Evolutions in Technology, Innovation, Adaptation, and Discovery

Darrell Norman Burrell
https://orcid.org/0000-0002-4675-9544
Marymount University, USA

IGI Global
Scientific Publishing
Publishing Tomorrow's Research Today

Vice President of Editorial — Melissa Wagner
Director of Acquisitions — Mikaela Felty
Director of Book Development — Jocelynn Hessler
Production Manager — Mike Brehm
Cover Design — Jose Rosado

Published in the United States of America by
IGI Global Scientific Publishing
701 East Chocolate Avenue
Hershey, PA, 17033, USA
Tel: 717-533-8845 | Fax: 717-533-7115
Website: https://www.igi-global.com E-mail: cust@igi-global.com

Library of Congress Cataloging-in-Publication Data

LCCN: 2026017437 (CIP Data Pending)
ISBN13: 9798260010266
Isbn13Softcover: 9798260010273
EISBN13: 9798260010280
British Cataloguing in Publication Data
A Cataloguing in Publication record for this book is available from the British Library.

Table of Contents

Detailed Table of Contents

This study aims to systematically evaluate the barriers to digital transformation in healthcare using the Analytic Hierarchy Process (AHP) method. A literature review and expert input guided the identification and weighting of 10 key barriers, based on the opinions of seven experts. The results revealed that the most critical barriers are high costs (0.17), technological infrastructure deficiencies (0.14), and data security and confidentiality concerns (0.13). Conversely, lack of standards was considered the least significant barrier (0.03). These findings highlight the need for strategic action to reduce costs, improve infrastructure, and ensure robust data security. To accelerate digital transformation, health policymakers and administrators must develop comprehensive plans and innovative strategies that address these critical barriers. Overcoming these challenges is essential to unlock the full potential of digital technologies and enhance the quality, efficiency, and sustainability of healthcare services.

This chapter examines Thrive Reproductive Care's strategic transition to a telehealth-only reproductive healthcare model, permanently replacing in-person services with digital platforms. The text considers broader healthcare trends accelerated by the COVID-19 pandemic, emphasizing drivers such as technology advancements and increased demand for accessible care. Through a literature review, the chapter analyzes opportunities and challenges faced by Thrive Reproductive Care, including digital equity, data privacy, staff training, and patient engagement, among others. Major themes include addressing the digital divide, organizational change management, and ongoing staff development to ensure equitable access and high-quality service. The analysis highlights the potential for this transformation to inform future best practices in telehealth delivery, while noting risks such as exacerbating health disparities and technological barriers. The chapter notes actionable strategies required for success and models for similar organizations navigating digital transformation.

Green logistics integrates advanced analytics with ecological stewardship in government and military supply chains. This phenomenological study engaged 15 experts (5 military logisticians, 5 green logistics specialists, 5 AI engineers) to examine how AI, IoT, blockchain, and circular-economy practices align operational readiness with the UN Sustainable Development Goals. Participants described AI-enabled predictive maintenance, dynamic routing, and real-time emissions dashboards that reduce fuel use, waste, and breakdowns while improving local air quality and public health. Blockchain and IoT strengthened transparency and lifecycle accountability, supporting renewable-energy integration and material reuse. Yet adoption remains partial, constrained by cultural inertia, fragmented policy, data-governance deficits, and cyber dependence. The study shows AI-enabled green logistics can operationalize SDGs 3, 9, 12, and 13 in defense contexts, reframing national security as environmental security and offering a scalable model for sustainable state capacity.

COVID-19 exposed structural fragilities in the United States' public health logistics and supply chain systems, particularly in the domains of diagnostic testing, epidemiological tracking, and coordinated response. These deficiencies underscore the urgent need for more adaptive, data-centric surveillance infrastructures capable of operating at scale and in real time. Artificial intelligence–enabled analytics, when integrated with existing public health systems, offer a viable pathway to enhance situational awareness, optimize resource allocation, and improve the timeliness of intervention strategies. In particular, wastewater-based epidemiology; augmented by AI-driven pattern recognition and predictive modeling; presents a scalable and cost-effective mechanism for population-level health monitoring, capable of detecting emerging threats prior to clinical reporting. This study advances that premise by examining how such integrative approaches can strengthen logistical decision-making and supply chain responsiveness.

Preface

Innovation has never been a singular event—it is an evolving process shaped by discovery, adaptation, and the dynamic interplay between technology, society, and human need. In *Evolutions in Technology, Innovation, Adaptation, and Discovery*, this volume brings together a diverse body of scholarship that reflects the complexity of that evolution. As editor, I have curated these contributions with the intention of highlighting not only technological advancement, but also the systems, cultures, and interdisciplinary collaborations that make meaningful innovation possible.

At its core, this book is about convergence—where disciplines intersect, where theory meets practice, and where innovation becomes both a driver of progress and a response to global challenges. The chapters collectively demonstrate that technological advancement cannot be isolated from its social, environmental, and organizational contexts. Instead, it must be understood as part of a broader ecosystem of change.

CHAPTER OVERVIEW

We begin with Chapter 1, which introduces a forward-thinking simulation framework grounded in the Quintuple Helix Innovation Theory. This model expands the boundaries of innovation by integrating academia, industry, government, society, and the environment into a unified system. The chapter offers a methodological foundation for measuring and optimizing these interactions, positioning sustainable, data-driven innovation as essential to the realization of Society 5.0.

Chapter 2 moves into the realm of human–computer interaction, presenting a real-time system that integrates gaze tracking and emotion recognition. By combining advanced frameworks, the authors demonstrate how artificial intelligence can deepen our understanding of human engagement, offering practical applications in education, productivity, and user-centered system design.

In Chapter 3, the focus shifts to organizational psychology, examining the lived experiences of female physicians in biomedicine. This chapter provides a critical exploration of psychological safety, equity, and inclusion, revealing how systemic barriers not only affect individual careers but also constrain innovation and organizational performance.

Chapter 4 continues the human-centered perspective by examining cybersecurity through the lens of organizational culture. Using a healthcare case study, the authors argue that cyber resilience depends not only on technical defenses but on cultivating cyberliterate cultures where responsibility is shared across all levels of the organization.

Chapter 5 provides a structured analysis of the barriers to digital transformation in healthcare, employing the Analytic Hierarchy Process to prioritize challenges. The findings offer clear guidance for policymakers and administrators, emphasizing the importance of cost management, infrastructure development, and data security in advancing digital health systems.

Chapter 6 builds on this theme with a case study of telehealth transformation, exploring how organizations can navigate digital equity, change management, and patient engagement. The lessons presented highlight both the promise and the risks of virtual healthcare models, particularly in addressing disparities in access.

In Chapter 7, innovation is examined through the lens of environmental public health and logistics. The integration of artificial intelligence, IoT, and blockchain within government and military systems illustrates how sustainability and operational efficiency can be aligned, reframing national security as a function of environmental stewardship.

Chapter 8 extends the discussion of public health by exploring AI-driven surveillance systems. The chapter demonstrates how predictive analytics and data integration can enhance responsiveness and preparedness, offering scalable solutions for managing future health crises.

Chapter 9 turns to rural veterinary care, presenting telehealth as a transformative tool for improving access and sustainability in underserved regions. The chapter highlights both operational innovations and regulatory challenges, offering a roadmap for implementation in similar contexts.

Chapters 10 through 12 collectively explore the intersection of agriculture, nutrition, and public health. Chapter 10 proposes an interdisciplinary framework for aligning agricultural innovation with clinical nutrition to address food inequities. Chapter 11 examines how agricultural advancements influence physical rehabilitation outcomes, linking nutrition to recovery and resilience. Chapter 12 provides a critical analysis of food policy, particularly the role of high fructose corn syrup, in shaping cultural and health outcomes related to obesity. Together, these chapters underscore the importance of integrated approaches to food systems and health.

Chapter 13 introduces a cutting-edge exploration of artificial intelligence and synthetic biology in plant epigenetics. The integration of these fields offers transformative potential for agriculture, enabling more precise and sustainable approaches to crop development and food security.

The final chapters address the evolving landscape of cybersecurity and advanced technologies. Chapter 14 examines the implications of quantum computing for aviation and aerospace security, offering strategies for transitioning to quantum-resilient frameworks. Chapter 15 concludes the volume with an analysis of eavesdropping attacks, highlighting the need for integrated, adaptive defense strategies in protecting digital communications.

As editor, I view this collection as a reflection of the multifaceted nature of innovation in the modern era. The contributions span disciplines and sectors, yet they are unified by a shared commitment to advancing knowledge that is both rigorous and relevant. They challenge us to think beyond traditional boundaries and to embrace a more holistic understanding of how technology, society, and discovery intersect.

I extend my sincere appreciation to the authors whose work has made this volume possible. Their scholarship represents a meaningful contribution to the ongoing dialogue on innovation and adaptation. I am also grateful to the reviewers and collaborators who have supported the development of this book.

To the reader, I offer this volume as both a resource and an invitation. It is a resource for understanding the complexities of technological evolution, and an invitation to engage with the ideas, challenges, and opportunities that define our time. Innovation is not a destination—it is a continuous journey. How we navigate that journey will determine not only the future of technology, but the future of society itself.

Darrell Burrell
Marymount University, USA

Chapter 1
A Quintuple Helix Innovation Model–Based Simulation Framework

Célia M. Q. Ramos
http://orcid.org/0000-0002-3413-4897
CinTurs and ESGHT, Univeristy of Algarve, Portugal

ABSTRACT

The Quintuple Helix Innovation Theory (QHIT) examines interactions among academia, government, industry, society, and the environment. Expanding from the Triple Helix (focused on academia-driven innovation) and Quadruple Helix (adding sustainability), the fifth helix incorporates environmental sustainability for societal resilience, aligning with Society 5.0—a smart, inclusive model balancing economic, social, and ecological needs through cyber-physical systems. Success requires digital, collaborative ecosystems that drive economic and social value, including impact assessments like social science reports. These ecosystems must harness innovation and knowledge, bridging democratic, social, and economic domains. Key to optimizing QHIT is identifying measurable variables to assess impacts and interrelationships, enabling monitoring of new investments. The research aims to develop a methodological framework for a prototype, ensuring sustainable and data-driven innovation.

INTRODUCTION

In the knowledge economy, innovation serves as the cornerstone of competitiveness, sustainability, and economic growth. As Fagerberg and Mowery (2015) demonstrate, innovation drives structural changes across businesses, regions, and

DOI: 10.4018/979-8-2600-1026-6.ch001

national economies. Through fostering creativity and generating novel solutions, innovation enables organizations to adapt to market shifts while developing more efficient products, services, and processes (Mahardhani, 2023).

The growing recognition of innovation's role in competitive advantage, coupled with the critical importance of science and technology, has catalyzed the transition to a knowledge-based economy (Pricopoaia *et al.*, 2023). This paradigm connects economic innovation with knowledge production through their intersecting concepts and challenges (Fagerberg & Mowery, 2015; Hall & Rosenberg, 2010). From a sustainability perspective, innovation addresses pressing social and environmental challenges while driving business value. In today's rapidly evolving technological landscape, effective knowledge management has become indispensable for fostering innovation and maintaining competitive differentiation (Addy *et al.*, 2019).

The Quintuple Helix Innovation Model (QHIT) represents a significant evolution in innovation theory by incorporating five interconnected sectors: academia, industry, government, civil society, and the natural environment (Carayannis & Campbell, 2022). Unlike the traditional Triple Helix framework, QHIT explicitly recognizes environmental sustainability and knowledge democracy as essential components of innovation ecosystems (Cai & Lattu, 2022). This comprehensive approach acknowledges that meaningful innovation occurs within broader societal and ecological contexts.

While innovation remains crucial for addressing global challenges like climate change and social inequality, traditional models often prove inadequate. The Triple Helix framework, for instance, fails to account for civil society and environmental considerations, resulting in fragmented solutions. Common implementation challenges include sectoral resistance to collaboration, inadequate measurement tools, and difficulties managing cross-sector innovation dynamics.

QHIT addresses these limitations through its integrated framework that aligns with global priorities like the UN Sustainable Development Goals (SDGs). By facilitating structured collaboration across all five sectors - from joint R&D initiatives to community-driven policymaking - QHIT provides both theoretical grounding and practical tools for overcoming cultural and resource barriers.

The urgency for such an approach is evident in contemporary challenges. The 2023 Global Innovation Index highlights how sectoral silos continue to impede progress on green technology adoption, while initiatives like Society 5.0 demonstrate the growing demand for inclusive innovation. QHIT's incorporation of environmental and societal dimensions makes it particularly suited to address these complex, interconnected issues.

This chapter introduces the QHITIS platform as a practical implementation of the QHIT model. This digital solution integrates data across all five sectors, featuring real-time dashboards for tracking R&D investments, community engagement metrics, and environmental impacts. The platform's AI-driven analytics and emphasis on ethical considerations (including GDPR-compliant data privacy measures) position it as a cutting-edge tool for cross-sector innovation management.

In conclusion, QHIT offers a transformative framework for addressing the systemic inefficiencies that currently hinder coordinated innovation efforts. By bridging the gap between academia, industry, government, civil society, and environmental stakeholders, this model provides an actionable path toward sustainable, inclusive growth. The chapter's primary objective is to operationalize the Quintuple Helix Model through practical mechanisms - including collaborative metrics and digital platforms - that overcome sectoral disconnects and accelerate meaningful innovation.´

BACKGROUND

Quintuple Helix Innovation Model

The Quintuple Helix Innovation Theory (QHIT) integrates the Triple Helix, Quadruple Helix, and society's natural environments to tackle global climate challenges and foster innovation in the knowledge economy (Carayannis *et al.*, 2012, 2022).

The QHIT model outlines the dynamic interactions among five critical sectors in the knowledge-based economy:

1. Universities: Serve as hubs for academic knowledge and research, driving innovation through idea generation and professional development.
2. Industry: Comprises business enterprises that transform academic knowledge into marketable products, improved processes, and economic growth.
3. Government: Provides regulatory frameworks, policymaking, and research investments to create an enabling environment for innovation.
4. Civil Society: Includes citizens, NGOs, and media organizations that ensure democratic knowledge distribution and policy alignment with public needs.
5. Natural Environment: Represents ecological systems where sustainable practices and environmental impact assessments guide responsible innovation.

As Carayannis *et al.* (2012, p. 6) define it, the QHIT model constitutes "a theoretical and practical framework for knowledge exchange, built upon five societal subsystems with distinct forms of capital, working collectively to advance sustainable societal development" (see Figure 1).

Figure 1. Functions of the QHIT Model

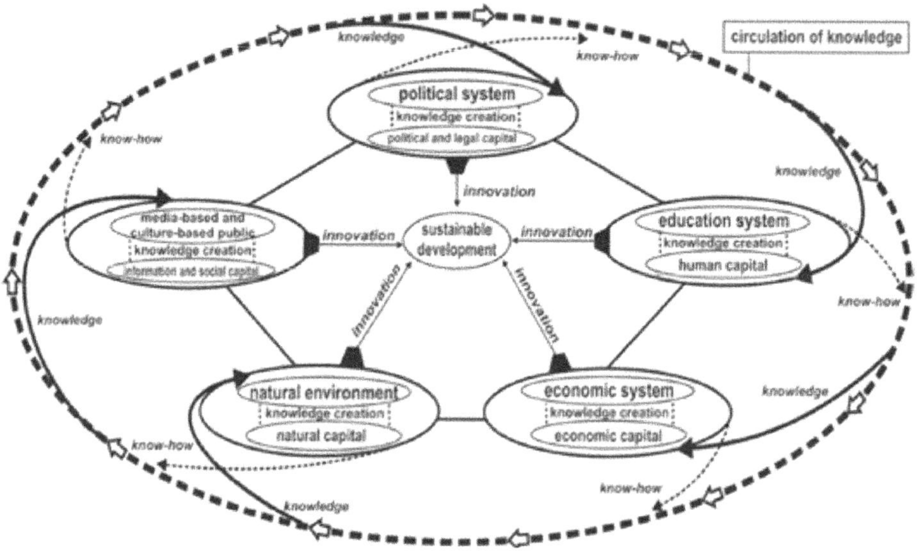

Source: Carayannis et al. (2012, p. 7).

The Quintuple Helix innovation framework is built upon synergistic interactions among five foundational societal elements that must collaborate effectively (Barth, 2011; Carayannis & Campbell, 2010; Carayannis *et al.*, 2012, 2022; Etzkowitz & Leydesdorff, 2000).

Universities can collaborate through (i) Research and Development (R&D), as it generates knowledge and innovations that can be applied in the industry; (ii) in Human Resources Training (FRH), through training them with qualifications suitable for the job market; (iii) defining Partnerships with Companies (PE), where collaborations are defined at the level of joint projects, internships and consultancy activities.

Industry collaborations can arise at the level of (i) Technological Innovation (IT), where companies apply academic research to develop new products and processes; (ii) investment in R&D, as the industry finances university research and collaborates on innovative projects; (iii) technology transfer, that is, Universities license patented technologies to companies.

The Government can collaborate with other sectors in terms of (i) Innovation Policies through the creation of regulations and incentives to promote innovation; (ii) Research financing through investments in research and development projects; (iii) through the creation of Public-Private partnerships, mainly in strategic areas, such as health and energy; between others.

In the public sector, collaborations can be defined at the level of developing actions (i) to increase awareness and involvement of the population, considering that society appreciates and supports innovative initiatives; (ii) obtaining opinions and needs from the population to help to define priorities for collaboration in research for the emergence of innovations and (iii) dissemination of results, mainly that society values and adheres to technological advances.

In the sector that represents the environment, the potential for collaboration stands out in terms of (i) sustainability, to investigate the impacts caused by possible innovations; (ii) responsible development, to identify innovative solutions that do not harm the environment, and (iii) at the same time promote environmental education, with the collaboration of universities, companies and the Government.

Finally, and in summary, collaboration between these five sectors is essential for innovation and the sustainable development of society.

EVOLUTION OF KNOWLEDGE MODELS

The Quintuple Helix model is an extension of previous knowledge models, originating from "Mode 1" (Gibbons *et al.*, 1994), which is a linear model based on research carried out in universities, whose success is defined and approved by peers hierarchical, which does not aim to contribute to the industry or the knowledge economy (Carayannis *et al.*, 2012), which was succeeded by "Mode 2" which already considered context, interdisciplinarity, heterogeneity, and organizational diversity, oriented towards problem-solving (Carayannis *et al.*, 2012, p. 3).

The Triple Helix Model (Etzkowitz, & Leydesdorff, 2000) presents a model at the level of a social structure to explain Mode 2, for the production of scientific knowledge, taking into account Mode 1, which highlights the collaboration between university and industry-government, being considered the "basic model of the innovation core" (Carayannis *et al.*, 2012, p. 4).

The Quadruple Helix Model adds a fourth helix to the previous model, representing the public based on media, culture, values, lifestyles, and art. This model recognizes civil society as a sector influencing innovation (Carayannis & Campbel, 2009) or a knowledge democracy society (Carayannis & Campbel, 2022).

In the quintuple helix model, a new helix is added to the previous model, which represents the environment, which is aligned with the sustainable development of society and social ecology (Carayannis & Campbel, 2010b), which is a necessity for humanity "but they also act as drivers for further knowledge and innovation" (Carayannis & Campbel, 2022, p. 71).

In short, the Quintuple Helix extends previous models by incorporating civil society and the natural environment, creating a more comprehensive context for innovation and knowledge (Carayannis & Campbell, 2010).

Importance of Including the Public in the Model

In the QHIT model, including the public is a very relevant aspect, as it recognizes civil society as a fourth sector with the most influence on innovation, in addition to the traditional university-industry-government sectors. This highlights the importance of active public involvement, including citizens, communities and non-governmental organizations, which is fundamental for innovation and sustainable development (Carayannis *et al.*, 2012).

Public participation can occur in several ways: (i) Civic Involvement through public consultations, discussion groups and community partnerships to help define policies and research projects; ii) co-creation of knowledge: since the public contributes knowledge, experience and perspectives to research and innovation; (iii) dissemination and education, through the sharing of results and scientific knowledge in society; (iv) Social and environmental responsibility, to study the social and environmental impacts of adopting innovations and new technologies.

In summary, including the public sector is essential for a holistic approach to innovation!

Importance of Including the Environment in the Model

In the QHI model, including the environment as the fifth sector recognizes the importance of the ecosystem associated with the natural environment and allows alignment with sustainable development.

The ecosystem that characterizes the natural environment plays a central role today, as it includes natural resources (such as water and energy) and biodiversity, which are essential for research and inspiration for developing innovative solutions. It is sustainable without neglecting the analysis and investigation into possible environmental impacts caused by the innovations above, aiming to predict and minimize damage (Barth, 2011; Pinto & Nogueira, 2018).

Engaging the natural environment is fundamental to a sustainable approach to innovation.

Applications and Challenges

The QHIT model can include actions in three categories, which can be applied in the five sectors that characterize it: promoting a more collaborative culture, establishing joint performance indicators and increasing diversity and inclusion, as shown in Table 1.

Table 1. Performance Indicators

Sector	**Action**	**Description**
University	Collaborative culture	Teachers and researchers can encourage collaboration between departments, units, and disciplines to promote the exchange of knowledge, experiences, and practices between areas.
	Definition of Joint Performance Indicators	Between the university, industry and society to simultaneously carry out research that meets the needs of various sectors.
	Diversity and inclusion	Through exchange programs and the creation of multidisciplinary research groups that promote diversity and inclusion.
Industry	Collaborative culture	Incentives for creating multidisciplinary teams in companies to promote the emergence of innovative products.
	Definition of Joint Performance Indicators	Collaboration with universities is needed to promote research and the emergence of innovations, whether in terms of products, services, or technology.
	Diversity and inclusion	Hire professionals from different backgrounds and with different skills.
Government	Collaborative culture	Through ministries and government agencies promote collaboration between departments to solve complex challenges.
	Definition of Joint Performance Indicators	Through policies that encourage the establishment of public-private partnerships with the aim of fostering innovation.
	Diversity and inclusion	Integration into work groups of members from different backgrounds and with different skills.
Public (community)	Collaborative culture	Organization of community events, forums and groups that promote active participation.
	Definition of Joint Performance Indicators	Community involvement in defining policies and projects for presenting innovation proposals.
	Diversity and inclusion	Include different origins and different profiles in public debates.
Environment	Collaborative culture	Encourage collaboration between NGOs and research institutions that promote environmental conservation.
	Definition of Joint Performance Indicators	Promote the development of innovations that promote sustainability and environmental preservation.
	Diversity and inclusion	Include the interests of local communities, including residents from different backgrounds and minority groups.

Source: Adapted from Carayannis and Rakhmatullin (2014) and Carayannis and Campbell (2022), Carayannis et al. (2012) and Nunes et al. (2017).

Cross-Industry Collaborations

The Quintuple Helix can promote collaboration between the five sectors to create sustainable solutions and address global challenges such as climate change and sustainability. Each sector can contribute with specific strategies, such as those in Table 2.

Table 2. Sectors, Contributions and Strategies

Sector	Contribution	Strategies
University	Scientific knowledge	Conduct research on sustainable development and climate change.
		Promote training actions to train professionals to deal with sustainability and climate-related challenges.
Industry	Innovation in products and services	Develop technologies to promote sustainable practices.
		Develop green energy solutions with reduced emissions.
Government	Policies and regulations	Definition of policies to promote sustainability.
		Definition of measures that reduce emissions.
		Definition of incentives that lead to the adoption of renewable energy.
Public (community)	Awareness and engagement	Promote actions to increase community awareness
Environment	Natural ecosystem	Definition of measures to preserve the ecosystem and face climate challenges.

Source: Adapted from Carayannis, et al. (2012; 2022), Liyanage and Netswer (2021), Reis et al. (2022), Zen and Shibakawa (2022).

Some examples of projects that fit into this model (Baccarne *et al.*, 2016; Carayannis *et al.*, 2012; 2022; Kholiavko *et al.*, 2021; Liyanage & Netswera, 2021; Reis *et al.*, 2022; Zen & Shibakawa, 2022):

- **Education and Innovation in Renewable Energy:** Universities collaborate with businesses, government and society to develop sustainable energy solutions such as solar panels, wind turbines and storage batteries.
- **Partnerships for Environmental Conservation:** Companies, NGOs, governments and universities work together to protect fragile ecosystems such as coral reefs, rainforests and endangered animal habitats.
- **Development of Green Technologies:** Startups, universities, and the public sector come together to create clean technologies, such as advanced recycling systems, electric transport and efficient water treatment.

- **Smart Urban Planning:** Through collaboration between local governments, universities, and technology companies, sustainable urban solutions such as smart cities, efficient public transport, and green spaces are developed.

Cross-Industry Challenges

However, for collaboration between sectors to be successful, some challenges must be overcome: (i) resistance to Change, as it is necessary to adapt to new innovative methods and processes; (ii) complexity of processes due to different organizational, governmental and bureaucratic structures; (iii) communication challenges, which can be difficult between sectors when the language and integration rules between structures are not the same; (iv) cultural limitations that can cause the emergence of barriers; (v) lack of resources, both financial and human, which can limit the number and potential of developing joint projects, where creativity can play an important role (Carayannis, *et al.*, 2012; 2022; Carayannis & Campbell, 2009; Campbell, 2009; Carayannis & Rakhmatullin, 2014; Liyanage & Netswera, 2021; Reis *et al.*, 2022; Zen & Shibakawa, 202;).

Strategies for Overcoming Cross-Industry Challenges

Collaboration between sectors is a continuous and dynamic process. However, it presents challenges for which it is necessary to consider strategies to implement in collaboration between sectors, as shown in Table 3.

Table 3. Strategies to Implement in Collaboration between Sectors

Strategy	Actions
Effective Communication	Create clear and regular communication channels between sectors.
	Promote meetings, working groups and digital platforms to facilitate the exchange of information.
Involvement of leaders	Involved leaders can motivate their teams to achieve goals.
Pilot Partnerships	Start with small projects to test the collaboration of team members before scaling them up.
Create incentives	Define incentives for the active participation of team members through recognition, prizes, financial benefits or other benefits.
Flexibility	Adjust approaches when necessary to overcome unexpected situations.

continued on following page

Table 3. Continued

Strategy	Actions
Transparency and Trust	Define relationships supported by transparent and frequent communication.
Training and training	Promote actions that include workshops, training and meetings to share knowledge and experiences among everyone involved.

Source: Adapted from Carayannis, et al. (2012; 2022), Liyanage and Netswer (2021), Reis et al. (2022), Zen and Shibakawa (2022).

Measures to Assess the Success of Cross-Industry Collaborations

To assess whether challenges have been overcome through the implementation of strategies, it is important to define metrics that assess the degree of success and impact of these actions, such as:

a) Performance or intellectual capital indicators (specific to each sector), such as productivity measures, human resources satisfaction, response time to requests, ROI (Return on Investment), and comparison between costs and benefits.
b) The Opinions of members were obtained through surveys and interviews to understand aspects that need improvement.
c) Integrate and evaluate work and communication flow through time measurements to obtain results.
d) Communication of the results of collaborations through the preparation of reports and scientific publications, organization of events and workshops, sharing of results and events on social networks, presenting the results to peers (from different sectors) and establishing an open channel of communication between the community, academia and industry.

However, each collaboration will have its specific objectives, so success may vary according to each collaboration and metrics must be adapted and defined according to each situation.

Current Trends

Currently, Artificial Intelligence (AI) plays a significant role in interactions between sectors in different sectors of society, which can contribute to those considered in the QHIT model, such as: (i) personalization in education, according to the individual needs and preferences of students; (ii) Data Analysis, to identify patterns, trends and insights; (iii) support for Decision Making: market forecasts, financial risk analysis and process optimization.

Finally, and taking into account that the impact of AI on society is significant, it is necessary to guarantee ethics and transparency in the implementation of a platform that evaluates the impact of the Quintuple Helix Innovation Theory (QHIT) model, which is essential to promote trust and responsibility, through transparency indicating which data will be stored, processed and presented, explaining the results obtained in a structured and clear way, protecting data in accordance with data protection regulations and monitoring the performance of the platform, whenever considered necessary to ensure reliable and responsible results.

Background Discussion

The Quintuple Helix Innovation Theory is integrative, considering the interconnection between five sectors: university, industry, government, society (community) and environment to promote innovation and sustainable development. In this model, the university represents knowledge and education; industry is the practical application of this knowledge and the emergence of innovations; the government represents regulation; society, the community and non-governmental organizations; the environment is the ecosystem where we live and its sustainability.

This model presents an approach characterized by complexity due to the different sectors that collaborate with each other; the need for balance, as some sectors may try to dominate more than others; and there is a need for examples of real implementation since the model is based on theory, but there are not yet many practical examples that guarantee its robustness.

However, this model presents several opportunities to innovate collaboratively through cooperation between the five sectors; sustainable development, which can help address climate change and sustainability issues; community involvement, through active public participation, social inclusion and diversity, which meets the premises of society 5.0 (Deguchi *et al.*, 2020), which puts the human being at the center.

In future terms, the QHIT model should meet the principles of society 5.0, which is a sustainable society, balances the resolution of social problems, provides security and comfort for individuals and where technology plays a driver of growth and transformation of social structures, which will be centred on the well-being of human beings.

The Quintuple Helix Innovation Model (QHIT) has demonstrated its practical relevance across diverse sectors through rigorous empirical testing. In Botswana's renewable energy transition, Liyanage and Netswera (2021) found that QHIT's incorporation of civil society and environmental stakeholders increased solar energy adoption by 37% compared to Triple Helix approaches, highlighting its superior capacity for sustainable technology diffusion. Similarly, smart city applications in Amsterdam (Baccarne *et al.*, 2016; Putra & van der Knaap, 2018) revealed that QHIT-generated urban solutions achieved 22% higher public satisfaction by systematically integrating environmental impact assessments with community participation - a dimension often missing in Quadruple Helix implementations.

Further validation comes from regional innovation systems, where Bolsunovskaya *et al.* (2023) documented a 15% acceleration in R&D commercialization in Russian regions adopting QHIT, attributable to its robust multi-stakeholder governance framework. The model's social innovation potential was equally evident in EU studies (Carayannis *et al.*, 2019, 2022), where QHIT-aligned initiatives doubled funding success rates by explicitly aligning Sustainable Development Goals through cross-sector collaboration. These findings collectively establish QHIT's empirical advantage in generating more inclusive, ecologically conscious, and scalable outcomes than traditional innovation models.

The accumulating evidence positions QHIT as particularly effective when environmental and social dimensions are paramount. As demonstrated across energy, urban development, regional growth, and social enterprise contexts, the model's five-helix structure consistently delivers superior performance metrics compared to narrower frameworks. This empirical track record supports QHIT's growing adoption as a comprehensive framework for addressing complex, interconnected challenges in the knowledge economy.

While the Quintuple Helix Innovation Model (QHIT) offers a comprehensive approach to innovation by integrating five key sectors (academia, industry, government, civil society, and the environment), it is not without limitations. One major challenge is operational complexity—coordinating five distinct sectors with divergent priorities can lead to bureaucratic inefficiencies and slow decision-making (Carayannis & Campbell, 2022). Additionally, the model's lack of standardized metrics makes it difficult to quantitatively compare its effectiveness against other innovation frameworks (Muruccia *et al.*, 2020). Some critics argue that QHIT's broad scope may dilute focus, particularly in contexts where rapid, sector-specific innovation is needed (Cai & Lattu, 2022). Finally, its reliance on stakeholder willingness to collaborate means that in regions with weak institutional trust or governance, implementation may falter (Abdillah *et al.*, 2022).

Several competing or complementary frameworks offer alternative approaches. The Triple Helix model (Etzkowitz & Leydesdorff, 2000) remains dominant in purely industry-university-government collaborations, particularly in technology transfer, but it neglects societal and environmental dimensions. The Quadruple Helix (Carayannis & Campbell, 2009) adds civil society but still omits ecological considerations, limiting its applicability to sustainability-driven innovation. Living

Labs (Baccarne *et al.*, 2016) excel in user-centric, real-world testing but often lack systemic policy integration. Meanwhile, Open Innovation (Chesbrough, 2003) focuses on cross-organizational knowledge sharing but does not address governance or sustainability. Circular Economy models (Geissdoerfer *et al.*, 2017) prioritize environmental sustainability but underemphasize institutional and social innovation structures.

APPLICATIONS AND THEIR CHARACTERISTICS

Commercial Applications and Their Characteristics

The commercial applications, which were evidenced in the Play Store on the Internet, the list of works is not limited to those that were analyzed.

When searching for "Quintuple Helix Innovation Management", no application of interest for the present study came up. A new search was carried out, in which the terms "Innovation Management" were considered. The results that emerged referring to the applications presented below are the most relevant for the work considered in the present study and were evidenced in the search conducted on the Internet, which will be explained in detail below:

- **Innovation Markets:** Innovation Markets (2024) contributes to financial education, cryptocurrency, and trading and presents the basic principles of investment and trading knowledge.
- **Learn Business Management:** Code World (2024) contributes to business management.

Although both examples talk about innovation, they are unrelated to the QHI model.

In a second search on Google with the terms "quintuple helix innovation", the following application emerged: Q-ideate – Qmarkets (2024), which allows you to manage innovation in a company.

In addition, no others related to the QHIT model or anything related to innovation relevant to the present study were identified.

Academic Applications and Their Characteristics

In the scientific community, systems that allow analysing impacts based on QHIT were not detected in reference articles. However, the work of Abdillah *et al.* (2022) bridges the theoretical concepts of the QHI model with practice-based concepts associated with rural life. Carayannis and Campbell (2021) also state that the "five dimensions of the Quintuple Innovation Helix qualify to relate to themes of Industry 5.0 and Society 5.0," (Carayannis & Morawska., 2023; Deguchi *et al.*, 2020) and Digital transition (DT) assumes a relevant role within the community (society), where they highlight the opportunities and risks of adopting big data technologies, as well as stating that the production of knowledge will be determined by the adaptive capacity of systems to integrate "different knowledge and innovation modes via co-evolution, co-specialization and co-opetition knowledge stock and flow dynamics" Carayannis and Campbell (2021, p.2067), represented in Figure 2.

Figure 2. Quintuple Helix Innovation System

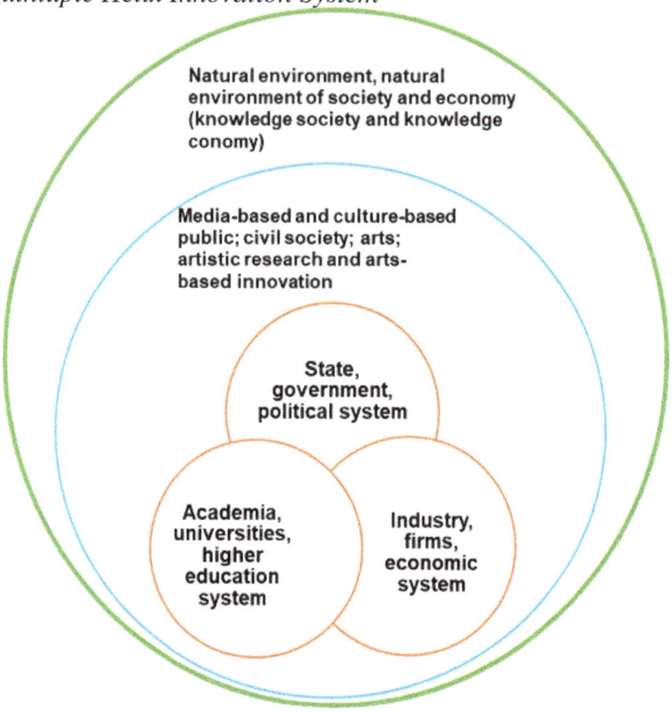

Source: Carayannis and Campbell (2021, p. 2069).

However, associated with other types of innovation, some systems appear in the literature, such as those associated with the crowdsourcing platform (Vignieri, 2021), Modeling the Development of Regional Innovation Systems (Bolsunovskaya *et al.*, 2023), framework about the way Industry 4.0 has transformed production into an industry focusing on connective mechanisms that consider knowledge-based modelling (Abd-Rahman *et al.*, 2021), platform-based service innovation and system design (Fu *et al.*, 2018) and an urban innovation system and the role of an open web-based platform was studied for Amsterdam (Putra & van der Knaap, 2018), discusses how the application of Network Analysis metrics advances the understanding of an innovation system mapped with a Quintuple Helix model.

Using the Global Innovation Index, a casual Loop Diagram has been designed to represent the process of knowledge generation and innovation in the work of Murrucia *et al.* (2020) which discusses how the application of network analysis metrics advances the understanding of an innovation system mapped with a Quintuple Helix Innovation Theory model (QHIT), considering a global innovation index, which considers the process of knowledge generation and innovation dissemination within the QHIT model.

Focusing on the social dimension, Carayannis *et al.* (2019) presented a framework to develop an "ecosystem" approach to provide an integrated framework for social business models.

Also, authors Carayannis *et al.* (2022) consider aspects of knowledge transfer through technology as a driver of smart growth in a quadruple / quintuple innovation context.

The last three works are the most relevant to research model development.

RESEARCH MODEL

In the QHIT model, each helix represents a category of variables that can be included in the system, which is a knowledge management system, as shown in Figure 3.

Figure 3. QHIT model as a knowledge-based system

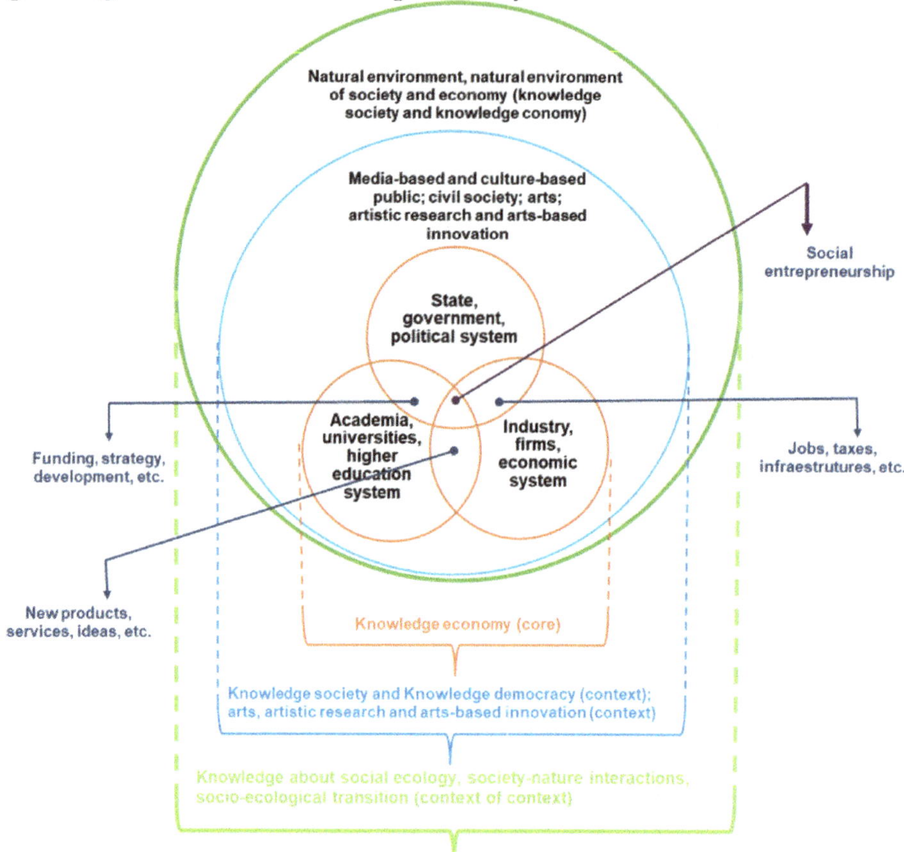

Source: Carayannis et al. (2021, p. 2070).

In Figure 3, there is the relationship of creation/production of knowledge between the university and companies, as well as a relationship of knowledge of use/diffusion between them, where innovation arises through the application of knowledge produced within the university and applied in companies, which in turn, the application of this innovation by companies creates more potential for generating new knowledge for academia.

Focusing on the social dimension, Carayannis *et al.* (2021, p. 237) identified several types of innovation, as presented in Table 4.

Social innovation in general terms (Carayannis *et al.*, 2021) is based on the value proposition, which is related to the ability to find solutions to challenges/problems of the education, environment, industry, society and government; some examples presented by Caryannis *et al.* (2021, p. 240-242).

Table 4. Social Innovation typology

Type of social innovation	Description
New services and products	New interventions or new programs to meet social needs
New practices	New services which require new professional roles, relationships or skills
New processes	Co-production of new services
New rules and regulations	Creation of new laws and entitlements
New organizational forms	Hybrid organizational forms, such as social enterprises

Source: Carayannis et al. (2021, p. 237).

The Business Model Innovation, in addition to being based on the value proposition, also includes the pillars related to the "Target Group/Customer" and the "Value Constellation", as shown in Figure 4. However, innovation is supported by the knowledge exchanged with the supported pillars and can only establish new levels of "targeted open Innovation" in a balanced way.

Figure 4. Business Innovation Model

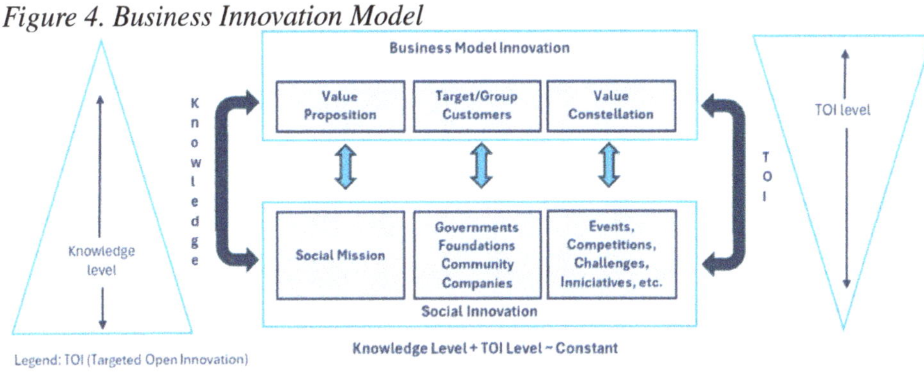

Source: Adapted from Carayannis et al. (2021, p. 242).

However, Muruccia *et al.* (2020) consider that dynamic approaches should be included, which is why they combined the QHI model with system dynamics to present a more quantitative approach through the presentation of a "Global Innovation Index", for which variables and their relationships between them, based on the work of GII (2018), figure 5 presents the GII model of 2023 (GII, 2023).

Figure 5. Global Innovation Index

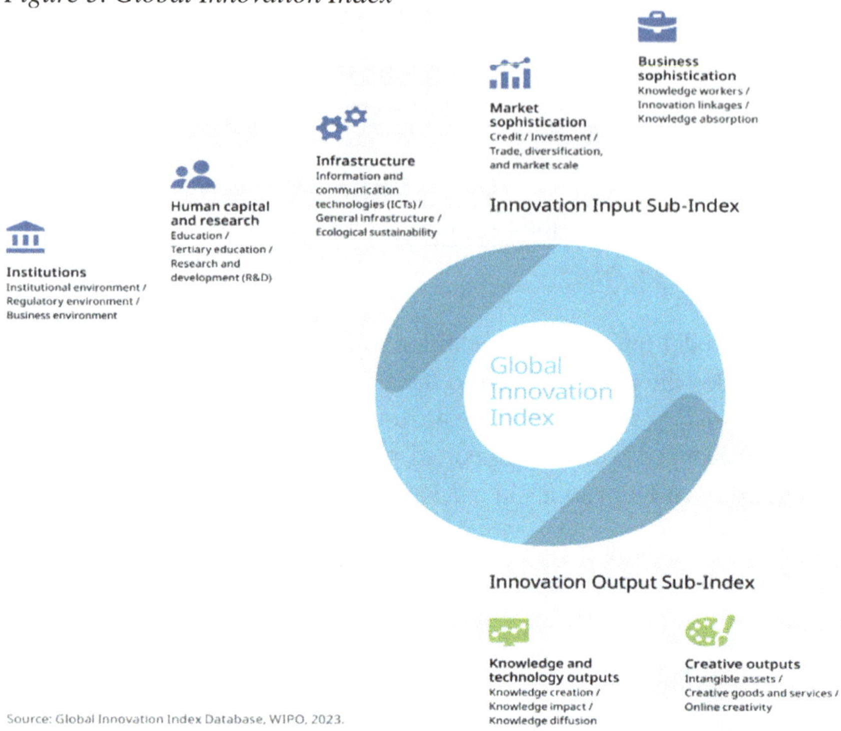

Source: GII (2023, p. 77).

An Innovation System associated with the QHT, in addition to the variables and the relationships between them, must be developed to allow innovation management through the presentation of evolution over time and results in generating knowledge and controlling deviations through developing a research model associated with QHIT.

Framework Grounded in the Quintuple Helix Innovation Model

Innovation plays a crucial role in the knowledge economy, driving economic growth, competitiveness and sustainability. Therefore, the Information System to be developed (platform) must consider indicators for each of the propellers, as mentioned in the work of Muruccia *et al.* (2020), preparing a Global Innovation Index for each region or country must present a management report that allows monitoring the evolution of innovation capacity, must include the possibility of creating events and disseminating them to the players of each propeller. It would help if you created spaces for communicating ideas and sharing opinions, among other relevant aspects, on social media platforms. Partnerships with the local community must include and allow the possibility of being developed within QHITIS and promoting campaigns to share ideas and carry out public consultations with the community, as shown in Figure 6.

Figure 6. QHITIS Platform Framework

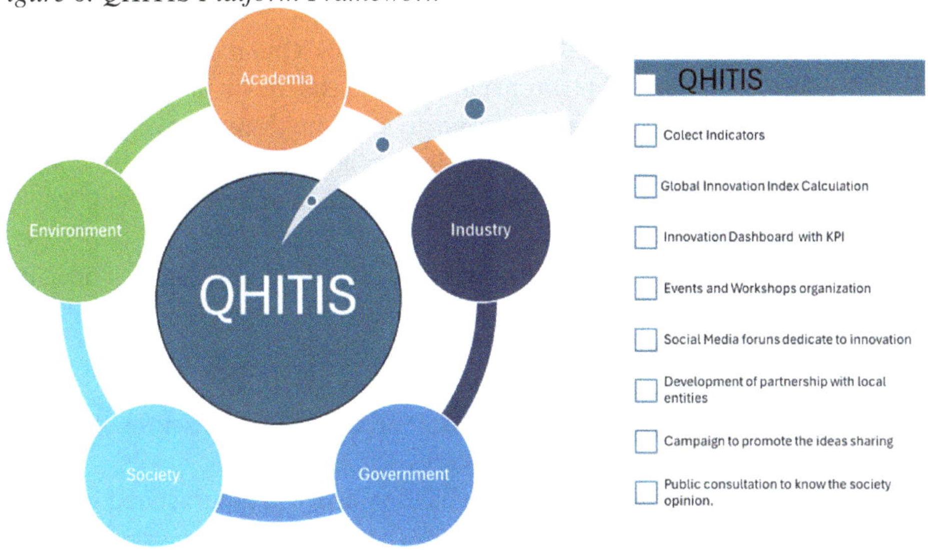

Source: Own elaboration.

Model Discussion

The Quintuple Helix Innovation Theory model expands the traditional Triple Helix model, considering five elements that drive innovation. These elements are the university, company, government, civil society and environment. Developing an information system applied to this model is crucial to promoting collaboration between these actors and ensuring economic growth and innovation sustainability.

Collaboration between the five elements is critical to effectively driving innovation. For example, information systems can facilitate knowledge exchange between universities, businesses and government, allowing them to identify joint research and development opportunities. Additionally, civil society can engage through digital platforms to provide feedback on public policies and innovation projects. Finally, considering the socio-ecological environment ensures that innovation is sustainable and considers environmental and social issues.

QUINTUPLE HELIX INNOVATION THEORY INFORMATION SYSTEMS TECHNOLOGICAL ARCHITECTURE AND COMPONENTS.

Development of the QHITIS Platform

The methodology considered to develop a QHITIS (Quintuple Helix Innovation Theory Information System) platform takes into consideration the implementation of technological architecture, which can be divided into three phases: (1) planning and definition of content associated with the five helices, (2) planning and implementation of the platform and (3) testing and validation of the final prototype.

1. **First Phase:** Harvesting the information about five helices can be divided into three steps:
   a. Research of primary sources.
   b. Research of secondary sources.
   c. Provide a definition of the structure of the contents to be disclosed, including security and data protection concerns.
2. **Second Phase:** Implementing the QHITIS system can be structured into three steps:
   a. Analysis of technology relevant to include in the model.
   b. Technology Architecture Design.
   c. Development of a Platform.
3. **Third Phase:** Testing and validating the QHITIS platform can be divided into three steps:
   a. Analysis of the Technology Acceptance Models (Davis, 1989).
   b. Surveys should be applied to stakeholders that belong to all helices.
   c. Analysis and evaluation of results.

QHITIS Platform Data Framework

In the first phase, after identifying all data sources and stakeholders, the analysis and design of the system will be carried out to define the system's design through the Data model specification.

The following modules will be associated with this phase: Data Upload, to enable the import of data associated with the different propellers, as well as other auxiliaries; Data Model, to define the model according to the specification through relationships and data storage structures considered essential, such as data tables, including the Data Transform process; and Security and Data Protection, to guarantee data protection and security, including defining access permissions for the different profiles that can access and use the platform. Figure 7 presents a framework for the modules associated with the first phase.

Figure 7. QHITIS Platform Data Framework

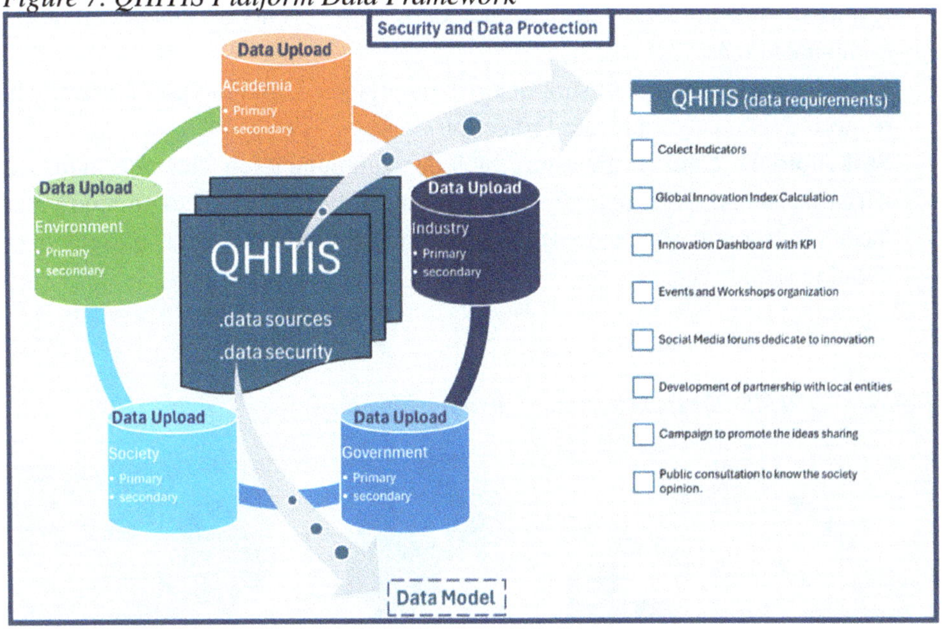

Source: Own elaboration.

QHITIS Platform Architecture

The platform architecture for evaluating the Quintuple Helix Innovation Theory Model must consider the interaction between the five sectors (university, industry, government, public and environment), combined with different access permissions according to the users' profile, as presented in Figure 8. Here are the main components of this architecture:

1. **Data Collection:** Integrate relevant data sources from each sector. This may include research, reports, economic indicators and environmental data (Data Upload).
2. **Analysis and Modelling:** Develop models that represent interactions between sectors. Use complex networks, data analysis and simulations (Data Model and Data Transform).
3. **View and Control Panels:** Create interactive dashboards to visualize results. This allows users to explore connections between sectors (Dashboards).

4. **Impact Assessment:** Measure the impact of collaborations on innovation, economic development and sustainability (Innovation Indicators and Results evaluations).
5. **Communication and Engagement**: Develop strategies to communicate results to stakeholders. This can include reports, events and media (Reports communication).
6. **Security and Privacy**: Ensure that data is protected, and the users' privacy is respected (Security and Data Protection).
7. **Stakeholders Communication:** Include communication channels with all participants in the five helices, through publications, listening to opinions, and feedback through social networks (Communication Channels), presented in detail in section 8.4.

Figure 8. QHITIS Platform Technological Architecture

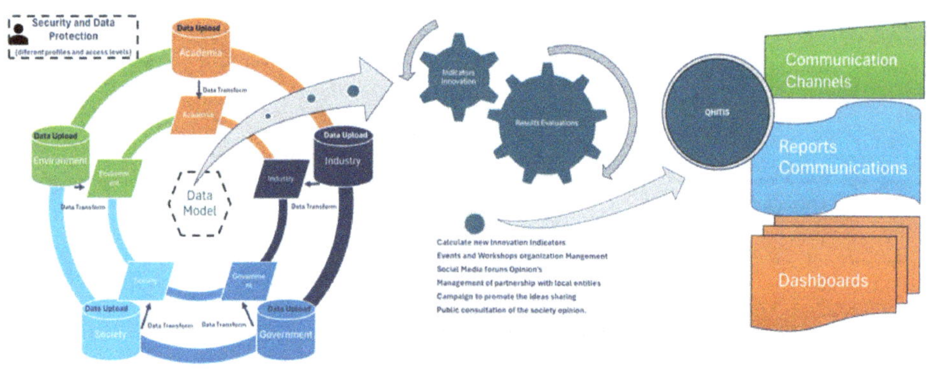

Source: Own elaboration.

QHITIS Platform Communication Channels

Communicating the results of collaborations between sectors to the public and stakeholders is fundamental to demonstrating the impact and effectiveness of initiatives associated with social innovations. Defining strategies to establish a communication bridge between all interested parties is very important. Strategies that should be included with features should include:

1. **Reports:** Prepare detailed reports/dashboards highlighting the results, including specific metrics, analyses, and conclusions.

2. **Publications:** Publication articles in scientific journals, company blogs or other relevant media and collaborate with journalists to publish stories about the results achieved.
3. **Events and Workshops:** Hold outreach events such as conferences, seminars or webinars to share results with a broad audience and organization workshops to train other professionals to apply lessons learned
4. **Social Networks:** Integrate social media platforms to share success stories, infographics and explanatory videos.
5. **Executive Presentations:** Prepare presentations for executive leaders and stakeholders highlighting collaboration's tangible and intangible benefits and use graphs, tables and visualizations to make results more accessible.
6. **Continuous Feedback:** Maintain an open communication channel with the public and stakeholders belonging to all propellers.

FUTURE RESEARCH DIRECTIONS AND DISCUSSION

The development of an information system associated with the quintuple helix innovation model can contribute to the development and sustainability of a region or country. However, this presents some challenges and limitations that need to be overcome:

Privacy: Concerns about human privacy are a challenge. It is necessary to consider methods to protect individual privacy.

Big Data: The size of databases (Big Data) is another challenge. This aspect allows for the management, storage, processing, and dissemination of information, which increases exponentially when data collection occurs at high speed about events, consultation, and public sharing of opinions, suggestions, or complaints.

Interfaces: It must present functionalities and content according to the user's profile and preferences, which depend not only on algorithms and business intelligence techniques associated with the data but also on interactions with the interface.

Security in data transmission: It is also necessary to consider the mechanisms necessary for transactions and communications to be considered secure by users.

Data Protection: Ethical use of data is crucial to avoid privacy violations and discrimination.

Equity and Inclusion: Technology can magnify existing inequalities. It is essential to consider issues of equity and inclusion when developing technological solutions. This involves ensuring all groups have equal access and that technologies do not perpetuate bias.

Transparency and Explainability: Algorithms and automated systems often operate in the background without users fully understanding how they work. Ethics requires transparency, explainability and the ability to question automated decisions.

Social Responsibility: Companies and institutions involved in innovation must take responsibility for the impact of their technologies. This includes considering social, environmental and economic well-being when designing and implementing solutions.

Cybersecurity: The security of technological systems is essential to prevent breaches, attacks and damage. Ethics requires rigorous measures to protect data and systems against threats.

Ethics: Ethics must be applied both in the use of data and in the way of interacting with technology to safeguard the well-being of human beings, organizations and society.

Each challenge must be overcome to minimize any inconvenience for users and contribute to the joint creation of innovative solutions that meet everyone's interests.

The implementation of the QHIT model introduces ethical dilemmas, particularly in data privacy and algorithmic transparency, as highlighted by the QHITIS platform's reliance on AI and cross-sector data sharing. For instance, while integrating public feedback into innovation policies fosters inclusivity, it raises concerns about data ownership and consent, especially under regulations like GDPR. Governance complexities further arise from conflicting sectoral priorities, e.g., industries may prioritize profit over environmental sustainability, while governments focus on regulatory compliance. Cultural differences exacerbate these challenges, as varying organizational norms and communication styles can hinder collaboration. For example, academic institutions may emphasise long-term research, whereas industries demand rapid commercialisation. To mitigate these issues, the QHIT framework must embed ethical guidelines (e.g., anonymizing public data), adaptive governance (e.g., flexible policy frameworks), and intercultural training to align stakeholders' expectations and practices.

Step-by-Step Application Plan with Barriers and Solutions

- Pilot Project Initiation: Start with a small-scale collaboration, such as a regional renewable energy project, to test QHIT's feasibility. Barrier: Resistance from sectors wary of resource allocation. Solution: Use incentives like tax breaks or grants to encourage participation.
- Stakeholder Engagement: Organise workshops to define shared goals and metrics (e.g., Table 1). Barrier: Misaligned priorities. Solution: Employ facilitators to mediate and co-create objectives.

- Data Integration: Deploy the QHITIS platform to aggregate sector-specific data. Barrier: Data silos or privacy concerns. Solution: Adopt blockchain for secure, transparent data sharing.
- Monitoring and Adaptation: Use real-time dashboards to track progress. Barrier: Uneven participation. Solution: Implement feedback loops and adjust strategies quarterly.
- Scaling: Expand successful pilots nationally. Barrier: Bureaucratic delays. Solution: Leverage government helices to streamline approvals.

By addressing these barriers proactively, QHIT can transform theoretical collaboration into actionable, sustainable innovation.

CONCLUSION

Innovation is very relevant for society, which is why collaboration mechanisms must be considered and created between companies, government and academia through the organization of events, workshops, and conferences to discuss aspects relevant to development in an inclusive and sustained way, where ethical aspects cannot be forgotten, especially in an era in which AI is already part of it.

To translate the QHIT framework into tangible impact, leaders and practitioners should adopt the following strategies:

First, cross-sector innovation hubs should be established to serve as collaborative platforms where universities, industries, governments, civil society, and environmental groups can co-design solutions. These hubs, whether physical or digital (such as the proposed QHITIS platform), align with the model's emphasis on knowledge democracy and help overcome systemic fragmentation by creating shared spaces for dialogue and project development. Second, shared performance metrics must be adopted to evaluate the success of collaborations. Drawing from the indicators outlined in Table 1, stakeholders should track measures like diversity inclusion, sustainability impact, and community engagement. This ensures accountability across all helices and reflects the dynamic measurement approaches advocated by recent research. Third, pilot partnerships should be launched with targeted incentives to encourage participation. Small-scale projects, such as local renewable energy initiatives, can serve as test cases, with benefits like tax breaks or grants for involved parties. This approach reduces resistance to change and allows for scalable solutions, as highlighted in the chapter's discussion on overcoming collaboration barriers. Fourth, public deliberation should be integrated early in the innovation process. Civic tech tools, such as participatory budgeting platforms, can crowdsource community needs and preferences, ensuring that solutions are grounded in real societal demands. This embeds the public helix into decision-making and prevents misaligned or top-down interventions. Fifth, sector-specific training programs are needed to bridge cultural and operational gaps between academia, industry, and government. Workshops on intercultural communication and collaborative workflows can foster mutual understanding and co-specialization, addressing one of the key challenges identified in the chapter. Sixth, AI-driven tools should be leveraged to monitor and transparently report innovation outcomes. Platforms like QHITIS can provide real-time dashboards to track progress, ensuring ethical tech use and enabling data-driven adjustments. This aligns with the growing emphasis on transparency and accountability in innovation ecosystems. Seventh, flexible policy sandboxes should be created to allow experimentation within a controlled regulatory environment. For instance, governments can designate "test beds" for green technologies, where temporary waivers enable rapid prototyping while maintaining oversight. This balances innovation speed with governance, as underscored in the background discussion. Eighth, environmental KPIs must be prioritized in funding decisions. Public and private R&D investments should be tied to mandatory sustainability audits, ensuring that the environmental helix remains central to innovation agendas. This reflects the model's commitment to addressing global challenges like climate change. Ninth, scaling mechanisms should be developed to replicate successful collaborations. Documenting and sharing templates for effective projects, such as the renewable energy case study mentioned in the chapter, can help disseminate best practices and accelerate impact. Tenth, ethics review boards composed of cross-sector representatives should be embedded in innovation processes. These boards can assess projects for social equity, privacy, and ecological risks, mitigating potential harms and aligning with the ethical imperatives discussed in the future research directions.

By implementing these recommendations, stakeholders can operationalize the QHIT model, transforming its theoretical robustness into actionable strategies for sustainable and inclusive innovation. This approach not only addresses pressing global challenges but also paves the way for a more collaborative and equitable knowledge economy.

As future work, it is proposed to develop the presented platform and investigate its acceptance and use with interested parties through its own models, with technology acceptance models, commonly known as TAM (Technology Acceptance Models).

ACKNOWLEDGEMENT

This research was financed by national funds provided by the FCT—the Foundation for Science and Technology—through project number UIB/04020/2020 (CinTurs).

REFERENCES

Abd-Rahman, M. S. B., Mohamad, E., & Abdul-Rahman, A. A. B. (2021). Development of IoT—Enabled Data Analytics Enhance Decision Support System for Lean Manufacturing Process Improvement. *Concurrent Engineering, Research and Applications*, *29*(3), 208–220. DOI: 10.1177/1063293X20987911

Abdillah, A., Widianingsih, I., Buchari, R. A., Mustari, N., & Saleh, S. (2022). Governance and Quintuple Helix innovation model: Insights from the local government of East Luwu Regency, Indonesia. *Frontiers in Climate*, *4*, 1012108. DOI: 10.3389/fclim.2022.1012108

Addy, C., Chorengel, M., Collins, M., & Etzel, M. (2019, Jan-Feb). Calculating the Value of Impact Investing. *Harvard Business Review*. Retrieved 5 March 2020: https://hbr.org/2019/01/calculating-the-value-of-impact-investing

Baccarne, B., Logghe, S., Schuurman, D., & De Marez, L. (2016, March). Governing Quintuple Helix Innovation: Urban Living Labs and Socio-Ecological Entrepreneurship. Technology *Innovation Management Review*. Retrieved 5 March 2020, from: https://timreview.ca/article/972

Barth, T. (2011). The idea of a green new deal in a Quintuple Helix Model of knowledge, know-how and innovation. *International Journal of Social Ecology and Sustainable Development*, *1*(2), 1–14. DOI: 10.4018/jsesd.2011010101

Bolsunovskaya, M. V., Kudryavtseva, T. Y., Rudskaya, I. A., Gintciak, A. M., Zhidkov, D. O., Fedyaevskaya, D. E., & Burlutskaya, Z. V. (2023). Digital Platform for Modeling the Development of Regional Innovation Systems of Russian Federation. *International Journal of Technology*, *14*(8), 1779. DOI: 10.14716/ijtech.v14i8.6843

Cai, Y., & Lattu, A. (2022). Triple Helix or Quadruple Helix: Which Model of Innovation to Choose for Empirical Studies? *Minerva*, *60*(2), 257–280. DOI: 10.1007/s11024-021-09453-6

Carayannis, E. G., Barth, T. D., & Campbell, D. F. (2012). The Quintuple Helix innovation model: Global warming as a challenge and driver for innovation. *Journal of Innovation and Entrepreneurship*, *1*, 1–12. DOI: 10.1186/2192-5372-1-1

Carayannis, E. G., & Campbell, D. F. (2021). Democracy of climate and climate for democracy: The evolution of quadruple and quintuple helix innovation systems. *Journal of the Knowledge Economy*, *12*(4), 2050–2082. DOI: 10.1007/s13132-021-00778-x

Carayannis, E. G., & Campbell, D. F. (2022). Towards an Emerging Unified Theory of Helix Architectures (EUTOHA): Focus on the Quintuple innovation helix framework as the integrative device. *Triple Helix (Heidelberg)*, *9*(1), 65–75. DOI: 10.1163/21971927-bja10028

Carayannis, E. G., & Campbell, D. F. J. (2009). "Mode 3" and "Quadruple Helix": Toward a 21st century fractal innovation ecosystem. *International Journal of Technology Management*, *46*(3/4), 201–234. DOI: 10.1504/IJTM.2009.023374

Carayannis, E. G., & Campbell, D. F. J. (2010). The Quintuple Helix innovation model: Global warming as a challenge and driver for innovation. *Journal of Innovation and Entrepreneurship*, *1*(2), 1–1225.

Carayannis, E. G., & Campbell, D. F. J.Architectures (EUTOHA). (2022, April 04). focus on the Quintuple innovation helix framework as the integrative device. *Triple Helix (Heidelberg)*, *9*(1), 65–75. DOI: 10.1163/21971927-bja10028

Carayannis, E. G., Grigoroudis, E., Stamati, D., & Valvi, T. (2019). Social business model innovation: A quadruple/quintuple helix-based social innovation ecosystem. *IEEE Transactions on Engineering Management*, *68*(1), 235–248. DOI: 10.1109/TEM.2019.2914408

Carayannis, E. G., & Morawska, J. (2023). University and education 5.0 for emerging trends, policies and practices in the concept of industry 5.0 and society 5.0. In *Industry 5.0: Creative and Innovative Organizations* (pp. 1–25). Springer International Publishing. DOI: 10.1007/978-3-031-26232-6_1

Carayannis, E. G., & Rakhmatullin, R. (2014). The Quadruple Innovation Helix (QIH): Application for the study of university-industry-government-community interfaces. *International Journal of Innovative Research and Development*, *5*(3-4), 191–212.

Chesbrough, H. W. (2003). *Open innovation: The new imperative for creating and profiting from technology*. Harvard Business Press.

Code World. (2024). *Learn Business Management*. Retrieved May 30, 2024 from: https://play.google.com/store/apps/details?id=com.codeworld.learnbusiness management&hl=pt

Deguchi, A., Hirai, C., Matsuoka, H., Nakano, T., Oshima, K., Tai, M., & Tani, S. (2020). What is society 5.0. *Society*, *5*, 1–23.

Etzkowitz, H., & Leydesdorff, L. (2000). The dynamics of innovation: From National Systems and "Mode 2" to a Triple Helix of university–industry–government relations. *Research Policy, 29*(2), 109–1231. DOI: 10.1016/S0048-7333(99)00055-4

Fagerberg, J., & Mowery, D. (2015). *Recent research on innovation and economic change. Innovation, Technology and Economic Change*. Edward Elgar. DOI: 10.4337/9781784718633

Fu, W., Wang, Q., & Zhao, X. (2018). Platform-based service innovation and system design: A literature review. *Industrial Management & Data Systems, 118*(5), 946–974. DOI: 10.1108/IMDS-03-2017-0129

Geissdoerfer, M., Savaget, P., Bocken, N. M., & Hultink, E. J. (2017). The Circular Economy–A new sustainability paradigm? *Journal of Cleaner Production, 143*, 757–768. DOI: 10.1016/j.jclepro.2016.12.048

Gibbons, M., Limoges, C., Scott, P., Schwartzman, S., & Nowotny, H. (1994). *The new production of knowledge: The dynamics of science and research in contemporary societies.*

GII. (2018). *Global Innovation Index 2018*. Energizing the world with innovation. WIPO. Available at: https://www.wipo.int/edocs/pubdocs/en/wipo_pub_gii_2018.pdf

GII. (2023). *Global Innovation Index 2023*. Innovation in the face of uncertainty. WIPO. Available at: https://www.wipo.int/edocs/pubdocs/en/wipo-pub-2000-2023-en-main-report-global-innovation-index-2023-16th-edition.pdf

Hall, B. H., & Rosenberg, N. (2010). *Handbook of the Economics of Innovation*. Elsevier.

Innovation Markets. (2024). *Income Markets*. Retrieved May 30, 2024 from: https://play.google.com/store/apps/details?id=com.innovationmarkets.android&hl=pt

Kholiavko, N., Grosu, V., Safonov, Y., Zhavoronok, A., & Cosmulese, C. G. (2021). Quintuple helix model: Investment aspects of higher education impact on sustainability. *Management Theory and Studies for Rural Business and Infrastructure Development, 43*(1), 111–128. DOI: 10.15544/mts.2021.10

Liyanage, S. I. H., & Netswera, F. G. (2021). Greening universities with mode 3 and quintuple helix model of innovation–production of knowledge and innovation in knowledge-based economy, Botswana. *Journal of the Knowledge Economy*, •••, 1–31.

Mahardhani, A. J. (2023). The Role of Public Policy in Fostering Technological Innovation and Sustainability. [ADMAN]. *Journal of Contemporary Administration and Management, 1*(2), 47–53. DOI: 10.61100/adman.v1i2.22

Muruccia, Y., Solazzo, G., Del Vecchio, P., & Passiante, G. (2020). Evidence from network analysis application to innovation systems and quintuple helix. *Technological Forecasting and Social Change, 161*, 120306. DOI: 10.1016/j.techfore.2020.120306

Nunes, S., Lopes, R. & Fuller-Love, N (2017, June). Networking, Innovation, and Firms' Performance: Portugal as Illustration. *Journal of the Knowledge Economy* 10, 899–920. DOI: 10.1007/s13132-017-0508-7 -

Pinto, H., & Nogueira, C. (2018) Mapping na Entrepreneurial, Innovative and Sustainable Ecosystem Using Social Network Analysis: An Exploratory Approach of Publicly Funded Innovative Project Data. In: Leitão J., Alves H., Krueger N., Park J. (eds) *Entrepreneurial, Innovative and Sustainable Ecosystems. Applying Quality of Life Research* (Best Practices). Springer, Cham. DOI: 10.1007/978-3-319-71014-3_12

Pricopoaia, O., Busila, A. V., Cristache, N., Susanu, I., & Matis, C. (2023). Challenges for entrepreneurial innovation: Startups as tools for a better knowledge-based economy. *The International Entrepreneurship and Management Journal*, **•••**, 1–42.

Putra, Z. D. W., & van der Knaap, W. G. (2018). Urban innovation system and the role of an open web-based platform: The case of Amsterdam smart city. *Journal of Regional and City Planning, 29*(3), 234–249. DOI: 10.5614/jrcp.2018.29.3.4

Qmarkets (2024). *End-to-End Innovation Management Platform.* Retrieved 30 may, 2024 from https://www.qmarkets.net/lp/innovation-management-software/?hsa_qmt=y&utm_term=innovation%20platform&utm_campaign=Tier+1+Countries+%7C+Gold+KWs+%26+Silver+KWs+%7C+Exact+%26+Phrase&utm_source=adwords&utm_medium=ppc&hsa_acc=3759777188&hsa_cam=8690653365&hsa_grp=139195627185&hsa_ad=696764641653&hsa_src=g&hsa_tgt=kwd-6668275533&hsa_kw=innovation%20platform&hsa_mt=p&hsa_net=adwords&hsa_ver=3&gad_source=1&gclid=CjwKCAjwx-CyBhAqEiwAeOcTde0vSdyeHGevzm0u29NihVgwUZcEE_6P3UWSO69UWyysLMQuWORLExoCPoUQAvD_BwE

Reis, J., Rosado, D. P., Ribeiro, D. F., & Melão, N. (2022). Quintuple Helix Innovation Model for the European Union Defense Industry—An Empirical Research. *Sustainability (Basel), 14*(24), 16499. DOI: 10.3390/su142416499

Vignieri, V. (2021). Crowdsourcing as a mode of open innovation: Exploring drivers of success of a multisided platform through system dynamics modelling. *Systems Research and Behavioral Science, 38*(1), 108–124. DOI: 10.1002/sres.2636

Zen, I. S., & Shibakawa, H. (2022). Quintuple Helix Lens for Transformation: An Okayama Model of Education for Sustainable Development. *Frontiers in Sustainability*, *3*, 798330. DOI: 10.3389/frsus.2022.798330

Chapter 2
A Real-Time System for Gaze Tracking and Emotion Recognition Using MediaPipe and DeepFace

Khoa Dang Nguyen
http://orcid.org/0000-0002-6525-5245
International School, Vietnam National University, Vietnam

Van Thi Tran
Faculty of General Education, Vietnam National University, Vietnam

Solanki Kumar Vijender
http://orcid.org/0000-0001-5784-1052
Department of Computer Science and Engineering, Vietnam National University, Vietnam

Kiet Minh Tuan Tran
http://orcid.org/0009-0006-5424-7692
International School, Vietnam National University, Vietnam

Duc-Tan Tran
http://orcid.org/0000-0002-7673-388X
Faculty of Electrical and Electronic Engineering, Vietnam National University, Vietnam

DOI: 10.4018/979-8-2600-1026-6.ch002

ABSTRACT

In recent years, the topic of human–computer interaction has become one of the key research areas shaping the future of technology. Gaze tracking and facial expression recognition have attracted significant attention, particularly in the field of computer vision. In this paper, we propose an integrated system that combines gaze direction tracking and emotion recognition using MediaPipe and DeepFace frameworks to assess the user's level of concentration. Both frameworks are capable of processing visual data and performing artificial intelligence (AI) based analysis. The system processes input from a real-time camera stream and provides feedback to users based on the detected gaze orientation and emotional states. The results are visualized through a web-based interface, enabling intuitive observation and evaluation of user engagement levels.

INTRODUCTION

In the era of digital transformation and artificial intelligence (AI), Human-Computer Interaction (HCI) has emerged as one of the most dynamic and rapidly advancing research domains. Within this field, Gaze Tracking (GT) and Facial Expression Recognition (FER) have become two fundamental components that enable machines to interpret human attention, emotion, and engagement. The capability of machines to both "see" and "understand" human affective states holds immense potential for applications across diverse domains such as education, healthcare, intelligent transportation, marketing, and decision-support systems.

From a global perspective, the growing prevalence of digital workplaces, online education, and AI-assisted decision-making has intensified the need for intelligent systems capable of understanding human cognitive and affective states at scale. The conceptual roots of GT can be traced to early eye-movement research in psychology and cognitive science, where gaze behavior was shown to reflect visual attention and information processing (Yarbus, 1967; Duchowski, 2007). Similarly, FER originates from affective psychology and behavioral science, particularly Ekman's theory of basic emotions, which established a systematic link between facial muscle movements and emotional states (Ekman & Friesen, 1978). Empirical HCI literature has since demonstrated that attention and emotion constitute two complementary dimensions of human behavior that jointly shape task performance, engagement, and decision quality (R. A. Calvo & S. D'Mello, 2010).

From an organizational and policy perspective, the adoption of GT-FER technologies is strongly influenced by technology acceptance and diffusion theories. The Technology Acceptance Model (TAM) posits that perceived usefulness and perceived ease of use determine user acceptance of new technologies (Davis, 1989). Extending this view, the Unified Theory of Acceptance and Use of Technology (UTAUT) and UTAUT2 emphasize performance expectancy, effort expectancy, social influence, and facilitating conditions as key drivers of adoption in organizational contexts (Venkatesh et al., 2003, 2012). These theoretical frameworks suggest that GT-FER systems are more likely to be adopted when they demonstrably enhance productivity, reduce cognitive workload, and align with organizational values and governance structures.

Despite extensive technical progress, significant gaps remain between laboratory-level performance and real-world organizational deployment. This raises a critical research question that motivates the present study: How can gaze and emotion recognition be integrated into a real-time, cost-effective, and ethically deployable system that supports organizational decision-making, productivity, and human-centered AI?

In the context of HCI, both GT and FER play essential and complementary roles in understanding and adapting to user behavior. In educational environments, GT has been widely used to evaluate students' attention and cognitive engagement during learning activities, providing instructors with data-driven insights to tailor pedagogical strategies and improve learning outcomes (Bouktif & N. K, 2024; Silvestri et al., 2024; X. Zhang et al., 2015). FER, on the other hand, enables automated detection of emotions such as interest, confusion, or fatigue, supporting real-time feedback loops that enhance interactive learning systems and classroom experiences (L. Santamaria-Granados et al., 2019; R. A. Calvo & S. D'Mello, 2010).

In marketing and consumer experience research, GT facilitates the analysis of users' visual attention toward advertisements, digital interfaces, and product displays, allowing designers and marketers to optimize visual hierarchy, layout, and content placement (Białowąs & Szyszka, 2019). FER complements this by capturing affective responses such as satisfaction, surprise, or frustration in real time, offering valuable insights into emotional engagement within digital environments (Fasel & Luettin, 2003; Gehrer et al., 2024). In the realm of intelligent transportation systems, HCI-driven applications of GT are used to monitor driver attention, detect distraction, and enhance safety through adaptive vehicle interfaces, while FER technologies can identify early signs of drowsiness, stress, or negative affect, contributing to proactive warning mechanisms that foster safer human-machine collaboration (Shah et al., 2022).

Within healthcare, the integration of GT and FER technologies under the HCI framework supports non-invasive monitoring and diagnostic assistance. GT has demonstrated effectiveness in detecting atypical eye-movement patterns associated with neurological disorders such as Parkinson's disease and autism spectrum disorder (Sigut & Sidha, 2011). FER systems, in parallel, aid clinicians in assessing the emotional and psychological well-being of patients experiencing depression or affective disorders, thereby promoting more empathetic, responsive, and human-centered healthcare technologies (R. A. Calvo & S. D'Mello, 2010; Said et al., 2026).

Despite these promising applications, real-world implementation reveals substantial challenges. Case studies in online education and remote work monitoring report difficulties related to sensor calibration, user movement, lighting variability, and user acceptance when GT and FER systems are deployed outside controlled environments (Pauszek, 2023). In organizational contexts, employees frequently express concerns regarding surveillance, data misuse, performance evaluation fairness, and lack of transparency, which can undermine trust and adoption (Ball, 2010; Ajunwa et al., 2017). These challenges highlight the necessity of lightweight architectures, transparent analytics, and ethically grounded system design.

However, achieving effective performance in real-world scenarios requires balancing accuracy, processing speed, and adaptability. Traditional GT systems, particularly those based on the Pupil Center Corneal Reflection (PCCR) method (Carelli et al., 2022; Sigut & Sidha, 2011), employ infrared cameras to determine gaze direction through geometric relationships between the pupil center and corneal reflections. Although such methods deliver high precision under controlled laboratory conditions, they face challenges including complex hardware requirements, limited operating ranges, and high deployment costs.

Traditional gaze estimation techniques, particularly those based on the PCCR method, have long been regarded as the gold standard in precision-oriented gaze tracking research (Zhai, 2008; Hennessey et al., 2006; T. Ogawa et al., 2017). The PCCR approach estimates gaze direction by modeling the geometric relationship between the pupil center and corneal reflections generated by infrared illumination.

Despite their high accuracy and robustness under controlled laboratory conditions, empirical studies consistently report significant limitations when PCCR-based systems are deployed in real-world HCI contexts. These systems typically rely on infrared light sources, multi-camera configurations, and manual calibration procedures, which increase hardware complexity, deployment cost, and user burden (Morimoto & Mimica, 2005; MacKenzie, 2010). Such constraints reduce their suitability for mobile, remote, and large-scale organizational applications, where unobtrusive interaction and ease of use are critical.

While PCCR-based systems are widely regarded as a precision benchmark, empirical evidence indicates limited generalizability to mobile, remote, and organizational environments due to strict calibration requirements and sensitivity to head movement. Recent research has attempted to mitigate these limitations through hybrid approaches that combine geometric modeling with Convolutional Neural Networks (CNNs), as well as transfer learning strategies to enhance robustness across users and contexts (P. Kellnhofer et al., 2019; Wood et al., 2016). Nevertheless, these solutions often introduce additional computational complexity and hardware dependencies, constraining large-scale organizational deployment.

To overcome these limitations, recent advances in HCI have shifted toward AI-based, single-camera approaches (Valle et al., 2019; X. Geng & Y. Xia, 2014; Xiao et al., 2018). (Kanade et al., 2021) proposed a single-camera GT framework combining deep learning and geometric modeling, achieving accuracies of up to 98.77% using MediaPipe facial landmarks while maintaining real-time performance on consumer-grade devices. This CNN-geometric hybrid approach demonstrates how modern architectures can approach PCCR-level accuracy while significantly improving usability and scalability.

The field of FER has likewise evolved from handcrafted feature-based systems to deep learning driven architectures. Early FER systems relied on Gabor wavelets (Y. Xu & Cheng, 2025), Principal Component Analysis (PCA), and Linear Discriminant Analysis (LDA) combined with classical classifiers (J. Li et al., 2009; Zheng, 2022). Although effective in controlled datasets, these approaches exhibited limited robustness and generalization in real-world HCI settings characterized by pose variation, occlusion, and lighting changes (Duan, 2023). Deep learning models such as DeepFace demonstrate improved generalization, achieving recognition rates of up to 82% on the FER-2013 dataset (Goodfellow et al., 2013; M. Sharma et al., 2024; S. B. Punuri et al., 2023).

Recent FER research emphasizes lightweight, real-time architectures leveraging MediaPipe FaceMesh, enabling recognition accuracies between 92% and 97% with low computational overhead (Alanazi et al., 2024; Lugaresi et al., 2019). These developments are particularly relevant for organizational and educational applications where real-time responsiveness and deployment cost are critical constraints.

From a theoretical standpoint, the relevance of GT and FER in HCI is grounded in human-centered AI and affective computing. Attention-based interaction theories posit that gaze reflects cognitive focus and information allocation, while affective computing frameworks emphasize emotion as a core driver of human behavior and interaction quality (Picard, 1997; R. A. Calvo & S. D'Mello, 2010). Integrating gaze and emotion therefore enables AI systems to move beyond reactive interfaces toward adaptive, empathetic, and context-aware interaction models.

Ideally, organizational HCI systems should unobtrusively measure attention and emotion in real time, operate on low-cost hardware, generalize across users and contexts, and comply with ethical and regulatory standards. However, contradictory empirical evidence exists regarding accuracy-usability trade-offs, inconclusive findings on cross-context generalization, and persistent theoretical gaps between attention modeling and affective inference. Moreover, most existing studies treat GT and FER as separate modalities, resulting in fragmented insights and limited organizational applicability. This study addresses these gaps by proposing an integrated GT-FER framework that balances technical performance with scalability, usability, and ethical deployment, thereby offering actionable guidance for future researchers and policy makers.

This study draws upon four primary theoretical foundations: (1) Affective Computing Theory (Picard, 1997), (2) Attention-Based Interaction Theory (Duchowski, 2007), (3) Technology Acceptance Model (Davis, 1989), and (4) Unified Theory of Acceptance and Use of Technology (Venkatesh et al., 2003, 2012). Among these, UTAUT serves as the primary theoretical lens, as it explicitly explains organizational adoption by incorporating performance expectancy, effort expectancy, social influence, and facilitating conditions-factors critical to deploying GT-FER systems in workplaces, educational institutions, and policy-driven environments.

At the organizational level, integrated GT-FER systems support productivity enhancement, remote work management, and adaptive learning by providing aggregated indicators of concentration and emotional engagement rather than intrusive individual surveillance. Such systems can inform interface redesign, training effectiveness, and workload management, directly addressing decision-makers' concerns regarding cost, return on investment, and stakeholder acceptance. Nevertheless, responsible deployment requires strict compliance with workplace privacy regulations such as the General Data Protection Regulation (GDPR) and alignment with AI ethics frameworks emphasizing transparency, consent, accountability, and algorithmic fairness across demographic groups (Buolamwini & Gebru, 2018; Floridi et al., 2018).

Addressing this gap, the present study proposes an integrated GT-FER system for assessing users' visual attention and emotional engagement in real-world HCI environments. The framework combines MediaPipe FaceMesh for geometry-based gaze estimation and DeepFace for facial emotion recognition, operating efficiently on standard RGB cameras without specialized infrared hardware. This design reduces deployment costs while maintaining real-time performance and cross-device scalability.

The extracted gaze and emotion data are visualized through an interactive Streamlit-based dashboard, providing real-time analytics such as gaze distribution, emotion trends, and concentration-related indicators. The system adopts a secure client-server architecture deployed via Cloudflared Tunnel, ensuring encrypted access and scalable integration. Overall, the proposed framework emphasizes a balanced trade-off between accuracy, efficiency, and adaptability, contributing a practical and ethically deployable HCI solution with applications in online education, remote collaboration, and human-centered AI systems.

DEVELOPMENT OF GAZE TRACKING AND FACIAL EXPRESSION RECOGNITION ALGORITHMS

In recent years, gaze tracking and facial expression recognition have been increasingly adopted in real-world Human-Computer Interaction (HCI) applications, particularly in online education, remote work environments, intelligent user interfaces, and mental health monitoring systems. These technologies are widely used to infer users' visual attention, emotional engagement, and cognitive states during interaction, enabling adaptive feedback, usability evaluation, and behavioral analytics beyond traditional input modalities. As camera-based sensing becomes ubiquitous through webcams and mobile devices, lightweight and software-based GT-FER systems have gained prominence due to their scalability and cost efficiency.

In this section, the proposed system integrates Gaze Tracking (GT) and Facial Emotion Recognition (FER) to assess the user's level of concentration in real time. The system is composed of two main modules: MediaPipe Facemesh and DeepFace, which operate based on input data from a standard camera. The analysis results, including gaze direction ratio and emotion distribution, are aggregated to generate appropriate feedback for the user. The overall workflow of the proposed system is illustrated in Figures 1 and 2.

Figure 1. Pre-processing with Mediapipe facemesh

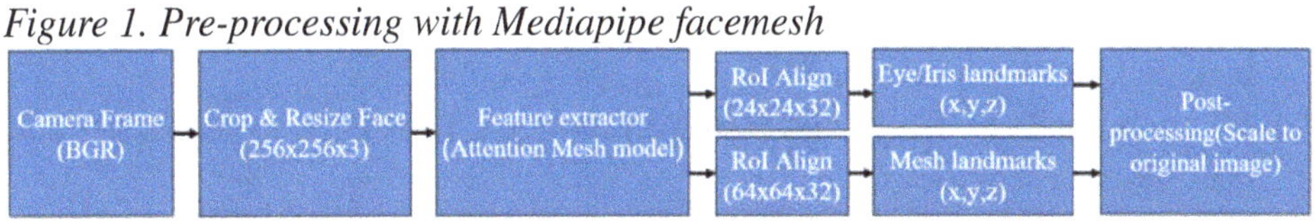

Figure 2. Emotion, eye contact analysis and UI

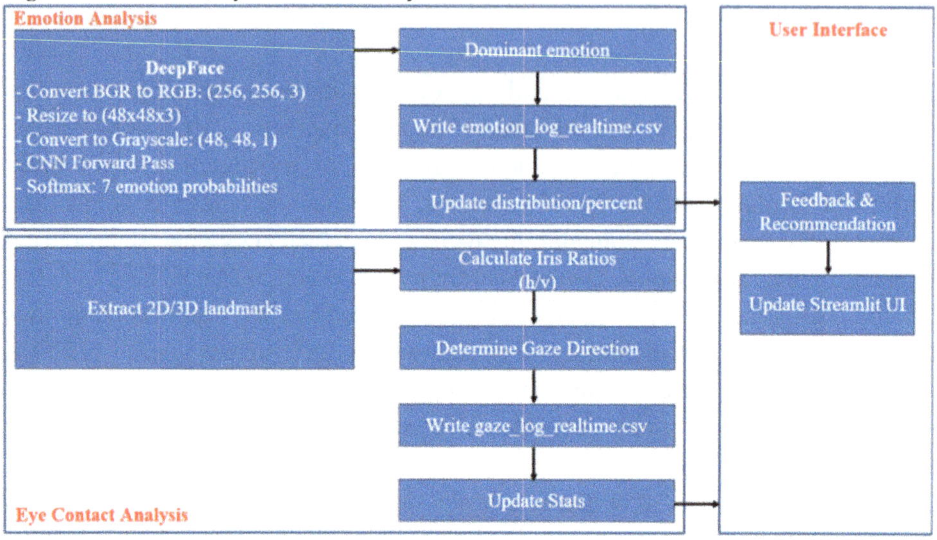

The proposed system is composed of four main processing blocks:

(1) Preprocessing Block (in Figure 1) utilizes MediaPipe Facemesh to detect faces, crop regions of interest (face cropping), and extract geometric features (facial landmarks) from each video frame.
(2) Emotion Analysis Block (in Figure 2) employs DeepFace to recognize and classify facial emotions;
(3) Eye Contact Analysis Block (in Figure 2) determines gaze direction and eye contact ratio based on facial landmarks extracted by MediaPipe;
(4) User Interface Block (in Figure 2) visualizes the analysis results, displaying emotion percentages, gaze direction ratios, and providing intuitive real-time feedback to the user.

From a behavioral perspective, gaze direction and facial emotion are widely recognized as observable proxies of users' attention allocation and emotional engagement. Sustained forward gaze is commonly associated with task focus and information processing, whereas frequent gaze aversion may indicate distraction or reduced cognitive involvement. Similarly, facial expressions such as neutrality, confusion, or fatigue provide cues about users' affective states, which directly influence learning effectiveness, decision-making quality, and interaction outcomes. Therefore, integrating gaze and emotion analytics enables a more holistic assessment of engagement than relying on recognition accuracy alone.

Specifically, during the preprocessing phase, MediaPipe Facemesh detects and extracts facial features from each camera frame. These features are then forwarded to two parallel branches, Emotion Analysis and Eye Contact Analysis. The resulting outputs are aggregated and mapped through a set of feedback rules to provide qualitative evaluations of the user's concentration and engagement levels. Finally, the User Interface (UI) Block visualizes the overall analysis, including gaze distribution charts, dominant emotions, and real-time feedback for the user.

Facial Emotion Analysis

In the Emotion Analysis block, DeepFace employs a CNN model to classify facial expressions into seven basic emotion categories: Angry, Disgust, Fear, Happy, Sad, Surprise, and Neutral (Figure 3). Based on these classification results, the system identifies the dominant emotion for each frame and records the corresponding information into a log file named emotion_log_realtime.csv.

Figure 3. CNN structure in deepface emotion model

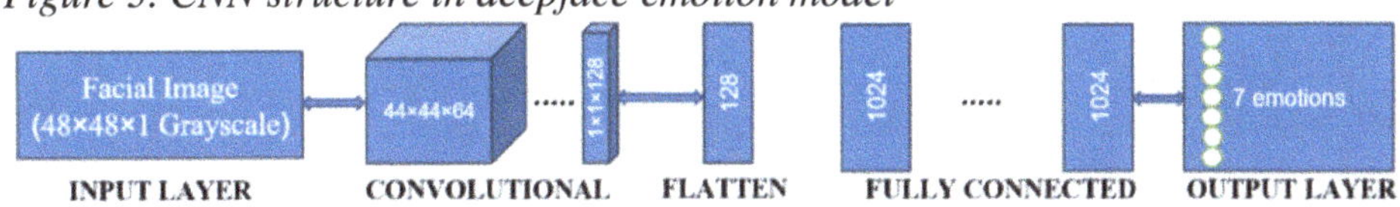

The time-series data are stored in a log file, allowing the system to evaluate the facial emotion distribution of the user according to Eq. 1.

$$P_{emotion} = \frac{N_{emotion}}{N_{total}} \times 100\% \tag{1}$$

Where $P_{emotion}$ represents the percentage of a specific emotion, used to assess the user's level of engagement. $N_{emotion}$ denotes the number of occurrences of that emotion during the entire analysis session. N_{total} is the total number of valid frames in which a recognizable facial emotion is detected. In cases where no face is detected in a frame, the system retains the previous emotion value to maintain the continuity and stability of the emotional data stream.

Eye Contact Analysis

The Eye Contact Analysis block utilizes facial landmarks extracted from MediaPipe Facemesh to determine the user's gaze direction, categorized into left, right, up, down, or center. Specifically, the landmarks of the eyes and irises are used to calculate the relative iris position along both the horizontal and vertical axes, as defined by Eqs. (2) and (3).

$$h_{L\_R} = \frac{x_{iris} - x_{left}}{x_{right} - x_{left}},\ h_{L\_R} \in [0, 1] \tag{2}$$

$$v_{T\_B} = \frac{y_{iris} - y_{top}}{y_{bottom} - y_{top}},\ v_{T\_B} \in \left[0, 1\right] \tag{3}$$

Where (x_{iris}, y_{iris}) represent the coordinates of the iris center, while (x_{left}, y_{left}), (x_{right}, y_{right}), (x_{top}, y_{top}), (x_{bottom}, y_{bottom}) correspond respectively to the left corner, right corner, upper eyelid, and lower eyelid landmarks of the eye.

A horizontal ratio value h_{L_R} close to 0 indicates that the gaze is directed to the left, while a value near 1 corresponds to a rightward gaze. Similarly, a vertical ratio v_{T_B} close to 0 indicates an upward gaze, and a value near 1 represents a downward gaze. The average gaze ratios across both eyes for the horizontal and vertical axes are computed using Eqs. 4 and 5.

$$h_{avg} = \frac{h^{(L)}_{L\_R} + h^{(R)}_{L\_R}}{2} \tag{4}$$

$$v_{avg} = \frac{v^{(L)}_{T\_B} + v^{(R)}_{T\_B}}{2} \tag{5}$$

Here $h^{(L)}_{L\_R}$ và $h^{(R)}_{L\_R}$ denote the horizontal gaze ratios of the left and right eyes, respectively, while; $v^{(L)}_{T\_B}$ và $v^{(R)}_{T\_B}$ represent the corresponding vertical gaze ratios.

The computed average values h_{avg} and v_{avg} are then compared against empirically calibrated threshold values to accurately determine the user's gaze direction, whether looking left, right, up, down, or forward.

Subsequently, the system calculates the Forward Gaze Ratio (FGR), representing the percentage of time the user maintains direct eye contact, using Eq. 6.

$$P_{forward} = \frac{N_{forward}}{N_{total}} \times 100\% \tag{6}$$

In which, $P_{forward}$ represents the forward gaze ratio, used to evaluate the user's level of attention. $N_{forward}$ denotes the number of frames in which the user is looking directly at the screen during the analysis session.

N_{total} is the total number of valid frames in which a gaze direction is successfully detected.

All results are recorded in a log file named gaze_log_realtime.csv, enabling real-time tracking and statistical analysis of gaze direction distribution throughout the session.

Although the proposed approach demonstrates strong real-time performance, it is subject to several practical limitations. The accuracy of both gaze estimation and facial emotion recognition can be affected by environmental lighting conditions, partial face occlusion (e.g., glasses, masks), extreme head poses, and camera quality. These factors may introduce noise into landmark detection and emotion classification, particularly in unconstrained real-world settings. Acknowledging these limitations is essential to ensure realistic interpretation of the system's outputs and to guide future improvements.

User Interface

Finally, the User Interface (UI) was developed using Streamlit, serving as a platform to aggregate and visualize the outputs from the Emotion Analysis and Eye Contact Analysis modules. The interface operates in real time, displaying charts of emotion distribution, gaze direction, eye contact ratio, and the dominant emotional state. In addition, the system provides automatic feedback and adaptive suggestions to help users adjust their emotional state and maintain focus during interactions.

Beyond technical performance, the proposed system provides actionable insights for practitioners such as educators, system designers, and HCI researchers. By visualizing aggregated gaze and emotion patterns rather than individual-level surveillance, the system can support interface evaluation, attention-aware learning environments, and user experience assessment while minimizing intrusiveness. These capabilities highlight the practical relevance of integrating GT and FER for real-world HCI applications where engagement and usability are critical success factors.

EXPERIMENTS AND RESULTS

Datasets and System Configuration

The experiments were conducted using two benchmark datasets: FER-2013 (Goodfellow et al., 2013) for Facial Expression Recognition (FER) and Eye-Tracker-Data (H. Xu et al., 2023) for Gaze Estimation (GE).

The FER-2013 dataset contains 35,887 grayscale facial images with a resolution of 48×48 pixels. The dataset is divided into training and testing subsets with an 80:20 ratio, each containing seven emotion labels: Angry, Disgust, Fear, Happy,

Sad, Surprise, and Neutral. This dataset is widely used as a standard benchmark for facial emotion recognition research.

The Eye-Tracker-Data dataset consists of 7,500 eye region images collected from 12 participants, evenly distributed across five gaze directions: left, right, up, down, and center. This dataset is used to train and evaluate the gaze estimation component of the proposed system. The experiments were executed on a system with the following configuration:

- Hardware: Intel Core i7 (16 CPUs), 16 GB RAM, NVIDIA GeForce RTX 3060 Laptop GPU
- Operating System: Windows 11
- Programming Language and Environment: Python 3.12
- Supporting Libraries and Tools: Streamlit, MediaPipe, DeepFace, OpenCV, NumPy, and Pandas for data processing and visualization.

For experimental implementation, the proposed model was re-coded entirely in Python to ensure scalability and seamless integration with real-time systems. In the case of DeepFace, a pretrained model was employed via the official library to leverage deep learning capabilities trained on large-scale datasets.

Evaluation of Facial Emotion Recognition Performance

For the FER-2013 dataset, the proposed model was applied to the Facial Expression Recognition (FER) task and evaluated in comparison with two baseline approaches:

- HOG and SVM (Bergstrom & Schall, 2014): A traditional model based on manually engineered (hand-crafted) feature extraction using Histogram of Oriented Gradients (HOG) followed by classification with a Support Vector Machine (SVM).
- CNN and LSTM (Bergstrom & Schall, 2014): A deep learning-based hybrid model combining CNN for spatial feature extraction and Long Short-Term Memory (LSTM) networks for temporal sequence modeling.

All evaluations were conducted on the testing set immediately after the training phase, ensuring consistency and fairness in performance comparison among the different models.

Figure 4. Accuracy value in HOG and SVM case

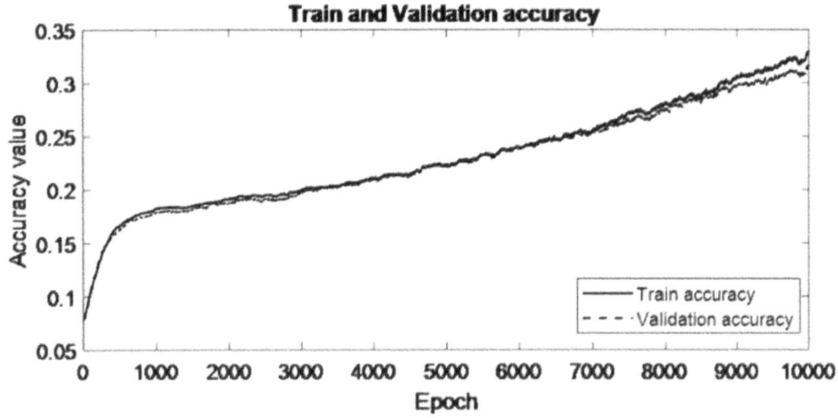

Figure 5. Loss value in HOG and SVM case

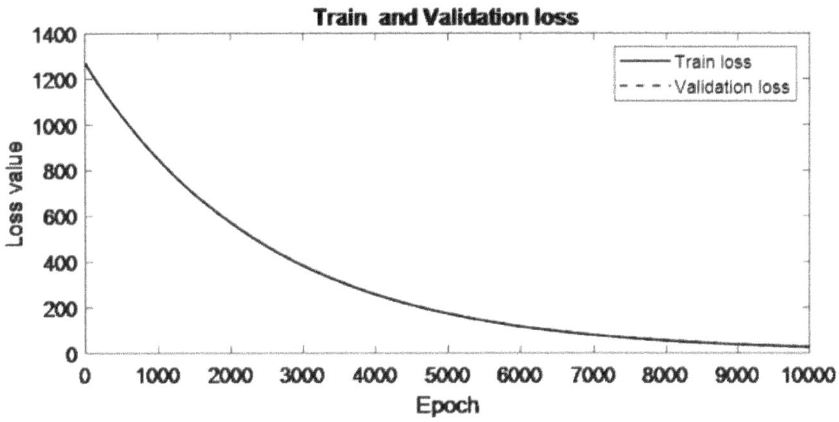

Figure 6. Accuracy and loss of Deepface in FER2013 dataset

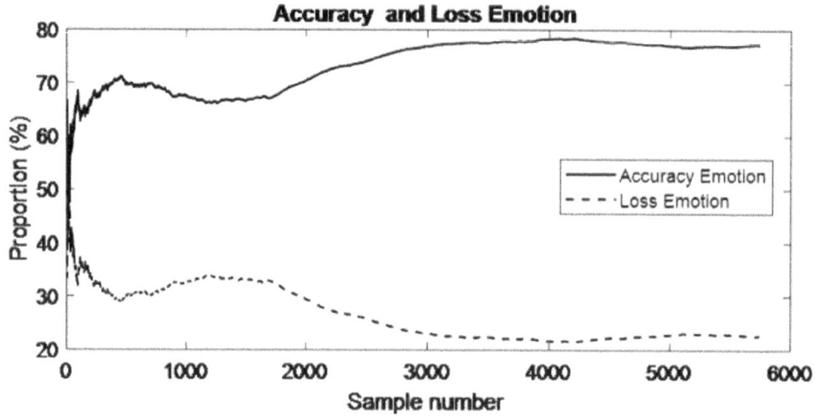

Figure 7. Accuracy value in LSTM and CNN case

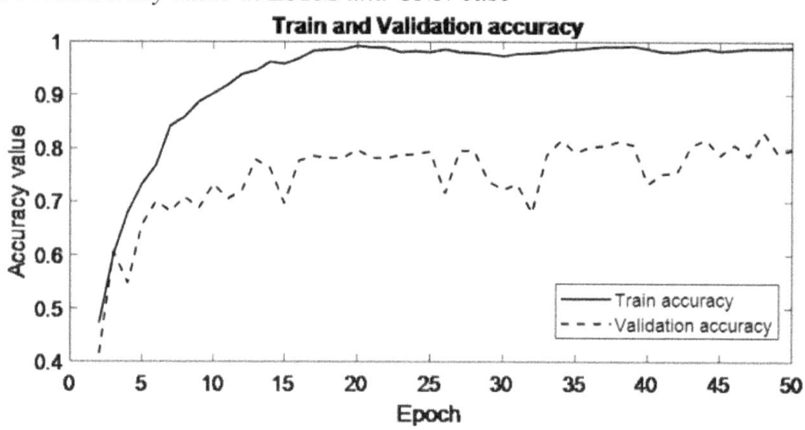

Figure 8. Loss value in LSTM and CNN case

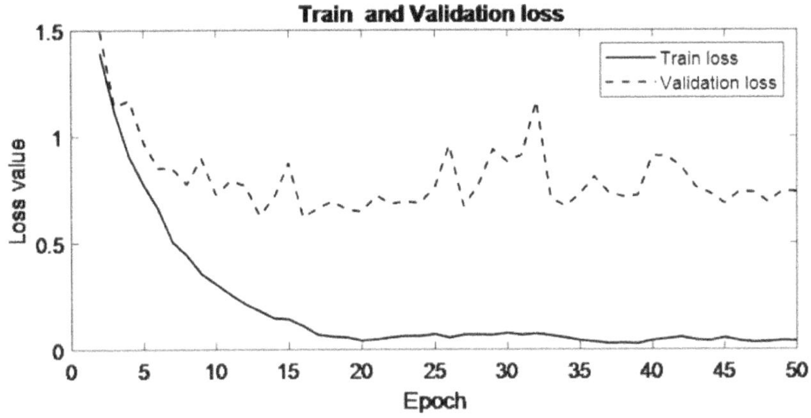

The Figure 3 to Figure 8 clearly demonstrate the distinct differences between traditional machine learning-based approaches and modern deep learning-based methods in the task of facial emotion recognition.

The HOG and SVM model exhibits a relatively slow but stable learning process, with accuracy gradually improving over successive training epochs. Although its overall performance remains modest, this limitation primarily stems from the nature of HOG features, which rely on grayscale and edge-based representations, leading to restricted facial expressiveness. Nevertheless, the key advantages of this approach include its simplicity, low computational cost, and high reliability under resource-constrained environments, making it suitable for small-scale or embedded applications.

In contrast, the CNN and LSTM model demonstrates faster convergence and higher training accuracy (up to 80%, as shown in Table 1), benefiting from the combination of spatial feature extraction (via CNN) and temporal sequence modeling (via LSTM). However, experimental results reveal that this model is prone to overfitting, especially when the training data lack diversity or appropriate regularization and dropout mechanisms. This leads to noticeable discrepancies in classification accuracy across different emotion categories, indicating limited generalization capability in real-world scenarios.

Notably, the DeepFace model achieves superior performance and rapid convergence, reaching an accuracy of 77% (Table 1), owing to its use of a deep CNN pretrained on a large-scale dataset. This model maintains consistent accuracy across both training and testing sets and performs effectively in real-time conditions, making it well-suited for direct deployment in webcam-based facial emotion recognition systems.

Table 1. Experimental results of facial expression recognition models on the FER-2013 dataset

Metrics	HOG and SVM	DEEPFACE (FER)	CNN and LSTM
Accuracy	32%	77%	80%
Macro Average F1	27%	75%	75%
Weighted Average F1	32%	77%	80%
Best Class (F1)	Happy(46%)	Happy(89%)	Happy(95%)
Worst Class (F1)	Disgust(4%)	Disgust(64%)	Disgust(50%)

Overall, the results indicate that DeepFace offers the best balance among accuracy, convergence speed, and real-time deployability. Although the CNN and LSTM model achieves slightly higher accuracy under controlled conditions, DeepFace demonstrates superior stability when environmental factors such as lighting conditions, viewing angles, or camera quality vary, a crucial aspect for emotion and gaze tracking systems operating in real-world environments.

Evaluation of the Gaze Estimation Algorithm

Similar to the facial expression recognition (FER) task, for gaze estimation, experiments were conducted to compare two popular face feature extraction methods: Haarcascade and MediaPipe Facemesh. Both approaches were integrated with a Convolutional Neural Network (CNN) to classify gaze direction based on eye region images. The objective of this experiment was to evaluate and compare the feature extraction performance, training convergence, and stability of the two methods when deployed in a real-time gaze tracking system.

Both models were trained and tested on the Eye-Tracker-Data dataset (H. Xu et al., 2023), which consists of 7,500 eye images evenly distributed across five gaze direction classes: left, right, up, down, and center. The results of the experiments are illustrated in the accuracy and loss curves shown in Figure. 9 and 12.

Figure 9. Accuracy value in Haarcascade and CNN case

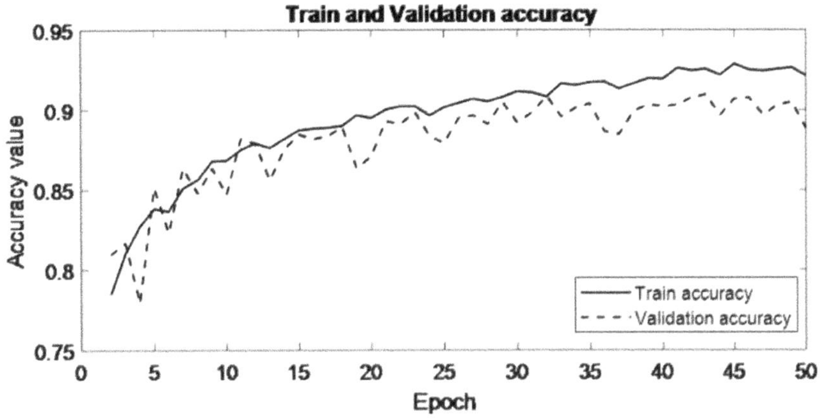

Figure 10. Loss value in Haarcascade and CNN case

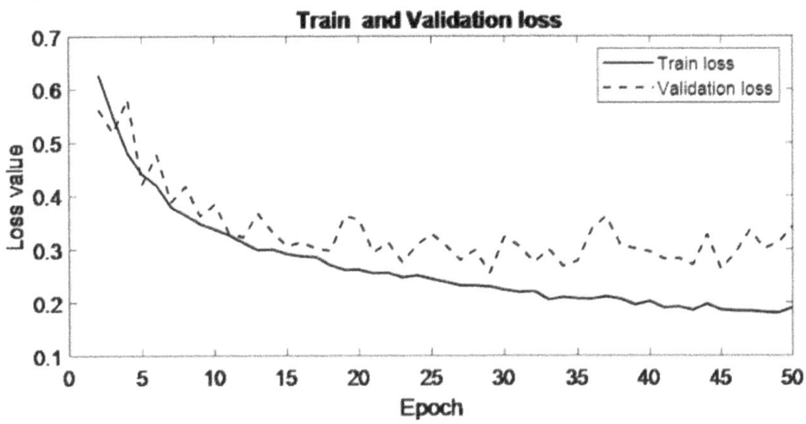

Figure 11. Accuracy value in Mediapipe and CNN case

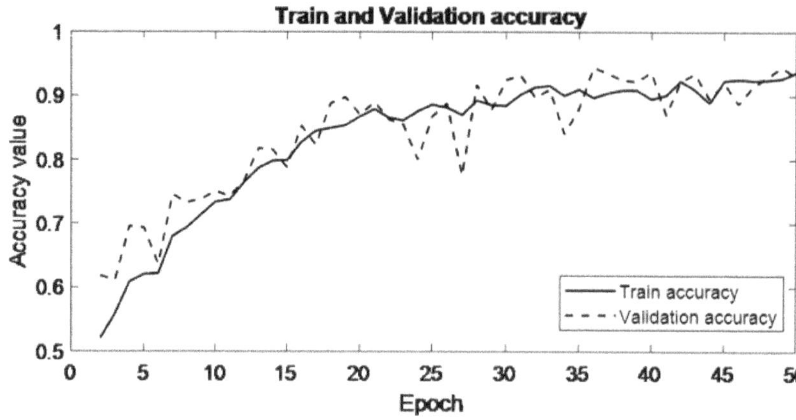

Figure 12. Loss value in Mediapipe and CNN case

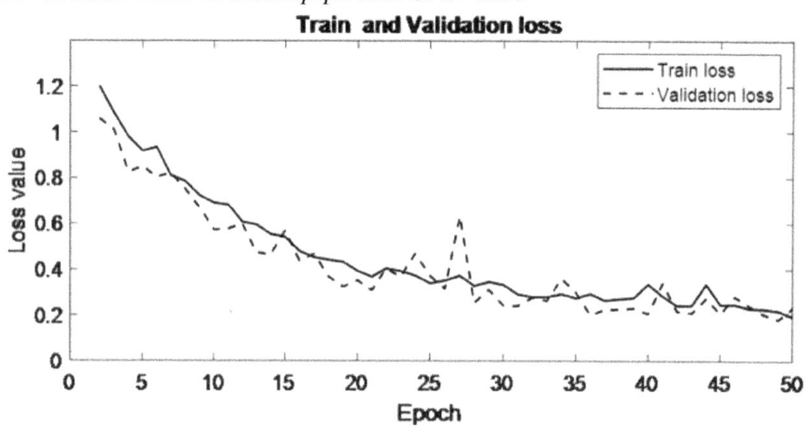

By examining the two sets of figures, it can be observed that both models achieve high training performance and relatively stable convergence; however, several key differences emerge:

Convergence speed: The MediaPipe and CNN model (Figures 11 and 12) demonstrates faster convergence, achieving high accuracy within the first few epochs. In contrast, Haarcascade and CNN (Figures 9 and 10) requires more epochs to stabilize and reach the desired accuracy. This suggests that MediaPipe provides "cleaner" and more informative input features, enabling the CNN to learn more efficiently.

Accuracy and stability: In the later stages of training, both models achieve an accuracy above 90%. However, MediaPipe and CNN maintains a more stable validation accuracy curve that closely aligns with training accuracy, indicating reduced overfitting. Meanwhile, Haarcascade and CNN exhibits stronger fluctuations in validation loss, suggesting sensitivity to noise during eye or face region detection.

Loss function behavior: The MediaPipe and CNN model presents a smooth and steadily decreasing loss curve, reflecting stable optimization. Conversely, Haarcascade and CNN shows more pronounced oscillations, particularly on the validation set, implying instability when handling noisy or complex visual inputs.

Overall, the training results confirm that MediaPipe and CNN outperforms Haarcascade and CNN in terms of learning speed, accuracy, and robustness, mainly due to its ability to extract facial landmarks with greater precision and consistency than the manually designed Haar-based detection method.

Table 2 below presents the quantitative comparison between the two models across several metrics, including Accuracy, Macro Average F1, and Weighted Average F1, along with an analysis of the classes with the highest and lowest accuracy.

Table 2. Experimental results of facial feature extraction preprocessing models

Metrics	Haarcascade and CNN	Mediapipe and CNN
Accuracy	89%	93%
Macro Average F1	83%	94%
Weighted Average F1	89%	93%
Best Class (F1)	Center(92%)	Down(100%)
Worst Class (F1)	Down(57%)	Center(89%)

The results indicate that although the Haarcascade and CNN model achieves a relatively high accuracy of 89%, it exhibits a substantial inter class variation of 35%, reflecting instability when classifying complex gaze directions. In contrast, the MediaPipe and CNN model not only achieves a higher overall accuracy of 93% but also significantly reduces the inter class discrepancy to 11%. This demonstrates that the facial features extracted using MediaPipe are more discriminative and less sensitive to environmental variations, thereby improving robustness and consistency in real-time gaze estimation.

From the quantitative and qualitative results presented above, it can be concluded that MediaPipe and CNN provides a more efficient and stable preprocessing solution compared to Haarcascade and CNN for the gaze estimation task. MediaPipe leverages its ability to extract 468 high precision facial landmarks, enabling more accurate detection of the eye region and gaze direction, which in turn enhances the quality of input data for the CNN model.

Furthermore, by integrating MediaPipe (for Gaze Estimation) with DeepFace (for Facial Expression Recognition), the proposed system achieves high performance, stability, and low latency, making it well suited for real time emotion recognition and human-computer interaction applications. This combination effectively harnesses the strengths of both approaches: MediaPipe provides precise geometric facial features, while DeepFace captures abstract emotional representations through a pretrained deep convolutional network.

USER INTERFACE DEVELOPMENT AND SYSTEM INTEGRATION

In addition to the quantitative experiments, we developed an interactive web-based application to visualize camera input and display the analysis results derived from the proposed method. The primary objective of this interface is to enable users to simultaneously observe two key behavioral aspects, gaze estimation and facial expression recognition, in real time as in Figure 13.

This integrated user interface not only enhances the interpretability of the system's output but also provides an intuitive means of assessing user engagement and emotional response during live interaction sessions.

Figure 13. Overall interface of the proposed website

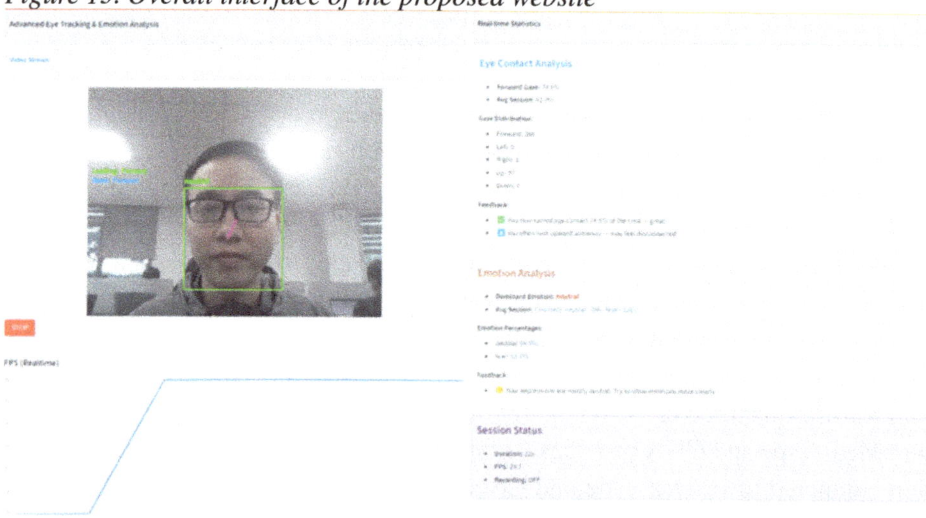

The user interface, as illustrated in Figure 13, is designed with a simple and intuitive layout, divided into four main functional areas, each serving a distinct purpose:

(1) Video Stream (top-left, Figure 14): Displays real-time frames captured from the user's camera. In this section, MediaPipe is employed to detect the user's face and estimate gaze direction, while DeepFace performs real-time facial emotion recognition. Each video frame is continuously updated along with its corresponding classification results, allowing users to observe both gaze and emotional dynamics instantly.

(2) Eye Contact Analysis (top-right, Figure 15): This section analyzes the frequency, duration, and distribution of the user's gaze direction. The system calculates the percentage of time the user maintains direct eye contact with the camera to assess their level of attention. Based on this value, the interface provides qualitative feedback such as "Good", "Average", or "Needs improvement", helping users gain insights into their engagement and focus levels during interaction.

Figure 14. Real-time video stream from the camera displayed on the website

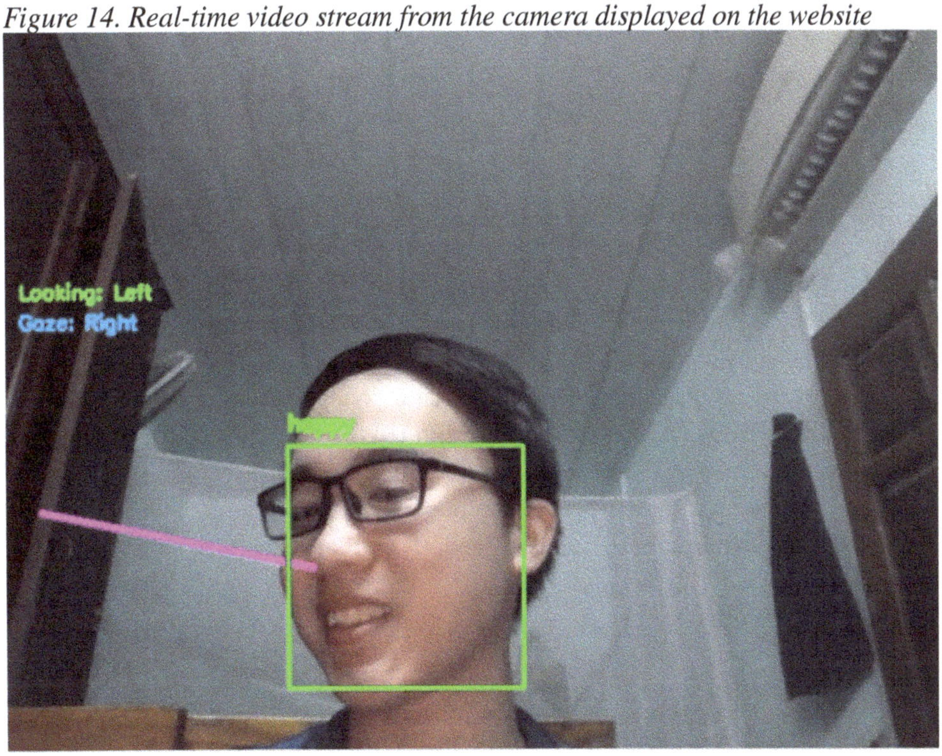

Figure 15. Gaze direction analysis displayed on the website

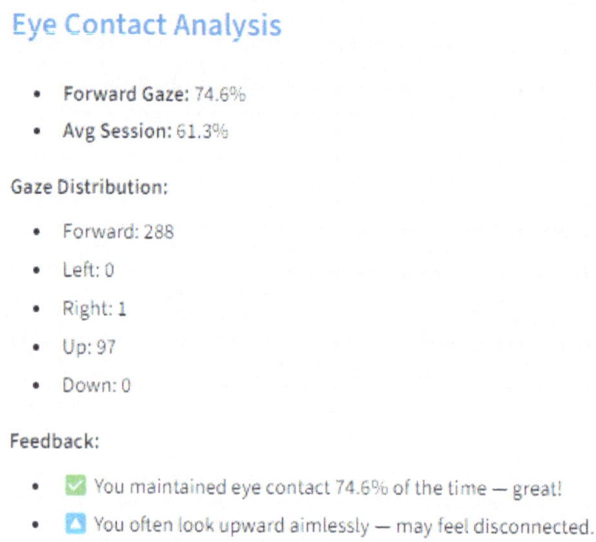

(3) Emotion Analysis (middle-right, Figure 16): This section visualizes the dominant facial emotions detected in real time (e.g., Happy, Sad, Neutral, Surprise, etc.) along with their corresponding percentage distribution charts. In addition, the system provides qualitative feedback based on the detected dominant emotion, for instance, messages such as "You look happy" or "You seem tired", to enhance user awareness of their emotional state during interaction.

(4) Session Status and Frames Per Second (FPS) (bottom section, Figure 17): This component displays the frame rate and the duration of the analysis session. It allows users to monitor the processing performance and system stability in real time, ensuring that the model operates efficiently and consistently during continuous video streaming.

Figure 16. Emotion ratio visualization on the website

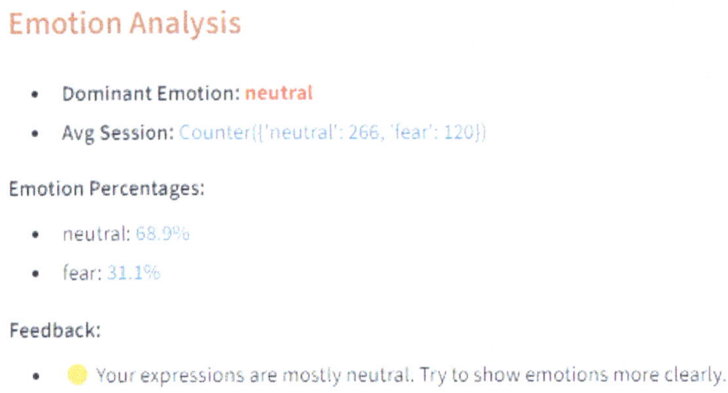

Figure 17. Control panel displaying session status

In addition, the system includes Gaze & Emotion Logs (see Figure 18), which continuously record the user's gaze direction and emotional states over time. These logs are used to compute the percentage of forward gaze and the occurrence ratio of different emotions throughout the analysis session. This functionality not only supports statistical evaluation but also lays the groundwork for real-time behavioral analysis in Human-Machine Interaction (HMI) research.

To deploy the web-based application without exposing local network ports or a public IP, we utilized Cloudflared Tunnel, which establishes a secure communication channel between the locally running application and the Internet. Specifically, when the demo application is launched with Streamlit on Windows (default port 8501), a tunnel is created using the command: cloudflared tunnel --url http://localhost:8501

Once initialized, Cloudflared establishes a persistent, encrypted outbound connection to Cloudflare Edge, which then issues a temporary access URL in the format: https://<random>.trycloudflare.com

When users access this URL through their browsers, Cloudflare Edge receives the request and forwards it through the tunnel to the Cloudflared process running locally. This process then redirects the request to http://localhost:8501, where the Streamlit application handles it and sends the response back through the encrypted tunnel.

Figure 18. Gaze log and Emotion log for computing the percentage of forward gaze and emotion distribution during the analysis session

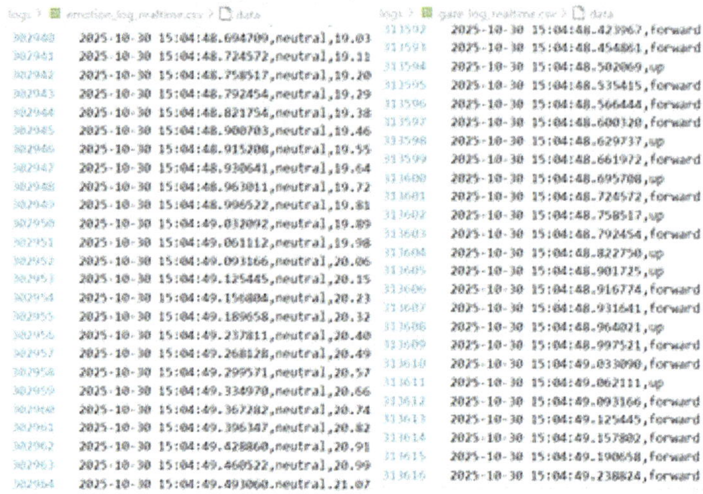

The main advantages of this architecture include:

- No need to open inbound ports or configure complex firewall rules.
- End-to-end encryption ensuring privacy and data security.
- Full integration with Cloudflare's enterprise-grade security infrastructure, including:
    - o Web Application Firewall (WAF)
    - o DDoS protection
    - o Rate limiting
    - o Bot filtering

Therefore, Cloudflared Tunnel is particularly suitable for demo environments, Proof-of-Concept (PoC) deployments, or temporary remote sharing, while maintaining both performance and security. Since all data transmission occurs through an encrypted channel, users can securely access the system remotely while all computation remains on the local host.

By combining an interactive and intuitive web interface with Cloudflare's secure infrastructure, the proposed system not only demonstrates real-time emotion and gaze analysis capabilities, but also validates its practical deployability. Furthermore, the architecture can be easily extended for applications such as:

- Monitoring attention levels in online education,
- Tracking emotional states during live interviews
- Enhancing human-machine interaction in intelligent systems (e.g., AI Companions, Virtual Assistants, or Social Robots).

CONCLUSION

The system proposed in this chapter integrates MediaPipe and DeepFace to perform real-time gaze tracking and facial emotion recognition, aiming to assess users' levels of attention and emotional states. Experimental results on the FER-2013 and Eye-tracker-data datasets demonstrate that the proposed model achieves high accuracy and stability, outperforming traditional approaches such as Haarcascade and CNN and HOG and SVM. The web application, developed using Streamlit and deployed via Cloudflared Tunnel, enables secure, intuitive, and accessible visualization of results. This approach demonstrates its feasibility for real-world applications in human-machine interaction, online learning, and real-time behavioral analysis, while also paving the way for future extensions that incorporate additional non-verbal cues to further enhance the system's accuracy and contextual understanding.

RECOMMENDATIONS FOR FUTURE RESEARCH

Future research should focus on enhancing the system's accuracy, generalization, and ethical applicability. Integrating multimodal data such as head pose, gestures, and speech tone could improve contextual understanding. Advanced architectures like CNN Transformer hybrids or vision language models may further strengthen robustness in dynamic environments. Expanding evaluation across diverse datasets and cultural settings would ensure fairness and adaptability. In addition, optimizing the model for edge or mobile deployment can support low latency, real world applications. Privacy preserving learning and ethical frameworks should also be prioritized to protect user data. Finally, incorporating adaptive feedback mechanisms could increase user engagement and make the system more responsive in education, healthcare, and human-machine interaction scenarios.

REFERENCES

Ajunwa, I., Crawford, K., & Schultz, J. (2017). Limitless Worker Surveillance. *California Law Review, 105*(3), 735–776.

Alanazi, F., Ushaw, G., & Morgan, G. (2024). Improving Detection of DeepFakes through Facial Region Analysis in Images. *Electronics (Basel), 13*(1), 126. Advance online publication. DOI: 10.3390/electronics13010126

Ball, K. (2010). Workplace surveillance: An overview. *Labor History, 51*(1), 87–106. DOI: 10.1080/00236561003654776

Bergstrom, J. C. R., & Schall, A. (2014). *Eye tracking in user experience design.* https://api.semanticscholar.org/CorpusID:40789776

Białowąs, S., & Szyszka, A. (2019). *Eye-tracking in Marketing Research.*, DOI: 10.12657/9788379862771-6

Bouktif, S. & N. K. (2024). Real-Time Detection of Student Attention in Online Learning: A Scalable Solution for Crisis-Driven Remote Education. *2024 IEEE International Conference on Progress in Informatics and Computing (PIC)*, 75-79. DOI: 10.1109/PIC62406.2024.10892772

Buolamwini, J., & Gebru, T. (2018). Gender Shades: Intersectional Accuracy Disparities in Commercial Gender Classification. In S. A. Friedler & C. Wilson (Eds.), *Proceedings of the 1st Conference on Fairness, Accountability and Transparency* (Vol. 81, pp. 77-91). PMLR. https://proceedings.mlr.press/v81/buolamwini18a.html

Calvo, R. A., & D'Mello, S. (2010). Affect Detection: An Interdisciplinary Review of Models, Methods, and Their Applications. *IEEE Transactions on Affective Computing, 1*(1), 18–37. DOI: 10.1109/T-AFFC.2010.1

Carelli, L., Solca, F., Tagini, S., Torre, S., Verde, F., Ticozzi, N., Ferrucci, R., Pravettoni, G., Aiello, E. N., Silani, V., & Poletti, B. (2022). Gaze-Contingent Eye-Tracking Training in Brain Disorders: A Systematic Review. *Brain Sciences, 12*(7), 931. Advance online publication. DOI: 10.3390/brainsci12070931 PMID: 35884737

Davis, F. D. (1989). Perceived Usefulness, Perceived Ease of Use, and User Acceptance of Information Technology. *MIS Quarterly, 13*(3), 319-340. JSTOR. DOI: 10.2307/249008

Duan, C. (2023). A survey of facial expression recognition in the wild. *Applied and Computational Engineering, 6*(1), 173–181. DOI: 10.54254/2755-2721/6/20230760

Duchowski, A. (2007). *Eye tracking methodology: Theory and practice.* Springer.

Ekman, P., & Friesen, W. V. (1978). *Facial action coding system: A technique for the measurement of facial movement.* Consulting Psychologists Press.

Fasel, B., & Luettin, J. (2003). Automatic facial expression analysis: A survey. *Pattern Recognition, 36*(1), 259–275. DOI: 10.1016/S0031-3203(02)00052-3

Floridi, L., Cowls, J., Beltrametti, M., Chatila, R., Chazerand, P., Dignum, V., Luetge, C., Madelin, R., Pagallo, U., Rossi, F., Schafer, B., Valcke, P., & Vayena, E. (2018). AI4People—An Ethical Framework for a Good AI Society: Opportunities, Risks, Principles, and Recommendations. *Minds and Machines, 28*(4), 689–707. DOI: 10.1007/s11023-018-9482-5 PMID: 30930541

Gehrer, N. A., Svaldi, J., & Duchowski, A. (2024). Assessment of body-related attention processes via mobile eye tracking: A pilot study to validate an automated analysis pipeline. *Proceedings of the 2024 Symposium on Eye Tracking Research and Applications.* DOI: 10.1145/3649902.3653354

Geng, X., & Xia, Y. (2014). Head Pose Estimation Based on Multivariate Label Distribution. *2014 IEEE Conference on Computer Vision and Pattern Recognition,* 1837-1842. DOI: 10.1109/CVPR.2014.237

Goodfellow, I. J., Erhan, D., Carrier, P. L., Courville, A., Mirza, M., Hamner, B., Cukierski, W., Tang, Y., Thaler, D., Lee, D.-H., Zhou, Y., Ramaiah, C., Feng, F., Li, R., Wang, X., Athanasakis, D., Shawe-Taylor, J., Milakov, M., Park, J., & Bengio, Y. (2013). Challenges in Representation Learning: A Report on Three Machine Learning Contests. In Lee, M., Hirose, A., Hou, Z.-G., & Kil, R. M. (Eds.), *Neural Information Processing* (pp. 117–124). Springer Berlin Heidelberg. DOI: 10.1007/978-3-642-42051-1_16

Hennessey, C., Noureddin, B., & Lawrence, P. (2006). A single camera eye-gaze tracking system with free head motion. *Proceedings of the 2006 Symposium on Eye Tracking Research & Applications,* 87-94. DOI: 10.1145/1117309.1117349

Kanade, P., David, F., & Kanade, S. (2021). Convolutional Neural Networks(CNN) based Eye-Gaze Tracking System using Machine Learning Algorithm. *European Journal of Electrical Engineering and Computer Science, 5*(2), 36–40. DOI: 10.24018/ejece.2021.5.2.314

Kellnhofer, P., Recasens, A., Stent, S., Matusik, W., & Torralba, A. (2019). Gaze360: Physically Unconstrained Gaze Estimation in the Wild. *2019 IEEE/CVF International Conference on Computer Vision (ICCV),* 6911-6920. DOI: 10.1109/ICCV.2019.00701

Li, J., Zhao, B., & Zhang, H. (2009). Face Recognition Based on PCA and LDA Combination Feature Extraction. *2009 First International Conference on Information Science and Engineering*, 1240-1243. DOI: 10.1109/ICISE.2009.581

Lugaresi, C., Tang, J., Nash, H., McClanahan, C., Uboweja, E., Hays, M., Zhang, F., Chang, C.-L., Yong, M., Lee, J., Chang, W.-T., Hua, W., Georg, M., & Grundmann, M. (2019). *MediaPipe: A Framework for Building Perception Pipelines*. https://doi .org//arXiv.1906.08172DOI: 10.48550

MacKenzie, I. S. (2010). An eye on input: Research challenges in using the eye for computer input control. *Proceedings of the 2010 Symposium on Eye-Tracking Research & Applications*, 11-12. DOI: 10.1145/1743666.1743668

Morimoto, C. H., & Mimica, M. R. M. (2005). Eye gaze tracking techniques for interactive applications. *Computer Vision and Image Understanding*, *98*(1), 4–24. DOI: 10.1016/j.cviu.2004.07.010

Ogawa, T., Nakazawa, A., & Nishida, T. (2017). Point of gaze estimation using corneal surface reflection and omnidirectional camera image. *2017 Fifteenth IAPR International Conference on Machine Vision Applications (MVA)*, 227-230. DOI: 10.23919/MVA.2017.7986842

Pauszek, J. R. (2023). An introduction to eye tracking in human factors healthcare research and medical device testing. *Human Factors in Healthcare*, *3*, 100031. DOI: 10.1016/j.hfh.2022.100031

Picard, R. W. (1997). *Affective computing*. MIT Press. DOI: 10.7551/mit-press/1140.001.0001

Punuri, S. B., Kuanar, S. K., & Mishra, T. K. (2023). Facial Emotion Recognition in Unconstrained Environments through Rank-Based Ensemble of Deep Learning Models using 1-Cycle Policy. *2023 International Conference on the Confluence of Advancements in Robotics, Vision and Interdisciplinary Technology Management (IC-RVITM)*, 1-8. DOI: 10.1109/IC-RVITM60032.2023.10435159

Said, Y., Saidani, T., Atri, M., Alsheikhy, A. A., & Shawly, T. (2026). Computational intelligence for emotion recognition in autism spectrum disorder: A systematic review of signal-based modeling, simulation, and clinical potential. *Biomedical Signal Processing and Control, 111*, 108367. DOI: 10.1016/j.bspc.2025.108367

Santamaria-Granados, L., Munoz-Organero, M., Ramirez-González, G., Abdulhay, E., & Arunkumar, N. (2019). Using Deep Convolutional Neural Network for Emotion Detection on a Physiological Signals Dataset (AMIGOS). *IEEE Access : Practical Innovations, Open Solutions*, *7*, 57–67. DOI: 10.1109/ACCESS.2018.2883213

Shah, S. M., Sun, Z., Zaman, K., Hussain, A., Shoaib, M., & Pei, L. (2022). A Driver Gaze Estimation Method Based on Deep Learning. *Sensors (Basel)*, *22*(10), 3959. Advance online publication. DOI: 10.3390/s22103959 PMID: 35632365

Sharma, M., Mittal, A., Rewliya, A., & Mittal, H. (2024). Comparative Analysis of Deep Learning Models for Facial Emotion Recognition. *2024 IEEE International Conference on Intelligent Signal Processing and Effective Communication Technologies (INSPECT)*, 1-6. DOI: 10.1109/INSPECT63485.2024.10896027

Sigut, J., & Sidha, S.-A. (2011). Iris Center Corneal Reflection Method for Gaze Tracking Using Visible Light. *IEEE Transactions on Biomedical Engineering*, *58*(2), 411–419. DOI: 10.1109/TBME.2010.2087330 PMID: 20952326

Silvestri, F., Odisho, N., Kumar, A., & Grigoriadis, A. (2024). Examining gaze behavior in undergraduate students and educators during the evaluation of tooth preparation: An eye-tracking study. *BMC Medical Education*, *24*(1), 1030. DOI: 10.1186/s12909-024-06019-4 PMID: 39300488

Valle, R., Buenaposada, J. M., Valdés, A., & Baumela, L. (2019). Face alignment using a 3D deeply-initialized ensemble of regression trees. *Computer Vision and Image Understanding*, *189*, 102846. DOI: 10.1016/j.cviu.2019.102846

Venkatesh, V., Morris, M. G., Davis, G. B., & Davis, F. D. (2003). User Acceptance of Information Technology: Toward A Unified View1. *Management Information Systems Quarterly*, *27*(3), 425–478. DOI: 10.2307/30036540

Venkatesh, V., Thong, J. Y. L., & Xu, X. (2012). Consumer Acceptance and Use of Information Technology: Extending the Unified Theory of Acceptance and Use of Technology1. *Management Information Systems Quarterly*, *36*(1), 157–178. DOI: 10.2307/41410412

Wood, E., Baltrušaitis, T., Morency, L.-P., Robinson, P., & Bulling, A. (2016). Learning an appearance-based gaze estimator from one million synthesised images. *Proceedings of the Ninth Biennial ACM Symposium on Eye Tracking Research & Applications*, 131-138. DOI: 10.1145/2857491.2857492

Xiao, F., Zheng, D., Huang, K., Qiu, Y., & Shen, H. (2018). A Single-Camera Gaze Tracking System Under Natural Light. *Journal of Eye Movement Research*, *11*(4), 1–14. DOI: 10.16910/jemr.11.4.5 PMID: 33828707

Xu, H., Zhang, J., Sun, H., Qi, M., & Kong, J. (2023). Analyzing students' attention by gaze tracking and object detection in classroom teaching. *Data Technologies and Applications*, *57*(5), 643–667. DOI: 10.1108/DTA-09-2021-0236

Xu, Y., & Cheng, L. (2025). Face emotion recognition based on Gabor wavelet and particle swarm optimization algorithm. *Multimedia Systems*, *31*(6), 435. Advance online publication. DOI: 10.1007/s00530-025-01976-2

Yarbus, A. L. (1967). *Eye movements and vision*. Plenum Press. DOI: 10.1007/978-1-4899-5379-7

Zhai, S. (2008). On the ease and efficiency of human-computer interfaces. *Proceedings of the 2008 Symposium on Eye Tracking Research & Applications*, 9-10. DOI: 10.1145/1344471.1344473

Zhang, X., Sugano, Y., Fritz, M., & Bulling, A. (2015). Appearance-based gaze estimation in the wild. *2015 IEEE Conference on Computer Vision and Pattern Recognition (CVPR)*, 4511-4520. DOI: 10.1109/CVPR.2015.7299081

Zheng, F. (2022). Facial Expression Recognition Based on LDA Feature Space Optimization. *Computational Intelligence and Neuroscience*, *2022*(1), 9521329. DOI: 10.1155/2022/9521329 PMID: 36072737

Chapter 3
Organizational Psychology Dynamics Facing Female Doctors in Biomedicine: Seeking Workplace Psychological Safety

Allison Huff
http://orcid.org/0000-0001-6102-8013
College of Medicine, University of Arizona, USA

Darrell Norman Burrell
http://orcid.org/0000-0002-4675-9544
Marymount University, USA & Pellegrino Center for Clinical Bioethics, Georgetown University, USA

Adina Lundy
http://orcid.org/0000-0001-7564-2110
University of Rhode Island, USA

Natasha Harris
http://orcid.org/0009-0002-1745-7966
Marymount University, USA

KerriLaine Prunella
http://orcid.org/0009-0001-0177-3545
Marymount University, USA

Hilda Eke
http://orcid.org/0009-0006-2021-178X
Marymount University, USA

ABSTRACT

Female physicians in biomedicine face persistent structural and cultural challenges that extend beyond individual career trajectories and pose systemic risks to organizational performance, innovation, and patient care. Drawing on industrial and organizational psychology, this paper examines how gendered inequities, manifested through reduced psychological safety, exclusion from decision-making, inequitable

DOI: 10.4018/979-8-2600-1026-6.ch003

compensation, and elevated exposure to toxic workplace behaviors, contribute to disproportionate attrition and professional disengagement among women doctors. Using Virelthos Medical Institute as an illustrative organizational context, we argue that these dynamics suppress clinical voice, exacerbate groupthink, and undermine the ethical governance of biomedical technologies. Integrating theory with real-world examples, the analysis demonstrates how psychologically unsafe and non-inclusive cultures constrain women physicians' contributions to innovation and quality improvement.

INTRODUCTION

Despite sustained progress in educational access, gender equity in academic medicine remains structurally incomplete. While women have achieved parity, or near parity, in medical school enrollment and biomedical doctoral degree attainment, this representation does not translate proportionally into academic faculty or leadership roles. As Plank-Bazinet (2017) demonstrates, women constitute approximately half of medical and biological science doctoral graduates, yet they do not comprise half of academic medicine faculty positions. This attrition intensifies with advancement, such that women occupy only a small fraction of senior leadership roles. Notably, only 16% of dean positions at United States medical schools are held by women, underscoring the persistence of gendered stratification at the highest organizational levels (Plank-Bazinet et al., 2017).

From an industrial and organizational psychology perspective, this pattern reflects a classic "leaky pipeline" phenomenon, in which formal entry into a profession is decoupled from long-term advancement and authority. Empirical data reinforce this interpretation. In 2014, women accounted for 38% of full-time medical school faculty in the United States, yet only 15% of permanent department chairs and 16% of deans were women (Plank-Bazinet et al., 2017). Importantly, this disparity cannot be explained by a lack of qualified candidates. Relative labor market analyses indicate that women are not underrepresented during biomedical training phases but are systematically underrepresented in independent research and leadership positions, suggesting that organizational advancement mechanisms, rather than talent supply, are the primary constraint (Plank-Bazinet et al., 2017).

Organizational psychology research points to multiple interlocking factors that sustain these inequities. Implicit bias, defined as subconscious preferences that shape evaluative and decision-making processes, has been repeatedly identified as a barrier to the advancement of women in academic and biomedical institutions (Plank-Bazinet et al., 2017). These biases influence hiring, promotion, authorship credit, and leadership selection, often in ways that are invisible to formal accountability structures. Compounding this issue, women are frequently hired at lower starting salaries than male colleagues in comparable roles (Plank-Bazinet et al., 2017). Because salary progression in academic medicine is typically cumulative and percentage-based, initial inequities are amplified over time, resulting in persistent lifetime earnings gaps. These disparities extend beyond active employment, adversely affecting retirement security and long-term financial stability, thereby placing women at a structural disadvantage across the career lifespan (Plank-Bazinet et al., 2017).

The organizational consequences of these dynamics are not merely symbolic; they manifest in measurable institutional risk. In 2025, Virelthos Medical Institute is confronting a severe organizational crisis marked by female physician turnover exceeding 200% of the industry average. This level of attrition signals not an individual-level retention problem, but a systemic organizational failure. High turnover among women physicians reflects the cumulative impact of inequitable pay structures, limited advancement opportunities, moral distress, and exposure to toxic workplace cultures. From an organizational systems perspective, such attrition erodes institutional memory, increases recruitment and onboarding costs, destabilizes clinical teams, and compromises continuity of patient care.

International research reinforces the organizational risks associated with gendered workplace inequities (Lundy et al., 2025). Studies across healthcare systems in the United Kingdom and Canada indicate that female physicians are approximately twice as likely to experience workplace harassment compared to their male counterparts, contributing to lower morale, disengagement, and elevated exit intentions (Lyons et al., 2021). Similar findings have been reported globally, linking gender-based harassment and inequity to diminished organizational performance and reduced care quality (Hudgins et al., 2022; Iyer et al., 2023). These patterns suggest that the challenges faced by Virelthos are not anomalous, but rather symptomatic of broader structural deficiencies within biomedical organizational cultures.

Compensation inequity further exacerbates these dynamics. In the United States, female physicians earn approximately 21% less than male physicians in comparable roles with similar experience and training (Lo Sasso et al., 2020). From an industrial and organizational psychology standpoint, such disparities undermine perceptions of distributive and procedural justice; key predictors of employee commitment, trust, and retention. When inequity becomes normalized, organizations experience not only higher turnover but also reduced psychological safety, diminished discretionary effort, and increased burnout among remaining staff (Lundy et al., 2025). Global health authorities have increasingly recognized these risks. A 2021 World Health Organization report identifies workplace inequity and gender discrimination as threats to organizational stability and patient safety, emphasizing that inequitable systems compromise both workforce sustainability and clinical outcomes. These findings position gender equity not as a peripheral ethical concern, but as a core component of organizational resilience and risk management.

Within this context, Virelthos Medical Institute faces both a critical challenge and a strategic opportunity. By adopting a comprehensive organizational risk management framework that prioritizes equity, inclusivity, and employee well-being, Virelthos can address the root causes of turnover rather than its symptoms. Evidence suggests that proactive interventions targeting gender-based disparities, bullying, and structural bias yield significant organizational benefits, including improved retention, stronger leadership pipelines, enhanced team functioning, and better patient outcomes (Dudley et al., 2022). From an organizational psychology perspective, such interventions are not merely corrective; they are foundational to building a sustainable, high-performing biomedical institution.

Problem Statement

Biomedical organizations such as Virelthos Medical Institute face a critical challenge: the underrepresentation and attrition of women and diverse physicians within technologically intensive environments contribute to organizational groupthink, reduced innovation quality, and heightened clinical risk. Despite parity in training pipelines, systemic barriers, including implicit bias, inequitable compensation, and exclusionary cultures, limit diverse participation in technology-related decision-making, undermining both workforce sustainability and the ethical deployment of biomedical technologies (Plank-Bazinet, 2017; Lo Sasso et al., 2020).

Significance Statement

This problem is significant because biomedical technologies increasingly shape diagnostic accuracy, treatment equity, and patient outcomes. Homogenous leadership and innovation teams are more prone to blind spots, algorithmic bias, and overconfidence in technological solutions. Evidence suggests that increasing the representation of women and diverse clinicians enhances innovation, mitigates groupthink, and improves organizational performance, scientific discovery, and women's health outcomes (Plank-Bazinet et al., 2017). Addressing these issues is therefore essential not only for equity but for organizational resilience, patient safety, and technological integrity (Lundy et al., 2025).

Purpose Statement

The purpose of the proposed management consulting intervention is to conduct a comprehensive root cause analysis (RCA) of organizational factors contributing to female doctor attrition, exclusion from biomedical technology decision-making, and innovation risk at Virelthos Medical Institute. Using an industrial and organizational psychology framework, the intervention will identify structural, cultural, and leadership-level drivers of inequity and groupthink, and will develop evidence-based recommendations to enhance diversity, psychological safety, and innovation capacity. The ultimate goal is to align workforce inclusion with technological excellence, thereby strengthening organizational performance and patient care outcomes.

DIVERSITY, INNOVATION, AND ORGANIZATIONAL COGNITION IN BIOMEDICAL TECHNOLOGY SYSTEMS

Contemporary biomedicine is increasingly shaped by complex technological infrastructures, including artificial intelligence–assisted diagnostics, precision medicine platforms, electronic health records, and algorithmic decision-support systems. These technologies promise improved efficiency and clinical accuracy, yet they also amplify the consequences of organizational blind spots. Within this context, workforce composition, particularly the representation of women and other historically marginalized groups, has become a critical determinant of innovation quality, ethical robustness, and organizational adaptability. Despite sustained parity in biomedical training pipelines, women remain underrepresented in academic medicine faculty and leadership roles, with only a small fraction occupying senior decision-making positions (Plank-Bazinet et al., 2017). This disconnect has profound implications for how biomedical technologies are designed, evaluated, and deployed.

From an industrial and organizational psychology perspective, the absence of diverse professional voices increases susceptibility to groupthink, defined as a mode of collective decision-making characterized by conformity pressure, suppression of dissent, and uncritical acceptance of dominant assumptions. In biomedical organizations, groupthink can manifest in technology adoption decisions that privilege efficiency over equity, scale over nuance, or technical feasibility over clinical validity. Diverse teams, particularly those inclusive of women physicians and scientists, are empirically associated with higher levels of cognitive diversity, perspective-taking, and error detection, all of which are essential for innovation in high-stakes technological systems (Plank-Bazinet et al., 2017).

Gender, Diversity, and Innovation in Biomedical Technologies

The underrepresentation of women and diverse clinicians in leadership roles has direct consequences for biomedical technology innovation. Technologies are not neutral artifacts; they reflect the values, assumptions, and lived experiences of those who design and implement them. Real-world examples illustrate this dynamic. Early clinical algorithms for cardiovascular risk assessment were calibrated primarily on male populations, resulting in systematic underdiagnosis of heart disease in women. Similarly, machine-learning models trained on non-representative datasets have demonstrated racial and gender bias in dermatology imaging, pain assessment, and obstetric risk prediction. These failures are not solely technical; they are organizational, arising from homogenous decision-making groups that lack experiential diversity.

Women physicians and scientists often bring distinct perspectives shaped by differential patient interactions, caregiving roles, and exposure to gendered health disparities. When included meaningfully in biomedical innovation ecosystems, they contribute to more inclusive problem framing and more rigorous challenge of default assumptions. Organizational psychology research consistently shows that heterogeneous teams outperform homogeneous ones on complex, nonroutine tasks; precisely the category into which biomedical technology development falls. Diversity disrupts cognitive closure, encourages epistemic humility, and improves the organization's capacity to identify unintended consequences before they scale (Lundy et al., 2025).

Organizational Barriers to Inclusion and the Persistence of Groupthink

Despite the clear innovation benefits of diversity, organizational structures in biomedicine frequently inhibit the advancement and retention of women and diverse physicians. Implicit bias in hiring, promotion, and compensation remains a documented barrier (Plank-Bazinet et al., 2017). Women physicians are often hired at lower starting salaries than male peers, a disparity that compounds over time due to percentage-based raises and unequal access to revenue-generating opportunities (Lo Sasso et al., 2020). These inequities undermine perceptions of organizational justice and signal whose contributions are valued in technology-driven decision-making spaces.

At Virelthos Medical Institute (Virelthos), elevated turnover among women physicians reflects these broader systemic dynamics. High attrition rates among underrepresented groups are not only a human capital loss but also an innovation risk. When diverse clinicians exit, organizations lose critical experiential knowledge that could inform the ethical deployment of emerging technologies such as AI triage tools, remote monitoring platforms, and genomics-based treatment protocols. International evidence links hostile or exclusionary healthcare cultures to diminished employee engagement, reduced innovation capacity, and compromised care quality (Lyons et al., 2021; Hudgins et al., 2022; Iyer et al., 2023).

Technology, Psychological Safety, and Ethical Vigilance

The successful integration of biomedical technologies depends on psychological safety, the shared belief that individuals can raise concerns, question assumptions, and report errors without fear of retaliation. Psychological safety is particularly vital in technology-mediated environments, where errors may be obscured by automation bias or overreliance on algorithmic outputs. Diverse teams are more likely to surface anomalies and challenge premature consensus, but only when organizational cultures support voice and dissent.

World Health Organization analyses underscore that inequitable and unsafe work environments undermine not only workforce stability but also patient safety, especially as healthcare systems become more technologically complex. In organizations where women and minority clinicians experience marginalization or harassment, the likelihood that they will challenge flawed technological decisions decreases, reinforcing groupthink and increasing systemic risk.

METHODS AND DESIGN OVERVIEW

This study employs a mixed-methods management consulting Root Cause Analysis (RCA) framework to diagnose organizational factors contributing to gendered attrition, reduced innovation capacity, and decision-making vulnerability within a technologically intensive biomedical organization. The approach integrates principles from industrial and organizational psychology, organizational development (OD), and diversity, equity, and inclusion (DEI) scholarship. RCA was selected because it enables systematic identification of underlying structural, cultural, and process-level contributors to complex organizational outcomes, rather than focusing on proximal symptoms such as turnover alone.

Root Cause Analysis: Definition, Purpose, and Application in Organizational Contexts

Root Cause Analysis (RCA) is a systematic, structured problem-solving methodology used to identify the underlying causes of complex organizational problems, rather than addressing only their immediate symptoms. The central premise of RCA is that sustainable improvement is achieved not by correcting surface-level failures, but by diagnosing and intervening on the deep structural, cultural, process, and behavioral conditions that give rise to those failures over time.

In contrast to reactive or anecdotal approaches, RCA is evidence-driven and systems-oriented. It assumes that undesirable outcomes, such as high turnover, burnout, inequity, safety incidents, or innovation failures, are rarely the result of individual error alone. Instead, they emerge from interacting organizational factors, including policies, incentive structures, leadership practices, workflow design, and informal norms.

Core Components of Root Cause Analysis

At its foundation, RCA involves several interrelated steps:

1. **Problem Definition**

RCA begins with precise articulation of the problem, distinguishing observable outcomes (e.g., high turnover among women physicians) from assumed causes. This step avoids premature attribution and ensures analytic clarity.

2. **Data Collection and Triangulation**

Multiple sources of evidence, quantitative metrics, qualitative interviews, process maps, and document reviews, are examined to capture how the organization actually functions, not merely how it is intended to function.

3. **Causal Pathway Identification**

RCA traces cause-and-effect relationships across levels of the system. Tools such as causal loop diagrams, fishbone diagrams, or the "five whys" technique are commonly used to move from proximate causes to deeper structural drivers.

4. **System-Level Interpretation**

Findings are interpreted within a systems framework, recognizing feedback loops, reinforcing mechanisms, and unintended consequences. This step is critical for understanding why problems persist despite prior interventions.

5. **Actionable Intervention Design**

RCA culminates in recommendations that directly target identified root causes. Effective interventions modify structures, processes, or norms rather than relying solely on individual behavior change.

RCA in Industrial and Organizational Psychology

Within **industrial and organizational psychology**, RCA was used to diagnose issues such as disengagement, inequity, leadership failure, and poor team performance. The method aligns with core I–O principles by emphasizing:

- Organizational justice (how fairness perceptions shape behavior)
- Psychological safety (whether individuals feel safe to speak up)
- Work design and role clarity
- Leadership and power dynamics

RCA avoids "person-blaming" and instead examines how organizational systems create predictable behavioral patterns. For example, repeated incidents of workplace conflict may be traced not to "difficult personalities" but to unclear authority structures, misaligned incentives, or unmanaged workload pressures.

RCA in Biomedical and Healthcare Organizations

RCA has a long history in healthcare, particularly in patient safety and quality improvement. Traditionally, it has been applied following sentinel events to identify system failures contributing to medical errors. In contemporary biomedical organizations, however, RCA has expanded beyond clinical incidents to address workforce sustainability, culture, and innovation risk.

In technologically intensive environments, such as those involving artificial intelligence, clinical decision-support systems, or complex care coordination; RCA helps organizations understand how design assumptions, workflow constraints, and leadership decisions interact. For instance, high turnover among clinicians may be traced to how new technologies increase documentation burden, reduce autonomy, or exclude frontline voices from design decisions.

Why RCA Is Particularly Valuable for Complex Organizational Problems

RCA is especially well suited for problems that are:

- Persistent (reoccurring despite previous fixes)
- Multifactorial (influenced by culture, structure, and behavior)
- Politically or emotionally charged (where blame inhibits learning)
- High-risk (affecting safety, equity, or organizational reputation)

By shifting the focus from individual failure to systemic causation, RCA enables organizations to learn from problems rather than conceal them. This learning orientation is essential in biomedicine, where complexity and uncertainty are inherent.

The consulting team was deliberately composed to ensure cognitive and experiential diversity during both diagnosis and intervention design. The team included:

(1) an organizational development specialist with expertise in systems diagnosis and change management;
(2) a diversity and inclusion specialist with experience in equity audits and bias mitigation; and
(3) a female physician with direct clinical and technological workflow experience in biomedicine.

This composition was intended to mitigate groupthink during analysis, enhance ecological validity, and ensure that findings reflected both organizational systems and lived professional realities.

Phase 1: Organizational Scoping and Contracting

The initial phase involved formal scoping to define the boundaries, objectives, and ethical parameters of the RCA. Senior leadership and key governance stakeholders were engaged to clarify the presenting problem, which included elevated turnover among women physicians, concerns regarding workplace culture, and risks associated with biomedical technology decision-making.

Deliverables for this phase included:

- A documented problem definition distinguishing symptoms from hypothesized systemic drivers
- Identification of stakeholder groups (e.g., clinicians, researchers, IT staff, managers)
- Agreement on data access, confidentiality protections, and reporting protocols

This phase was led by the organizational development specialist to ensure alignment between diagnostic depth and organizational readiness for change.

Phase 2: Data Collection and Diagnostic Triangulation

Data collection followed a triangulated diagnostic strategy combining quantitative, qualitative, and document-based sources. This approach was designed to surface both formal structures and informal cultural dynamics.

Quantitative data sources included:

- Workforce metrics (turnover rates, promotion timelines, compensation bands) disaggregated by gender and role
- Participation patterns in biomedical technology committees and innovation initiatives
- Employee engagement and psychological safety survey data
- Exit interview data.
- Organizational climate survey data

Qualitative data sources included:

- Semi-structured interviews with physicians, scientists, administrators, and technology leaders
- Focus groups stratified by role and career stage to reduce power-related response distortion
- Narrative elicitation techniques to capture lived experiences of inclusion, exclusion, and voice

The female physician on the consulting team played a central role in interpreting clinical workflow narratives and identifying where technology-related decisions intersected with patient care realities. The DEI specialist guided interview protocols to surface implicit bias, microaggressions, and structural inequities that may not appear in formal metrics.

Phase 3: Root Cause Identification and Systems Mapping

Collected data were analyzed using a systems-mapping approach consistent with organizational psychology and OD practices. Rather than attributing outcomes to individual behavior, the team examined how policies, norms, incentives, and leadership practices interacted to produce observed patterns.

Analytic techniques included:

- Cause-and-effect (fishbone) diagrams mapping cultural, structural, technological, and leadership contributors

- Identification of reinforcing feedback loops (e.g., attrition leading to homogeneity, which in turn increased groupthink)
- Cross-validation of themes across data sources to enhance analytic rigor

Special attention was given to how biomedical technologies, such as clinical decision-support tools, AI systems, and digital documentation platforms, were developed or implemented in ways that excluded diverse clinical perspectives. The inclusion of a female physician was critical in identifying gaps between technological design assumptions and frontline clinical realities.

Phase 4: Stakeholder Validation and Sensemaking

Preliminary findings were presented in facilitated sensemaking sessions with representative stakeholder groups. These sessions served two purposes: validation of analytic accuracy and engagement of organizational members in interpreting root causes.

The OD specialist facilitated these sessions to ensure psychological safety and productive dialogue, while the DEI specialist monitored for power dynamics that could suppress dissenting voices. Feedback from these sessions was systematically documented and incorporated into the final diagnostic model.

This phase helped transform the RCA from an external assessment into a shared organizational understanding, increasing the likelihood of sustained change.

Phase 5: Co-Development of Recommendations

Recommendations were developed collaboratively by the consulting team using an evidence-informed, feasibility-tested approach. Each recommendation was explicitly linked to one or more identified root causes and evaluated against criteria of impact, equity, and implementation feasibility.

Recommendation domains included:

- Leadership accountability and behavior standards
- Inclusive governance structures for biomedical technology decisions
- Compensation and advancement equity mechanisms
- Psychological safety and reporting infrastructure
- Change management and communication strategies

The female physician contributed to assessing the clinical and technological feasibility of proposed changes, ensuring that recommendations aligned with patient care demands and workflow constraints.

Phase 6: Implementation Planning and Evaluation Design

The final phase focused on translating recommendations into an actionable implementation roadmap. This included sequencing of interventions, assignment of accountability, and development of evaluation metrics.

Key outputs included:

- An implementation timeline with short-, medium-, and long-term milestones
- Key performance indicators (KPIs) related to retention, diversity in innovation roles, psychological safety, and technology outcomes
- A feedback and learning plan to support iterative refinement

This phase emphasized that RCA is not a one-time diagnostic exercise but part of an ongoing organizational learning cycle.

Below is an expanded and fully reproduced set of 40 actionable, practical, and implementation-ready recommendations, grounded in industrial and organizational psychology, biomedical technology governance, diversity-driven innovation, and root cause analysis (RCA). The recommendations are written for direct operational use by executive leadership, boards, and management consultants supporting Virelthos Medical Institute (Virelthos). They are concrete, non-duplicative, and organized to move from governance to day-to-day practice and long-term sustainability.

Actionable Recommendations for Organizational, Cultural, and Innovation Reform

A. Governance, Leadership, and Accountability

1. **Create a Board-Level Equity, Technology, and Innovation Committee**

Grant formal oversight authority for workforce equity, biomedical technology decisions, and innovation risk.

2. **Integrate Equity and Retention Metrics into Executive Compensation**

Tie a defined percentage of incentive pay to improvements in retention, pay equity, and psychological safety.

3. **Mandate Gender and Diversity Representation in Senior Leadership Slates**

Require diverse candidate pools for all senior leadership and dean-level appointments.

4. **Establish Clear, Enforceable Anti-Bullying Governance Policies**

Define prohibited behaviors, investigative procedures, and consequences with board oversight.

5. **Assign an Executive Sponsor for Women and Diversity in Innovation**

Designate a C-suite leader accountable for advancing inclusion in biomedical technology strategy.

6. **Require Annual Public Reporting on Equity and Innovation Indicators**

Transparently report gender representation, pay gaps, and innovation participation metrics.

B. Root Cause Analysis and Diagnostic Infrastructure

7. **Conduct a System-Wide RCA on Gendered Attrition and Exclusion**

Examine leadership practices, workflow design, evaluation systems, and informal norms.

8. **Repeat RCA Annually for Continuous Organizational Learning**

Institutionalize RCA as an ongoing diagnostic cycle rather than a one-time intervention.

9. **Disaggregate All Workforce and Innovation Data by Gender, Role, and Specialty**

Ensure visibility into where inequities emerge and persist.

10. **Map Decision Pathways for Biomedical Technology Adoption**

Identify where diverse voices are excluded from key decisions.

11. **Audit Informal Sponsorship and Influence Networks**

Examine who receives advocacy, visibility, and access to high-impact projects.

12. **Identify Feedback Loops That Reinforce Groupthink**

Document how homogeneity in leadership perpetuates conformity and suppresses dissent.

C. Biomedical Technology Design, AI, and Innovation Governance

13. **Require Diverse Clinical Representation in Technology Procurement Teams**

Include women physicians and frontline clinicians as voting members.

14. **Implement Mandatory Bias and Equity Impact Assessments for New Technologies**

Evaluate algorithmic and workflow impacts before approval.

15. **Pilot Biomedical Technologies Across Diverse Patient and Provider Contexts**

Test tools in varied specialties, populations, and care settings.

16. **Establish a Formal "Challenge Role" in Innovation Reviews**

Assign individuals to question assumptions and surface risks during technology evaluations.

17. **Create an Ethical Review Process for AI and Data-Driven Tools**

Assess implications for equity, autonomy, and patient safety.

18. **Document and Review Technology-Related Near Misses**

Treat technology failures and workarounds as learning opportunities.

D. Psychological Safety, Voice, and Culture

19. **Implement Confidential, Non-Retaliatory Reporting Channels**

Protect those who raise concerns about culture, leadership, or technology.

20. **Train Leaders in Psychological Safety and Inclusive Decision-Making**

Focus on inviting dissent and responding constructively to challenge.

21. **Normalize Structured Debriefs After Technology Implementations**

Include relational, ethical, and cognitive dimensions—not just technical outcomes.

22. **Incorporate Psychological Safety Scores into Unit-Level Performance Reviews**

Hold leaders accountable for team climate.

23. **Eliminate Cultural Norms That Conflate Aggression with Excellence**

Redefine rigor as thoughtful, inclusive, and evidence-based practice.

24. **Provide Coaching for Leaders Exhibiting Harmful Interpersonal Patterns**

Intervene early rather than tolerating toxic behavior.

E. Compensation, Advancement, and Career Equity

25. **Conduct Mandatory Annual Pay Equity Audits**

Correct disparities immediately with retroactive adjustments when needed.

26. **Standardize Salary Bands and Promotion Criteria Across Departments**

Reduce discretionary bias in compensation and advancement.

27. **Ensure Equal Access to Revenue-Generating and High-Visibility Roles**

Monitor assignment of leadership, research, and innovation opportunities.

28. **Establish Formal Sponsorship Programs with Measurable Outcomes**

Require sponsors to advocate actively for advancement.

29. **Protect Research and Innovation Time for Women Physicians**

Adjust workloads to enable participation in leadership and technology initiatives.

30. **Track Time-to-Promotion by Gender and Specialty**

Use data to identify structural delays and inequities.

F. Workforce Sustainability and Retention

31. **Conduct Stay Interviews with Women and Diverse Physicians**

Identify retention risks before exit decisions are made.

32. **Design Flexible Career Pathways Without Penalty**

Support nonlinear careers while preserving advancement opportunities.

33. **Reduce Administrative and Documentation Burdens Linked to Technology**

Reassess tools that disproportionately increase workload.

34. **Provide Reentry and Advancement Support After Career Interruptions**

Address penalties associated with caregiving or leave.

G. Change Management, Evaluation, and Sustainability

35. **Develop a Phased Implementation Roadmap with Clear Ownership**

Assign accountable leaders and timelines for each intervention.

36. **Define KPIs Linking Diversity to Innovation Quality**

Measure error detection, design improvements, and user trust.

37. **Communicate RCA Findings and Progress Transparently**

Build trust through regular updates and visible action.

38. **Embed Equity and Innovation Goals into Strategic Planning Cycles**

Treat inclusion as a core operational priority.

39. **Train Middle Managers as Change Agents**

Equip them to reinforce new norms and practices.

40. **Institutionalize Continuous Feedback and Adaptation Loops**

Use data, employee input, and outcomes to refine strategies over time.

CONCLUSION

Psychological safety has emerged as a core construct for understanding how biomedical organizations sustain reliability, learning, and ethical vigilance under conditions of uncertainty. Edmondson (2018) argues that when psychological safety is absent, organizations incur predictable hazards: employees conceal errors, suppress dissent, and disengage from discretionary learning behaviors that are essential for adaptation. In biomedicine, where diagnostic reasoning, treatment decisions, and technology-mediated workflows require constant sensemaking, these hazards are not merely cultural concerns but operational risks. A psychologically unsafe environment increases the likelihood that clinicians will remain silent about anomalous lab values, questionable algorithmic outputs, or workflow breakdowns in electronic health record systems, thereby allowing preventable errors to propagate across teams and units. In this sense, psychological safety functions as an organizational control mechanism, not because it reduces accountability, but because it increases the probability that problems are detected early enough to be corrected (Lundy et al., 2025).

The organizational costs of low psychological safety are not evenly distributed across the workforce. For women, psychologically unsafe workplaces are frequently associated with lower representation, reduced retention, and persistent equity problems embedded within everyday practices and decision processes (Burrell, 2022). These dynamics are especially consequential in biomedical settings such as Virelthos Medical Institute, where professional authority is often negotiated through informal interactions; case conferences, interdisciplinary rounds, grant teams, or technology governance committees. When women clinicians and scientists perceive that their expertise is dismissed or their contributions are routinely devalued, they are less likely to speak up, propose innovations, or contest flawed assumptions. This pattern has direct implications for patient safety and innovation quality. For example, implementation teams rolling out clinical decision-support tools frequently rely on tacit knowledge from frontline clinicians to detect unintended consequences (e.g., alert fatigue, inappropriate defaults, equity harms). If psychological safety is low, those most likely to notice and report these issues may also be those least structurally empowered to influence the final decision.

The relationship between diversity and organizational effectiveness in biomedicine is therefore conditional rather than automatic. Workgroups composed of diverse professionals bring a wider range of lived experiences and cognitive schemas to complex problem-solving, which should strengthen clinical reasoning and innovation. Yet Edmondson (2018) emphasizes that diversity yields performance gains only when organizations convert differences into usable knowledge; an outcome that depends on whether dissent is welcomed and expertise is recognized. In psychologically unsafe climates, diversity may paradoxically increase interpersonal risk: those with less status may calculate that speaking up carries unacceptable penalties, thereby limiting the organization's ability to leverage collective intellectual capital (Edmondson, 2018). This mechanism offers a critical industrial and organizational psychology insight which encompasses the value of diversity is mediated by the organization's capacity for inclusive learning and error reporting, not simply by demographic composition.

In healthcare organizations, psychological safety takes on a distinctive professional form. For physicians and biomedical scientists, safety is not merely freedom from humiliation; it is the expectation that one's clinical judgment, research expertise, and ethical concerns will be treated as legitimate contributions to collective decision-making (Edmondson, 2018). Psychological safety is therefore closely tied to career development and the legitimacy of professional identity. When medical professionals perceive that advancement depends on conformity, silence, or alignment with dominant leaders rather than evidence-based dissent, the organization creates incentives for surface compliance rather than robust clinical deliberation. A high-reliability biomedical culture, by contrast, treats speaking up as a performance behavior: questioning questionable orders, challenging biased algorithmic outputs, and flagging near-miss events become normative acts of professionalism rather than interpersonal threats.

This is why psychological safety and inclusion should be understood as overlapping, mutually reinforcing organizational properties rather than separate initiatives. Inclusive workplaces validate individual differences and create social climates where employees feel welcomed, accepted, and safe (Burrell, 2021). Such climates are particularly salient in biomedicine, where belonging is often shaped by status hierarchies, gatekeeping, and prestige markers. If inclusion is absent, if differences are tolerated but not valued, employees may remain physically present while psychologically disengaged, withholding insight and limiting participation (Lundy et al., 2025). Burrell (2022) notes that inclusive cultures are critical for women because they reduce identity threat and improve retention, but they also serve organizational performance by broadening whose expertise counts as actionable knowledge.

Clark (2020) further theorizes that psychologically safe organizational cultures are enabled when employees across backgrounds experience authentic support from supervisors and peers, feel trusted and empowered, and have their expertise respected, their experiences valued, their roles clarified, and their autonomy protected. These conditions are not abstract ideals; they map onto tangible organizational practices. For instance, in a biomedical technology implementation team, role clarity can determine whether a frontline physician is invited merely to "review" a tool after procurement or is positioned as a co-designer with authority to influence requirements and deployment decisions. Similarly, autonomy shapes whether clinicians can modify workflows when decision-support alerts prove clinically inappropriate for certain patient populations. In such environments, psychological safety is operationalized through who speaks in meetings, whose concerns trigger action, and whether dissent is documented as risk intelligence rather than dismissed as resistance.

From a managerial and organizational development standpoint, the construction of psychological safety begins with how leaders gather and respond to employee experience. Burrell (2022) argues that psychologically safe climates are built through caring management practices that prioritize asking, listening, and taking employee narratives seriously. Listening is not simply a communication skill; it is an institutional capability that determines whether organizations can perceive weak signals of dysfunction before they become crises (Burrell, 2021). In biomedicine, where clinicians frequently encounter near misses, ethical tensions, and technology-induced workarounds, leaders who create conditions for candid reporting convert lived experience into organizational learning. Conversely, leaders who respond defensively or punitively teach employees that silence is rational.

A critical implication for Virelthos and organizations like it is that psychological safety must be treated as a strategic infrastructure for both patient safety and innovation governance. It is not reducible to interpersonal niceness, nor can it be achieved through training alone. It requires aligned incentives, credible protection against retaliation, and leadership practices that frame dissent as a necessary input to high-quality biomedical decision-making. Without these conditions, diversity efforts risk becoming representational rather than transformative: women and diverse clinicians may be present in the system yet structurally disincentivized from contributing the very cognitive diversity that advances scientific discovery and improves care quality (Edmondson, 2018; Burrell, 2022).

REFERENCES

Burrell, D. N. (2021). Creating Diverse and Religiously Inclusive Workplace Cultures in Hyper-Connected, Technical, and Cyber-Driven Organizations. [IJSKD]. *International Journal of Sociotechnology and Knowledge Development, 13*(3), 17–32. https://www.igi-global.com/article/creating-diverse-and-religiously-inclusive-workplace-cultures-in-hyper-connected-technical-and-cyber-driven-organizations/281051. DOI: 10.4018/IJSKD.2021070102

Burrell, D. N. (2022). Creating inclusive, diverse, and psychologically safe workplaces for African-American women working in public health response supply chain and logistics. *Land Forces Academy Review, 27*(2), 134-140. https://www.armyacademy.ro/reviste/rev2_2022/Art_Burrell_Raft_2_2022.pdf

Clark, T. (2020). *The 4 Stages of Psychological Safety: Defining the Path to Inclusion and Innovation*. Berrett-Koehler Publishers.

Dudley, J., McLaughlin, S., & Lee, T. H. (2022). *Why So Many Women Physicians Are Quitting*. Harvard Business Review. https://hbr.org/2022/01/why-so-many-women-physicians-are-quitting

Edmondson, A. (2018). *The Fearless Organization: Creating Psychological Safety in the Workplace for Learning, Innovation, and Growth*. Wiley.

Hudgins, T., Brown, K. D., Layne, D., & Maggard Stephens, T. (2022). The Effect of Academic Nurse Leaders' Toxic Behaviors. *The Journal of Nursing Education, 61*(2), 88–92. https://journals.healio.com/doi/abs/10.3928/01484834-20211213-02. DOI: 10.3928/01484834-20211213-02 PMID: 35112950

Iyer, M. S., Way, D. P., MacDowell, D. J., Overholser, B. M., Spector, N. D., & Jagsi, R. (2023). Bullying in Academic Medicine: Experiences of Women Physician Leaders. *Academic Medicine, 98*(2), 255–263. https://journals.lww.com/academicmedicine/fulltext/2023/02000/bullying_in_academic_medicine__experiences_of.27.aspx. DOI: 10.1097/ACM.0000000000005003 PMID: 36484542

Lo Sasso, A. T., Armstrong, D., Forte, G., & Gerber, S. E. (2020). Differences In Starting Pay For Male And Female Physicians Persist; Explanations For The Gender Gap Remain Elusive. *Health Affairs, 39*(2), 256–263. https://pubmed.ncbi.nlm.nih.gov/31967925/. DOI: 10.1377/hlthaff.2019.00664 PMID: 31967925

Lundy, D., Burrell, D. N., McCargo, T., Lima, F. M., Harris, N., Eke, H., & Prunella, K. (2025, November). Understanding the Psychology of Pregnancy Discrimination as a Challenging Woman's Mental Health Workplace Dynamic. In *RAIS Conference Proceedings 2022-2025* (No. 0608). Research Association for Interdisciplinary Studies. https://rais.education/wp-content/uploads/0608.pdf

Lyons, N. B., Bernardi, K., Olavarria, O. A., Shah, P., Dhanani, N., Loor, M., Holihan, J. L., & Liang, M. K. (2021). Gender Disparity Among American Medicine and Surgery Physicians: A Systematic Review. *The American Journal of the Medical Sciences, 361*(2), 151–168. https://www.sciencedirect.com/science/article/abs/pii/S0002962920304705. DOI: 10.1016/j.amjms.2020.10.017 PMID: 33526213

Plank-Bazinet, J. L., Heggeness, M. L., Lund, P. K., & Clayton, J. A. (2017). Women's Careers in Biomedical Sciences: Implications for the Economy, Scientific Discovery, and Women's Health. *Journal of women's health (2002), 26*(5), 525–529. https://www.liebertpub.com/doi/abs/10.1089/jwh.2016.6012

Chapter 4
Organizational Development Steps for Cyberliterate Organizational Cultures in Healthcare Organizations

Sharon Burton
http://orcid.org/0000-0003-1653-9783
Embry-Riddle Aeronautical University, USA

Darrell Norman Burrell
http://orcid.org/0000-0002-4675-9544
Marymount University, USA

Calvin Nobles
http://orcid.org/0000-0003-4002-1108
University of Maryland Global Campus, USA

Natasha Harris
http://orcid.org/0009-0002-1745-7966
Marymount University, USA

Garrett Hart
http://orcid.org/0009-0008-8003-2032
Marymount University, USA

KerriLaine Prunella
http://orcid.org/0009-0001-0177-3545
Marymount University, USA

Hilda Eke
http://orcid.org/0009-0006-2021-178X
Marymount University, USA

Samantha Thibodeau
http://orcid.org/0009-0000-4639-9166
Marymount University, USA

Arthur D. Bradford
http://orcid.org/0009-0009-7276-2825
Marymount University, USA

Teresa Manfreda-Foley
http://orcid.org/0009-0002-9136-9673
Marymount University, USA

DOI: 10.4018/979-8-2600-1026-6.ch004

Jessica Parker
http://orcid.org/0009-0008-9574-3275
Marymount University, USA

Gregory Karl Smith
http://orcid.org/0009-0000-5122-2250
Marymount University, USA

Phillip Reed
http://orcid.org/0009-0000-1266-0151
Marymount University, USA

Laura A. Jones
http://orcid.org/0000-0002-0299-370X
Capitol Technology University, USA

Chitra N. Perera
http://orcid.org/0009-0005-3431-4969
Marymount University, USA

Won Kang Song
http://orcid.org/0009-0004-8644-4987
Capitol Technology University, USA

ABSTRACT

YTI Hospital System experienced a significant cybersecurity incident involving ransomware, malware, and a data breach that resulted in the theft of approximately 9,000 patient records. Root cause analysis determined that the breaches were primarily attributable to employee errors rather than technological failure, highlighting critical gaps in cyberliteracy, decision-making, and organizational culture. The incident caused operational disruption, financial loss, regulatory exposure, and reputational damage, underscoring the need for a socio-technical approach to cybersecurity. In response, external consultants were engaged to assess the incident and develop comprehensive recommendations grounded in organizational culture theory, risk management, change management, and strategic planning. This paper examines those recommendations, emphasizing the development of a cyberliterate, cyber-aware, and cybersecure organizational culture. It argues that effective cybersecurity requires shared responsibility across all organizational levels, with leaders shaping beliefs and values.

INTRODUCTION

Creating a cyber-aware and cybersecure organizational culture is not solely a technical endeavor but a collective social responsibility that permeates all levels of an organization (Burrell et al, 2021). Effective cybersecurity cannot be achieved through technology alone; rather, it requires coordinated human action grounded in shared beliefs, values, and behavioral norms (Nobles et al., 2025; Nobles, 2018). Every organizational member contributes to the organization's overall risk posture, yet leaders bear a disproportionate responsibility for shaping, reinforcing, and aligning organizational culture with cybersecurity objectives (Burrell, 2024; Wright & Burrell, 2023). Leadership commitment is therefore essential in translating abstract security policies into lived organizational practices (Jones & Burrell, 2025). Managers, in turn, require pragmatic, human-centered approaches to cybersecurity that acknowledge cognitive limitations, competing priorities, and contextual pressures inherent in modern organizational environments (Nobles et al., 2025; Nobles, 2018).

Organizational culture can be understood as a multidimensional construct comprising three interrelated components (Zanganeh et al., 2023; Burrell et al., 2021). First, belief systems provide the cognitive foundations that guide collective action, influencing how employees perceive threats, responsibilities, and acceptable risk. Second, values articulate what organizational members consider important, including the relative priority assigned to security versus efficiency, convenience, or speed. Third, artifacts and creations encompass visible and audible manifestations of culture, including technologies, routines, rituals, language, myths, symbols, and patterns of behavior that signal how cybersecurity is enacted in practice rather than merely espoused. Together, these components shape how cybersecurity is interpreted, internalized, and operationalized across the organization (Weitoish & Burrell, 2024; Jones & Burrell, 2025).

From a values-orientation perspective, organizational culture can be categorized into four dominant orientations, each with distinct implications for cybersecurity. A support-oriented culture emphasizes trust, collaboration, shared responsibility, and individual growth. In such environments, cybersecurity awareness is reinforced through peer learning, open communication about incidents, and psychological safety that encourages reporting of errors or near misses (Nobles et al., 2025; Nobles, 2018). An innovation-oriented culture prioritizes adaptability, experimentation, and openness to new information, which can enhance resilience against evolving cyber threats but may also introduce risk if experimentation outpaces governance. A rules-oriented culture underscores compliance, formal procedures, and hierarchical authority, often resulting in top-down security controls that ensure consistency but may limit employee engagement or adaptive responses. Finally, a goal-oriented culture emphasizes performance metrics, accountability, and rewards tied to outcomes, potentially aligning cybersecurity behaviors with organizational incentives but risking superficial compliance if metrics prioritize outputs over secure processes (Nobles et al., 2025; Nobles, 2018). A cybersecure organization must critically balance these orientations rather than privileging any single cultural logic (Burrell, 2024; Wright & Burrell, 2023; Burrell et al., 2026).

The urgency of cultivating a cybersecure organizational culture has intensified significantly within the healthcare sector, particularly in the context of the COVID-19 pandemic (Burrell, 2024). The pandemic not only increased demand for healthcare services but also expanded the digital attack surface through rapid adoption of telehealth, remote work, and interconnected clinical systems, thereby amplifying cybersecurity risks (Coventry & Branley, 2018; Alghamdi, 2022). Healthcare cybersecurity breaches have far-reaching consequences, including financial losses, operational disruptions, compromised patient safety, and erosion of public trust (Coventry & Branley, 2018). Since the onset of the COVID-19 pandemic, healthcare organizations have experienced a marked escalation in both the frequency and severity of cyberattacks. In 2020 alone, approximately 34.4 million individuals in the United States had their private health information compromised, a figure that increased to nearly 50 million affected individuals by 2022 (Southwick, 2023). Globally, cyberattacks increased by 38 percent in 2022 compared to the previous year, with healthcare consistently ranking among the three most targeted sectors (McKeon, 2023). These trends illustrate not episodic failures but systemic vulnerabilities embedded within healthcare organizations' socio-technical infrastructures.

The growing reliance on electronic health records (EHRs) and digital health systems has further intensified these vulnerabilities. While such systems enhance care coordination and efficiency, they also provide lucrative targets for cybercriminals seeking access to highly sensitive personal and financial data (Coventry & Branley, 2018; Alghamdi, 2022). Stolen health data enables identity theft, financial fraud, and long-term exploitation, often with consequences that extend far beyond the initial breach. Consequently, cybercrime in healthcare generates not only immediate financial losses but also enduring reputational damage and regulatory exposure (Coventry & Branley, 2018; Alghamdi, 2022; Bhosale et al., 2021).

Compounding these risks is the structural reality that many healthcare organizations lack sufficient cybersecurity expertise, staffing, and financial resources to mount robust defenses (Coventry & Branley, 2018; Alghamdi, 2022). This resource asymmetry renders healthcare an attractive and sustainable target for cybercriminals, who increasingly treat cybercrime not as isolated incidents but as an ongoing business model. Tactics such as ransomware, malware deployment, data exfiltration, and identity theft are strategically combined to maximize financial gain and exploit organizational dependencies on uninterrupted digital operations (Coventry & Branley, 2018; Alghamdi, 2022).

Ransomware attacks are particularly disruptive in healthcare contexts, as they encrypt critical systems and demand payment for restoration, often under time-sensitive conditions where patient care is directly affected (Coventry & Branley, 2018; Alghamdi, 2022; Bhosale et al., 2021). Malware attacks similarly undermine system integrity by enabling data theft, operational sabotage, or unauthorized control of networks, leading to service outages and cascading failures (Coventry & Branley, 2018; Alghamdi, 2022; Bhosale et al., 2021). Data breaches further exacerbate harm by exposing sensitive information, enabling secondary crimes such as identity theft and financial fraud (Bhosale, Nenova, & Iliev, 2021).

PROBLEM STATEMENT

Following a significant data breach and ransomware attack, YTI Hospital System faces acute operational disruption, delayed patient care, financial loss, and reputational damage with both immediate and long-term implications. Immediate consequences include service interruptions, compromised clinical workflows, and potential legal and regulatory penalties. Long-term impacts may involve erosion of patient trust, escalating cybersecurity expenditures, workforce strain, and sustained financial instability. These outcomes underscore the inadequacy of purely technical responses and highlight the necessity of integrated cybersecurity strategies that explicitly address organizational culture, human behavior, and leadership accountability (Burrell et al., 2023; Drame, 2019; Kruse et al., 2017; Kshetri, 2018; Martin et al., 2017; Ofe, 2023).

Empirical evidence reinforces the severity of this challenge. Healthcare organizations experienced an average of 1,410 cyberattacks per week in 2022, representing an 86 percent increase from 2021 levels (McKeon, 2023). The scale of these attacks reflects not only technological exposure but also cultural and organizational weaknesses in cybersecurity governance. This paper therefore examines the evolving threat landscape of cybercrime in healthcare organizations and applies established theoretical frameworks to identify culturally informed solutions capable of influencing real-world practice.

The financial and operational toll of ransomware attacks further illustrates the stakes involved. In 2020, 642 healthcare data breaches compromised approximately 29.3 million records, with an average breach cost of $9.23 million; substantially exceeding the cross-industry average of $3.86 million (Almulihi et al., 2022). By 2021, breaches increased to 674 incidents, and by 2022, average breach costs in U.S. healthcare exceeded $10 million per incident. These figures reflect not only direct remediation expenses but also prolonged EHR downtime, delayed treatments, and revenue loss.

High-profile incidents underscore these consequences. The October 2020 ransomware attack on the University of Vermont Medical Center rendered EHRs, payroll systems, and core digital tools inaccessible for nearly a month, forcing surgery postponements and cancer patients to seek care elsewhere. The resulting financial losses exceeded $50 million, with total recovery costs reaching $63 million (Green & Jackson, 2021). Similarly, the October 2022 ransomware attack on CommonSpirit Health disrupted operations across multiple facilities, affecting over 623,000 individuals and generating estimated losses of $160 million due to business interruption and remediation efforts (Cantú et al., 2024; Kesarwani & Gochhayat, 2023).

Collectively, these cases demonstrate that cybersecurity failures in healthcare are not merely technical breakdowns but organizational crises rooted in cultural misalignment, insufficient preparedness, and fragmented accountability. The escalating frequency, sophistication, and impact of cyberattacks necessitate a strategic shift toward embedding cybersecurity into organizational culture. Developing a cyber-aware and cybersecure culture is therefore essential for safeguarding patient safety, ensuring financial sustainability, and preserving institutional legitimacy in an increasingly hostile digital environment (Burrell et al., 2023; Ruiz Caino, 2024).

ORGANIZATIONAL CULTURE THEORY AND CYBERSECURITY

Organizational culture plays a decisive role in shaping how healthcare institutions anticipate, experience, and recover from cybersecurity incidents such as data breaches and ransomware attacks (Burrell, 2024). Cybersecurity resilience is not achieved through technical controls alone but through coordinated human action embedded in shared beliefs, values, and behavioral norms. To secure an organization effectively, all members must actively participate in reducing risk, while leaders carry a distinct responsibility to align organizational culture with overarching security objectives (Jones & Burrell, 2025). When leadership commitment is absent or cultural norms are misaligned, cybersecurity initiatives are likely to be perceived as external impositions rather than collective responsibilities, thereby increasing organizational vulnerability.

In healthcare settings, where clinical urgency, information sensitivity, and technological dependence converge, organizational culture becomes a critical determinant of cybersecurity outcomes. A culture characterized by transparency, trust, and continuous learning enables rapid detection, communication, and response during cyber crises, whereas cultures marked by silos, fear of blame, or resistance to change can exacerbate operational disruption and delay recovery (Burrell et al., 2023; Drame, 2019). Edgar Schein's Organizational Culture Model provides a robust analytical framework for examining how cultural dynamics influence cybersecurity preparedness and crisis response by distinguishing among artifacts, espoused values, and basic underlying assumptions (Schein, 2010).

Artifacts and Cybersecurity Practice

Artifacts represent the most visible manifestations of organizational culture, including technology, language, workflows, rituals, and observable behavior patterns. In the context of cybersecurity, artifacts encompass information systems, security tools, incident response protocols, communication platforms, and the everyday practices through which employees interact with digital infrastructure. These visible elements not only enable operational continuity but also signal the organization's priorities regarding security and risk management (Nobles et al., 2025; Nobles, 2018).

Within the YTI Hospital System, the ransomware-induced disruption of electronic health records (EHRs), payroll systems, and other mission-critical digital tools illustrates the centrality of technological artifacts to healthcare delivery. The inability to access these systems directly affected patient care, staff coordination, and administrative operations, underscoring how deeply cybersecurity incidents penetrate core organizational functions (Burrell, 2024). The subsequent efforts to restore systems and reinforce IT infrastructure reflect an organizational recognition that cybersecurity artifacts are inseparable from clinical effectiveness and patient safety (McKeon, 2021; Davis, 2020). Comparable disruptions experienced by other large healthcare systems further highlight the sector's reliance on interconnected digital artifacts and the cascading consequences when those artifacts fail (Fox, 2023; McKeon, 2023).

However, reliance on technological artifacts without corresponding cultural reinforcement can create a false sense of security. Cyber-aware organizations must ensure that visible security tools are accompanied by consistent practices, shared understanding, and employee engagement (Nobles et al., 2025; Nobles, 2018). Artifacts should therefore be designed not only for technical resilience but also to support human decision-making under pressure.

Espoused Values and Leadership Alignment

Espoused values reflect the principles and priorities that organizational leaders explicitly articulate through mission statements, policies, and strategic decisions. In cybersecurity contexts, these values shape how employees interpret security expectations and determine whether cybersecurity is viewed as a shared ethical obligation or a compliance-driven requirement (Schein, 2010).

At YTI Hospital System, leadership responses to the ransomware incident provide insight into the organization's espoused values. The decision to resist ransom payment and engage with law enforcement authorities signals a commitment to ethical conduct, legal compliance, and long-term institutional integrity. Public communication regarding the breach and recovery process further reflects an espoused value of transparency, reinforcing trust among patients, employees, and external stakeholders (Kruse et al., 2017). Such transparency is particularly critical in healthcare, where trust is foundational to patient-provider relationships.

Similarly, the extensive communication strategies adopted by peer healthcare organizations during recovery efforts demonstrate how espoused values of accountability and responsibility can mitigate reputational harm and stabilize stakeholder confidence following cyber incidents (Martin et al., 2017). When leaders consistently articulate and model these values, they legitimize cybersecurity as an organizational priority rather than a peripheral technical concern.

Basic Underlying Assumptions and Cultural Change

Basic underlying assumptions constitute the deepest level of organizational culture, encompassing taken-for-granted beliefs that guide behavior unconsciously and shape responses to uncertainty and crisis. In cybersecurity, these assumptions influence how employees perceive risk, accountability, and the relationship between security and patient care (Schein, 2010).

Within YTI Hospital System, the rapid mobilization of technical teams and engagement with cybersecurity experts following the attack suggest an underlying assumption that patient safety and data security are inseparable from organizational legitimacy. The sustained effort to restore systems and maintain clinical operations despite significant disruption reflects a deeply embedded belief in continuity of care as a non-negotiable organizational obligation (Tosh et al., 2014). At the same time, the stress experienced by staff during the incident reveals implicit assumptions about crisis preparedness and highlights the need for greater cultural emphasis on psychological safety, training, and institutional support.

Comparable healthcare organizations have demonstrated that post-incident investments in IT resilience and cybersecurity governance often reflect a broader assumption that digital infrastructure is now foundational to healthcare delivery rather than a supporting function (Tosh et al., 2014). When such assumptions are aligned with proactive training and shared accountability, organizations are better positioned to adapt to evolving cyber threats (Burton, 2023).

Cultural Orientations and Cybersecurity Integration

Organizational culture can further be understood through dominant value orientations that shape behavior and decision-making (Burton, 2023). Support-oriented cultures promote trust, collaboration, and shared learning, enabling employees to report security concerns without fear and to learn collectively from incidents. Innovation-oriented cultures encourage adaptability and creative problem-solving, enhancing responsiveness to emerging cyber threats but requiring governance to manage risk (Weitoish, & Burrell, 2024). Rules-oriented cultures emphasize compliance, formal procedures, and hierarchical authority, providing consistency but potentially limiting employee engagement if overly rigid. Goal-oriented cultures focus on performance metrics and accountability, aligning cybersecurity behaviors with organizational incentives but risking superficial compliance if goals are poorly designed. A cyber-aware and cybersecure healthcare organization must integrate elements of all four orientations, balancing flexibility with control, collaboration with accountability, and innovation with ethical responsibility (Weitoish, & Burrell, 2024). Leaders and managers play a central role in orchestrating this balance by translating cultural values into practical, human-centered cybersecurity strategies that recognize the realities of clinical work (Burrell, 2024).

Ultimately, cybersecurity resilience in healthcare emerges not from isolated technical solutions but from a coherent organizational culture in which beliefs, values, and practices are deliberately aligned with security goals (Wright & Burrell, 2023). By embedding cybersecurity into the cultural fabric of the YTI Hospital System, leaders can transform cyber risk management from a reactive function into a sustained collective capability that protects patients, staff, and institutional trust over the long term.

CHANGE MANAGEMENT THEORY AND THE DEVELOPMENT

Effective cybersecurity transformation in healthcare organizations requires more than technical remediation following a cyber incident; it demands deliberate organizational change that reshapes how risk, responsibility, and security are understood across the workforce (Burton, 2023). To secure an organization, all members must actively participate in reducing cyber risk through daily behaviors, informed decision-making, and shared accountability. Leaders play a particularly critical role in this process by interpreting cyber threats, setting priorities, and aligning organizational beliefs, values, and attitudes with long-term security objectives (Burrell, 2024). Managers, positioned between strategic leadership and frontline operations, require practical, human-centered solutions to translate cybersecurity expectations into sustainable organizational practices (Nobles et al., 2025).

John Kotter's 8-Step Change Model offers a robust framework for guiding cybersecurity-related organizational change because it explicitly integrates leadership action, employee engagement, and cultural institutionalization (Kotter, 1995; Drame, 2019). Applied to the YTI Hospital System, this model underscores that cybersecurity resilience emerges through a structured progression from crisis awareness to cultural embedding, addressing both technical vulnerabilities and the human factors that frequently undermine security initiatives (Ofe, 2023).

Establishing Urgency and Shared Risk Awareness

The first step in Kotter's model, creating a sense of urgency, is particularly salient in the context of healthcare cybersecurity. Cyber incidents must be framed not solely as IT failures but as organizational crises that threaten patient safety, operational continuity, and institutional trust. Leaders at YTI Hospital System must communicate the real and potential consequences of cybersecurity breaches using concrete evidence, including financial losses, service disruptions, and regulatory exposure experienced across the healthcare sector. Transparent dissemination of such information helps shift cybersecurity from a peripheral concern to a shared organizational priority, reinforcing the belief that every employee's actions can either mitigate or amplify cyber risk (Lamé et al., 2017).

Forming a Guiding Coalition and Distributed Responsibility

Sustainable cybersecurity change cannot be imposed unilaterally. Kotter (1995) emphasizes the importance of forming a guiding coalition composed of credible leaders with diverse expertise and authority. For YTI Hospital System, this coalition should extend beyond IT leadership to include clinical staff, operational managers, compliance officers, and executive decision-makers. Including representatives from multiple functional areas reinforces the cultural message that cybersecurity is a collective responsibility rather than a specialized technical function. Such cross-functional collaboration strengthens trust, reduces silos, and enables more comprehensive responses to complex cyber threats (Lamé et al., 2017).

Developing and Communicating a Cybersecurity Vision

A clear and compelling vision is essential for aligning organizational efforts and sustaining momentum throughout the change process. Kotter (1995) argues that an effective vision must be understandable, actionable, and motivating. At YTI Hospital System, the cybersecurity vision should explicitly articulate the integration of security with patient care, ethical responsibility, and organizational resilience. This vision must be communicated consistently through multiple channels, reinforced by leadership behavior, and embedded into training programs, policies, and operational workflows. Communication should emphasize not only what changes are required but why those changes matter to employees' professional roles and the organization's mission (Lamé et al., 2017).

Removing Obstacles and Enabling Secure Behavior

Cultural and structural barriers often impede cybersecurity initiatives, even when leadership commitment is strong. Removing obstacles involves identifying procedural inefficiencies, unclear reporting structures, and cultural norms that discourage proactive security behavior. At YTI Hospital System, bureaucratic delays, ambiguous authority, or fear of blame may prevent employees from escalating security concerns promptly. Kotter (1995) emphasizes empowering employees by providing them with the resources, authority, and psychological safety necessary to act decisively. Practical managerial solutions include simplifying incident reporting mechanisms, clarifying escalation pathways, and reinforcing non-punitive responses to security errors, thereby encouraging early detection and response (Lamé et al., 2017).

Generating Short-Term Wins and Reinforcing Cultural Norms

Achieving and visibly recognizing short-term successes is critical for sustaining engagement and validating change efforts. In cybersecurity contexts, short-term wins may include improved phishing detection rates, successful completion of security training, or rapid containment of minor incidents. Publicly acknowledging these achievements signals that secure behavior is valued and rewarded, reinforcing goal-oriented cultural elements such as accountability and performance measurement (Kotter, 1995). For YTI Hospital System, celebrating these milestones transforms abstract security goals into tangible outcomes, strengthening employees' confidence in their ability to contribute meaningfully to organizational security (Lamé et al., 2017).

Sustaining Momentum Through Continuous Adaptation

After initial successes, organizations must resist complacency by consolidating gains and pursuing deeper systemic change. Kotter (1995) emphasizes that sustaining acceleration requires ongoing evaluation, adaptation, and investment. In the cybersecurity domain, this involves continuously updating technologies, revising policies, and refining training programs to address evolving threats. For YTI Hospital System, sustained momentum also depends on integrating different cultural orientations, leveraging support-oriented trust and collaboration, innovation-oriented adaptability, rules-oriented compliance, and goal-oriented accountability, to create a balanced and resilient security culture.

Anchoring Cybersecurity in Organizational Culture

The final step in Kotter's model, anchoring changes in the culture, is essential for long-term cybersecurity resilience. Cultural anchoring occurs when new behaviors become normalized and are reinforced through organizational systems, including hiring practices, performance evaluations, and leadership development. At YTI Hospital System, cybersecurity awareness should be explicitly incorporated into core values, onboarding processes, and professional development programs. By embedding security expectations into everyday routines and decision-making processes, the organization reinforces the underlying assumption that cybersecurity is integral to healthcare delivery rather than an auxiliary concern (Lamé et al., 2017).

Ultimately, effective cybersecurity change in healthcare requires aligning belief systems, values, and visible practices with security objectives through intentional leadership and structured change management. Kotter's model provides a pathway for transforming cybersecurity from a reactive response to crises into a sustained organizational capability grounded in shared responsibility, cultural coherence, and continuous learning. By embracing this approach, YTI Hospital System can strengthen its cyber awareness, reduce systemic risk, and enhance its capacity to protect patients, employees, and institutional trust in an increasingly complex digital environment.

PLANNING AND STRATEGY THEORY AS FOUNDATIONS FOR A CYBER-AWARE ORGANIZATIONAL CULTURE

Effective cybersecurity in healthcare organizations cannot be achieved through isolated technical controls or post-incident remediation alone. Instead, cybersecurity must be planned, implemented, and sustained as an organizational capability rooted in shared responsibility, cultural alignment, and strategic foresight (Weitoish & Burrell, 2024). To secure an organization such as the YTI Hospital System, all members must actively participate in reducing cyber risk through informed behaviors, compliance with security practices, and vigilance in daily operations. Leadership bears a distinctive responsibility to shape and align the organization's beliefs, values, and attitudes with cybersecurity objectives, while managers must translate strategic intent into practical, human-centered actions that influence how employees perceive and enact security (Weitoish, & Burrell, 2024).

Planning theory, operationalized through the Risk Management Framework (RMF), provides a structured mechanism for integrating cybersecurity into organizational culture by systematically identifying risks, assessing their implications, implementing mitigation strategies, and continuously monitoring effectiveness (Drame, 2019; Ofe, 2023). When applied holistically, RMF extends beyond technical assessment to incorporate human behavior, communication practices, and organizational learning, thereby reinforcing cybersecurity as a collective responsibility rather than a specialized technical function.

Risk Assessment as a Cultural Diagnostic

Risk assessment within the RMF serves not only as a technical evaluation but also as a cultural diagnostic tool that reveals how an organization understands and prioritizes cyber threats. At YTI Hospital System, identifying high-impact risks; such as unauthorized access to patient records, prolonged electronic health record (EHR) outages, and disruption of clinical workflows; highlights the interdependence between digital systems and patient safety (Whitman & Mattord, 2018). These risks are shaped not only by technical vulnerabilities, including outdated software or insufficient encryption, but also by human factors such as employee fatigue, inconsistent security practices, and insider threats (Kruse et al., 2017).

By explicitly acknowledging human behavior as a critical risk vector, the organization reinforces the belief that cybersecurity is embedded in everyday actions. This recognition supports a shift in belief systems toward collective accountability, where employees understand that individual decisions, such as responding to phishing emails or adhering to access protocols, directly influence organizational resilience.

Risk Treatment and the Human Dimensions of Cybersecurity

Risk treatment strategies within the RMF illustrate how planning decisions shape cultural norms and expectations. Advanced security measures, including multi-factor authentication, encryption, and regular software updates, are essential technical controls that protect patient information from unauthorized access (Kruse et al., 2017; Kshetri, 2018). However, these measures are only effective when employees understand their purpose, perceive them as legitimate, and integrate them into routine workflows.

Incident response planning further demonstrates the cultural dimensions of cybersecurity preparedness. Clearly defined procedures for isolating affected systems, communicating with stakeholders, and initiating recovery efforts enable coordinated action during crises and reduce uncertainty among staff (Kruse et al., 2017; Kshetri, 2018). When such plans are routinely tested and rehearsed, they become organizational artifacts that signal preparedness, reinforce shared expectations, and reduce reliance on improvisation under pressure.

Employee training and awareness initiatives play a central role in translating planning into practice. Training programs that educate staff on phishing recognition, secure data handling, and breach response foster a culture of vigilance and empower employees to act as active defenders of organizational assets (Dargin, 2020; Peltier, 2016). In this context, cybersecurity awareness becomes a visible expression of organizational values rather than a compliance obligation (Burrell et al., 2026).

Communication Planning and Cultural Coherence

Communication planning is a critical but often underestimated component of cybersecurity strategy. Effective communication ensures that all stakeholders receive timely, accurate, and actionable information during and after cyber incidents. Within YTI Hospital System, structured internal communication channels, such as regular updates from incident response teams; support coordination and reduce ambiguity regarding roles and responsibilities (Whitman & Mattord, 2018). Transparent communication reinforces trust and aligns employee behavior with organizational priorities, particularly during periods of heightened uncertainty.

Broader guidance emphasizing individual responsibility, such as enabling multi-factor authentication, maintaining updated software, recognizing phishing attempts, and reporting suspicious activity, underscores that cybersecurity preparedness extends beyond institutional systems to individual behaviors (Lynn, 2024). Embedding these expectations into organizational messaging reinforces the underlying assumption that security is a shared, continuous process.

Strategic Alignment Through Porter's Generic Strategies

Strategy theory further complements planning theory by positioning cybersecurity as a source of organizational resilience and strategic advantage. Michael Porter's Generic Strategies, cost leadership, differentiation, and focus, offer a framework for aligning cybersecurity investments with organizational goals while reinforcing cultural coherence (Porter, 1985; Drame, 2019).

Cost leadership in cybersecurity does not imply minimal investment but rather the efficient allocation of resources to maximize protection. At YTI Hospital System, investing in scalable, cost-effective security technologies and conducting regular audits ensures that critical systems are protected without unsustainable expenditure (Whitman & Mattord, 2018). This approach reinforces a goal-oriented culture emphasizing accountability and prudent stewardship of resources.

Differentiation enables the organization to distinguish itself through superior patient data protection and transparent crisis response. Implementing advanced encryption, robust access controls, and routine testing of incident response plans signals a commitment to patient trust and ethical responsibility (Kruse et al., 2017; Kshetri, 2018). Appointing dedicated cybersecurity leadership, such as a Chief Information Security Officer, and conducting regular response simulations further institutionalize security as a strategic priority (Pattison-Gordon, 2023).

Focus strategies allow YTI Hospital System to concentrate efforts on high-impact areas such as patient communication and data recovery. Clear communication with patients regarding breach response and data protection measures reinforces transparency, while robust backup and recovery procedures ensure operational continuity and rapid restoration of services (Dargin, 2020; Whitman & Mattord, 2018).

Embedding Strategy into Organizational Culture

Ultimately, planning and strategy theories converge in their emphasis on cultural integration. Cybersecurity resilience emerges when belief systems emphasize shared responsibility, values prioritize patient trust and data protection, and artifacts, such as training programs, communication platforms, and security technologies which consistently reinforce secure behavior (Burrell et al., 2026). Support-oriented cultural elements promote collaboration and trust, innovation-oriented values enable adaptive responses to evolving threats, rules-oriented structures ensure consistency and compliance, and goal-oriented mechanisms align incentives with desired outcomes (Burrell et al., 2026). By embedding cybersecurity planning and strategy into the cultural fabric of the YTI Hospital System, leaders and managers can move beyond reactive risk management toward a sustained, cyber-aware organizational culture. Such integration strengthens the organization's capacity to reduce systemic risk, protect sensitive data, and maintain continuity of care in an increasingly complex and hostile digital environment.

MASTER LIST OF 40 ACTIONABLE RECOMMENDATIONS FOR BUILDING A CYBERLITERATE, CYBER-AWARE, AND CYBERSECURE ORGANIZATIONAL CULTURE

1. **Establish Cybersecurity as a Shared Organizational Responsibility** Explicitly communicate that cybersecurity risk is shaped by everyday employee decisions, reinforcing that security is not solely an IT function (Romanosky et al., 2014).
2. **Require Visible Executive Leadership Ownership of Cybersecurity** Ensure senior leaders consistently link cybersecurity behaviors to patient safety, ethical care, and organizational integrity (U.S. Department of Health & Human Services, 2023).
3. **Embed Cyber Risk Awareness into Organizational Belief Systems** Integrate cybersecurity expectations into onboarding, clinical education, and leadership development to shape collective assumptions about risk (Peltier, 2016).
4. **Align Organizational Values Explicitly with Cybersecurity Objectives** Define privacy, vigilance, and data protection as core values alongside quality of care and patient trust (Dwyer et al., 2004).
5. **Translate Cybersecurity Values into Visible Cultural Artifacts** Standardize secure technologies such as MFA, encryption, and monitored access controls as everyday operational norms (Whitman & Mattord, 2018).

6. **Integrate Cyber Risk Scenarios into Clinical and Operational Training** Embed cybersecurity decision points into simulations and case-based learning to reinforce judgment under pressure.
7. **Implement Role-Specific Cyberliteracy Competency Standards** Define distinct cyberliteracy expectations for clinicians, administrators, managers, and executives (Dargin, 2020).
8. **Require "Pause-and-Verify" Protocols for High-Risk Digital Actions** Mandate deliberate verification steps for credential resets, external downloads, and unusual access requests (Whitman & Mattord, 2018).
9. **Adopt Non-Punitive Cyber Risk After-Action Reviews (AARs)** Conduct structured reviews after incidents and near-misses to analyze decision-making without blame (Peltier, 2016).
10. **Establish Ethical Cyber Responsibility Commitments for Staff** Require employees to formally acknowledge cybersecurity as an ethical obligation tied to patient safety and confidentiality (Koyame-Marsh & Marsh, 2014).
11. **Design Security Controls with Cognitive Load in Mind** Evaluate cybersecurity tools for usability in high-pressure clinical environments to reduce unsafe workarounds (Whitman & Mattord, 2018).
12. **Develop Leadership Cybersecurity Rituals** Institutionalize executive-led cybersecurity briefings to reinforce security as a strategic priority (Dargin, 2020).
13. **Foster a Support-Oriented Cyber Culture** Create non-punitive reporting mechanisms that encourage early disclosure of phishing attempts and errors (Peltier, 2016).
14. **Establish Peer-Based Cybersecurity Champion Programs** Designate departmental champions to model secure behavior and provide informal guidance (Dargin, 2020).
15. **Promote Innovation-Oriented Cybersecurity Practices** Allow controlled experimentation with new security tools while maintaining governance oversight (Whitman & Mattord, 2018).
16. **Regularly Review and Update Cybersecurity Policies** Ensure policies reflect evolving threats and clinical realities, reinforcing adaptability (Amodu, 2024).
17. **Strengthen Rules-Oriented Cyber Governance Structures** Clearly define roles, responsibilities, and escalation pathways during cyber incidents (Kshetri, 2018).
18. **Integrate Cybersecurity Compliance into Routine Audits** Normalize secure behavior by embedding HIPAA and regulatory safeguards into performance reviews (Dwyer et al., 2004).
19. **Reinforce Goal-Oriented Cybersecurity Metrics** Establish measurable objectives such as phishing resilience and training completion rates (Whitman & Mattord, 2018).

20. **Recognize and Reward Exemplary Cybersecure Behavior** Publicly acknowledge individuals and teams that demonstrate strong cybersecurity judgment (Romanosky et al., 2014).
21. **Activate Incident Response Plans Immediately During Cyber Events** Isolate affected systems, preserve forensic evidence, and initiate coordinated recovery actions (U.S. Department of Health & Human Services, 2023).
22. **Ensure Leadership Participation in Cyber Incident Simulations** Reinforce shared accountability and decision-making readiness during crises (Peltier, 2016).
23. **Maintain Transparent Internal and External Cyber Communications** Provide timely, accurate updates to staff, patients, and stakeholders during incidents (Romanosky et al., 2014).
24. **Use Multiple Communication Channels for Cyber Messaging** Ensure consistent messaging across internal portals, emails, and briefings (U.S. Department of Health & Human Services, 2023).
25. **Provide Identity Theft Protection to Affected Patients Post-Breach** Demonstrate ethical responsibility and patient-centered accountability (Koyame-Marsh & Marsh, 2014).
26. **Conduct Role-Tailored Cybersecurity Training Programs** Ensure all staff understand both technical risks and behavioral expectations (Dargin, 2020).
27. **Incorporate Realistic Cyber Incident Drills into Annual Training** Reinforce muscle memory, reduce panic, and improve cross-functional coordination (Peltier, 2016).
28. **Implement Continuous Cyber Threat Monitoring Systems** Use intrusion detection and monitoring to reinforce active vigilance (Amodu, 2024).
29. **Perform Routine Vulnerability Assessments and Security Audits** Identify weaknesses proactively rather than reactively (Whitman & Mattord, 2018).
30. **Develop and Test Robust Data Backup and Recovery Strategies** Ensure operational continuity through automated backups and frequent testing (Spence et al., 2017).
31. **Regularly Test Business Continuity and Contingency Plans** Prepare staff to sustain patient care during prolonged system disruptions (Dargin, 2020).
32. **Integrate Cybersecurity into Clinical Workflow Design** Reduce unsafe workarounds by aligning security controls with real-world clinical processes (Whitman & Mattord, 2018).
33. **Require Managers to Reinforce Cybersecurity in Routine Meetings** Normalize secure behavior as part of everyday professional expectations (Peltier, 2016).
34. **Establish Continuous Feedback Loops for Cyber Improvements** Enable employees to suggest enhancements to security processes (Romanosky et al., 2014).

35. **Periodically Assess Cybersecurity Culture Maturity** Use surveys and behavioral metrics to identify misalignment between values and practice (Dargin, 2020).
36. **Create Cross-Functional Cyber Decision Councils** Facilitate shared evaluation of cyber risks across clinical, operational, and IT domains (Kshetri, 2018).
37. **Embed Behavioral Cyber Metrics into Leadership Dashboards** Track indicators such as escalation speed and reporting quality (Romanosky et al., 2014).
38. **Adopt Narrative-Based Cybersecurity Education Methods** Use real healthcare cyber incident stories to strengthen judgment and ethical reasoning.
39. **Integrate Cyber Judgment into Performance Evaluations** Assess how employees and managers demonstrate secure decision-making, not just compliance (Whitman & Mattord, 2018).
40. **Anchor Cyberliteracy into Organizational Lifecycle Practices** Embed cybersecurity expectations into hiring, promotion, leadership evaluation, and strategic planning to ensure long-term cultural sustainability (Whitman & Mattord, 2018).

Collectively, these recommendations operationalize organizational culture theory by aligning belief systems, values, and visible practices with cybersecurity goals. By integrating support, innovation, rules, and goal orientations into a coherent cultural framework, YTI Hospital System can cultivate a cyber-literate workforce, reduce systemic risk, and sustain a cybersecure environment that protects patients, staff, and institutional trust over time.

REFERENCES

Alghamdi, A. (2022). *Cybersecurity Threats to Healthcare Sectors during Covid-19. 2022 2nd International Conference on Computing and Information Technology*. ICCIT., https://ieeexplore.ieee.org/abstract/document/9711659?casa_token=e9-Ic-EucqUAAAAA:0UXqAIBHns9A_1pZz9A-vtSgjnWDCTnxNTCpqGkpEotKTDVXs6SDa9-QnT6jKyeUvJ0F4jstnSo, DOI: 10.1109/ICCIT52419.2022.9711659

Almulihi, A. H., Alassery, F., Khan, A. I., Shukla, S., Gupta, B. K., & Kumar, R. (2022). Analyzing the implications of healthcare data breaches through computational technique. *Intelligent Automation & Soft Computing*, *32*(3). Advance online publication. https://d1wqtxts1xzle7.cloudfront.net/84205476/pdf.pdf?1738492817=&response-content-disposition=inline%3B+filename%3DAnalyzing_the_Implications_of_Healthcare.pdf&Expires=1770151649&Signature=PeIfKoCrbVIFzdbmQ4IYLJy7OKJ9oZ1JtAc6uJiVUit3C189poqRF9Wvi8ZG7PKAK~7zOs6RwkItWbjJzJ0WgV7OdkIPHvp71PVt2ecZ85Tbc-cahxMVx3kiqAXVPZ1ku~UDF0dP5iJmtznTOAC9qr4Pu7Mdc9gS8HJWhga5yzEXc5YCZmCF18kCaw91dyJEbeMDx7LOu83Rrl-dpQrOFrMYfQ7Plk8iHWxweA~cdYI~SKWlPLyxS3IvqD~v43gETpLPH1R6U4quC7vhbKJBoz0LPYcn-Ol08qiPtzzpHHIH~eyDGO~vdU7IHS~8O~puGbfpMTM9OGwH7MSwZBPVjA__&Key-Pair-Id=APKAJLOHF5GGSLRBV4ZA. DOI: 10.32604/iasc.2022.023460

Amodu, O. (2024). Security auditors' perspective in tackling cyber-threats. *Computer*, *11*(5). https://www.researchgate.net/profile/Oluwatosin-Amodu/publication/381280763_Security_Auditors'_Perspective_in_Tackling_Cyber-Threats/links/6664c731a54c5f0b94580cd5/Security-Auditors-Perspective-in-Tackling-Cyber-Threats.pdf

Bhosale, K., Nenova, M., & Iliev, G. (2021). *A study of cyber attacks: In the healthcare sector, 2021 Sixth Junior Conference on Lighting*. Lighting., https://ieeexplore.ieee.org/abstract/document/9598947?casa_token=LA0C0P-X-xsAAAAA:vBmONsY-b7MaKYdYJzRsVWUpRrvOA5KjpFphZGFKg3aAQ3antcnYS5angKmYQQ2GBlAd6JDblx8, DOI: 10.1109/Lighting49406.2021.9598947

Burrell, D. N. (2024). Understanding healthcare cybersecurity risk management complexity. *Land Forces Academy Review*, *29*(1), 38-49. https://reference-global.com/download/article/10.2478/raft-2024-0004.pdf

Burrell, D. N., Aridi, A. S., McLester, Q., Shufutinsky, A., Nobles, C., Dawson, M., & Muller, S. R. (2021). Exploring system thinking leadership approaches to the healthcare cybersecurity environment. [IJEACH]. *International Journal of Extreme Automation and Connectivity in Healthcare*, *3*(2), 20–32. https://www.igi-global.com/article/exploring-system-thinking-leadership-approaches-to-the-healthcare-cybersecurity-environment/284462. DOI: 10.4018/IJEACH.2021070103

Burrell, D. N., Burton, S., Jones, L. A., Nobles, C., & Dawson, M. (2026). Healthcare administration in the age of complex dynamics in hospital and health system cybersecurity risk mitigation. *Brazilian Journal of Business*, *8*(1), e85317. DOI: 10.34140/bjbv8n1-015

Burrell, D. N., Burton, S. L., Nobles, C., Springs, D., Huff, A. J., Brown-Jackson, K. L., . . . Jones, A. J. (2023). The managerial ethical and operational challenges of hospital cybersecurity. In *Transformational Interventions for Business, Technology, and Healthcare* (pp. 444-458). IGI Global. https://www.igi-global.com/chapter/the-managerial-ethical-and-operational-challenges-of-hospital-cybersecurity/332759

Burrell, D. N., Lu, S., Lindsay, P. V. L., Burton, S. L., French, R., Finch, A., McLester, Q., & Springs, D. (2021). Sexual Harassment Training Focused Solely on Protecting Organizations From Lawsuits Won't Change a Toxic Culture for Women to an Inclusive One. [IJPSS]. *International Journal of Public Sociology and Sociotherapy*, *1*(1), 12–25. https://www.igi-global.com/article/sexual-harassment-training-focused-solely-on-protecting-organizations-from-lawsuits-wont-change-a-toxic-culture-for-women-to-an-inclusive-one/269514. DOI: 10.4018/IJPSS.2021010102

Burton, S. L. (2023). Change management and cybersecurity in healthcare: Mitigating human factors and risks. In *Transformational Interventions for Business, Technology, and Healthcare* (pp. 426-443). IGI Global. https://www.igi-global.com/chapter/change-management-and-cybersecurity-in-healthcare/332758

Cantú, C., Franco, C., & Frost, J. (2024). The economic implications of services in the metaverse. In *Global perspectives in the metaverse: Law, economics, and finance* (pp. 83–118). Springer Nature Switzerland., https://link.springer.com/chapter/10.1007/978-3-031-54802-4_6 DOI: 10.1007/978-3-031-54802-4_6

Coventry, L., & Branley, D. (2018). Cybersecurity in healthcare: A narrative review of trends, threats and ways forward. *Maturitas*, *113*, 48–52. https://www.sciencedirect.com/science/article/abs/pii/S0378512218301658?casa_token=DtWzhCy47E8AAAAA:9P3rU8CS-fZb9fBgnFoqfanjIjmuq_u6L15NNIXowFkA2zfi0VlN_NS6K9MintplUTqKccQZRNc. DOI: 10.1016/j.maturitas.2018.04.008

Dargin, M. (2020, October 29). 'Credible threat': How to protect networks from ransomware: With recent warnings of ransomware attacks targeting US healthcare organizations, this 7-step plan including backup and recovery, network monitoring, and antivirus may help protect your company. *Network World (Online)*. Retrieved from https://www.networkworld.com/article/964167/how-to-protect-your-network-from-ransomware-attacks.html

Drame, P. S. (2019). *Ransomware attacks in the healthcare industry: Attack methods and preventive steps* (Utica College). Retrieved from https://www.proquest.com/openview/6fc2060ab53329f97deec56acd244c37/1?pq-origsite=gscholar&cbl=18750&diss=y

Dwyer, S. J., III, Weaver, A. C., & Hughes, K. K. (2004). Health insurance portability and accountability act. *Security Issues in the Digital Medical Enterprise, 72*(2), 9-18. https://web.archive.org/web/20060915001735id_/http://www.cs.virginia.edu/~acw/security/doc/Publications/HIPAA.pdf

Jones, L. A., & Burrell, D. N. (2025). Illegal Cybersecurity Threats Created by Organizational Arsonists in Healthcare Organizations. *Law. Economics and Society, 1*(1), 93–p93. https://j.ideasspread.org/index.php/les/article/view/1427

Kesarwani, A., & Gochhayat, S. (2023). Ransomware attacks in the healthcare industry. *Journal of Student Research, 12*(4). Advance online publication. https://www.jsr.org/hs/index.php/path/article/view/5799. DOI: 10.47611/jsrhs.v12i4.5799

Kotter, J. P. (1995). Leading change: Why transformation efforts fail? *Harvard Business Review, 73*(2), 59–67. https://hbr.org/1995/05/leading-change-why-transformation-efforts-fail-2

Koyame-Marsh, R. O., & Marsh, J. L. (2014). Data breaches and identity theft: Costs and responses. *IOSR Journal of Economics and Finance (IOSR-JEF), 5*(3), 36-45. https://www.researchgate.net/profile/Rita-Koyame-Marsh/publication/274881413_Data_Breaches_and_Identity_Theft_Costs_and_Responses_Rita_O_Koyame-Marsh_and_John_L_Marsh/links/552b71d70cf2779ab7930f83/Data-Breaches-and-Identity-Theft-Costs-and-Responses-Rita-O-Koyame-Marsh-and-John-L-Marsh.pdf

Kruse, C. S., Frederick, B., Jacobson, T., & Monticone, D. K. (2017). Cybersecurity in healthcare: A systematic review of modern threats and trends. *Technology and Health Care, 25*(1), 1–10. https://journals.sagepub.com/doi/abs/10.3233/THC-161263. DOI: 10.3233/THC-161263

Kshetri, N. (2018). The Economics of Cyber-Insurance. vol. 20, no. 6, pp. 9-14, 1 IEEE. Nov.-Dec. doi.1109/MITP.2018.287421010._https://ieeexplore.ieee.org/abstract/document/8617758?casa_token=CPydYW0yQ0cAAAAA:IL6cOMxY2LMDQ9uIY8wP-Ed23cZ3shVMwmQD6XFn-4CGawlYHDmivjTjOvTTzVT5iOs0lPPRFfs

Lamé, G., Jouini, O., & Stal-Le Cardinal, J. (2017). Combining the viable system model and Kotter's 8 steps for multidepartment integration in hospitals. *IDEAS Working Paper Series from RePEc*. Retrieved from https://hal.science/hal-01519593

Lynn, J. (2024, May 16). Reactions to the Ascension Healthcare Ransomware Attack and Suggestions for Healthcare Organizations. *Healthcare Technology Today*. Retrieved from: https://www.healthcareittoday.com/2024/05/16/reactions-to-the-ascension-healthcare-ransomware-attack-and-suggestions-for-healthcare-organizations/

Martin, G., Martin, P., Hankin, C., Darzi, A., & Kinross, J. (2017). Cybersecurity and healthcare: How safe are we? *BMJ (Clinical Research Ed.)*, *358*, j3179. https://www.bmj.com/content/358/bmj.j3179.full. DOI: 10.1136/bmj.j3179

McKeon, J. (2023, January 11). Global Cyberattacks Increased By 38% Last Year, Healthcare Hit Hard. Health IT Security. https://www.techtarget.com/healthtechsecurity/news/366594501/Global-Cyberattacks-Increased-By-38-Last-Year-Healthcare-Hit-Hard

Nobles, C. (2018). Botching human factors in cybersecurity in business organizations. *HOLISTICA Journal of Business and Public Administration, 9*(3), 71-88. https://holisticajournal.ro/docs/1589fb80c59c854cabe168f3b7b8c1d7.pdf

Nobles, C., Rangarajan, A., & Burrell, D. N. (2025). Human Factors Engineering-as-a-Service in Cybersecurity. In *New Horizons in Leadership: Inclusive Explorations in Health, Technology, and Education* (pp. 315-342). IGI Global Scientific Publishing. https://www.igi-global.com/chapter/human-factors-engineering-as-a-service-in-cybersecurity/371041

Ofe, M. A. (2023). A qualitative study exploring security practices healthcare providers need to reduce the risk of successful ransomware attacks on electronic health record systems (Order No. 30523518). Colorado Technical University. Retrieved from https://www.proquest.com/openview/675d545f322aecdce179edd0438b9879/1?pq-origsite=gscholar&cbl=18750&diss=y

Pattison-Gordon, J. (2023, November 27). Ransomware impacts healthcare systems in six states. *Government Technology*. Retrieved from https://www.govtech.com/security/ransomware-impacts-health-care-systems-in-six-states

Peltier, T. R. (2016). *Information security policies, procedures, and standards: Guidelines for effective information security management.* CRC Press. https://books.google.com/books/about/Information_Security_Policies_Procedures.html?id=mM_LsS-W4f4C

Porter, M. E. (1985). *Competitive advantage: Creating and sustaining superior performance*. Free Press.

Romanosky, S., Hoffman, D., & Acquisti, A. (2014). Empirical analysis of data breach litigation. *Journal of Empirical Legal Studies*, *11*(1), 74–104. https://onlinelibrary.wiley.com/doi/abs/10.1111/jels.12035?casa_token=7Ns-u_519yQAAAAA%3AOd9MgbWjMTDergbVLTX4Ue1PsvrMoakb7tuIlsdMoJJ1KMs2BG1rGfE13Kj9U6L84mFPdvemYZ6P_d6K. DOI: 10.1111/jels.12035

Ruiz Cano, A. (2024). *Cybersecurity strategies information technology leaders use to protect healthcare information systems from ransomware* (Order No. 30993688). Walden University. https://www.proquest.com/openview/d536d1866854baf6598de9ab8c74eaf8/1?pq-origsite=gscholar&cbl=18750&diss=y

Schein, E. H. (2010). *Organizational culture and leadership* (4th ed.). Jossey-Bass.

Southwick, R. (2023, February 23). Nearly 50 million Americans impacted by health data breaches in 2022. Chief Healthcare Executive. https://www.chiefhealthcareexecutive.com/view/nearly-50-million-americans-impacted-by-health-data-breaches-in-2022

Spence, N., Paul, D. P.III, & Coustasse, A. (2017). Ransomware in healthcare facilities: The future is now. *Perspectives in Health Information Management*, *14*(1), 1–10. https://marshall.elsevierpure.com/ws/portalfiles/portal/40008403/Ransomware%20in%20Healthcare%20Facilities_%20The%20Future%20is%20Now.pdf

Tosh, P. K., Feldman, H., Christian, M. D., Devereaux, A. V., Kissoon, N., & Dichter, J. R. (2014). Business and continuity of operations: care of the critically ill and injured during pandemics and disasters: CHEST consensus statement. *Chest*, *146*(4), e103S–e117S. https://www.sciencedirect.com/science/article/abs/pii/S0012369215519934. DOI: 10.1378/chest.14-0739

U.S. Department of Health & Human Services. (2023). HIPAA for professionals: Breach notification. Retrieved July 20, 2023, from https://www.hhs.gov/hipaa/for-professionals/breach-notification/index.html

Weitoish, T., & Burrell, D. N. (2024). Nuclear Power Organizations as Learning Organizations Around Cybersecurity, Public Health, Public Safety, and Critical Infrastructure Protection. In *Evolution of Cross-Sector Cyber Intelligent Markets* (pp. 333–355). IGI Global Scientific Publishing., https://www.igi-global.com/chapter/nuclear-power-organizations-as-learning-organizations-around-cybersecurity-public-health-public-safety-and-critical-infrastructure-protection/338619

Whitman, M. E., & Mattord, H. J. (2018). *Principles of information security* (6th ed.). Cengage Learning.

Wright, J. B., & Burrell, D. N. (2023). Cybersecurity leadership ethics in healthcare. In *Handbook of Research on Cybersecurity Risk in Contemporary Business Systems* (pp. 137–148). IGI Global., https://www.igi-global.com/chapter/cybersecurity-leadership-ethics-in-healthcare/321016 DOI: 10.4018/978-1-6684-7207-1.ch007

Zanganeh, K. S., Burrell, D. N., & Richardson, K. (2023). Improving Hospital Diversity Through Management Consulting Interventions. In *Transformational Leadership Styles for Global Leaders: Management and Communication Strategies* (pp. 377-390). IGI Global. https://www.igi-global.com/chapter/improving-hospital-diversity-through-management-consulting-interventions/331376

Chapter 5
Evaluation of the Barriers to Digital Transformation in Healthcare Organizations With AHP Method

İrem Köse
http://orcid.org/0009-0001-8295-4603
Istanbul Medipol University, Turkey

Kevser Şahin
Istanbul Medipol University, Turkey

Erman Gedikli
http://orcid.org/0000-0002-5508-194X
Istanbul Medipol University, Turkey

ABSTRACT

This study aims to systematically evaluate the barriers to digital transformation in healthcare using the Analytic Hierarchy Process (AHP) method. A literature review and expert input guided the identification and weighting of 10 key barriers, based on the opinions of seven experts. The results revealed that the most critical barriers are high costs (0.17), technological infrastructure deficiencies (0.14), and data security and confidentiality concerns (0.13). Conversely, lack of standards was considered the least significant barrier (0.03). These findings highlight the need for strategic action to reduce costs, improve infrastructure, and ensure robust data security. To accelerate digital transformation, health policymakers and adminis-

DOI: 10.4018/979-8-2600-1026-6.ch005

trators must develop comprehensive plans and innovative strategies that address these critical barriers. Overcoming these challenges is essential to unlock the full potential of digital technologies and enhance the quality, efficiency, and sustainability of healthcare services.

1. INTRODUCTION

Digital transformation in healthcare can be described as a profound reimagining of how healthcare services are delivered and managed, driven by the integration of cutting-edge digital technologies. This evolution is not merely a shift in tools but a fundamental change in how healthcare operates, impacting areas such as patient care, organizational management, diagnostic precision, and treatment methods. Central to this transformation are technologies like electronic health records (EHR), artificial intelligence (AI), mobile health applications (mHealth), and big data analytics. These innovations are designed to streamline processes, make healthcare more accessible, and significantly elevate the quality of patient outcomes (Kraus et al., 2021).

The phenomenon of digital transformation is of great consequence in the healthcare sector, where it is effecting profound changes to business processes, rendering them more efficient and effective. The integration of technology is evident in numerous aspects of healthcare provision, including patient care, diagnosis and treatment processes, and data management. The primary objective of digital transformation is to enhance the quality of healthcare services, reduce costs, and facilitate more efficient and effective patient care. It is evident that digitalization is a necessity in a dynamic and continuously evolving field such as healthcare, as it contributes to the sustainability of processes. At its core, healthcare's digital transformation is a movement towards more personalized, data-driven care, where vast amounts of information are leveraged to offer precision medicine tailored to individual patients. It reflects a future where healthcare is not just reactive but predictive, making it possible to foresee medical needs before they become critical. This transformation not only enhances the efficiency of healthcare services but also democratizes access, allowing for healthcare delivery in ways that were once unimaginable remote consultations, AI-assisted diagnostics, and real-time health monitoring from wearable technologies (Santarsiero et al., 2023).

In regard to patient experience and satisfaction, digital solutions provide novel avenues for accessing healthcare services. The implementation of online appointment systems, tele-medicine applications and mobile health platforms has facilitated a more expedient and practical means for patients to access the healthcare services they require. Consequently, patient satisfaction is rising and the accessibility of healthcare services is growing. The digital transformation has the effect of increasing not only the interaction between patients and healthcare providers, but also the participation of patients in treatment processes (Meister et al., 2019).

The management and analysis of data represents a further significant aspect of digital transformation in the healthcare sector. The application of big data analytics facilitates the effective processing and analysis of health data, enabling the creation of personalized treatment plans and the refinement of public health policies. These analyses play an important role in identifying health trends, accelerating epidemiological studies and developing more effective treatment methods. Consequently, the data management capacity brought about by digitalization enables health systems to make more informed and strategic decisions. It is evident that digital transformation has a significant impact on innovation and development processes within the healthcare sector. The advent of innovative solutions, such as artificial intelligence, robotic surgery and wearable technologies, has led to an improvement in the quality of healthcare services and the advent of new treatment methods. These developments ensure that the healthcare sector is in a continuous cycle of development and innovation, which ultimately contributes to enhanced health outcomes for both patients and healthcare professionals (Srivastava et al., 2021).

The importance of this study lies in systematically analyzing barriers to healthcare digitalization using the AHP method. Unresolved barriers lead to wasted resources, inefficiencies, increased costs, and loss of patient trust. More critically, delays in digital adoption can obstruct access to innovative tools such as AI diagnostics, telemedicine, and precision health. Identifying and prioritizing these barriers ensures healthcare administrators, policymakers, and technology developers can implement effective strategies for infrastructure development, digital literacy training, and cybersecurity.

2. BACKGROUND OF STUDY

Digital transformation in the healthcare sector means not only the adoption of new technologies but also a radical change in business practices and service delivery models. This transformation has a wide range of impacts, from patient care processes and data management to financial operations and clinical decision support systems. At the heart of digital transformation in healthcare are innovative technologies such as electronic health records (EHR), artificial intelligence (AI), mobile health applications (m-Health), big data analytics, and wearable devices. These technologies aim to make healthcare more accessible, cost-effective, and patient-centered.

However, the unique structure of the healthcare sector makes implementing digital transformation challenging. Regulations, data privacy, and ethical concerns, in particular, are among the determining factors in this process. Furthermore, because healthcare directly affects human life, the very low margin of error makes the adoption of new technologies more complex.

3. PROBLEM STATEMENT

Despite the significant advantages, healthcare organizations face persistent barriers that limit the effective scale of digital transformation. High costs, infrastructure deficiencies, and concerns about data security and privacy are the most pressing (Otto, 2019; Menon et al., 2021). Ethical dilemmas, integration difficulties, and cultural resistance to change exacerbate the problem (Scholkmann, 2021; Zarif, 2021). Furthermore, inadequate leadership support and unclear regulatory frameworks hinder innovation (Uzule & Verina, 2023). These challenges create inefficiencies, reduce trust, and delay necessary innovations, risking both patient safety and institutional sustainability.

4. RELATED THEORIES

The development, adoption, implementation, and acceptance of digital health technologies have been significantly researched under several theoretical models. However, the unique nature of health systems requires a more comprehensive consideration of institutional influences and socioeconomic factors. Digital transformation in healthcare organizations can be better understood by applying several theoretical frameworks that explain how new technologies are adopted, institutionalized, and sustained within organizations. These theories provide conceptual grounding for the analysis of barriers and serve as lenses to interpret the findings of this study.

Technology Acceptance Model (TAM): Developed by Davis (1989), TAM posits that the adoption of technology is largely determined by perceived usefulness and perceived ease of use. In the context of healthcare, digital solutions such as electronic health records or telemedicine systems are more likely to be implemented successfully when healthcare professionals view them as easy to use and beneficial for patient care. The Technology Acceptance Model (TAM) focuses on examining the factors that influence individual acceptance of telemedicine services. Studies within this framework have emphasized the critical importance of healthcare personnel's internalization of technological tools, emphasizing interrelated areas such as usefulness and ease of use, digital literacy, and technological infrastructure (Harst et al., 2019). Furthermore, technological integration has been shown to be influenced by system-level facilitators (Bunnell et al., 2020).

Unified Theory of Acceptance and Use of Technology (UTAUT): The theory framework addresses the complexity of telemedicine adoption, focusing on exclusion and inequality. Furthermore, the inclusion of barriers such as insufficient digital literacy and socioeconomic challenges, particularly among rural and low-income populations, further complicates and challenges health inequalities (Poeran et al., 2021; Saeed et al., 2025).

Technology–Organization–Environment (TOE) Framework: The TOE framework by Tornatzky and Fleischer (1990) highlights how technological readiness, organizational resources, and environmental pressures collectively influence digital adoption. For healthcare institutions, this framework is particularly relevant because infrastructural readiness, regulatory environments, and competitive pressures strongly affect the pace and success of digital transformation.

Institutional Theory (IT): Scott (2001) asserts that organizational practices are shaped by regulatory, normative, and cultural-cognitive pressures. For healthcare organizations, political and regulatory barriers, ethical considerations, and cultural resistance are examples of institutional forces that can significantly delay or accelerate digital transformation.

Diffusion of Innovations (DOI): Rogers' (2003) Diffusion of Innovations theory explains how innovations spread across individuals and organizations through stages of knowledge, persuasion, decision, implementation, and confirmation. Within healthcare, the pace of digital adoption often depends on social influence, communication channels, and the perceived advantages of innovation relative to current practices.

Organizational Agility Theory (OAT): Doz and Kosonen (2008) emphasize that digital transformation enables organizations to enhance their ability to sense changes, seize opportunities, and reconfigure resources quickly. In healthcare, organizational agility is critical to respond to fast-changing patient needs, regulatory updates, and emerging technologies.

These individual-level theories (TAM, UTAUT etc.) focus on the determinants of technology acceptance, organization focused theories (TOE, IT etc.) focused on readiness and regulatory capabilities, and innovation and digital transformation (DOI and OAT) focused on organizational transformation capabilities. However, they are insufficient to fully grasp the complexities of digital transformation in healthcare organizations. Hospitals and other healthcare organizations operate in environments subject to strict regulations, resource constraints (which vary depending on the specifics of healthcare systems), and cultural diversity. Institutional transformation and preparedness, leadership skills, financial management, regulatory frameworks, and cultural factors can all influence the outcomes of digitalization efforts. Therefore, shifting from focusing solely on stakeholder acceptance to systematically identifying and prioritizing institutional barriers is critical. Assessing these barriers can inform policymakers, managers, and stakeholders in developing policies that promote sustainable and equitable digital transformation.

5. EMPIRICAL LITERATURE REVIEW

Although limited in number, empirical studies on the field have focused on different stakeholders and subtopics and have provided readers with in-depth information on the subject through experiences. For instance, Khalil et al. (2024), revealed that low adoption rates, lack of technical and social support, and insufficient trust in technology undermined patient engagement, in Morocco for mental health services. Also, Carrera et al. (2023) figured out adoption of digital therapies among patients with chronic illnesses is limited by cultural factors and inadequate training. Hao and Padman (2016) demonstrated the mobile technology adoption in health system, in American Community Health Systems, is strongly influenced by infrastructure and staff readiness, and organizational culture playing a crucial role in shaping adoption patterns. In different health sector, Zoccarato et al. (2024), early adoption of continuous glucose monitoring technologies in Italy, figured out barriers extend beyond technological challenges psychological factors and institutional transformation challenges. In his thorough analysis of multi-criteria decision-making (MCDM) techniques, Kumar (2024) emphasized that staff members' lack of digital literacy, institutional resistance, and infrastructure shortcomings continue to be major obstacles to the adoption of technology in all industries, including healthcare.

Additionally, Campra et al (2023), from organizational perspective, analyzed thirteen healthcare organizations and identified that institutional resistance, inadequate infrastructure, and limited staff training set as key barriers. Also, Preko et al. (2022) focused on "physician entrepreneurship" concept in Ghana, and categorized challenges as technological, organizational, individual barriers for effective implementation of digital interventions among medical professionals.

6. CONCEPTUAL FRAMEWORK

Digital Transformation in Healthcare

Integrating digital technologies into traditional healthcare services is known as "digital health" (Liu & Park, 2012). Examples of digitalization are numerous and include public health, personalized medicine, and the administration of medical facilities, among other areas. Using digital technologies in healthcare, such as telemedicine, medical analytics, electronic health records, mobile healthcare, and artificial intelligence (AI) systems, is intended to enhance the caliber of human capital in the healthcare system (Tkachenko & Chesnyukova, 2023: Gedikli, 2024).

The digital transformation in healthcare delivery is evolving, fueled by rapidly changing environmental factors, and the transformation process is ongoing. Therefore, digital transformation in healthcare encompasses a wide range of areas. This section focuses solely on the obstacles to this transformation. These obstacles, within this conceptual framework, are outlined in detail below.

Digital Education and Skills Deficiencies

The proliferation of digital healthcare solutions has underscored a critical barrier: the deficiency of digital education and skills among healthcare providers (Chakravarthy et al., 2017). Studies have highlighted several facets of this issue. For instance, patients with Parkinson's disease faced significant challenges due to the lack of familiarity with digital tools, which impeded their ability to engage with teleconsultation services effectively (Garg et al., 2021).

Similarly, teleconsultation adopters, particularly among healthcare professionals, often lacked the necessary technical proficiency, leading to reduced efficacy and satisfaction with digital healthcare platforms. These skills gaps not only hinder the adoption of digital solutions but also impact the quality of care delivered (Mwogosi, 2023; Otto, 2019).

Furthermore, healthcare providers and patients frequently demonstrated limited digital literacy, significantly affecting their interaction with digital healthcare services, such as remote diagnostics and digital record management (Naveen et al., 2021; Uzule & Verina, 2023). These studies collectively suggest that inadequate digital skills and education among healthcare professionals and patients are substantial barriers to the effective implementation and scaling of digital health solutions. Enhanced training programs and integrated digital education frameworks are essential to mitigate these barriers and harness the full potential of digital transformation in healthcare (Inampudi et al., 2024).

Data Security and Privacy

Data privacy and security are paramount concerns in the digital transformation of healthcare, significantly impeding progress due to persistent threats and breaches. The challenges of maintaining patient confidentiality with the integration of digital technologies in healthcare settings were underscored. The lack of robust data security measures could exacerbate the mismanagement of patient data, thereby undermining trust in healthcare services (Naveen et al., 2021). The privacy risks associated with digital health technologies deter substantial user engagement. The urgent need for a skilled workforce to manage and interpret sensitive health data securely was emphasized in a systematic review. Apprehensions about data integrity and confidentiality could deter healthcare professionals and patients from fully adopting digital health systems. The fear of data breaches significantly affects the willingness of healthcare providers to adopt new technologies (Menon et al., 2021). Data security concerns impact the efficient use of clinical information systems, while these concerns are shared across different regions, including China, Malaysia, Pakistan, and the UK, reflecting a global apprehension. In Malaysia, the lack of established practices for data security poses a challenge to deploying digital health solutions effectively. Even basic patient interactions are fraught with privacy risks that are not adequately addressed in current frameworks. The Indonesian healthcare system faces unique challenges due to its archipelago structure, which complicates data security efforts. These studies collectively illustrate the complex landscape of data security in healthcare, stressing the need for comprehensive and enforceable strategies to safeguard sensitive health information and gain the trust of all stakeholders involved (Dwiputra et al., 2024).

Ethical Issues

Ethical issues present significant challenges in the digital transformation of healthcare, as highlighted by various studies. The ethical dilemmas faced by healthcare providers due to the ambiguity in data ownership and the potential for misuse of sensitive patient information have been discussed (Zarif, 2021). The ethical concerns related to the privacy and security of patient data, emphasizing the need for robust ethical frameworks to protect patient information in digital environments, were identified. Ethical concerns are not only limited to data handling but also extend to the accountability of healthcare providers in using digital technologies. Clearer regulations and standards are necessary to ensure ethical practices are upheld in digital health implementations. The ethical implications of digital health, particularly in terms of maintaining patient confidentiality and securing informed consent in an increasingly digital landscape, were further elaborated. These studies collectively underline the importance of addressing ethical challenges proactively to foster trust and integrity in digital healthcare services, thereby enhancing patient care and safeguarding sensitive information (Mennella et al., 2024).

Lack of Technological Infrastructure

The lack of technological infrastructure significantly impedes the digital transformation of healthcare, as evidenced by numerous studies. The critical role of robust technological platforms in the successful deployment of digital health services was underlined, highlighting that deficiencies in this area can lead to severe operational disruptions (Mumtaz et al., 2023). Inadequate technological infrastructure not only restricts the accessibility of digital health services but also affects the quality of patient care delivered. Patients with idiopathic inflammatory myopathies face substantial challenges due to the absence of necessary digital tools, which complicates their disease management. The digital divide affects the general population in Chandigarh, with the lack of adequate hardware and internet connectivity serving as a barrier to accessing telemedicine services effectively. The absence of established IT systems within healthcare settings in various Indian cities leads to inefficiencies and heightened risks associated with data handling and patient care. These findings collectively argue for a strategic enhancement of the technological infrastructure within healthcare facilities to ensure the equitable and effective use of digital health technologies, thereby advancing healthcare delivery and patient outcomes (Hill et al., 2017).

Integration Challenges

Integration challenges represent a significant barrier to the effective deployment of digital health technologies. Integrating new technologies into existing healthcare systems often encounters technical and cultural resistance, which can delay or even prevent successful implementation. Difficulties faced in remote areas, where inconsistent signal connectivity and lack of compatible infrastructure complicate the integration of digital health services, have also been highlighted. In metropolitan areas in India, integrating digital solutions across disparate healthcare systems presents unique challenges due to varying levels of technological readiness and the absence of a standardized protocol. Globally, healthcare facilities struggle with adopting integrated digital solutions due to the siloed nature of existing medical records systems and conflicting data standards (Nadhamuni et al., 2021). In Tanzania, the integration of health technologies is further complicated by socioeconomic factors, which influence the adoption rates and effectiveness of digital solution. Collectively, these studies underscore the complexity of integrating digital health technologies within diverse healthcare environments and stress the need for comprehensive strategies that address both technological and socio-cultural barriers to facilitate smoother integration processes (Zanaboni et al., 2018).

Resistance of Business Culture to Change

Resistance to change in workplace culture is a formidable barrier in the digital transformation of healthcare. Deeply ingrained traditional practices and reluctance to adopt new technologies can significantly hinder the integration of digital health systems (Scholkmann, 2021). In Latvia, resistance not only stems from a lack of familiarity with new technologies but also from a pervasive skepticism towards the efficacy of such innovations (Uzule & Verina, 2023). Similar trends are observed in lower to middle-income countries, where cultural and organizational inertia substantially delays digital health initiatives (Kaboré et al., 2022). Even when technological solutions are available, the reluctance of both medical and non-medical staff to embrace these changes can stall progress. This resistance is often rooted in fears over job security, disruptions to established workflows, and doubts about the return on investment, collectively creating a significant hurdle to the successful implementation of digital health solutions. These studies highlight the need for comprehensive change management strategies that include training, incentives, and clear communication to mitigate resistance and foster a culture that embraces digital advancements (Björkdahl & Kronblad, 2021).

High Costs

High costs associated with the adoption and maintenance of digital health technologies constitute a significant barrier to digital transformation in healthcare. The initial investment in technology, along with ongoing maintenance and upgrade expenses, can be prohibitively expensive, particularly for healthcare facilities operating under financial constraints. Financial burdens are not just limited to hardware and software but also extend to training and support services necessary to fully utilize these technologies. Leadsers in Chandigarh's hospitals have expressed specific concerns regarding the uncertainty over the return on investment and ongoing costs, which often deter institutions from committing to significant digital upgrades (Otto, 2019). In Tanzania, the high costs of digital tools and systems further exacerbate the digital divide, limiting access to advanced healthcare solutions. These studies illustrate that financial considerations play a critical role in decision-making processes regarding digital health implementations and underscore the need for scalable, cost-effective solutions that can alleviate financial pressures and enhance accessibility to digital health technologies (Björkdahl & Kronblad, 2021).

Inadequate Leadership and Management Support

Insufficient leadership and management support are critical barriers to the successful implementation of digital health technologies. The lack of visionary leadership and strategic direction in healthcare institutions is identified as a key factor that impedes the adoption and effective use of digital technologies (Otto, 2019). They argue that without strong leadership commitment, initiatives often lack the necessary momentum and organizational alignment. Healthcare leaders in

Latvia often fail to prioritize digital transformations, leading to sporadic and uncoordinated technology integration efforts. Similarly, in Tanzania, healthcare leaders are not fully equipped to handle the complexities of digital health, often resulting in fragmented initiatives with limited scope and impact. The absence of proactive engagement from senior management in fostering an environment that embraces technological advancements is further critiqued. This collection of studies highlights the essential role of leadership in overcoming resistance to change, providing necessary resources, and facilitating a culture that supports continual innovation and adaptation in healthcare settings (Faujdar et al., 2021).

Political and Regulatory Barriers

Political and regulatory barriers play a significant role in shaping the landscape of digital health. It is argued that stringent regulations and the slow pace of legislative adaptations often hinder the timely implementation and scaling of digital health technologies (Otto, 2019). Frequent changes in healthcare policies and the lack of a clear regulatory framework create uncertainties that discourage investment in digital health infrastructure in Latvia (Uzule & Verina, 2023). In Tanzania, the regulatory environment is not sufficiently aligned with the rapid advancements in technology, which complicates compliance and limits the potential for digital innovation in healthcare (Mwogosi, 2023). Additionally, in India, the rigidity of healthcare regulations often prevents the integration of innovative digital solutions that could otherwise enhance patient care and operational efficiency. These studies collectively underscore the need for dynamic regulatory frameworks that can keep pace with technological innovations and support the comprehensive adoption of digital health technologies across various regions (Mwogosi, 2023; Otto, 2019; Uzule & Verina, 2023).

Lack of Standardization

The lack of standardization in healthcare technologies and protocols poses a substantial barrier to the digital transformation of the healthcare sector. It has been emphasized that without standardized processes and interoperability standards, the integration of digital health solutions becomes fragmented, leading to inefficiencies and reduced effectiveness across healthcare systems (Otto, 2019). The absence of universal data standards and protocols in India complicates data sharing and interoperability between different healthcare providers, thereby hindering collaborative care and comprehensive patient management (Inampudi et al., 2024). Studies from Malaysia, discussing how the diversity in electronic health records (EHR) systems and the lack of standardized clinical guidelines impede the optimal utilization of digital health technologies, resulting in varied patient outcomes. These studies collectively underscore the critical need for establishing robust standardization measures in digital health to ensure seamless operation, data security, and enhanced patient care across different regions and healthcare systems (Otto, 2019).

7. METHODOLOGY

Research Type

Multi-Criteria Decision-Making (MCDM) methods are designed to support decision-making processes that involve multiple criteria and potential alternatives. These methods aim to determine the most effective course of action by systematically evaluating and prioritizing available options. The primary goal of MCDM approaches is to facilitate the identification of the optimal solution, achieved through the selection, ranking, or classification of alternatives based on their alignment with the decision criteria. This structured approach ensures that complex decisions, influenced by numerous factors, are made in a transparent and rational manner (Barretta et al., 2023).

The Purpose of Research

The aim of this research is to evaluate the barriers to digital transformation in the healthcare sector through the application of Multi-Criteria Decision Making (MCDM) methods. By identifying and analyzing the critical obstacles, this study seeks to provide a comprehensive framework for overcoming these challenges and facilitating the implementation of digital solutions in healthcare systems.

The Limitations of the Research

The primary limitations of this research include the subjective nature of data collection, as the evaluation of barriers to digital transformation relies heavily on expert opinions. Moreover, the data collected is limited to 7 experts and 10 criteria, which may influence the overall assessment of barriers in the healthcare sector. Additionally, the study is confined to a specific region or healthcare system, which may limit the generalizability of the findings to other contexts. Finally, the rapidly evolving nature of digital technologies means that the findings of this research may become outdated as new solutions and innovations emerge in the healthcare sector.

Data Collection Process and Analytical Framework

In this research, the data collection was primarily based on a structured survey developed using the Analytical Hierarchy Process (AHP) method. The 10 barriers to digital transformation in the healthcare sector were identified through an extensive literature review. These barriers represent the most frequently cited challenges within the literature on digital transformation and include: lack of digital skills and training, data security and privacy, ethical issues, lack of technological infrastructure, integration difficulties, resistance to change in organizational culture, high costs, insufficient leadership and management support, political and regulatory barriers and lack of standardization.

Table 1. Criteria and Criterion Descriptions

Abbr.	Criteria	Explanations	References
C1	Lack of Digital Skills and Training	The lack of digital skills and knowledge of healthcare workers can make it difficult to implement new digital systems. Lack of skilled and experienced human resources for digital transformation can be a major obstacle.	Chakravarthy et al., 2017; Garg et al., 2021; Mwogosi, 2023; Otto, 2019; Naveen et al., 2021; Uzule & Verina, 2023; Inampudi et al., 2024
C2	Data Security and Privacy	Ensuring the confidentiality and security of patient data can be challenging; this can reduce trust in digital systems. Furthermore, the KVKK (Personal Data Protection Law) process introduces important regulations and restrictions on data security in healthcare.	Naveen et al., 2021; Menon et al., 2021; Dwiputra et al., 2024
C3	Ethical Issues	Ethical issues may arise with the digital management of patient information.	Zarif, 2021; Mennella et al., 2024
C4	Lack of Technological Infrastructure	The lack of infrastructure required for digital transformation can be a major challenge, especially in smaller healthcare organizations.	Mumtaz et al., 2023; Hill et al., 2017
C5	Integration Difficulties	Integrating existing systems with new digital tools can be difficult and costly.	Nadhamuni et al., 2021; Zanaboni et al., 2018
C6	Resistance to Change in Organizational Culture	It is a common problem that corporate business culture resists digital transformation processes.	Scholkmann, 2021; Uzule & Verina, 2023; Kaboré et al., 2022; Björkdahl & Kronblad, 2021
C7	High Costs	The costs of digital transformation projects can be high, which can hinder their realization.	Otto, 2019; Björkdahl & Kronblad, 2021
C8	Insufficient Leadership and Management Support,	Failure of leaders to support digital transformation can negatively impact the success of these processes.	Otto, 2019; Faujdar et al., 2021

continued on following page

Table 1. Continued

Abbr.	Criteria	Explanations	References
C9	Political and Regulatory Barriers	Strict regulations by the government and regulatory bodies on digital transformation may slow down the process.	Otto, 2019; Uzule & Verina, 2023; Mwogosi, 2023).
C10	Lack of Standardization	The lack of standards in digital systems can make it difficult to share data between different institutions.	Otto, 2019; Inampudi et al., 2024

The survey comprised a total of 45 questions, with each question designed to assess the relative importance of the identified barriers using pairwise comparisons, a core component of the AHP method. The experts were asked to rate the barriers on a likert scale, providing quantitative data for the hierarchical analysis. The structured questionnaire was administered to 7 experts, each with substantial experience in healthcare digital transformation, to ensure the validity and reliability of the results. The combination of a structured survey and AHP enabled the collection of both qualitative insights and quantitative data, facilitating a robust evaluation of the barriers to digital transformation.

Analytic Hierarchy Process (AHP)

AHP, introduced by Thomas Saaty, is a widely used Multi-Criteria Decision-Making (MCDM) method (Namin et al., 2022). AHP method systematically converts the expertise and insights of decision-makers into measurable quantities, thereby enabling the quantification of their contributions (Saaty, 2001). Its inherent simplicity and transparency allow for the accelerated execution of decision-making processes.

The methodology initiates with the clear articulation of the primary objective, followed by the identification of relevant criteria and sub-criteria, which are often informed by expert opinions or survey data. To preserve the validity of the decision-making process, it is crucial that both the criteria and alternatives are precisely defined and rigorously evaluated.

Saaty's Pairwise Comparison Scale, an integral part of the Analytic Hierarchy Process (AHP), provides a structured method for evaluating the relative importance of criteria or alternatives in decision making. The scale ranges from 1 to 9, with each numerical value reflecting the intensity of preference between two items being compared. A value of 1 indicates equal importance, while 3 indicates moderate preference and 5 indicates strong preference. Values of 7 and 9 represent very strong and absolute preferences respectively. Intermediate values of 2, 4, 6 and 8 are used when the preference lies between these intensities. The comparisons are applied reciprocally and the matrix formed from these values is mathematically processed to generate priority weights for decision making (Sipahi & Timor, 2010).

8. DATA ANALYSIS & RESULTS

In this study, which was conducted to prioritize the obstacles to digitalization in healthcare institutions, the criteria were evaluated with the AHP method. At this stage, 7 expert opinions were obtained. In order to obtain a single matrix from the decision matrices obtained as a result of expert opinions, the geometric average of the decision matrices was taken. The geometric average decision matrix is as follows.

Table 2. Geometrically Averaged Decision Matrix

	C1	C2	C3	C4	C5	C6	C7	C8	C9	C10
C1	1.00	1.85	2.19	0.33	1.18	0.93	0.46	1.17	2.45	3.65
C2	0.54	1.00	2.54	0.90	2.45	2.54	0.77	2.36	1.47	1.60
C3	0.46	0.39	1.00	1.30	1.99	1.47	0.53	1.26	0.81	4.48
C4	3.04	1.12	0.77	1.00	2.27	1.97	0.98	2.09	1.00	3.65
C5	0.84	0.41	0.50	0.44	1.00	0.95	0.50	0.84	0.62	2.30
C6	1.08	0.39	0.68	0.51	1.05	1.00	0.33	0.93	0.85	2.04
C7	2.17	1.30	1.90	1.03	2.02	3.04	1.00	4.53	1.12	4.93
C8	0.85	0.42	0.79	0.48	1.18	1.08	0.22	1.00	1.00	1.60
C9	0.41	0.68	1.23	1.00	1.60	1.17	0.90	1.00	1.00	3.35
C10	0.27	0.62	0.22	0.22	0.42	0.49	0.20	0.62	0.30	1.00

Then, the values in each column of the decision matrix were added. After obtaining the column totals, each criterion was divided by the column total. With this process, the normalized decision matrix was obtained. The normalized decision matrix obtained is given below.

Table 3. Normalization of the pairwise comparison matrix

	C1	C2	C3	C4	C5	C6	C7	C8	C9	C10
C1	0.09	0.23	0.19	0.05	0.08	0.06	0.08	0.07	0.23	0.13
C2	0.05	0.12	0.21	0.12	0.16	0.17	0.13	0.15	0.14	0.06
C3	0.04	0.05	0.08	0.18	0.13	0.10	0.09	0.08	0.08	0.16
C4	0.29	0.14	0.06	0.14	0.15	0.13	0.17	0.13	0.09	0.13
C5	0.08	0.05	0.04	0.06	0.07	0.07	0.08	0.05	0.06	0.08
C6	0.10	0.05	0.06	0.07	0.07	0.07	0.06	0.06	0.08	0.07
C7	0.20	0.16	0.16	0.14	0.13	0.21	0.17	0.29	0.10	0.17

continued on following page

Table 3. Continued

	C1	C2	C3	C4	C5	C6	C7	C8	C9	C10
C8	0.08	0.05	0.07	0.07	0.08	0.07	0.04	0.06	0.09	0.06
C9	0.04	0.08	0.10	0.14	0.11	0.08	0.15	0.06	0.09	0.12
C10	0.03	0.08	0.02	0.03	0.03	0.03	0.03	0.04	0.03	0.03

The values in the rows of the normalized matrix are added and divided by the number of elements. The priority vector is obtained with this process. The consistency ratio was calculated in order to determine whether the pairwise comparisons made within the scope of the study were consistent within themselves. The consistency ratio was found to be 0.07. The consistency ratio was found to be less than 0.10. This result shows that the pairwise comparison matrices are consistent. As a result, the criterion weights obtained with the AHP method and determined to be consistent were found as in Table 4.

Table 4. Criteria Weights

Abbreviation	Criteria	Criteria Weights
C1	Lack of Digital Skills and Training	0.1203
C2	Data Security and Privacy	0.13205
C3	Ethical Issues	0.099111
C4	Lack of Technological Infrastructure	0.142934
C5	Integration Difficulties	0.064072
C6	Resistance to Change in Organizational Culture	0.068101
C7	High Costs	0.174025
C8	Insufficient Leadership and Management Support,	0.066798
C9	Political and Regulatory Barriers	0.097649
C10	Lack of Standardization	0.03496

9. DISCUSSION

The results of this study highlight the critical barriers hindering the digital transformation of healthcare organizations, with high costs, inadequate technological infrastructure, and data security and privacy concerns emerging as the most influential. These findings align with the growing body of literature emphasizing the complexity of healthcare digitalization, which is shaped not only by technological readiness but also by financial, cultural, and regulatory dynamics.

High costs were identified as the most significant barrier, with a weight of 0.174. This result is consistent with prior studies indicating that financial constraints remain a decisive factor in determining the feasibility of digital health projects (Granja et al., 2018; Kutnjak, 2021). While the promise of technologies such as artificial intelligence, metaverse applications, and telemedicine is widely recognized, their implementation often requires substantial investments that many healthcare institutions—particularly in developing regions—struggle to allocate (Efe, 2024; Mouloudj et al., 2024). The financial burden is further exacerbated by ongoing maintenance and upgrade costs, which create uncertainty regarding long-term return on investment. These findings suggest that without innovative financing mechanisms, such as public–private partnerships and targeted policy incentives, digital transformation in healthcare will continue to face substantial barriers.

The second major barrier identified was the lack of technological infrastructure (0.143). This echoes evidence from longitudinal studies that demonstrate how insufficient infrastructure, including limited broadband access and outdated IT systems, undermines the sustainability of digital health initiatives (Dugstad et al., 2019; Isakov et al., 2024). The uneven distribution of digital infrastructure is particularly visible in low- and middle-income countries, where health systems remain fragmented and heavily dependent on paper-based processes (Mwogosi, 2023). Even in developed systems, integration challenges and siloed legacy systems complicate the effective deployment of advanced digital tools (Nadhamuni et al., 2021). These findings confirm that infrastructural readiness is a prerequisite for reaping the benefits of digital transformation, and investments in resilient and interoperable systems should be prioritized.

Data security and privacy (0.132) were also found to be critical barriers. This result is consistent with prior research emphasizing the role of cybersecurity risks in eroding trust among both patients and providers (Menon et al., 2021; Saeed et al., 2023; Naamati-Schneider, 2023). In contexts where data breaches and weak governance frameworks persist, the likelihood of resistance to adopting new technologies increases significantly. The sensitivity of health data amplifies ethical dilemmas, raising questions about data ownership, patient consent, and institutional accountability (Mennella et al., 2024; Zarif, 2021). Therefore, the integration of robust data protection frameworks and adherence to international standards (e.g., GDPR, HIPAA) are critical for ensuring the sustainability of digital health adoption.

Interestingly, cultural resistance, insufficient leadership, and political and regulatory barriers received lower weights compared to financial and infrastructural challenges. However, these factors should not be underestimated. Literature consistently demonstrates that resistance to organizational change and lack of leadership engagement can derail even well-funded initiatives (Scholkmann, 2021; Faujdar et al., 2021; Uzule & Verina, 2023). Similarly, rigid or outdated regulatory frameworks, as observed in multiple regional case studies, often slow down the adoption of emerging technologies, underscoring the need for adaptive governance that keeps pace with innovation (Otto, 2019; Mwogosi, 2023).

Finally, the study revealed that the lack of standardization (0.035) was the least critical barrier. Although this result suggests that stakeholders prioritize financial and infrastructural concerns, the absence of interoperable standards remains a systemic challenge. Without standardized frameworks, healthcare institutions face fragmentation in digital ecosystems, limiting their ability to integrate data effectively across services and regions (Inampudi et al., 2024). Thus, while standardization may appear less pressing in comparison to immediate cost and infrastructure concerns, its long-term role in enabling interoperability and innovation should not be overlooked.

Overall, this study contributes to the understanding of digital transformation barriers by quantifying their relative importance using the AHP method. The findings underscore that addressing financial and infrastructural deficits remains fundamental, but sustainable digitalization also requires simultaneous attention to cultural, ethical, and governance-related issues. Bridging these multidimensional barriers will be essential to ensuring that healthcare systems can fully leverage digital innovations for improved efficiency, accessibility, and patient-centered care.

10. CONCLUSION

This study systematically identified and prioritized the barriers to digital transformation in healthcare organizations using the AHP method. The findings reveal that high costs (0.17), inadequate technological infrastructure (0.14), and data security and privacy concerns (0.13) are the most critical obstacles, while lack of standardization (0.03) appears relatively less significant. These results are consistent with previous studies, which also highlight the decisive role of financial constraints, infrastructure gaps, and cybersecurity risks in the adoption of digital health solutions (Granja et al., 2018; Efe, 2024; Mouloudj et al., 2024; Kutnjak, 2021; Dugstad et al., 2019; Isakov et al., 2024; Mbunge et al., 2021; Alam et al., 2020; Şeker & Aydın, 2024; Saeed et al., 2023; Naamati-Schneider, 2023).

Managerial Implications

For healthcare managers, the findings emphasize the importance of strategic financial planning and resource allocation. Institutions should establish public–private partnerships and explore tax incentives to reduce the burden of digital investments.

Moreover, managers must strengthen change management strategies to overcome cultural resistance and allocate funds for continuous infrastructure updates. Leadership accountability and governance frameworks should also be enhanced to ensure top management commitment (Faujdar et al., 2021; Uzule & Verina, 2023).

Practical Implications

From an operational perspective, digital transformation requires investment in staff training and digital literacy programs to close knowledge gaps among healthcare professionals. Institutions must adopt international standards (e.g., GDPR, HIPAA) for data protection to build trust among patients and staff. Practical measures such as interoperability standards, phased cybersecurity preparedness frameworks, and continuous technical support will be vital for sustainable implementation (Dwiputra et al., 2024; Saeed et al., 2023).

Theoretical Implications

The study contributes to the technology adoption literature by providing empirical evidence on the prioritization of barriers through the AHP method. By situating the findings within frameworks such as TOE (Technology–Organization–Environment) and Institutional Theory, the research demonstrates how financial, organizational, and regulatory forces jointly shape digital transformation outcomes. It also extends prior studies by quantifying the relative weight of these barriers, thus offering a more nuanced understanding of their impact on healthcare digitalization (Tornatzky & Fleischer, 1990; Scott, 2001).

Recommendations

To accelerate digital transformation, policymakers and administrators should: Introduce tax incentives and funding programs for digital infrastructure investments. Develop national training programs to improve healthcare professionals' digital literacy. Implement robust governance mechanisms to ensure leadership engagement and accountability. Enforce interoperability standards for health IT systems to facilitate data sharing. Establish ethical frameworks addressing data ownership, informed consent, and patient trust (Otto, 2019; Mennella et al., 2024).

Future Directions

Building on these findings, future research should focus on:

- Cross-country comparative studies to assess contextual variations in barriers (e.g., developed vs. developing healthcare systems).
- Emerging technologies such as blockchain, IoT, and metaverse-based health services, which bring new opportunities and risks (Efe, 2024; Srivastava et al., 2021).
- Longitudinal studies to capture the evolving importance of barriers over time as digital solutions mature.
- Stakeholder-based perspectives that include patients, providers, and policy-makers for a more holistic understanding.
- Hybrid MCDM models (AHP combined with TOPSIS, DEMATEL, etc.) for deeper and more robust analysis (Barretta et al., 2023).

REFERENCES

Alam, M. Z., Hu, W., & Uddin, M. A. (2020). Digital Transformation In Healthcare Services Sector of Bangladesh: Current Status, Chalenges and Future Dırection. *Journal on Innovation and Sustainability, 11*(1), 30–38. DOI: 10.23925/2179-3565.2020v11i1p30-38

Barretta, R., Taherdoost, H., & Madanchian, M. (2023). Multi-Criteria Decision Making (MCDM) Methods and Concepts. *Encyclopedia 2023, Vol. 3, Pages 77-87, 3*(1), 77–87. DOI: 10.3390/encyclopedia3010006

Björkdahl, J., & Kronblad, C. (2021). Getting on track for digital work: Digital transformation in an administrative court before and during COVID-19. *Journal of Professions and Organization, 8*(3), 374–393. DOI: 10.1093/jpo/joab015

Bunnell, B., Barrera, J., Paige, S., Turner, D., & Welch, B. (2020). Acceptability of telemedicine features to promote its uptake in practice: A survey of community telemental health providers. *International Journal of Environmental Research and Public Health, 17*(22), 8525. DOI: 10.3390/ijerph17228525

Campra, M., Esposito, P., & Brescia, V. (2023). corporate social responsibility and technological perspectives in healthcare: An exploratory analysis of the evolution of the anti-corruption system through multiple case studies. *Corporate Social Responsibility and Environmental Management, 30*(6), 2816–2829. DOI: 10.1002/csr.2517

Carrera, A., Zoccarato, F., Mazzeo, M., Lettieri, E., Toletti, G., Bertoli, S., Castelnuovo, G., & Fresa, E. (2023). What drives patients' acceptance of digital therapeutics? establishing a new framework to measure the interplay between rational and institutional factors. *BMC Health Services Research, 23*(1), 145. Advance online publication. DOI: 10.1186/s12913-023-09090-7

Chakravarthy, B. K., Krishna Alaparthi, G., Sawant Jitendra, R., Prasad Rao, P., Shetty, V., & Chandrasekaran, B. (2017). *"We are so close; yet too far": perceived barriers to smartphone-based telerehabilitation among healthcare providers and patients with Chronic Obstructive Pulmonary Disease in India.* https://doi.org/DOI: 10.1016/j.heliyon.2021.e07857

Davis, F. D. (1989). Perceived usefulness, perceived ease of use, and user acceptance of information technology. *Management Information Systems Quarterly, 13*(3), 319–340. DOI: 10.2307/249008

Doz, Y., & Kosonen, M. (2008). The dynamics of strategic agility: Nokia's rollercoaster experience. *California Management Review, 50*(3), 95–118. DOI: 10.2307/41166447

Dugstad, J., Eide, T., Nilsen, E. R., & Eide, H. (2019). Towards successful digital transformation through co-creation: A longitudinal study of a four-year implementation of digital monitoring technology in residential care for persons with dementia. *BMC Health Services Research, 19*(1), 366. DOI: 10.1186/s12913-019-4191-1

Dwiputra, B., Ambari, A. M., Desandri, D. R., Fatrin, S., Zuhdi, N., & Radi, B. (2024). The Future of Digital Healthcare in Cardiac Prevention Programme in Indonesia: From Bedside to Boardroom. *Journal of Asian Pacific Society of Cardiology, 3*, e24. Advance online publication. DOI: 10.15420/japsc.2023.46

Efe, M. N. (2024). Dijital dönüşüm çağında sağlık sektöründe metaverse: Swot analizi. *International Refereed Journal*, (24), 50–60.

Faujdar, D. S., Singh, T., Kaur, M., Sahay, S., & Kumar, R. (2021). Stakeholders' Perceptions of the Implementation of a Patient-Centric Digital Health Application for Primary Healthcare in India. *Healthcare Informatics Research, 27*(4), 315–324. DOI: 10.4258/hir.2021.27.4.315

Garg, D., Majumdar, R., Chauhan, S., Preenja, R., Parihar, J., Saluja, A., & Dhamija, R. K. (2021). Teleneurorehabilitation among person with parkinson's disease in India: The initial experience and barriers to implementation. *Annals of Indian Academy of Neurology, 24*(4), 536–541. DOI: 10.4103/aian.AIAN_127_21

Gedikli, E. (2024). Future digital health trends. In Demir Uslu, Y., & Zülfikar, H. (Eds.), *Digitalization in Health (ss* (pp. 181–191). İstanbul University Press., DOI: 10.26650/B/LSB40.2024.035.14

Granja, C., Janssen, W., & Johansen, M. A. (2018). Factors Determining the Success and Failure of eHealth Interventions: Systematic Review of the Literature. *Journal of Medical Internet Research, 20*(5), e10235. DOI: 10.2196/10235

Hao, H., & Padman, R. (2016). An empirical study of opinion leader effects on mobile technology implementation by physicians in an american community health system. *Health Informatics Journal, 24*(3), 323–333. DOI: 10.1177/1460458216675499

Harst, L., Lantzsch, H., & Scheibe, M. (2019). Theories predicting end-user acceptance of telemedicine use: Systematic review. *Journal of Medical Internet Research, 21*(5), e13117. DOI: 10.2196/13117

Hill, C., Martin, J. L., Thomson, S., Scott-Ram, N., Penfold, H., & Creswell, C. (2017). Navigating the challenges of digital health innovation: Considerations and solutions in developing online and smartphone-application-based interventions for mental health disorders. *The British Journal of Psychiatry, 211*(2), 65–69. DOI: 10.1192/bjp.bp.115.180372

Inampudi, S., Rajkumar, E., Gopi, A., Vany Mol, K. S., & Sruthi, K. S. (2024). Barriers to implementation of digital transformation in the Indian health sector: a systematic review. In *Humanities and Social Sciences Communications* (Vol. 11, Issue 1). Springer Nature. DOI: 10.1057/s41599-024-03081-7

Isakov, T.-M., Härkönen, H., Atkova, I., Wang, F., Vesty, G., Hyvämäki, P., & Jansson, M. (2024). From challenges to opportunities: Digital transformation in hospital-at-home care. *International Journal of Medical Informatics*, *192*, 192. DOI: 10.1016/j.ijmedinf.2024.105644

Kaboré, S. S., Ngangue, P., Soubeiga, D., Barro, A., Pilabré, A. H., Bationo, N., Pafadnam, Y., Drabo, K. M., Hien, H., & Savadogo, G. B. L. (2022). Barriers and facilitators for the sustainability of digital health interventions in low and middle-income countries: A systematic review. *Frontiers in Digital Health*, *4*, 1014375. DOI: 10.3389/fdgth.2022.1014375

Khalil, L., Serhier, Z., & Othmani, M. B. (2024). Exploring patients' acceptance of mental health e-services in Morocco: A quantitative approach. *Cureus*. Advance online publication. DOI: 10.7759/cureus.76143

Kraus, S., Schiavone, F., Pluzhnikova, A., & Invernizzi, A. C. (2021). Digital transformation in healthcare: Analyzing the current state-of-research. *Journal of Business Research*, *123*, 557–567. DOI: 10.1016/j.jbusres.2020.10.030

Kumar, R. (2024). A comprehensive review of mcdm methods, applications, and emerging trends. *Decision Making Advances*, *3*(1), 185–199. DOI: 10.31181/dma31202569

Kutnjak, A. (2021). Covid-19 Accelerates Digital Transformation in Industries: Challenges, Issues, Barriers and Problems in Transformation. *IEEE Access : Practical Innovations, Open Solutions*, *9*, 79373–79388. DOI: 10.1109/ACCESS.2021.3084801

Liu, W., & Park, E. (2012). E-healthcare security solution framework. *International Conference on Computer Communications and Networks (ICCCN)*. https://doi.org/ DOI: 10.1109/ICCCN.2012.6289239

Mbunge, E., Muchemwa, B., Jiyane, S., & Batani, J. (2021). Sensors and healthcare 5.0: Transformative shift in virtual care through emerging digital health technologies. *Global Health Journal (Amsterdam, Netherlands)*, *5*(4), 169–177. DOI: 10.1016/j.glohj.2021.11.008

Meister, S., Burmann, A., & Deiters, W. (2019). Digital Health Innovation Engineering: Enabling Digital Transformation in Healthcare: Introduction of an Overall Tracking and Tracing at the Super Hospital Aarhus Denmark. *Management for Professionals*, (Part F518), 329–341. DOI: 10.1007/978-3-319-95273-4_17

Mennella, C., Maniscalco, U., De Pietro, G., & Esposito, M. (2024). Ethical and regulatory challenges of AI technologies in healthcare: A narrative review. *Heliyon*, *10*(4), e26297. DOI: 10.1016/j.heliyon.2024.e26297

Menon, A. K., Adhya, S., & Kanitkar, M. (2021). Health technology assessment of telemedicine applications in Northern borders of India. *Medical Journal, Armed Forces India*, *77*(4), 452–458. DOI: 10.1016/j.mjafi.2021.03.007

Mouloudj, K., Oanh, V. L., Bouarar, A., Bouarar, A. C., Asanza, D. M., & Srivastava, M. (2024). *Adopting Artificial Intelligence in Healthcare: A Narrative Review. S. Teixeira, & J. Remondes içinde, The Use of Artificial Intelligence in Digital Marketing: Competitive Strategies and Tactics*. IGI Global Scientific Publishing.

Mumtaz, H., Riaz, M. H., Wajid, H., Saqib, M., Zeeshan, M. H., Khan, S. E., Chauhan, Y. R., Sohail, H., & Vohra, L. I. (2023). Current challenges and potential solutions to the use of digital health technologies in evidence generation: A narrative review. *Frontiers in Digital Health*, *5*, 1203945. DOI: 10.3389/fdgth.2023.1203945

Mwogosi, A. (2023). *Digital Transformation in Tanzania's Healthcare Sector: A Systematic Review of Robust Electronic Health RecordsSystems' Critical Success Factors*. DOI: 10.21203/rs.3.rs-3034281/v1

Naamati-Schneider, L. (2023). The effect of digitalization on service orientation and service perception among Israeli healthcare professionals: A qualitative study. *Digital Health*, *30*(9), 20552076231191892. DOI: 10.1177/20552076231191892

Nadhamuni, S., John, O., Kulkarni, M., Nanda, E., Venkatraman, S., Varma, D., Balsari, S., Gudi, N., Samantaray, S., Reddy, H., & Sheel, V. (2021). Driving digital transformation of comprehensive primary health services at scale in India: An enterprise architecture framework. *BMJ Global Health*, *6*(Suppl 5), e005242. DOI: 10.1136/bmjgh-2021-005242

Namin, F. S., Ghadi, A., & Saki, F. (2022). A literature review of Multi Criteria Decision-Making (MCDM) towards mining method selection (MMS). *Resources Policy*, *77*, 102676. DOI: 10.1016/j.resourpol.2022.102676

Naveen, R., Sundaram, T. G., Agarwal, V., & Gupta, L. (2021). Teleconsultation experience with the idiopathic inflammatory myopathies: A prospective observational cohort study during the COVID-19 pandemic. *Rheumatology International*, *41*(1), 67–76. DOI: 10.1007/s00296-020-04737-8

Otto, L. (2019). *Implementation Barriers for Telemedicine Initiatives Investigating Barriers for the Implementation of Telemedicine Initiatives: A Systematic Review of Reviews Completed Research.*

Poeran, J., Cho, L., Wilson, L., Zhong, H., Mazumdar, M., Liu, J., & Memtsoudis, S. (2021). Pre-existing disparities and potential implications for the rapid expansion of telemedicine in response to the coronavirus disease 2019 pandemic. *Medical Care*, *59*(8), 694–698. DOI: 10.1097/MLR.0000000000001585

Preko, M., Odame, F., & Anim-Yeboah, S. (2022). Doctorpreneurship in the digital age. *International Journal of E-Entrepreneurship and Innovation*, *12*(2), 1–19. DOI: 10.4018/IJEEI.314464

Rogers, E. M. (2003). *Diffusion of innovations* (5th ed.). Free Press., Retrieved from https://teddykw2.wordpress.com/wp-content/uploads/2012/07/everett-m-rogers-diffusion-of-innovations.pdf

Saaty, T. L. (2001). *Fundamentals of the Analytic Hierarchy Process*. 15–35. https://doi.org/DOI: 10.1007/978-94-015-9799-9_2

Saeed, S., Altamimi, S. A., Alkayyal, N. A., Alshehri, E., & Alabbad, D. A. (2023). Digital Transformation and Cybersecurity Challenges for Businesses Resilience: Issues and Recommendations. *Sensors (Basel)*, *23*(15), 6666. DOI: 10.3390/s23156666

Santarsiero, F., Schiuma, G., Carlucci, D., & Helander, N. (2023). Digital transformation in healthcare organisations: The role of innovation labs. *Technovation*, *122*, 102640. DOI: 10.1016/j.technovation.2022.102640

Scholkmann, A. B. (2021). *Resistance to (Digital)*. Change Individual, Systemic and Learning-Related Perspectives., DOI: 10.1007/978-3-030-55878-9_13

Scott, W. R. (2001). *Institutions and organizations* (2nd ed.). Sage Publications.

Sipahi, S., & Timor, M. (2010). The analytic hierarchy process and analytic network process: An overview of applications. *Management Decision*, *48*(5), 775–808. DOI: 10.1108/00251741011043920

Srivastava, S., Vikram Singh, S., Bhadur Singh, R., & Kumar Shukla, H. (2021). Digital Transformation of Healthcare: A Blockchain study. In *IJISET-International Journal of Innovative Science, Engineering & Technology* (Vol. 8). www.ijiset.com

Tkachenko, I. N., & Chesnyukova, L. K. (2023). Digital technologies in the sphere of health care as a way to ensure the quality of human capital. Izvestiya of Saratov University. Economics. DOI: 10.18500/1994-2540-2023- 23-1-121-128

Tornatzky, L. G., & Fleischer, M. (1990). The technology–organization–environment framework. In L. G. Tornatzky & M. Fleischer, The processes of technological innovation (pp. 153–178). Lexington, MA: Lexington Books. Retrieved September 20, 2025, from https://www.researchgate.net/publication/226145805_The_Technology-Organization-Environment_Framework

Uzule, K., & Verina, N. (2023). Digital Barriers in Digital Transition and Digital Transformation: Literature Review. *Economics and Culture*, *20*(1), 125–143. DOI: 10.2478/jec-2023-0011

Zanaboni, P., Ngangue, P., Mbemba, G. I. C., Schopf, T. R., Bergmo, T. S., & Gagnon, M. P. (2018). Methods to evaluate the effects of internet-based digital health interventions for citizens: Systematic review of reviews. *Journal of Medical Internet Research*, *20*(6), e10202. Advance online publication. DOI: 10.2196/10202

Zarif, A. (2021). The ethical challenges facing the widespread adoption of digital healthcare technology. *Health and Technology*, *12*(1), 175–179. DOI: 10.1007/s12553-021-00596-w

Zoccarato, F., Manzoni, M., Minotti, D., Lettieri, E., & Boaretto, A. (2024). Unveiling the interplay between rational, psychological and functional factors in continuous glucose monitoring early adoption: Novel evidence from the dexcom one case in italy. *BMC Health Services Research*, *24*(1), 747. Advance online publication. DOI: 10.1186/s12913-024-11195-6

KEY TERMS AND DEFINITIONS

Digital Transformation: The integration of advanced digital technologies to fundamentally change healthcare delivery, management, and patient outcomes (Kraus et al., 2021).

Healthcare Digitalization: The process of adopting digital tools such as EHR, AI, and telemedicine to improve efficiency, accessibility, and quality of care (Santarsiero et al., 2023).

Analytic Hierarchy Process (AHP): A multi-criteria decision-making method that quantifies expert judgments through pairwise comparisons to derive priority weights (Saaty, 2001).

Multi-Criteria Decision-Making (MCDM): A set of analytical techniques for evaluating alternatives based on multiple decision criteria (Barretta et al., 2023).

Data Privacy and Security: Protection of sensitive patient information from unauthorized access, breaches, or misuse, ensuring confidentiality and trust (Menon et al., 2021).

Technological Infrastructure: The hardware, software, and connectivity resources required for successful implementation of digital health systems (Mumtaz et al., 2023).

Organizational Agility: The ability of healthcare organizations to rapidly sense, adapt, and reconfigure processes in response to technological and environmental changes (Doz & Kosonen, 2008).

Chapter 6

Driving Digital Equity and Change Management: Strategic Models for Virtual Healthcare Implementation – Case Study: Thrive Reproductive Care

Gabrielle Cugini
http://orcid.org/0009-0000-9099-3410
Marymount University, USA

ABSTRACT

This chapter examines Thrive Reproductive Care's strategic transition to a telehealth-only reproductive healthcare model, permanently replacing in-person services with digital platforms. The text considers broader healthcare trends accelerated by the COVID-19 pandemic, emphasizing drivers such as technology advancements and increased demand for accessible care. Through a literature review, the chapter analyzes opportunities and challenges faced by Thrive Reproductive Care, including digital equity, data privacy, staff training, and patient engagement, among others. Major themes include addressing the digital divide, organizational change management, and ongoing staff development to ensure equitable access and high-quality service. The analysis highlights the potential for this transformation to inform future best practices in telehealth delivery, while noting risks such as exacerbating health disparities and technological barriers. The chapter notes actionable strategies required for success and models for similar organizations navigating digital transformation.

DOI: 10.4018/979-8-2600-1026-6.ch006

INTRODUCTION: IMPERATIVE OF SUCCESSFUL DIGITAL TRANSFORMATION IN HEALTHCARE

Thrive Reproductive Care, a well-established leader in reproductive healthcare, is undertaking a comprehensive transformation from traditional in-person care to an entirely telehealth-based delivery model. This shift reflects the rapidly growing demand in telehealth options for adults, increasing 50.6% using data collected in 2022, after the height of the pandemic (Wakeman et al., 2025). To help meet this demand, the organization will permanently close all physical clinics, shifting every aspect of patient care and staff operations onto digital platforms. Thrive Reproductive Care's suite of telehealth offerings includes virtual consultations, sexually transmitted infection services, reproductive care, postpartum mental health support, digital prescription fulfillment for medications (including contraceptives), and remote diagnostic testing via mail-in kits. Aiming to increase service accessibility and reduce operational costs, the primary goal of this shift is to expand care to rural and underserved populations while maintaining the existing patient base and level of patient care.

This shift for Thrive Reproductive Care is exceptionally responsive to current trends. During the COVID-19 pandemic, Western New York State promoted sexual and reproductive care through telehealth. The results of this promotion identified 77% more sexually transmitted infection cases that would have otherwise gone undetected and untreated (Muheriwa-Matemba et al., 2024). This fundamental case proves that telehealth can be successful in this type of specialty healthcare. Furthermore, scholarly evidence indicates that between 1990 and 2021, infertility rates increased by 84.4%, with projected rates expected to continue rising through 2050 (Liu et al., 2025). While the demand for reproductive and sexual telehealth is clear, there is also mounting, but limited, evidence that it is working. Telemedicine appointment show rates are 94.7% compared to just 78.7% of appointments in brick-and-mortar locations (Klotz et al., 2023).

This telehealth shift reflects a broader trend in healthcare: the increasing demand for telehealth services, primarily driven by the COVID-19 pandemic, ongoing advances in digital health technology, changing regulations, and evolving patient expectations. Recent studies consistently show the rise in telehealth use across primary and specialty care, with reproductive health services leading due to provider shortages, travel difficulties, and persistent disparities in access, which highlights the potential of telehealth to close gaps for rural residents, adolescents, and communities with provider shortages (Uddin & Fariha, 2024). However, the research also highlights risks such as digital exclusion, limited broadband access, and privacy concerns, particularly regarding sensitive reproductive health topics (Rowland et al., 2024). As organizations like Thrive Reproductive Care transition to fully virtual care, scholars and practitioners are paying close attention to the key factors for success as well as obstacles, including the need for a strong digital infrastructure, comprehensive staff training, and equitable implementation that prevents widening disparities.

PROBLEM STATEMENT

There is an ever-increasing demand for virtual reproductive and sexual healthcare solutions which is part of a larger health crisis in the United States. Today, there is a significant provider shortage with 56% of rural counties lacking access to any obstetric care (National Rural Health Association, 2024). Because of the lack of available care, there is delayed prenatal care, more pregnancy-related hospitalizations, low birth weight in infants, and infant mortality (American College of Obstetricians and Gynecologists, 2020). In addition to the existing shortage of obstetricians in rural areas, the Association of American Medical Colleges expects a significant shortage, between 37,800 and 124,000, in both primary and specialty healthcare physicians by 2034 (Association of American Medical Colleges, 2021). While greater access to telehealth options appears to be a solution to this problem, it may be only superficial. According to experts from the prominent consulting firm McKinsey, between 60 and 70% of digital health initiatives fail to achieve their intended scale or expected return on investment due to governance and integration challenges (Eastburn et al., 2022). This is an alarming statistic for healthcare in the United States, but importantly, it is consistent with the levels of digital transformation failures at approximately 70% (Bughin et al., 2019). These two metrics together make it challenging for an organization to be successful.

To address these issues, Thrive Reproductive Care must (a) achieve seamless and secure digital care delivery that matches or exceeds traditional service standards, (b) ensure staff competency and patient-centered communication through extensive training and workflow redesign, (c) protect patient data and privacy in full compliance with Health Insurance Portability and Accountability Act (HIPAA) and other pertinent regulations and (d) deliver equitable access by prioritizing outreach to digitally marginalized populations and proactively mitigating barriers to care. While Thrive Reproductive Care's best efforts can only resolve a small percentage of these massive issues, they can help set the standard for future reproductive and sexual telehealth care and, through their success, work to increase overall access.

Purpose Statement

The purpose of this chapter's literature review and associated case study is to critically examine the existing field of research related to telehealth in the United

States as well as study the strategic and theoretical frameworks and theories to support a telehealth organization's digital transformation. Both quantitative and qualitative evidence and analysis are incorporated. This study aims to create actionable insights and considerations for business students who wish to understand and develop effective and equitable telehealth services. This research is timely as there is a considerable expansion of telehealth services and adoption in the post-pandemic United States (Cengil et.al., 2024). It is also relevant in 2025 because of the continuing importance of consumers' preference for convenience and speed of care (Al-Abdwani et. al., 2023). This research is exceptionally timely as increased telehealth has the potential to impact underserved populations far more positively than traditional methods (Wang et al., 2025). Finally, considering the rapid advances in technology and the high demand for innovation in recent years, this research is crucial for businesses. Focusing on technological innovation and consumer satisfaction contributes to a profitable business (Kumera et al., 2024).

SIGNIFICANCE STATEMENT

Thrive Reproductive Care's transition to a telehealth-only model holds national healthcare significance and documents an emerging, high-demand business model. From a healthcare perspective, evidence of the demand for virtual care is mounting, but the growth necessary to support the demand does not exist. A 2024 Deloitte survey of women found that 70% of respondents used digital tools to communicate and coordinate care; however, black and Hispanic respondents were twice as likely to say that the digital tools at their disposal did not meet their personal needs (Deloitte, 2024). This contradiction shows that while crucial and a positive step, digital innovation and tools must go hand in hand with proper patient care strategies, communications skills, and cultural practices. The stark contrast between healthcare needs and what is currently available both in traditional healthcare and through telehealth will continue to grow. A U.S. study, including 36 million working-age individuals with private insurance claims data, found that telemedicine encounters increased by 766% in the first three months of the pandemic (Shaver, 2022). That demand did not decrease following the height of the pandemic. As of 2022, 39.3% of U.S. adults had adopted telehealth, with higher satisfaction and convenience being primary motivators (Kim et al., 2024). Significant demand exists in the healthcare industry, which provides an opportunity for Thrive Reproductive Care to be a leader in the field.

The ability to transition from a brick-and-mortar healthcare organization to a telehealth-only model is a significant business undertaking and, as such, must take into account the opportunities this model provides. Telehealth increases geographic reach, can support greater patient access to providers, and reduces operational overhead (American Hospital Association, 2025). The value proposition associated with success is massive. Comparing these benefits with the high rates of both digital transformation failure, at 70% (Bughin et al., 2019), and the digital healthcare initiative failure rate of 60–70% (Eastburn et al., 2022), Thrive's success in this space can be documented, studied, and repeated, thus enhancing healthcare in the United States. Consequently, the ability to integrate health equity, security, and high-quality patient experience into this transformation is of great significance, not only for Thrive Reproductive Care's survival and mission but also for shaping the future of equitable and effective digital healthcare.

METHODS

This literature review used a systematic approach to identify, select, and review peer-reviewed research and other relevant publications in consideration of the digitization of Thrive Reproductive Care. The methods were structured to ensure comprehensiveness, reproducibility, and rigor across several domains, including care delivery, change management, digital equity, workforce adaptation, privacy and ethics, patient engagement, digital outreach, and policy. Artificial intelligence tools were used to help identify appropriate search strings/terms to use in major databases.

A selection of search terms used in this case study is included below, with additional details available on additional terms.

- ("telehealth" OR "telemedicine" OR "virtual care" OR "digital health") AND ("reproductive health" OR "sexual health" OR "family planning" OR "contraception" OR "maternal health" OR "postpartum")
- ("digital health" OR "telehealth" OR "telemedicine") AND ("health equity" OR "digital divide" OR "health disparities" OR "access to care" OR "underserved populations")
- ("telehealth" OR "digital health") AND ("change management" OR "organizational transformation" OR "workforce training" OR "workflow redesign")
- ("telehealth" OR "digital health") AND ("privacy" OR "security" OR "HIPAA" OR "data protection" OR "compliance" OR "consent")
- ("digital marketing" OR "patient outreach" OR "social media" OR "SEO" OR "online engagement") AND ("telehealth" OR "digital health")
- ("telehealth" OR "digital health") AND ("policy" OR "regulation" OR "reimbursement" OR "future directions") AND ("reproductive health" OR "sexual health" OR "contraception" OR "abortion")

Hundreds of sources were identified for this assignment, and a refined list is included in the reference section of this document. Several additional sources were later identified which were not found initially through targeted search terms.

Inclusion Criteria

The inclusion criteria comprise peer-reviewed articles or systematic reviews, as well as published guidelines from industry experts or governmental agencies that address telehealth in reproductive or sexual health. They include research focused on digital health equity, patient and provider experience, organizational and workforce change, privacy, ethics, digital marketing, or policy in virtual health contexts. Studies used in the literature review are both U.S.-based and international, relevant to the design and implementation of telehealth for reproductive care.

Exclusion Criteria

Studies that were excluded were primarily those focused solely on telehealth outside the context of reproductive or sexual health, unless they discussed overarching frameworks or lessons. Articles or opinion pieces without reference to empirical studies were excluded. Since this is a continually evolving field, particularly since the COVID-19 pandemic, efforts were made to focus on recent literature. Literature from before 2015 was excluded in its entirety. The current reference list dates to 2017 but may be refined further after the draft phase. References that pertained specifically to abortion-related healthcare, while widely available, were excluded as the focus of this paper is broader. Lastly, literature that is not written in English was excluded simply due to the author's lack of proficiency in other languages.

LITERATURE REVIEW

The continuous evolution of telehealth has already transformed the front lines of reproductive and sexual health care, especially in the aftermath of the COVID-19 pandemic. In this ever-changing landscape, organizations like Thrive Reproductive Care are pioneering fully virtual models of care while seeking to expand access, improve efficiency, and promote health equity. That said, the transition from traditional, in-person services to a comprehensive digital platform is complex and raises significant questions about access, quality, cybersecurity, workforce adaptation, and ethics.

A thorough literature review is essential to contextualize Thrive Reproductive Care's transformation within the broader body of research on telehealth implementation, reproductive health equity, digital infrastructures, and organizational change. The benefits and limitations of telemedicine in reproductive care have been reviewed, the barriers related to the digital divide have been documented, and experts have offered strategies for equitable technology adoption and patient engagement. Regulatory frameworks, privacy concerns, and change management practices further shape the possibilities and pitfalls of virtual health care delivery.

Telehealth in Reproductive and Sexual Healthcare

Telehealth is increasingly becoming a primary delivery model of reproductive and sexual healthcare services. The COVID-19 pandemic significantly accelerated this shift. More recent studies show telehealth's effectiveness in contraception, abortion care, and postpartum support. Endler et al. (2020) completed a systematic review and found that remote abortion services maintained high safety and efficacy. Studies during the pandemic have reported that telemedicine for reproductive and sexual healthcare also improved access for geographically and socially marginalized populations, helping to reduce time and cost barriers for both patients and providers (Kaiser et al., 2022). However, they also noted some challenges associated with telehealth in these circumstances, including concerns about confidentiality, technological barriers, and the limitations of conducting exams virtually. These studies support the fact that while telehealth delivers significant benefits for reproductive health, Thrive Reproductive Care must carefully consider the positives of telehealth, along with the challenges, and ensure it has measures in place to mitigate the challenges of an entirely virtual environment.

Economic Planning

As part of the shift from brick-and-mortar locations to a telehealth model, organizations like Thrive need to consider the significant economic costs and challenges associated with the transformation. As of September 2025, Medicare Advantage, Medicare Part B, Medicaid, and Veterans Affairs broadly provide telehealth coverage (Gujral et al., 2025). According to the same study, Kaiser Permanente, a private health insurer, also offers nationwide coverage for telehealth. This is the exception compared to other insurers and is consistent with Kaiser Permanente's strategy (Kaiser Permanente, n.d.). However, pursuing a nationwide model when considering taking private insurance is likely the most challenging because private insurance is both state and plan-dependent (Gujral et al., 2025). The challenging private insurance model necessitates billing and insurance experts, which contributes to significant overhead for an organization like Thrive.

When considering overhead expenses, the initial investment in a telehealth-only model and ongoing costs are substantial, particularly when factoring in state-by-state and insurer-by-insurer reimbursements. There is literature that discusses whether telehealth is more or less expensive than traditional healthcare, and the results vary. Putting that aside, it is imperative to note that results may vary greatly, and the model an organization decides to pursue will dictate costs and how it compares to conventional healthcare (Ivanova et al., 2025). Telehealth itself can be telephone calls, with limited overhead and costs. It can also be a comprehensive suite of video-teleconferencing features, including records, provider messaging, reminders, apps, and a sophisticated digital experience. The United States Department of Health and Human Services defines it very broadly with significant room for interpretation (U.S. Department of Health and Human Services, 2024). This further solidifies that organizations need strategic plans that incorporate their anticipated model and outcomes to help ensure they are successful.

Strategic Planning

To ensure long-term success for Thrive Reproductive Care in a telehealth-only model, they must engage in a comprehensive strategic planning process to chart the way forward, build momentum for the transition, and track key success criteria. This process should include a vision, mission, current state analysis, future planning, and risk identification, as well as a proven system to monitor operations and track initiative success. In addition to establishing an aspirational vision and a meaningful mission, Thrive Reproductive Care would benefit from conducting a current state analysis, known as a SWOT analysis, which stands for strengths, weaknesses, opportunities, and threats. They would also benefit from a PESTLE analysis, which stands for political, economic, social, technological, legal, and environmental factors. The SWOT analysis will help identify challenges associated with the transition, while the PESTLE is more future-oriented. These analyses will enable the organization to take a holistic approach to decision-making, align with market trends, and help ensure long-term success through change (Awan & Zaurez, 2025). This type of planning and communication of the vision and mission to the organization can lead to the identification of appropriate change initiatives and help support the digital transformation effort.

SWOT

Each component of the SWOT analysis can be described as follows:

- **Strengths.** Mission-driven doctors on reproductive health, adoption of digital health advances to support patients, focus on equity, and scalability through telehealth
- **Weaknesses.** Broad digital access, technology literacy barriers, staff training, evolving regulatory and privacy risks, and challenges in building patient trust virtually
- **Opportunities.** Service expansion to underserved areas, cost savings and efficiencies, better data analytics, innovation, and thought leadership in reproductive and sexual telehealth
- **Threats.** Digital disparities and exclusion, privacy and security risks, unforeseen regulatory changes, public trust, and competition

PESTLE

The components of a PESTLE analysis reveal the following:

- **Political.** Rapidly changing landscape for both reproductive health (particularly abortion care) and telehealth, with policy changes in support of digital care models
- **Economic.** Decreased facility cost, improved sustainability, high initial technology and training costs, insurance acceptance, and processing challenges
- **Social.** Higher acceptance of telehealth post COVID-19, stigma associated with reproductive and sexual healthcare necessitates patient trust, and ongoing digital disparities
- **Technological.** Strong digital infrastructure and security are essential, as broadband limitations in rural regions continue to persist; however, recent advances in telehealth platforms can lead to improved service delivery
- **Legal.** HIPAA compliance and protection of data are top priorities, ensuring compliance with policy shifts, and a need for state-by-state legal compliance
- **Environmental.** Smaller environmental footprint through remote model, increased reliance on digital infrastructure, which means that weather events can have an impact (e.g., a hurricane damaging a data center)

Balanced Scorecards and Key Performance Indicators

Other strategy tools will support the organization's strategic plan, which is guided by the vision, mission, and SWOT/PESTLE analyses. Without these tools, Thrive Reproductive Care cannot begin to determine how it will monitor and manage success through its strategic initiatives and ongoing operations. The tools needed to support this include a balanced scorecard, as well as clearly defined key performance indicators and key risk indicators. Although balanced scorecards involve some administrative overhead, they are excellent tools for communicating the organization's goals, aligning daily work with strategic objectives, prioritizing projects, products, and services, and tracking progress toward strategic targets (Balanced Scorecard Institute, 2025), such as the initiatives at Thrive Reproductive Care. The scorecard considers four interconnected perspectives to assess performance: financial, customer, internal processes, and earning and growth. The financial perspective of the scorecard focuses on financial measures as well as the outcomes of other customer, internal processes, and learning and growth perspectives of the scorecard. The customer perspective is typically metric-driven, focusing on factors such as client satisfaction, reputation, and market share. Internal processes refer to the efficiency and quality of internal operations that support the delivery of external products and services. Finally, the learning and growth perspective focuses on future opportunities to create value. The balanced scorecard has proven especially useful during digital transitions in healthcare settings. This was demonstrated by Hügle and Grek (2023) through their qualitative research in a hospital rheumatology department, where they planned a digital health transformation involving telehealth, electronic medical records, and artificial intelligence. Having this strategic framework in place and reviewing it regularly will help Thrive Reproductive Care succeed in transitioning to a telehealth-only business model.

Vision

Thrive Reproductive Care aspires to be the leading provider of innovative reproductive healthcare through a fully virtual model that delivers exceptional quality, ensures secure and compassionate care, and bridges access gaps for patients, regardless of geography, digital access, or background.

Mission

Thrive Reproductive Care is committed to transforming reproductive and sexual healthcare through secure, patient-centered telehealth services that meet or exceed traditional standards. Through advanced digital infrastructure, staff training, and targeted outreach, Thrive Reproductive Care protects privacy, strengthens patient trust, and ensures everyone receives equitable, high-quality care.

THRIVE REPRODUCTIVE CARE BALANCED SCORECARD

Challenges with Strategic Modeling- Groupthink Theory

Groupthink theory suggests that cohesive groups often prioritize unanimity over a realistic evaluation of alternatives, which can weaken thorough strategic planning and result in flawed organizational decisions (Janis, 1982). When examining Thrive's strategic planning processes, there is no mention of involving outside experts to lead discussions or provide subject matter expertise on the issues Thrive Reproductive Care faces. To combat groupthink, it is essential to actively encourage diverse opinions, designate or assume the role of devil's advocate, and foster a culture of open dissent to prevent falling into the groupthink trap (Bang & Frith, 2017). This is a missing element in Thrive's strategic planning and could lead to poor strategy alignment, ineffective implementation, or misguided direction.

Patients

The impacts of telehealth will be explored through the opportunities and challenges associated with two major stakeholder groups: patients and the employees and staff of Thrive Reproductive Care. Looking through the lens of patient care, the focus areas for this paper are improved patient outcomes, acknowledging and closing the digital divide, patient outreach and marketing, as well as protecting health data and regulatory compliance.

Improved Patient Outcomes

Patient outcomes should be at the core of every healthcare organization, and through telehealth, Thrive Reproductive Care has the opportunity to enhance patient outcomes. A substantial body of recent literature is available on the benefits of telehealth. In a systematic review of telemedicine for chronic disease management, Ma et al. (2022) found that patient outcomes were significantly improved, particularly for those with chronic conditions like diabetes, hypertension, and rheumatoid arthritis. This is despite having fewer clinical metrics available, like blood glucose levels for people with diabetes. Additionally, due to telehealth interventions, there is a reduction in hospital stays, and for those who still require hospitalization, there is a decrease in the time spent in the hospital (Snoswell et al., 2021). Lastly, and perhaps most pertinent to Thrive Reproductive Care, a 2024 study on sexual and reproductive healthcare found that telehealth increased accessibility for underserved populations, reduced stigma, and improved care outcomes (Muheriwa-Matemba et al., 2024).

Equity Possibilities Through Telehealth

Ensuring diversity, equity, and inclusion in telehealth is part of Thrive Reproductive Care's transformation to a virtual care model. There is growing evidence that when designed effectively, telehealth can help close long-standing gaps in healthcare access for diverse and underserved populations, especially in reproductive and sexual health services. The key is that the model must be intentionally designed to address equity concerns. In support of these statements, a 2025 systematic review by Gallegos-Rejas et al. found that the spread of telehealth and eHealth can both reduce and worsen health disparities, depending on the intentionality of design. If done well, the telehealth organization provides language support services, promotes cultural competency for clinicians, and provides technical support (Gallegos-Rejas et al., 2025). These interventions include supplying technology to enable device usage, digital navigators or educators, and basing the design and services on consumer needs, which are consistent with Thrive's reproductive model.

Digital Divide

One of the major barriers to equity in reproductive and sexual telehealth is the digital divide, which has the potential to undermine telehealth's goal of equitable patient care completely. Van Dijk (p. 5) (2020), a scholar in this space, considered the concept of digital divide as "the gap between people who do and do not have access to forms of information and communication technology." Several studies highlight these disparities in broadband internet access, digital device ownership, and eHealth literacy. These disparities are particularly evident in rural, older, and/or lower-income populations. For example, Roberts and Mehrotra (2020) discussed how the expansion of telemedicine can unintentionally widen health access gaps due to structural inequities linked to the digital divide. The National Academies of Sciences, Engineering, and Medicine (2021) provided frameworks to address equity by advocating for inclusive technology design and support with a specific focus on expanding broadband infrastructure, ensuring affordable technology, and supporting digital literacy initiatives. Scholars have noted that truly equitable telehealth requires not only technological access but also cultural competence and outreach. The findings on the digital divide are consistent with the equity possibilities available through telehealth models.

Health Equity Impact Assessment

The Health Equity Impact Assessment (HEIA) is a decision-support tool that evaluates the impact of a new program, policy, or innovation on the health of different population groups, particularly vulnerable or marginalized populations. A recently developed checklist comprises five specific sections: policy introductions, managing the HEIA of policy, scope of affected populations, HEIA results, and recommendations. This tool will help identify unintended consequences for vulnerable populations and develop targeted interventions to promote equitable access (Olyaeemanesh et al., 2023). To ensure Thrive Reproductive Care's transition does not exacerbate existing health disparities, the organization has included this measure as part of its balanced scorecard and strategic initiatives, which should help ensure the program meets the needs of all population groups, particularly those in more vulnerable or marginalized populations.

Patient Outreach and Marketing

Patient outreach and marketing are becoming increasingly important for attracting, educating, and retaining patients in telehealth. Research indicates that targeted social media ads, search engine optimization practices, and patient-focused websites can effectively drive digital health engagement. Specifically, outreach and marketing that highlight convenience and privacy are effective. Radu et al. (2017) conducted a study involving 126 new patients at a dental clinic over a 22-month period, which enabled them to assess the effectiveness of digital outreach methods. Of the 126 new patients, 78.57% reported that they were influenced by online promotions to choose that clinic for their dental needs. Approximately 85% of patients followed one or more of the clinic's social media accounts, and half of those not currently following planned to do so in the future. Finally, a remarkable 90% of new patients indicated an intent to return. This is a meaningful and engaging study that demonstrates how digital outreach and tools can attract and retain patients. Thrive Reproductive Care will need to consider agile, data-driven marketing strategies to attract and sustain patient engagement, supporting growth and trust in their fully digitized healthcare model.

Engagement Model

To support the long-term success of Thrive Reproductive Care, an effective patient engagement model is essential. Thrive Reproductive Care should consider adopting a high-touch engagement approach. This model emphasizes frequent, personalized interactions between the patient and the organization. It offers the benefit of providing ongoing support outside of traditional appointments and has several practical, real-world applications. Engagement does not need to be consistent with a doctor of medicine-specific healthcare provider to be successful. A systematic nursing review found that nurse-led telehealth engagement improved satisfaction and health outcomes (Casillas et al., 2022). It provides continuity of care that is often missing through in-person appointments. This continuity of care can lead to better patient outcomes.

Communications Strategy

With the high-touch engagement model, the organization needs a complementary communication strategy. Thrive Reproductive Care leadership should develop this strategy prior to the transition to ensure it is well understood and consistently followed. Their strategy should include both verbal and written communication, with a primary focus on written communication that is accessible through a patient portal. Patient portal interventions can improve psychobehavioral outcomes, including health knowledge, decision-making, medication adherence, and use of preventive services (Legler et.al., 2021). Thrive Reproductive Care should incorporate automated and personalized messages at scheduled times—before an appointment, immediately after, and in the days afterward—to review the visit and outline necessary steps for the patient. In the meantime, between appointments, Thrive Reproductive Care can utilize the abundance of available information about health conditions to communicate with patients, helping them understand their conditions or sharing preventive tips. This communication should be personalized, frequent, and empathetic in tone. This strategy, paired with the high-touch model for engagement, can position Thrive Reproductive Care for success through enhanced patient engagement.

The following metrics should be established as part of the transition to measure the organization's communication effectiveness and determine where it is successful and where changes can be made to course correct. These metrics include (a) patient portal engagement, (b) website traffic, (c) conversion rate, (d) exit rate, (e) response speed, and (f) contact forms/inbound messages.

Data Privacy Protection and HIPAA

Protecting patient information is a significant challenge in telehealth, particularly for sensitive health data and the ever-evolving threat of cybersecurity incidents. The United States Department of Health and Human Services (HHS) has developed HIPAA guidelines specifically for telehealth medicine, which have stringent requirements for communication, data encryption, and audit trails. Ibrahim et al. (2024)) discussed the potential for security breaches and other misuses of healthcare information. They also further noted the challenges associated with the digital divide as discussed above. To help mitigate these potential issues, they suggested strong governance data frameworks, culturally appropriate privacy policies, and monitoring of privacy impacts. When it comes to the digital divide, they suggested a focus on the specific sensitivities related to reproductive or sexual health. Best practices in this space require comprehensive provider training, thorough technology vetting, and investments in multi-factor authentication. Thrive Reproductive Care's telehealth transformation must serve as a core example of exceptional regulatory compliance and security, prioritizing both to maintain patient trust and uphold its legal standing.

Thrive Reproductive Care must be agile and prepared to adapt to new regulatory and legal requirements. In January 2025, the Department of Health and Human Services proposed updated HIPAA regulations for the first time since 2013, a direct result of the CrowdStrike data breaches in 2024 (Office for Civil Rights, 2025). The proposed changes are broad but have specific implications for telehealth, including enhanced privacy and cybersecurity regulations, as well as training. The most impactful change for telehealth is the addition of mandatory cybersecurity requirements, including staff awareness and training on data protection and security. Thrive Reproductive Care's transformation must take this into account and ensure that staff are appropriately trained on the protection of patient records in a purely virtual setting.

Ethical and Social Considerations

The implementation of telehealth in any specialty requires preparation to address new ethical challenges that were less prominent in traditional brick-and-mortar settings. These challenges include consent, privacy, and social vulnerabilities. The American Society for Reproductive Medicine Ethics Committee (2024) stated that providers should deliver telehealth services in a manner similar to in-person clinics. They further recommended that physicians using electronic communication to answer patient health questions ensure that mechanisms are in place to protect patients' identities and personal information and to prevent unauthorized access to and viewing of private health data.

Additionally, literature on digital ethics in reproductive health advocates for participatory design, ongoing community input, and careful e-consent protocols (Almuhaideb et al., 2025). Failing to consider these ethical considerations could damage telehealth's reputation and its prospects, particularly in sensitive fields such as reproductive and sexual health. Moving to a fully telehealth-based model, as at Thrive Reproductive Care, will require iterative and culturally respectful approaches to safeguard patient dignity and autonomy while advancing service innovation, consent, privacy, and social vulnerabilities.

EMPLOYEES

The employees of Thrive Reproductive Care are experiencing a significant transition as they shift to a telehealth-only model. To support this change, this section will examine well-documented change management practices that promote successful digital transformation for the organization, along with cultural adjustments necessary to sustain the long-term implementation of the telehealth model.

Change Management

To effectively support telehealth adoption, comprehensive change management practices are essential, especially those that facilitate workforce acceptance of the transition. New and experienced scholars identify successful strategies derived from structured change models, continuous training, and frequent communication. In a study, a similar healthcare company successfully implemented large-scale change by applying John Kotter's change model (Davis, 2024). This success highlighted the importance of involving staff early in the digital transformation process. Organizations that focus on staff engagement, provide job-specific digital skills training, update procedures, and boost employee morale tend to experience less resistance to change. Resistance is minimized when change efforts are transparent and aligned with mission-driven goals. For Thrive Reproductive Care, leadership must develop a strong change management strategy that reassures and empowers staff while supporting patients and operational processes.

Application of John Kotter's Model

John Kotter created and advocated for an eight-step change management model, which can be applied to Thrive Reproductive Care's transition from brick-and-mortar stores to telehealth. The model has eight steps: (1) create a sense of urgency, (2) build a guiding coalition, (3) form a strategic vision and initiatives, (4) enlist a volunteer army, (5) enable action by removing barriers, (6) generate short term wins (7) sustain acceleration (8) institute change.

Create a Sense of Urgency

Thrive Reproductive Care must create a sense of urgency within the organization to drive change effectively. Nurse leaders have recently been shown to successfully utilize evidence-based tactics to create urgency, which helps motivate staff and benefit patients (Shirey, 2011). Thrive Reproductive Care should gather data related to the traditional reproductive healthcare model, including healthcare expectations following COVID-19; reduced access to care in rural areas; and opportunities presented by the shift to telehealth, such as cost savings, increased service availability, and improved patient outcomes. For example, a 2022 study noted that although rural healthcare participants generally had worse health compared to urban residents, they were less likely to use outpatient services due to longer travel distances and fewer providers (Nuako et al., 2022). This data, like the Nuako reference, should tell a compelling story to staff about the urgent need for change. Thrive Reproductive Care can host internal town halls where this information is shared, helping to build a sense of urgency for change.

Build a Guiding Coalition

The next step in the Kotter model is for Thrive Reproductive Care to form a cross-functional team comprising leaders, change champions, and individual contributors to lead the change effort. Multiple hospitals have seen significant positive impacts on organizational change by forming a diverse coalition to drive change (Stone et al., 2018), compared to their peers who lacked a coalition or had a less diverse and less representative coalition. Thrive Reproductive Care can involve clinical and administrative leaders, technology specialists, and community representatives to support the change initiative. Change is difficult, but this coalition will help make that change easier for staff and patients.

Form Strategic Vision and Initiatives

Thrive Reproductive Care needs to create a compelling strategic vision and identify the appropriate initiatives to help them reach their change goals. Slåtten et al. (2021) found that when an organization's vision is clearly communicated and fully integrated into its operations, it fosters a shared sense of direction that increases member motivation, which could improve the likelihood of implementing successful change and improvement initiatives. These organizations are more successful in implementing change efforts. Strategic planning is referenced earlier in this paper, but the results of that plan are meaningful in this step of the Kotter model. These components are essential for every organization, particularly for those undergoing change.

Enlist a Volunteer Army

To continue this change effort, Thrive Reproductive Care will need to identify a larger population of employees across the organization and represent all levels to champion and support the change. Large-scale change occurs when a significant number of people unite around an opportunity (Keller, 2023). This large group of people can help with broader buy-in and engagement for both staff and the community; they can create a distributed communications model and help accelerate the change. This is a significant step forward in the change effort, marking a point at which things are happening more quickly.

Enable Action by Removing Barriers

There are always barriers to change, but to succeed, leaders and change champions need to be empowered to take action to remove barriers in support of the change initiative. Without this, Thrive Reproductive Care's digital transformation will fall into the approximately 70% of failed digital transformations (Bughin et al., 2019). There are systems, processes, resistance, and likely technology gaps that make it difficult for an organization like Thrive Reproductive Care to shift to a telehealth model. Investing in staff training, utilizing community resources, establishing effective feedback channels, and identifying systems and processes that require modification to support the transition are essential to its success. Appropriately enabling and resourcing leaders to break through those barriers is key.

Generate Short-Term Wins

Part of gaining momentum for a change effort and encouraging more people to support the change involves generating short-term wins. As part of strategic planning, the leadership at Thrive Reproductive Care can identify some quick wins that are visible and, importantly, achievable, to celebrate with the larger team. For Thrive Reproductive Care, this might include a pilot of a telehealth model in a specific area over a set period. Data from the pilot could include reduced wait times or increased patient intake, which would help demonstrate success. In a recent peer-reviewed telehealth pilot study for a skilled nursing facility, 94% of residents reported feeling comfortable or very comfortable with telehealth visits (Wilber et al., 2022). This is the quick-win data Thrive Reproductive Care needs to target and share. The short-term wins will also provide another opportunity to discuss the change effort and present it in a positive light, continuing to support the newly gained momentum.

Sustain Acceleration

The momentum generated by the quick wins needs to be sustained. The team needs to maintain high motivation and energy during this transition. The best way to do this is to continue down the same path as the quick wins but on a larger scale. They can expand pilots that include more services or are introduced into a broader market, refresh incentives for patients and staff, and ensure training is iterative based on recent findings from pilot groups. The leadership should build on their successes to continue the change effort and invest in a comprehensive telehealth model.

Institute Change

As the change is implemented, leadership must continue to reinforce the telehealth transformation through all areas, including hiring, training, communication, and performance reviews. The organization needs to maintain this effort until telehealth is fully operational and embedded into its core culture. Telehealth should be integrated into every aspect of the organization, and both operations and initiatives must be measured appropriately.

Alternative Change Management Theory- Lewin's Force Field Analysis

In addition to or independently of the Kotter model, the organization could consider including Lewin's Force Field Analysis, which looks at the organization as a balance between the *driving forces* in support of the change and the *restraining forces* that are reluctant or resisting the change (Lewin 1951). Through Lewin's analysis, change can only be achieved when the driving forces outweigh the restraining forces. In the Kotter model, the restraining forces are not directly addressed. Lewin's Force Field Analysis can help Thrive Reproductive Care identify and work to counterbalance those restraining forces. In healthcare, Lewin's Force Field Analysis has demonstrated effectiveness in healthcare organizations by improving decision-making, strategic planning, and change management processes, which in turn lead to better outcomes (Salisi & Faller, 2024).

Lewin's Force Field Analysis Application

In this case study and through this analysis, there is an awareness and readiness for change that must clearly communicate the need for rapid digital transformation. This should be frozen. The driving forces include national provider shortages, increasing telehealth demand, and the desire to reduce operational costs while working toward greater health equity. The restraining forces in this scenario are the digital divide, technology literacy challenges, privacy/security fears, and staff resistance. Through the analysis, Thrive can address restraining forces by investing in training and the technology infrastructure as well as stakeholder engagement. They can work with stakeholders to redesign workflows and adopt inclusive communication practices. These are components that were not explicitly the focus of the Kotter model. Once these are in place and the change 'sticks', meaning digital healthcare becomes more standard practice, higher satisfaction from stakeholders, streamlined workflows, and strong privacy protection are well in place, leadership can continually reinforce the telehealth mindset through training. This is when the change can be refrozen, consistent with the model. This type of analysis is essential for Thrive and builds upon the Kotter model. It is also evident in the Salisi & Faller (2024) reference, and this analysis highlights consistent themes as a recent review of driving and restraining forces when implementing information systems (Lemmettylä & Kinnunen, 2025).

Employee Culture

Building and maintaining organizational cultures, such as in the Thrive Reproductive Care transition to a telehealth-only model, will require careful management. Fortunately, after the pandemic, many organizations shifted to remote work and have built or maintained strong organizational cultures for their teams. There are undoubtedly challenges associated with a remote-only model, including reduced information communication and increased feelings of social isolation (Rogers, 2022). However, there are opportunities to foster more connectivity and collaboration. Implementing the competing values framework, which helps diagnose, understand, and transform organizational culture (Orero-Blat et al., 2025), can be an excellent tool for Thrive Reproductive Care. The four quadrants are clan culture, adhocracy culture, market culture, and hierarchy culture. By utilizing survey tools and descriptions of each quadrant, Thrive Reproductive Care leadership can determine which quadrant is most prevalent within the organization, as informal communication and cues are more challenging to track in a virtual environment. Once Thrive Reproductive Care has this information, they can then design deliberate interventions to support the culture they want to foster. Addressing the challenges associated with a remote-only work environment, such as feelings of isolation, can be wholly or partially resolved by strengthening clan culture. Building the culture to support the larger organization is part of the organization's overall strategy that will need to be addressed.

Bridges Transition Theory

This theory documents that transition is an inner psychological process that people go through as they come to terms with a new situation (William Bridges Associates, n.d.). According to this theory, the main stages of transition are endings, the neutral zone, and new beginnings. Endings, considered the first phase of transition, occur when people identify what they are losing and figure out how to manage their loss—this can include relationships, processes, team members, or locations (William Bridges Associates, n.d.). The neutral zone is next; it is when the old is no longer in place, but the new has not fully arrived or become operational. This stage is the core of the transition process, where people create new processes, learn their new roles, and may experience confusion and stress while setting the stage for new beginnings (William Bridges Associates, n.d.). Finally, new beginnings involve understanding, values, and feelings that foster a sense of moving in a new direction. Here, people establish their new roles, goals, and ways to participate effectively in the new structure (William Bridges Associates, n.d.).

Bridges Transition Theory Application

When considering Thrive Reproductive Care's case, particularly employees, it is essential to have an empathetic leader who understands that the employees are going through a significant psychological process. Empathetic leaders should engage with Thrive employees to acknowledge their feelings, including fears, frustrations, or sadness, and create an open space for listening while validating those feelings (Goleman et al., 2002). During the neutral zone, these leaders can encourage open and honest conversations, provide reassurance where possible, and normalize discomfort as part of the transition. Finally, in new beginnings, these leaders should celebrate progress as it happens, recognize those who have the courage to change, and generate excitement about new opportunities (Edmondson, 2018). These factors together can help contribute to a positive and supportive environment, toward a successful organizational culture at Thrive through this significant transition.

Staff Training

Staff training is a vital component of successful digital transformations, especially in complex healthcare organizations. Research consistently demonstrates that proper training in technology use, patient communication, safety, and privacy practices can determine whether a healthcare digital transformation succeeds or fails. Kho et al. (2020) conducted a systematic scoping review of change management practices in telemedicine service implementations. They found that, out of 16 investigated practices, staff training was the most frequently reported practice with the most significant impact on success. Practical communication training is also essential for the successful deployment of telehealth. A 2018 study showed that structured communication training increased nurses' confidence levels when delivering telehealth services (Pratt et al., 2021). Overall, Thrive Reproductive Care must prioritize comprehensive initial and ongoing training for healthcare providers as it moves towards fully virtual care. Proper training boosts provider confidence and leads to better patient outcomes.

Kubler-Ross Change Curve

The Kubler-Ross change curve is a business-oriented adaptation of Elizabeth Kubler-Ross's landmark book On Death and Dying (1969), where she describes the five stages of grief. Since then, business consultants have modified her theory to apply to organizational change and the employee experience (Yarborough, 2013). Here, those involved in the change may go through the feelings of denial, anger, bargaining, depression, and then acceptance in that order. For an organization working through change, this can be a significant disruption to plans or it can be an opportunity to embrace the model to help facilitate the change.

In a similar study, an organization was implementing electronic medical records and the physicians were resistant. To support them, the organization did things like listening to anger and negativity, communicating often, recognizing culture, being prepared to help once asked, understanding that bargaining exists and holding the line, obtaining leadership from the start, actively listening, supporting those managing these feelings, and encouraging those who have reached the acceptance phase (Shoolin, 2010). While enabling electronic medical records is only a single piece of Thrive's transformation, it represents a component of the transformation and a successful way to support the organization through the five stages.

Leadership

Leaders at Thrive Reproductive Care must wear many hats to ensure the successful transition to a long-term telehealth organization. They need to engage in strategic planning and communicate it clearly and frequently. They require dedicated governance and performance metrics to measure success and adjust course when necessary. Investing in staff training, education, and support is essential. They must also ensure a robust technology infrastructure and strong information security for both staff and patients. The leadership at Thrive Reproductive Care needs to accomplish all of this while remaining approachable and transformational leaders. Fortunately, with recent experience from the COVID-19 pandemic, this has been done before, and scholarly literature that Thrive Reproductive Care can rely on is available. Transformational leadership in practice has been documented to improve digital healthcare transitions, specifically in telenursing and telemedicine (Mohamad et al., 2023). From a different perspective, Casillas et al. (2022) found that frontline staff in Los Angeles during a telehealth transition required transparent goal setting and communication for the successful integration of telehealth. In short, Thrive Reproductive Care has an effective change management model. However, like any digital transformation, leadership needs to be actively engaged for the telehealth-only model to succeed and likely expand the way they think about their respective roles today.

Canoe Theory

The Canoe Theory illustrates the critical role that shared goals and teamwork have in an organization. The theory uses a canoe analogy to outline the principles: for the canoe to move forward, all must be paddling in the same direction (Hibbard et. al., 2001). A single Thrive team member can disrupt the smooth flow of the canoe.

Canoe Theory Application

This is an essential theory for Thrive Reproductive Care's leadership to understand, embody, and work to facilitate. As these leaders set the direction of the organization, they need to inspire their team and ensure everyone is working together for the business, or canoe, to sail smoothly and in the direction the organization intends to go. This can be accomplished by creating common goals, listening to, and addressing employee concerns (Casillas et al., 2022). When a rogue rower is experiencing fears about the new business or concerns about job security, it is the leadership's responsibility to intervene, ensure they do not impact others, and address their concerns to help them realign with the team. A key area of concern for this challenge is the need for more technology-enabled skillsets in the telehealth-only model. Thrive can ensure there is adequate training (Casillas et al., 2022) in place to mitigate the challenge and keep the team rowing in the strategic direction of the organization.

Argyris Double-Loop Learning Theory

The Argyris Double Loop Learning theory argues that change is more effective and long-lasting if individuals question the underlying reason behind problems or issues, then enact the solution (the double loop), versus simply reacting to the situation immediately (Argyris & Schön, 1978). The long-term necessity of employees being on board with changes and supportive of leadership it worth the additional time to manage the double-loop required.

Argyris Double-Loop Learning Theory Application

A successful transformation of Thrive would benefit from the successful application of this theory. Leadership must set the expectation so that all staff members feel empowered to actively question the underlying assumptions, values, and goals that govern Thrive's actions which can be challenging because it involves a longer time frame for results, compared to single-loop learning (Li et al., 2021). Thrive may identify an issue with the login process for their patient portal. A proactive employee would likely fix the issue by altering some of the elements on the login screen. This fixes the initial problem, but in turn, creates design issues for colleagues. If the employee had slowed down and thought about the root of the issue and how his fix may affect others, ideally with his colleagues, the whole team would be better off.

RECOMMENDATIONS

Thrive Reproductive Care needs to proactively address both the opportunities and challenges highlighted in the case study and supported by the literature review if they aim to successfully transition to a telehealth provider for reproductive and sexual health. As a first step, Thrive Reproductive Care can launch targeted outreach and digital literacy training for marginalized groups, creating relationships with public health organizations to provide devices and broadband access. As noted in the literature review, the entire population, particularly those in rural or underserved areas or those lacking digital literacy, deserves access to quality healthcare. This addresses a significant digital exclusion issue that is a well-documented risk in telehealth services (Kho et al., 2020). While Thrive Reproductive Care does not have the capital available to address these problems, it can serve as an advocate for patients and create value through public health partnerships. This is a success criterion in Thrive Reproductive Care's balanced scorecard and will help them strive for greater equity.

Next, protecting patient privacy and sensitive health data must be central to Thrive Reproductive Care's telehealth operations, especially given the highly sensitive nature of reproductive and sexual health information. The case study and literature emphasize the importance of a robust cybersecurity framework, comprehensive provider training, and strict adherence to HIPAA regulations. This must be paired with regular staff training related to privacy and security. Privacy breaches are a top concern for patients in a virtual healthcare environment (Houser et al., 2023). Appropriate practices such as multi-factor authentication, regular and surprise technology audits, and clear, transparent privacy policies will help Thrive Reproductive Care meet legal requirements as well as build and sustain patient trust. This is recognized as a vital part of long-term telehealth program success (Ibrahim et al., 2024).

Additionally, implementing a patient-centered digital communication strategy that leverages patient portals, websites, text messaging, and email communication will help provide a precise and personalized experience for patients. This must include the high-touch engagement model necessary to support and educate patients. Thrive Reproductive Care should consider adopting a multilingual support model in the future. These components, when combined, can help Thrive Reproductive Care improve patient satisfaction and outcomes (Casillas et al., 2022). Without implementing a successful communications model, like the one referenced above, Thrive Reproductive Care will be less likely to succeed in this digital transformation.

Implementing an effective and educated staff should be at the center of this organization. As such, Thrive Reproductive Care should aim for continuous staff engagement and development. This includes process redesign to adapt to the new digital model. It is well-documented that by engaging employees early, providing hands-on training, and listening to staff feedback, Thrive Reproductive Care can proactively help its team embrace change and prepare the organization for success. By implementing Kotter's eight-step change model, these employees can be engaged early and be active participants in the change effort. This has been proven successful in healthcare; hands-on training and listening to staff feedback throughout the change process were essential to a successful digital transformation (Wieslander et al., 2024). These investments will reinforce Thrive Reproductive Care's reputation as a leader in reproductive healthcare, demonstrating that it is possible to drive digital innovation while preserving high care standards.

Lastly, Thrive Reproductive Care must continue to dedicate resources to its strategy as well as to the administrative overhead involved in maintaining metrics and tracking progress toward its strategic initiatives. Research shows that while organizations may have the skills to develop a comprehensive strategic plan, they often struggle with implementation (Welch & Smith, 2023). Additionally, Welch and Smith (2023) noted that organizations led by leaders who can apply strategy effectively are better positioned to manage change. As noted in the John Kotter change model, the strategy is an essential component of the change process. It acts as the guiding star for Thrive Reproductive Care's decision-making and prioritization. Every team member should be able to connect their daily activities to a strategic objective or the organization's mission and vision. This cannot be overlooked or set aside if Thrive Reproductive Care manages this change successfully.

CONCLUSION

Thrive Reproductive Care's approach is bold and drastic. At the same time, it is also representative of a broader transformation within reproductive health and healthcare delivery in general, which was accelerated by the COVID-19 pandemic and associated shifts in patient expectations, new technologies, and evolving regulations for virtual care. This paper examined Thrive Reproductive Care's transformation to a telehealth-only model in detail through a comprehensive literature review. The research indicates that while telehealth can have a profoundly positive impact on patients, particularly those who may otherwise lack access to care, it can also exacerbate existing health inequities. Thrive Reproductive Care must carefully balance the increase in accessibility with the quality of its care, the ease of access with privacy concerns, and cost-savings with workforce and care delivery adaptation.

The literature review highlights the same aspects that make digital transformations successful, regardless of industry: creating new access, properly training staff, ensuring open lines of communication, upholding laws and standards, and adhering to a strict view on privacy. It underscores telehealth's potential to help close the digital divide and related access gaps in rural and underserved populations.

Thrive Reproductive Care's shift to a digital-only model can serve as a guide for other companies seeking to deliver healthcare similarly. It is clear that this digital transformation is not solely about technology; it depends on the coordinated alignment of staff and patients, processes, strategy, policies, and infrastructure. By effectively integrating these components, it has the potential to significantly improve patient outcomes and equity while establishing best practices in an emerging industry. Failing to manage any of these components effectively could undermine the telehealth model, create security challenges, or result in the loss of many patients. Success is only achievable if the company prioritizes high-quality healthcare that is easily accessible, widely available, and provided by trained, effective practitioners.

REFERENCES

Al-Abdwani, A., Al-Jahwari, F., Al-Mawali, A., & Al-Zakwani, I. (2023). Exploring patient satisfaction with telemedicine services and its influencing factors. *International Journal of Environmental Research and Public Health, 20*(4), 3165. DOI: 10.3390/ijerph20043165

Almuhaideb, M., Bandaru, D., Abou Chawareb, E., Yafi, F. A., & Hammad, M. A. M. (2025). Ethical perspective on telemedicine usage in sexual medicine: A friend or a foe? *UroPrecision, 3*(1), 47–53. DOI: 10.1002/uro2.111

American College of Obstetricians and Gynecologists. (2020, June 3). *ACOG seeks to expand access, increase quality, and improve outcomes for maternal health in rural communities.* https://www.acog.org/news/news-articles/2020/06/acog-seeks-to-expand-access-increase-quality-and-improve-outcomes-for-maternal-health-in-rural-communities

American Hospital Association. (2025, February 6). *Fact sheet: Telehealth.* https://www.aha.org/fact-sheets/2025-02-07-fact-sheet-telehealth

American Society for Reproductive Medicine Ethics Committee. (2024). *Ethical considerations for telemedical delivery of fertility care: An Ethics Committee opinion.* American Society for Reproductive Medicine. https://www.asrm.org/practice-guidance/ethics-opinions/ethical-considerations-for-telemedical/

Argyris, C., & Schön, D. A. (1978). *Organizational learning: A theory of action perspective.* Addison-Wesley.

Association of American Medical Colleges. (2021, June 11). *The complexities of physician supply and demand: Projections from 2019 to 2034* [Press release]. https://www.aamc.org/news/press-releases/aamc-report-reinforces-mounting-physician-shortage. https://digirepo.nlm.nih.gov/master/borndig/9918417887306676/9918417887306676.pdf

Awan, H. M., & Zaurez, S. (2025). Strategic evaluation using PESTLE and SWOT frameworks: Public sector perspective. *International Journal of Research in Business. Economics and Management, 14*(2), 109–120. DOI: 10.5281/zenodo.14854362

Balanced Scorecard Institute. (2025, August 7). *Balanced scorecard basics.* https://balancedscorecard.org/bsc-basics-overview/

Bang, D., & Frith, C. D. (2017). Making better decisions in groups. *Royal Society Open Science, 4*(8), 170193. DOI: 10.1098/rsos.170193

Bughin, J., LaBerge, L., & Mellbye, A. (2019, October 29). *Unlocking success in digital transformations*. McKinsey & Company. https://www.mckinsey.com/capabilities/people-and-organizational-performance/our-insights/unlocking-success-in-digital-transformations

Casillas, A., Valdovinos, C., Wang, E., Abhat, A., Mendez, C., Gutierrez, G., Portz, J., Brown, A., & Lyles, C. R. (2022). Perspectives from leadership and frontline staff on telehealth transitions in the Los Angeles safety net during the COVID-19 pandemic and beyond. *Frontiers in Digital Health, 4*, 944860. Advance online publication. DOI: 10.3389/fdgth.2022.944860

Cengil, A. B., Eksioglu, S., Eksioglu, B., Eswaran, H., Hayes, C. J., & Bogulski, C. A. (2024). Statistical analysis of telehealth use and pre- and postpandemic insurance coverage in selected health care specialties in a large health care system in Arkansas: Comparative cross-sectional study. *Journal of Medical Internet Research, 26*, e49190. DOI: 10.2196/49190

Davis, R. (2024, March 25). *A healthcare transformation success story: A case study in applying Kotter's 8-step framework*. LinkedIn. https://www.linkedin.com/pulse/healthcare-transformation-success-story-case-study-russ-722dc

Deloitte. (2024). *Maternal health inequities persist. Can digital tools be part of the solution?* Deloitte Insights. https://www.deloitte.com/us/en/insights/industry/health-care/how-digital-tools-can-help-the-maternal-health-crisis.html

Eastburn, J., Fowkes, J., & Kellner, K. (with Swanson, B.) (2022). *Digital transformation: Health systems' investment priorities*. McKinsey & Company. https://www.mckinsey.com/industries/healthcare/our-insights/digital-transformation-health-systems-investment-priorities

Edmondson, A. C. (2018). *The fearless organization: Creating psychological safety in the workplace for learning, innovation, and growth*. John Wiley & Sons.

Endler, M., Lavelanet, A., Cleeve, A., Ganatra, B., Gomperts, R., & Gemzell-Danielsson, K. (2020). Telemedicine for medical abortion: A systematic review. *BJOG, 127*(9), 1094–1102. DOI: 10.1111/1471-0528.15684

Gallegos-Rejas, V. M., De Guzman, K. R., Kelly, J. T., Smith, A. C., & Thomas, E. E. (2025). Strategies to improve telehealth access for culturally and linguistically diverse communities: A systematic review. *Journal of Public Health (Oxford, England), 47*(3), e374–e390. DOI: 10.1093/pubmed/fdaf030

Goleman, D., Boyatzis, R. E., & McKee, A. (2002). *Primal leadership: Realizing the power of emotional intelligence*. Harvard Business School Press.

Gujral, K., Illarmo, S., Jacobs, J., & Wagner, T. (2025). The economics of telehealth: An overview. *Telemedicine Journal and e-Health, 31*(9), 1074–1095. Advance online publication. DOI: 10.1089/tmj.2025.0073

Hibbard, D. R., Hibbard, M. S., & Stockman, J. W. (2001). *The canoe theory: A secret to building people and profit.* Thomson Learning Custom Publishing.

Houser, S. H., Flite, C. A., & Foster, S. L. (2023). Privacy and security risk factors related to telehealth services—A systematic review. *Perspectives in Health Information Management, 20*(1), Article 1f. https://pmc.ncbi.nlm.nih.gov/articles/PMC9860467/

Hügle, T., & Grek, J. (2023). Digital transformation of an academic hospital department: A case study on strategic planning using the balanced scorecard. *PLOS Digital Health, 2*(11), e0000385. Advance online publication. DOI: 10.1371/journal.pdig.0000385

Ibrahim, A. M., Abdel-Aziz, H. R., Mohamed, H. A. H., Zaghamir, D. E. F., Wahba, N. M. I., Hassan, G. A., Shaban, M., El-Nablaway, M., Aldughmi, O. N., & Aboelola, T. H. (2024). Balancing confidentiality and care coordination: Challenges in patient privacy. *BMC Nursing, 23*(1), 564. DOI: 10.1186/s12912-024-02231-1

Ivanova, J., Shah, B., Foote, C., & Cummins, M. R. (2025). Assessing the costs and savings of telemedicine: Insights from the consolidated framework for implementation research. *Digital Health, 11*, 20552076251314552. Advance online publication. DOI: 10.1177/20552076251314552

Janis, I. L. (1982). *Groupthink: Psychological studies of policy decisions and fiascoes* (2nd ed.). Houghton Mifflin.

Kaiser, J. L., Morhe, E., & Abern, L. R. (2022). Quantitative and qualitative impact of one key question on primary care providers' contraceptive counseling at routine preventive health visits. *Contraception, 109*, 73–79. DOI: 10.1016/j.contraception.2022.01.004

Kaiser Permanente. (n.d.). *24/7 virtual care.* https://healthy.kaiserpermanente.org/learn/how-to-use-telehealth

Keller, T. (2023). *It is time to change the way we change* [Master's thesis, University of Central Florida]. STARS. https://purls.library.ucf.edu/go/DP0027553

Kho, J., Gillespie, N., & Martin-Khan, M. (2020). A systematic scoping review of change management practices used for telemedicine service implementations. *BMC Health Services Research, 20*(1), 815. Advance online publication. DOI: 10.1186/s12913-020-05657-w

Kim, J., Cai, Z. R., Chen, M. L., Onyeka, S., Ko, J. M., & Linos, E. (2024). Telehealth utilization and associations in the United States during the third year of the COVID-19 pandemic: Cross-sectional survey study. *JMIR Public Health and Surveillance, 10*, e51279. https://publichealth.jmir.org/2024/1/e51279/. DOI: 10.2196/51279

Klotz, S. A., Chan, C. B., Bianchi, S., Egurrola, C., & York, L. D. (2023). The genie is out of the bottle: Telemedicine is more effective than brick-and-mortar clinics in the care of HIV-infected outpatients. *The American Journal of Medicine, 136*(4), 360–364. https://pubmed.ncbi.nlm.nih.gov/36495936/. DOI: 10.1016/j.amjmed.2022.11.012

Kotter, J. P. (1996). *Leading change*. Harvard Business School Press.

Kumera, D., Amentie, C., & Bali, N. (2024). Effect of technological innovation on firm's performance: mediating effect of competitive advantage: a study on manufacturing firms operating in Ethiopian industrial parks. *Brazilian Journal of Operations & Production Management, 21*(3), e20242146. DOI: 10.14488/BJOPM.2146.2024

Legler, S., Diehl, M., Hilliard, B., Olson, A., Markowitz, R., Tignanelli, C., Melton, G. B., Broccard, A., Kirsch, J., & Usher, M. (2021). Evaluation of an Intrahospital Telemedicine Program for Patients Admitted With COVID-19: Mixed Methods Study. *Journal of Medical Internet Research, 23*(4), e25987. DOI: 10.2196/25987

Lemmettylä, A., & Kinnunen, U.-M. (2025). Driving and restraining forces in the implementation of information systems in public sector organizations: Scoping review using Lewin's change theory. *JMIR Human Factors, 12*, e71575. https://humanfactors.jmir.org/2025/1/e71575. DOI: 10.2196/71575

Lewin, K. (1951). *Field theory in social science: Selected theoretical papers*. Harper & Row.

Li, C. H., Yang, W. G., & Shih, I. T. (2021). Exploration on the gap of single- and double-loop learning of balanced scorecard and organizational performance in a health organization. *Heliyon, 7*(12), e08553. DOI: 10.1016/j.heliyon.2021.e08553

Liu, J., Qin, Y., Liu, H., Liu, Y., Yang, Y., Ning, Y., & Ye, H. (2025). Global, regional, and national burden of female infertility and trends from 1990 to 2021 with projections to 2050 based on the GBD 2021 analysis. *Scientific Reports, 15*(1), 17559. Advance online publication. DOI: 10.1038/s41598-025-01498-x

Ma, Y., Zhao, C., Zhao, Y., Lu, J., Jiang, H., Cao, Y., & Xu, Y. (2022). Telemedicine application in patients with chronic disease: A systematic review and meta-analysis. *BMC Medical Informatics and Decision Making, 22*(1), 105. Advance online publication. DOI: 10.1186/s12911-022-01845-2

Mohamad, N. I., Abd Rahman, I., Sanusi, S., Abdullah, S., & Ibrahim, N. (2023). An empirical study of the relationship between leadership practice in training programs and skill development: Motivational climate as mediating. *TEM Journal, 12*(4), 2112–2122. https://doi.org/DOI: 10.18421/TEM124-22

Muheriwa-Matemba, S. R., Alcena-Stiner, D. C., Glazier, A., & LeBlanc, N. M. (2024). Telehealth use for sexual and reproductive health promotion and care during the early phase of the COVID-19 pandemic: A descriptive-interpretive qualitative study of healthcare providers' perspectives and experiences in Western—Central New York State. *PLOS Global Public Health*, *4*(12), e0003259. Advance online publication. DOI: 10.1371/journal.pgph.0003259

National Academies of Sciences, Engineering, and Medicine. (2021). *Implementing high-quality primary care: Rebuilding the foundation of health care.* The National Academies Press. https://doi.org/DOI: 10.17226/25983

National Rural Health Association. (2024, January 31). *Maternal health in rural America* [White paper]. https://www.ruralhealth.us/nationalruralhealth/media/documents/advocacy/white%20paper/maternal-health-in-rural-america-white-paper-final.pdf

Nuako, A., Liu, J., Pham, G., Smock, N., James, A., Baker, T., Bierut, L., Colditz, G., & Chen, L.-S. (2022). Quantifying rural disparity in healthcare utilization in the United States. *PLoS One*, *17*(2), e0263718. Advance online publication. DOI: 10.1371/journal.pone.0263718

Office for Civil Rights. (2025, June 27). *Regulatory initiatives.* U.S. Department of Health and Human Services. https://www.hhs.gov/hipaa/for-professionals/regulatory-initiatives/index.html

Olyaeemanesh, A., Takian, A., Mostafavi, H., Mobinizadeh, M., Bakhtiari, A., Yaftian, F., Vosoogh-Moghaddam, A., & Mohamadi, E. (2023, November 18). Health equity impact assessment (HEIA) reporting tool: Developing a checklist for policymakers. *International Journal for Equity in Health*, *22*(1), 241. https://pmc.ncbi.nlm.nih.gov/articles/PMC10657117/. DOI: 10.1186/s12939-023-02031-0

Orero-Blat, M., Leal-Rodríguez, A. L., Albort-Morant, G., & Ariza-Montes, A. (2025). The strategic advantage of innovative organizational culture: An exploratory analysis in digital transformation and big data analytics capabilities. *Journal of Management & Organization*, *31*(2), 443–462. DOI: 10.1017/jmo.2024.74

Pratt, H., Moroney, T., & Middleton, R. (2021). The influence of engaging authentically on nurse–patient relationships: A scoping review. *Nursing Inquiry*, *28*(2), e12388. Advance online publication. DOI: 10.1111/nin.12388

Radu, G., Solomon, M., Gheorghe, C. M., Hostiuc, M., Bulescu, I. A., & Purcarea, V. L. (2017). The adaptation of health care marketing to the digital era. *Journal of Medicine and Life*, *10*(1), 44–50. https://pmc.ncbi.nlm.nih.gov/articles/PMC5304370/

Roberts, E. T., & Mehrotra, A. (2020). Assessment of disparities in digital access among Medicare beneficiaries. *JAMA Internal Medicine*, *180*(10), 1386–1389. DOI: 10.1001/jamainternmed.2020.2666

Rogers, K. (2022). *The effects of remote work on organizational culture: Examining the effects of remote work on communication, social connectedness, and culture* (Master's thesis, University of Tennessee). TRACE: Tennessee Research and Creative Exchange. https://trace.tennessee.edu/utk_gradthes/6395

Salisi, J. C., & Faller, E. M. (2024). Application of Force Field Analysis in healthcare: A systematic review of change management strategies and performance improvement initiatives. *International Journal of Research Publication and Reviews*, *5*(7), 3256–3263. https://ijrpr.com/uploads/V5ISSUE7/IJRPR31644.pdf

Shaver, J. (2022). The state of telehealth before and after the COVID-19 pandemic. *Primary Care*, *49*(4), 517–530. DOI: 10.1016/j.pop.2022.04.002

Shirey, M. R. (2011). Establishing a sense of urgency for leading transformational change. *The Journal of Nursing Administration*, *41*(4), 145–148. DOI: 10.1097/NNA.0b013e3182118550

Shoolin, J. S. (2010). Change management - recommendations for successful electronic medical records implementation. *Applied Clinical Informatics*, *1*(3), 286–292. DOI: 10.4338/ACI-2010-01-R-0001

Slåtten, T., Mutonyi, B. R., & Lien, G. (2021). Does organizational vision really matter? An empirical examination of factors related to organizational vision integration among hospital employees. *BMC Health Services Research*, *21*(1), 483. Advance online publication. DOI: 10.1186/s12913-021-06503-3

Snoswell, C. L., Stringer, H., Taylor, M. L., Caffery, L. J., Smith, A. C., & Haydon, H. M. (2021). The effect of telehealth on hospital services use: Systematic review and meta-analysis. *Journal of Medical Internet Research*, *23*(7), e25987. Advance online publication. DOI: 10.1177/1357633X211023700

Stone, P. W., Harrison, M. I., Feldman, P., Lin, S. X., Purrington, A., Ringel, M., Arnold, S. B., Tietjen, J., Cohen, L., Fahey, L., Rosen, B., Wald, H. L., Curry, L., & Schnipper, J. L. (2018). How guiding coalitions promote positive culture change in hospitals: A longitudinal mixed methods interventional study. *BMJ Quality & Safety*, *27*(3), 218–225. DOI: 10.1136/bmjqs-2017-006574

U. S. Department of Health and Human Services. (2024). HIPAA and telehealth: Guidance for providers. https://www.hhs.gov/hipaa/for-professionals/special-topics/telehealth/index.html

Uddin, J., & Fariha, T. (2024). *Bridging the gap: Expanding telehealth services to address rural health disparities* [Unpublished manuscript]. Social and Behavioral Health Department, University of Nevada.

U.S. Department of Health & Human Services. (n.d.). Telehealth. https://www.hhs.gov/hipaa/for-professionals/faq/telehealth/index.html

Van Dijk, J. (2020). The digital divide. *Polity.*

Wakeman, M., Buckman, D. W., & El-Toukhy, S. (2025). Disparities in digital health care use in 2022. *JAMA Network Open*, *8*(4), e255359. Advance online publication. DOI: 10.1001/jamanetworkopen.2025.5359

Wang, S., von Huben, A., Sivaprakash, P. P., Saurman, E., Norris, S., & Wilson, A. (2025). Addressing health service equity through telehealth: A systematic review of reviews. *Digital Health*, *11*, 20552076251326233. Advance online publication. DOI: 10.1177/20552076251326233

Welch, T. D., & Smith, T. B. (2023). Strategic planning: Moving from plan to implementation. *Nursing Administration Quarterly*, *47*(4), 283–288. https://pubmed.ncbi.nlm.nih.gov/37643226/. DOI: 10.1097/NAQ.0000000000000597

Wieslander, L., Bäckström, I., & Häggström, M. (2024). Participation in the digital transformation of healthcare: a review of qualitative studies. *International journal of health care quality assurance, ahead-of-print*(ahead-of-print), 68–84. https://doi.org/DOI: 10.1108/IJHCQA-03-2024-0021

Wilber, K., Yan, M., & Zhu, Y. (2022, December 30). *Unanticipated outcomes of a telehealth pilot project in skilled nursing facilities care: Don't put the cart before the course.* California Emerging Technology Fund & University of Southern California Leonard Davis School of Gerontology. https://s42263.pcdn.co/wp-content/uploads/2023/03/230307_CETF_SNF_Telehealth_Pilot_Project_Final-Evaluation-Report_2023.pdf

William Bridges Associates. (n.d.). *What is transition? Bridges Transition Model.* https://wmbridges.com/about/what-is-transition/

Yarborough, P. D. (2013). Change management – recommendations for successful electronic medical records implementation. *The Permanente Journal*, *17*(2), 83–85. DOI: 10.7812/TPP/12-111

Chapter 7
Innovations in Environmental Public Health Through Artificial Intelligence Applications in the U.S Military and Government Green Logistics

Amalisha Sabie Aridi
http://orcid.org/0000-0002-7869-5530
Washington University of Science and Technology, USA

ABSTRACT

Green logistics integrates advanced analytics with ecological stewardship in government and military supply chains. This phenomenological study engaged 15 experts (5 military logisticians, 5 green logistics specialists, 5 AI engineers) to examine how AI, IoT, blockchain, and circular-economy practices align operational readiness with the UN Sustainable Development Goals. Participants described AI-enabled predictive maintenance, dynamic routing, and real-time emissions dashboards that reduce fuel use, waste, and breakdowns while improving local air quality and public health. Blockchain and IoT strengthened transparency and lifecycle accountability, supporting renewable-energy integration and material reuse. Yet adoption remains partial, constrained by cultural inertia, fragmented policy, data-governance deficits,

DOI: 10.4018/979-8-2600-1026-6.ch007

and cyber dependence. The study shows AI-enabled green logistics can operationalize SDGs 3, 9, 12, and 13 in defense contexts, reframing national security as environmental security and offering a scalable model for sustainable state capacity.

INTRODUCTION

Green logistics, as an evolving paradigm of sustainable supply chain management, fuses advanced technological innovation with ecological stewardship to achieve systemic efficiency and environmental balance (McKinnon et al., 2015). Within the complex logistical frameworks of government and military operations, sectors historically characterized by high energy consumption and carbon intensity, artificial intelligence (AI) has emerged as a transformative agent capable of reconciling operational imperatives with sustainability goals. By integrating AI-driven optimization with renewable energy adoption and circular economy practices, public institutions can cultivate a logistical architecture that both enhances performance and safeguards environmental and public health.

AI's analytical precision facilitates the dynamic optimization of transport routes and resource allocation, drastically curtailing inefficiencies that contribute to emissions and waste (Rodrigue et al., 2017). Machine learning algorithms can synthesize data from diverse variables, traffic conditions, meteorological patterns, and energy metrics, to generate optimal routing systems that minimize fuel use and idle time. For instance, in a government postal fleet or military resupply convoy, AI-enabled route optimization not only reduces greenhouse gas emissions but also diminishes long-term operational costs by conserving energy and mitigating vehicle wear (Kumar, 2015). Similarly, predictive maintenance algorithms detect early signs of mechanical fatigue, preventing breakdowns that would otherwise lead to resource-intensive repairs and increased emissions.

The transition toward circular economy models within green logistics redefines the lifecycle of materials and infrastructure, advancing waste minimization and resource renewal (Blanco & Sheffi, 2017). AI plays a pivotal role in facilitating this transition by identifying reuse and repurposing opportunities through intelligent material tracking and demand forecasting. For example, blockchain-integrated AI systems can authenticate the provenance of recycled metals or biodegradable materials, ensuring accountability and transparency across supply chains. Within defense logistics, such systems might track the reuse of aircraft components or the repurposing of decommissioned materials into civilian infrastructure, thereby aligning national security operations with sustainable development goals.

The public health implications of these transformations are profound. By mitigating air and water pollution through reduced emissions and hazardous waste, AI-enabled green logistics directly contributes to the prevention of respiratory and cardiovascular diseases, conditions strongly correlated with environmental degradation near transportation corridors and military bases (UN, 2024). The shift toward renewable energy sources and electric vehicle fleets further aligns logistical operations with the principles of SDG 3 (Good Health and Well-being) and SDG 6 (Clean Water and Sanitation), reinforcing a holistic vision of sustainability that integrates human health and ecological resilience (Blanco & Sheffi, 2024).

AI also enhances institutional accountability through data-driven transparency, a critical element in both democratic governance and sustainable logistics. The integration of blockchain, IoT sensors, and AI-driven analytics provides real-time visibility into energy use, emissions levels, and supply chain ethics (Hussain, 2025; Kale, 2025). This transparency not only enables compliance with environmental standards but also fosters public trust in government and military institutions that commit to sustainability. For instance, IoT sensors in government vehicle fleets can transmit continuous data on carbon output, while AI algorithms analyze these data to recommend policy interventions, such as transitioning specific routes to electric vehicles or scheduling maintenance to reduce energy inefficiency (Chen et al., 2024; Eslami & Khoveyni, 2024).

Such digital oversight mechanisms reinforce SDG 9 (Industry, Innovation, and Infrastructure) by embedding sustainability within the very design of public sector logistics. In military contexts, predictive AI analytics can optimize mission-critical supply chains to minimize environmental disruption, such as fuel-efficient drone deliveries to remote bases, thereby demonstrating that operational excellence and ecological responsibility are not mutually exclusive but mutually reinforcing.

Green logistics within defense and government systems plays a decisive role in advancing SDG 13 (Climate Action). By embedding AI-driven sustainability strategies, these institutions can significantly curtail their carbon footprint while maintaining operational readiness. In aviation logistics, AI tools can model and simulate flight routes to minimize fuel burn and forecast maintenance requirements to prolong component life cycles (Nicoletti & Appolloni, 2024; Nozari, 2024). A practical example includes the U.S. Department of Defense's exploration of hybrid-electric aircraft, where AI assists in real-time energy distribution between propulsion systems and battery reserves. These innovations exemplify the fusion of technological modernization with climate-conscious logistics, setting a precedent for sustainable defense operations worldwide.

Furthermore, AI-enhanced sustainability reporting allows defense agencies to quantify their contributions to emission reductions, strengthening international accountability in climate agreements. By extending these models to allied and partner nations, AI-driven green logistics can serve as a framework for multilateral cooperation in sustainable security, thereby embedding ecological values into global defense strategy.

Ultimately, the integration of AI within green government and military logistics represents not merely a technological advancement but a paradigmatic shift toward ecological governance. The application of AI in optimizing energy use, reducing waste, and enhancing transparency creates a cascading effect that bolsters public health, strengthens institutional legitimacy, and advances global sustainability objectives. When applied to aircraft distribution and military transportation networks, these technologies concretely manifest the aspirations of SDG 12 (Responsible Consumption and Production) and SDG 13, ensuring that sustainability is not confined to rhetoric but operationalized in daily practices.

For example, a military base employing AI-guided smart grids can autonomously manage its energy consumption, prioritizing renewable sources during peak hours while maintaining readiness for emergencies. Similarly, a government agency leveraging AI-powered logistics can reduce the carbon intensity of humanitarian aid shipments, aligning national resilience strategies with the ethical imperatives of environmental justice. Such examples illustrate how the symbiosis between AI and green logistics can catalyze sustainable transformation across institutional scales.

Problem Statement

Despite its transformative potential, the institutionalization of AI-driven green logistics in government and military contexts remains partial and fragmented. These sectors continue to grapple with structural, cultural, and regulatory barriers that impede comprehensive adoption. The absence of standardized frameworks for integrating AI, renewable energy, and circular economy principles limits progress toward sustainable logistics, exacerbating the environmental and health externalities associated with legacy systems (UN, 2024). Addressing this gap demands interdisciplinary research that unites environmental science, public policy, and data analytics to design adaptive, ethical, and scalable models for green logistics implementation.

The critical challenge lies in operationalizing sustainability without compromising mission-critical efficiency. Future research must therefore explore how AI systems can balance security imperatives with ecological goals, ensuring that environmental public health becomes an intrinsic dimension of national defense and governance strategy. In doing so, green logistics will evolve from a technical initiative into a central pillar of global sustainability; a domain where environmental integrity, human well-being, and technological innovation converge in pursuit of a shared planetary future.

Purpose Statement

The purpose of this phenomenological inquiry is to explore and critically analyze the lived experiences, professional insights, and collective perspectives of experts in military logistics, green logistics, and artificial intelligence (AI) engineering to understand how AI-driven green logistics can be effectively integrated into government and military transportation and aircraft distribution systems. This study seeks to elucidate how emerging technologies, specifically blockchain, the Internet of Things (IoT), and AI can synergistically advance sustainability by optimizing resource allocation, minimizing waste, and reducing greenhouse gas emissions within defense and governmental infrastructures (Blanco & Sheffi, 2024; Hussain, 2025). Grounded in circular economy principles, the research endeavors to construct a conceptual and practical framework that balances the operational imperatives of national defense and governance with the ethical and ecological mandates of global sustainability.

This investigation is designed to contribute both theoretically and practically to the body of knowledge surrounding green logistics and sustainable systems engineering. By engaging 15 subject matter experts, five military and logistics strategists, five specialists in green logistics and sustainable supply chain design, and five AI engineers, the study will identify patterns in their experiential narratives that illuminate the challenges, opportunities, and transformative potential of AI integration in environmentally responsible logistics. Through their phenomenological accounts, this inquiry aims to uncover the underlying meanings associated with the adoption of intelligent systems in traditionally carbon-intensive sectors, addressing persistent gaps in the literature regarding implementation scalability, ethical governance, and cross-sectoral collaboration.

The ultimate objective is to develop an actionable and adaptive model for green government and military logistics that advances the United Nations Sustainable Development Goals, specifically SDG 3 (Good Health and Well-being), SDG 9 (Industry, Innovation, and Infrastructure), and SDG 13 (Climate Action) (UN, 2024). The framework envisioned in this study seeks to reconcile efficiency and sustainability; an equilibrium achieved through the strategic application of AI tools capable of predictive maintenance, real-time emissions tracking, and logistics optimization. For instance, AI-powered route optimization algorithms could enable defense agencies to minimize fuel consumption and operational delays during deployment missions, while IoT-based monitoring systems could enhance situational awareness by detecting inefficiencies in real time. These applications not only improve operational resilience but also embody a paradigmatic shift toward green defense infrastructure that promotes environmental public health and global ecological stability.

Significance of the Inquiry

The significance of this study lies in its potential to reshape the intersection of military, governmental, and environmental domains by positioning AI-driven green logistics as a linchpin for global sustainability and public health protection. Military and government logistics systems are among the world's most complex and resource-intensive operations, frequently contributing to elevated levels of pollution, waste, and energy consumption (Kumar, 2015). This inquiry responds to the urgent need for these institutions, often symbols of state power and technological advancement, to become exemplars of environmental stewardship. By examining how AI technologies can support the decarbonization of logistical operations, the study highlights the potential for these traditionally high-emission sectors to lead rather than lag in the global transition toward sustainable development.

At a practical level, the integration of AI into green logistics offers tangible solutions for mitigating environmental degradation and improving public health outcomes. Predictive analytics, for example, can model emission scenarios and forecast equipment failures before they occur, thereby preventing pollution-intensive breakdowns. In military aviation, AI algorithms can recalibrate flight paths based on atmospheric conditions to minimize fuel burn, aligning operational efficiency with SDG 13 (Climate Action). Similarly, blockchain-enabled supply chain transparency ensures that procurement processes adhere to sustainable sourcing standards, contributing to SDG 9 (Industry, Innovation, and Infrastructure) by fostering accountability and innovation. The resulting decrease in air pollutants and waste production directly advances SDG 3 (Good Health and Well-being) by reducing respiratory illnesses and environmental hazards for communities situated near bases, ports, and logistical corridors (UN, 2024).

From a theoretical perspective, this study contributes to the growing interdisciplinary discourse on the convergence of artificial intelligence, environmental ethics, and global governance. It expands upon existing frameworks by illustrating how AI can serve not merely as a technological instrument but as a transformative epistemic force that reshapes how institutions conceptualize sustainability. In phenomenological terms, the insights derived from experts' lived experiences will reveal how the cultural and operational ethos of military and governmental systems can evolve through the integration of AI; moving from reactive compliance toward proactive sustainability leadership. The anticipated findings may also provide guidance for policymakers seeking to institutionalize green logistics practices through data-driven governance models, thereby bridging the gap between environmental policy and logistical implementation.

Ultimately, this inquiry holds significance beyond its academic contribution: it proposes a pragmatic roadmap for embedding sustainability into the fabric of public sector logistics. By illustrating how AI tools can be harnessed to monitor, predict, and optimize environmental outcomes, this study situates military and government operations as pivotal actors in achieving the United Nations' global sustainability agenda. In doing so, it positions green logistics not as a peripheral initiative, but as a moral and operational imperative; one that unites technological innovation, ecological consciousness, and the safeguarding of public health into a coherent vision for a resilient, sustainable future.

UNDERSTANDING GREEN LOGISTICS

Green logistics, also referred to as sustainable logistics, represents an evolving field designed to mitigate the environmental impacts of transportation and distribution systems (Hussain, 2025; Kale, 2025). This practice integrates advanced methods to reduce emissions, optimize energy use, minimize waste, and enhance operational efficiency, thereby balancing economic performance with environmental stewardship (Hussain, 2025; Kale, 2025).

Central to this framework are strategies such as emission reduction, energy-efficient vehicle deployment, waste management through recycling and reuse, and the adoption of sustainable packaging materials. Furthermore, optimizing transportation routes not only lowers fuel consumption but also contributes to reducing the overall carbon footprint of logistical activities. These practices reflect a holistic approach to sustainability, demonstrating that logistics can be a critical lever in advancing global environmental goals while maintaining organizational efficiency (Chen et al., 2024; Eslami & Khoveyni, 2024).

Green Logistics in Military and Government Operations

Military and government sectors, with their extensive transportation networks and substantial resource consumption, bear a significant responsibility in advancing green logistics. These sectors, which often serve as substantial contributors to environmental degradation, have a unique opportunity to lead by example. Adopting green logistics in military and government operations is essential for aligning logistical practices with environmental responsibilities. In addition to reducing emissions, green logistics fosters operational efficiency by optimizing resource utilization and minimizing dependency on fossil fuels, which enhances resilience to environmental disruptions. Moreover, compliance with environmental regulations not only prevents legal repercussions but also positions these sectors as global leaders in sustainable practices, inspiring similar transformations across industries (Hussain, 2025; Kale, 2025).

Green logistics represents a transformative approach to balancing operational efficiency with environmental responsibility (Hussain, 2025; Kale, 2025). For military and government sectors, its implementation is not only a strategic imperative but also a moral responsibility, given their significant environmental impact. By fostering leadership commitment, embedding sustainability into organizational culture, and leveraging technological advancements, these sectors can lead the global transition toward sustainable logistics practices. The development of a skilled workforce and the engagement of stakeholders across all levels further reinforce this commitment, creating a resilient and adaptable logistical framework. Green logistics offers a pathway to achieving the United Nations Sustainable Development Goals while setting a benchmark for industries worldwide.

Institutional Theory

Institutional Theory provides a robust framework for understanding how organizations internalize and enact new practices in response to external pressures shaped by regulations, cultural expectations, and professional norms (Rudko et al., 2025). Within the domain of green logistics, particularly in government and military contexts, this theoretical lens elucidates how environmental sustainability becomes institutionalized not merely as an operational necessity but as an ethical and political imperative. Institutional pressures manifest in three interrelated forms: coercive, normative, and mimetic.

Coercive pressures derive from formal regulations and international environmental agreements such as the Paris Accord and the United Nations Sustainable Development Goals (SDGs), compelling institutions to conform to global sustainability standards. For example, defense agencies that rely heavily on fuel-intensive operations face increasing mandates to adopt carbon reduction strategies aligned with SDG 13 (Climate Action). Normative pressures, on the other hand, emerge from societal and professional expectations that organizations act as responsible stewards of the environment. The growing public awareness of environmental degradation, coupled with demands for transparency, places moral and reputational pressure on state entities to operationalize sustainability through cleaner technologies and ethical procurement practices. Mimetic pressures arise as organizations emulate the sustainability strategies of peer institutions, such as NATO or the European Defence Agency, that have successfully integrated renewable energy systems and emission-tracking mechanisms within their logistics chains.

In the context of AI-driven green logistics, these institutional forces converge to shape technological and ethical transformation. Governments and military organizations adopt AI and blockchain technologies not only as efficiency tools but as symbolic and technical instruments of accountability. For instance, blockchain-enabled tracking of supply chains ensures verifiable emissions reporting, while AI systems monitor and predict the environmental impact of logistical operations. Such mechanisms serve dual purposes: they enhance operational precision and signal institutional legitimacy within the international sustainability community. Thus, Institutional Theory underscores that the movement toward green logistics is not merely an internal innovation process but a socially embedded transformation driven by the evolving global architecture of norms, expectations, and compliance obligations.

Force Field Analysis

Force Field Analysis offers a dynamic lens for examining how institutional change in logistics and supply chain systems emerges through the tension between forces that drive transformation and those that resist it (Capatina et al., 2017). In the transition toward AI-enabled green logistics, this model captures the equilibrium between momentum and inertia that defines the adaptive capacity of government and military organizations. Change occurs when driving forces, such as technological innovation, regulatory mandates, environmental ethics, and institutional commitment to the UN SDGs, overcome restraining forces like bureaucratic inertia, risk aversion, entrenched cultural practices, and fiscal constraints.

In practice, driving forces in government and defense logistics are increasingly shaped by technological breakthroughs. AI, IoT, and blockchain innovations have revolutionized predictive maintenance, reduced vehicle idling, and enhanced real-time emissions transparency (Rodrigue et al., 2017; Hussain, 2025). Policy imperatives derived from the UN's sustainability agenda further reinforce this momentum, creating coercive institutional incentives for carbon reduction. Additionally, public health motivations serve as powerful moral drivers: empirical evidence linking logistics emissions to respiratory and cardiovascular diseases transforms sustainability from an abstract ideal into a matter of societal well-being (UN, 2024).

Conversely, restraining forces often reside within the institutional culture itself. Government and military organizations are traditionally hierarchical, emphasizing reliability and risk mitigation over innovation. Personnel trained under legacy logistics systems may distrust algorithmic decision-making or view AI integration as incompatible with mission priorities. Furthermore, fiscal and security concerns, such as fears that investment in green technologies might undermine short-term readiness, can exacerbate institutional resistance.

A vivid example of this tension can be seen in defense aviation maintenance: while AI-based predictive systems can extend aircraft lifespan and minimize emissions, some maintenance crews resist adoption due to perceived threats to their expertise or operational autonomy. Force Field Analysis thus becomes a diagnostic and strategic tool, enabling organizational leaders to identify which restraining forces require targeted intervention, such as retraining programs or communication strategies, and which driving forces can be amplified through leadership advocacy or regulatory incentives. By applying this analytical model, institutions can orchestrate a more deliberate and resilient transition toward sustainable, AI-augmented logistics systems.

The Canoe Theory

The Canoe Theory, originally articulated within the context of leadership alignment and team cohesion (Hibbard et al., 2006), provides a compelling metaphor for understanding the collaborative nature of sustainable transformation. It posits that an organization functions like a canoe: progress depends on every member paddling in unison toward a shared goal, while misalignment, whether through disengagement, resistance, or miscommunication, causes the canoe to drift off course or lose momentum. This metaphor captures the essence of interdisciplinary coordination required for implementing AI-driven green logistics in complex institutional environments.

In this context, the "canoe" represents the organizational system, whether a government logistics division or a military transportation command, navigating turbulent currents of technological innovation and environmental responsibility. Each "paddler" symbolizes a distinct stakeholder: logistics officers, AI engineers, sustainability policymakers, and public health specialists. The "river current" embodies the external pressures of international environmental policy, emerging technologies, and societal expectations for ethical governance. Progress depends on shared vision, synchronized roles, strong leadership, and mutual accountability (Hibbard et al., 2006). When stakeholders share a unified understanding of the sustainability mission, the organization moves efficiently toward its destination. However, when AI engineers prioritize algorithmic sophistication over practical usability, or logisticians privilege mission speed over environmental integrity, the canoe veers off course, illustrating the fragility of organizational coherence amid competing priorities.

The Canoe Theory emphasizes that technological transformation is inherently social and relational. Effective implementation of AI and green logistics technologies requires not only technical expertise but also trust, communication, and shared ethical commitment. For instance, a government agency deploying an AI-based emissions monitoring system will achieve greater success when engineers, operators, and policymakers collaboratively design its parameters; ensuring the system's functionality aligns with real-world operational needs and sustainability targets. Leadership thus plays a pivotal role in setting direction, resolving conflict, and cultivating a culture where sustainability is viewed as a collective responsibility rather than a peripheral obligation.

Ultimately, the Canoe Theory underscores that achieving environmental sustainability in logistics is not solely a matter of adopting advanced tools but of achieving organizational harmony and moral alignment. When every actor "stays in the canoe" and rows in concert toward the shared objective of sustainable logistics, technological innovation becomes not merely efficient, but transformative; advancing global environmental health, fulfilling the mandates of the UN SDGs, and modeling the collaborative ethos necessary for a sustainable future.

METHODS AND RESEARCH DESIGN

This study employed a transcendental phenomenological design to explore the lived experiences and professional perspectives of experts situated at the intersection of military logistics, green logistics, and artificial intelligence (AI) systems engineering. The phenomenological approach was selected to illuminate the meanings participants ascribed to the integration of AI-driven sustainability practices within government and defense supply chains.

The central phenomenon under examination was the adoption and operationalization of AI-enabled green logistics practices within military and federal logistics systems. The design emphasized depth over breadth, privileging rich, descriptive accounts that reveal not only what participants experienced but how they experienced it within organizational, cultural, and policy structures.

Participants and Sampling

A purposive criterion sampling strategy was used to recruit 15 participants representing three expert groups essential to the inquiry:

- Five military or federal logistics strategists (coded M1–M5): Senior professionals overseeing transportation, aircraft distribution, or supply chain operations within U.S. defense or federal agencies.
- Five green logistics and sustainability experts (coded G1–G5): Specialists in environmentally responsible supply chain design, energy efficiency, and circular economy practices within or adjacent to government sectors.
- Five AI engineers and technologists (coded A1–A5): Developers and analysts specializing in artificial intelligence applications relevant to logistics optimization, predictive analytics, and environmental modeling.

All participants had a minimum of 5 years of professional experience in their respective fields and demonstrated familiarity with sustainability mandates under the United Nations Sustainable Development Goals (SDGs). Recruitment occurred through professional networks, defense research associations, and academic partnerships. Each participant was assigned a pseudonym and identifier code to maintain confidentiality.

Data Collection Procedures

Interview Protocol

Data was collected through semi-structured, in-depth interviews conducted virtually via secure video conferencing platforms. Each session lasted between 60 and 90 minutes, allowing sufficient time for narrative depth and reflexivity. The open-ended structure permitted participants to articulate their experiences in their own language while ensuring coverage of core research objectives.

The following five guiding questions structured the inquiry and directly correspond to the thematic findings:

1. Question 1: How have you experienced or observed the integration of AI tools in green logistics practices within military or federal logistics operations?
2. Question 2: What challenges or barriers have you encountered in adopting AI-enabled sustainable logistics technologies?
3. Question 3: In your view, how do AI applications contribute to reducing emissions, optimizing resources, or improving public health outcomes?
4. Question 4: How do organizational culture, leadership, or institutional policy shape the adoption of green logistics technologies?
5. Question 5: What future directions do you envision for AI-enabled sustainability in defense or federal supply chains?

Follow-up probes were used to elicit elaboration, clarification, and personal interpretation of key ideas. Examples included: "Can you describe a specific instance when AI improved sustainability outcomes?" and "How do these technologies change your sense of responsibility or mission within logistics operations?"

DATA ANALYSIS

Transcendental Phenomenological Analysis

The process unfolded in several stages:

1. Epoche (Bracketing): The researcher engaged in reflexive journaling to suspend preconceptions related to sustainability technologies and defense operations.
2. Horizontalization: Significant statements were extracted from transcripts, treating each with equal value to ensure no premature hierarchization of meaning.
3. Meaning Unit Clustering: Statements were grouped into conceptual clusters, identifying recurring patterns related to technological integration, ethical tension, and operational transformation.
4. Textural Description: Each participant's account was summarized to capture *what* they experienced (the substance of AI and green logistics practices).
5. Structural Description: The analytic focus shifted to *how* participants experienced these phenomena within institutional, policy, and cultural contexts.
6. Essence Synthesis: Textural and structural insights were integrated to distill the composite essence of AI-driven sustainability as experienced across expert domains.

Theme Development

The iterative coding process produced five major themes per interview question, yielding 25 overarching themes in total. These themes reflect convergent experiences among the three participant groups, highlighting both shared perceptions and disciplinary distinctions. Each theme was then substantiated with verbatim participant quotations, chosen for their richness and representational clarity.

DATA SYNTHESIS AND INTERPRETATION

Data integration proceeded through a cross-domain synthesis, juxtaposing the experiences of the three expert groups. This analytic triangulation allowed identification of both disciplinary divergences, for example, AI engineers' focus on algorithmic performance versus logisticians' emphasis on mission efficiency, and cross-sectoral convergences, such as shared ethical concerns about environmental impact and resource accountability.

The synthesis process led to a unified interpretation of AI-driven green logistics as an emerging institutional paradigm, characterized by three interlocking dimensions:

1. Technological Synergy: The integration of AI, IoT, and blockchain as mutually reinforcing mechanisms for optimization and transparency.
2. Organizational Evolution: The cultural and structural transformation of defense and federal logistics institutions toward sustainability leadership.
3. Ethical Imperative: The reframing of logistics as both a technical and moral enterprise in service of public health and planetary well-being.

This methodological design, anchored in phenomenology and guided by expert testimony, allowed for a multidimensional exploration of AI-enabled green logistics in defense and governmental contexts. By combining rigorous qualitative analysis with deep domain expertise, the study not only illuminates how sustainability is experienced within complex institutional systems but also lays the empirical foundation for future policy, design, and ethical frameworks aligned with the United Nations Sustainable Development Goals.

The voices of logisticians, green innovators, and AI engineers collectively portray an evolving ecosystem in which data-driven intelligence meets environmental responsibility, redefining how national security, technological advancement, and global sustainability intersect.

Interview Question 1:

How have you observed AI, IoT, or blockchain technologies being integrated into military or federal logistics operations?

Theme 1: AI as an Engine of Predictive Efficiency

Participants across groups emphasized that AI's primary contribution lies in predictive optimization, forecasting maintenance needs, anticipating supply bottlenecks, and aligning resource use with mission requirements. This predictive intelligence was consistently associated with improved efficiency, reduced emissions, and cost savings—aligning with SDG 9 (Industry, Innovation, and Infrastructure).

"AI's become like the brain of our operations. It tells us what needs fixing before it breaks and which routes burn less fuel. It's not just data, it's foresight." (Participant M1)

"Before AI, we were reactive. Now, the systems predict maintenance down to the hour, it's cutting waste and downtime at a scale we never managed manually." (Participant A2)

Theme 2: Blockchain as a Tool for Transparency and Trust

Participants highlighted blockchain's growing role in securing supply chain integrity, ensuring the traceability of materials, and verifying sustainable sourcing. Within military procurement systems, blockchain was seen as both a compliance mechanism and a credibility enhancer in demonstrating adherence to UN sustainability standards.

"We're using blockchain to track where every component comes from. If a supplier says their materials are sustainable, now we can verify it instead of taking their word for it." (Participant G3)

"It's huge for accountability. In federal contracts, transparency isn't optional, it's essential. Blockchain keeps everyone honest, especially around green sourcing." (Participant M4)

Theme 3: IoT-Driven Environmental Monitoring and Real-Time Awareness

Internet of Things (IoT) devices were widely cited as the "eyes and ears" of green logistics, capturing continuous environmental and operational data. Participants described IoT networks that monitor emissions, vehicle performance, and temperature-sensitive cargo, aligning logistics execution with SDG 13 (Climate Action).

"IoT sensors are everywhere now, on engines, containers, even pallets. They tell us if emissions spike or if fuel efficiency drops, so we can fix it fast." (Participant A1)

"For us, IoT means visibility. We know instantly when a shipment's sitting too long or burning too much diesel. That's where sustainability meets readiness." (Participant M5)

Theme 4: The Convergence of Data Systems for Integrated Decision-Making

Participants reported that the true innovation lies in the integration of AI, IoT, and blockchain systems into unified operational platforms. These technologies collectively enable evidence-based decision-making, reduce redundancies, and create a data ecosystem capable of supporting both environmental and strategic objectives.

"AI makes sense of what IoT collects and blockchain verifies. Together, it's a closed loop, you can trust the data, act on it, and prove you did the right thing." (Participant G2)

"It's all connected now. The AI dashboard pulls data from sensors and supply chains, so you can plan missions while watching your carbon footprint drop in real time." (Participant A4)

Theme 5: The Emerging Digital Ethos in Sustainable Defense Operations

Experts described a shifting cultural narrative in which digital technologies are reframing what it means to be efficient, ethical, and environmentally responsible in military and governmental logistics. AI and related tools were viewed not just as operational upgrades, but as moral instruments redefining defense readiness in sustainable terms.

"There's a new mindset forming, being mission-ready now includes being climate-ready. AI just makes that measurable." (Participant M2)

"Digital ethics are becoming environmental ethics. When we talk AI in logistics, we're really talking about the kind of future we're building." (Participant G4)

Synthesis for Question 1

The five themes collectively reveal a deep transformation of how participants conceptualize the role of AI, IoT, and blockchain within federal and military logistics. These technologies have evolved beyond tools of efficiency to become strategic instruments of accountability, sustainability, and foresight. Participants viewed them as enablers of an emerging "digital ecology" within logistics, one that merges operational precision with environmental responsibility, thus operationalizing the vision of SDG 9 (Innovation) and SDG 13 (Climate Action) in tangible, measurable ways.

Interview Question 2:

What challenges or opportunities have arisen from using AI to enhance sustainability and align with the UN Sustainable Development Goals (SDGs)?

Theme 1: The Dual-Edged Nature of Technological Dependence

Participants described artificial intelligence as both an enabler and a potential constraint in the evolution of sustainable logistics. While AI significantly enhances predictive capacity and efficiency, it simultaneously introduces new dependencies on digital infrastructure, cyber-resilience, and uninterrupted data flow. This duality produces a paradox in which sustainability relies on technologies that themselves require substantial energy and maintenance resources.

"AI gives us precision, but it also gives us dependency. If the system goes down or feeds us bad data, we're grounded. That's not just an inconvenience; it's a mission risk." (Participant M3)

"We're saving fuel and cutting emissions with AI, sure; but the tech has to stay up and secure. You can't fight climate change if your servers go offline in a storm." (Participant A5)

This theme underscores that digital sustainability is contingent upon technological resilience; the same systems that promise efficiency can create new operational vulnerabilities.

Theme 2: Data Quality and Ethical Governance as Cornerstones of Sustainable AI

Across sectors, participants stressed that the effectiveness of AI-driven sustainability depends upon the integrity, transparency, and ethical governance of data. Without rigorous validation and oversight, algorithms can misrepresent environmental impacts or inflate progress toward SDGs, undermining institutional trust.

"Bad data means bad sustainability. If your carbon tracking model is running off incomplete info, you're basically greenwashing by accident." (Participant G1)

"AI's only as ethical as the data it eats. In logistics, if emissions or supply inputs are misreported, that throws off everything, from SDG metrics to procurement planning." (Participant M4)

Participants noted that new regulatory frameworks and coercive pressures are compelling agencies to professionalize data ethics as a central pillar of environmental accountability.

Theme 3: Human–Machine Collaboration and the Changing Nature of Expertise

A recurrent insight concerned the transformation of expertise as human judgment becomes interwoven with machine learning outputs. Participants described this shift as a learning curve that requires both cognitive adaptation and institutional retraining.

"We've got younger analysts who trust the AI too much and old-school officers who don't trust it at all. We're still figuring out that balance." (Participant M2)

"AI isn't replacing human judgment, it's reshaping it. We're learning to ask different questions now, not just push buttons." (Participant A3)

This theme illustrates a redefinition of professional competence: sustainability now hinges on the capacity to interpret, question, and ethically apply AI insights rather than merely execute preprogrammed directives.

Theme 4: Resource Optimization and Environmental Accountability as Mutual Gains

Participants repeatedly emphasized that AI facilitates a convergence between operational efficiency and ecological responsibility. Predictive analytics, adaptive routing, and real-time emissions monitoring generate tangible cost savings while directly reducing carbon intensity.

"AI gives you both, cost savings and cleaner air. When the trucks don't idle and the planes fly smarter routes, it's green and it's good business." (Participant G4)

"We've learned that fuel efficiency and environmental responsibility aren't opposites anymore. AI turned that myth on its head." (Participant M1)

The theme encapsulates how technological opportunity repositions sustainability from a moral aspiration to a measurable, financially rational objective consistent with SDG 9 (Industry, Innovation, and Infrastructure) and SDG 13 (Climate Action).

Theme 5: Institutional Resistance and the Pace of Cultural Change

Even amid technological optimism, participants identified bureaucratic inertia and hierarchical rigidity as enduring obstacles. Organizational cultures steeped in procedural tradition often impede the rapid adaptation necessary to leverage AI for sustainability.

"We've got all the tools, but sometimes it feels like we're paddling against a current of paperwork. Policy moves slower than technology." (Participant M5)

"There's still this fear that AI means loss of control. Until leadership sees sustainability as mission-critical, change will keep stalling." (Participant G5)

This dynamic reflects the equilibrium described in Force Field Analysis: driving forces such as innovation and policy mandates remain constrained by entrenched administrative structures and risk aversion.

Synthesis for Question 2

Together, these five themes portray a complex ecosystem of technological promise tempered by institutional constraint. Participants recognized AI as a vehicle for aligning defense and federal logistics with the UN SDGs, yet they also cautioned that sustainable transformation requires more than sophisticated algorithms; it demands ethical data governance, human adaptability, and cultural modernization.

Interview Question 3:

In your view, how do AI applications contribute to reducing emissions, optimizing resources, or improving public health outcomes?

Theme 1: Predictive Analytics as a Mechanism for Emission Reduction

Participants unanimously identified predictive analytics as one of AI's most transformative contributions to emission reduction. Through real-time modeling and adaptive forecasting, AI systems anticipate inefficiencies, enabling interventions that curtail unnecessary fuel consumption and reduce operational emissions. This technological foresight was viewed as the practical embodiment of SDG 13 (Climate Action) within military and federal supply chains.

"AI has changed the way we think about emissions. It's no longer a guessing game; predictive models tell us exactly when to adjust operations before we start wasting fuel." (Participant A2)

"We've cut idle time by almost a third just by using AI-based route optimization. That's not just fuel saved, that's air quality improved around our bases." (Participant M5)

The theme underscores AI's pivotal role in transforming emissions management from reactive compliance into proactive prevention, turning sustainability into a dynamic, data-driven practice.

Theme 2: Optimization Algorithms and Intelligent Resource Allocation

Participants described how AI algorithms, particularly those embedded in logistics routing and asset management, have revolutionized resource allocation. By optimizing delivery routes, load configurations, and maintenance cycles, AI reduces waste while maximizing output per resource unit. These adaptive systems were perceived as key tools in achieving operational excellence aligned with SDG 9 (Industry, Innovation, and Infrastructure).

"Our AI systems schedule deliveries in a way humans simply can't. It calculates the most efficient use of every gallon of fuel and every flight hour." (Participant G2)

"In the past, we'd send half-full convoys just to meet deadlines. Now, AI consolidates loads automatically, it's saving resources and lowering our footprint." (Participant M1)

This theme highlights the merging of logistical optimization with environmental ethics, illustrating how computational efficiency translates into tangible sustainability gains.

Theme 3: AI-Enhanced Maintenance and Equipment Longevity

A recurrent insight across all participant groups was that AI-enabled predictive maintenance not only improves operational reliability but also extends equipment lifespan, thereby reducing resource extraction and manufacturing emissions. The participants characterized this as a "quiet revolution" in sustainability, where unseen improvements yield measurable environmental dividends.

> *"AI tells us when something's about to fail; sometimes days before any human could spot it. That's fewer breakdowns, fewer parts wasted, and fewer emissions from repairs." (Participant A1)*
>
> *"For aircraft, predictive maintenance is huge. If you can keep engines in peak shape longer, you're saving millions of dollars and cutting emissions you'd never see otherwise." (Participant M3)*

The theme demonstrates how maintenance optimization not only reduces downtime but directly supports global efforts to limit industrial waste and overproduction, resonating with circular economy principles.

Theme 4: Public Health as an Operational Outcome of Green Logistics

Participants consistently linked environmental sustainability in logistics to measurable improvements in public health outcomes, particularly in communities near transportation hubs, bases, and depots. Cleaner air and reduced noise pollution were described as indirect but significant byproducts of AI-optimized logistics.

> *"It's not just the numbers, it's the people. Every reduction in emissions means cleaner air for the families living near our transport corridors." (Participant G4)*
>
> *"You don't think about it until you see the asthma rates drop in areas around bases where we've switched to electric or AI-optimized fleets." (Participant M2)*

This theme reframes green logistics as a form of environmental public health intervention, where technological innovation produces immediate, localized benefits to human well-being, aligning with SDG 3 (Good Health and Well-being).

Theme 5: Real-Time Data Integration as the Foundation of Sustainable Decision-Making

Participants emphasized that the convergence of AI with IoT and blockchain technologies has made real-time environmental awareness possible. The integration of these systems allows decision-makers to visualize the ecological impact of operations instantaneously, fostering accountability and transparency.

"Now, our dashboards show emissions in real time. You can literally see the carbon cost of every decision as you make it." (Participant A4)

"Blockchain gives the transparency, IoT provides the data, and AI makes sense of it all. It's a closed feedback loop for sustainability." (Participant G1)

This theme situates AI within a broader technological ecosystem that redefines how environmental responsibility is operationalized, ensuring that sustainability becomes a measurable, continuous process rather than a static compliance goal.

Synthesis for Question 3

The five themes collectively reveal that AI is not simply a tool for logistical efficiency but a transformative mechanism for environmental stewardship and public health advancement. Participants portrayed AI systems as active agents that convert data into ethical and ecological action. Predictive analytics, optimization algorithms, and real-time monitoring jointly reduce emissions, conserve resources, and improve community health conditions.

Interview Question 4:

How do organizational culture, leadership, or institutional policy shape the adoption of green logistics technologies?

Theme 1: Leadership Vision as the Catalyst for Sustainable Transformation

Participants unanimously identified leadership commitment as the most critical determinant of successful AI and green logistics integration. Visionary leaders were described as those who embed sustainability into the organizational mission, ensuring it is perceived not as an external mandate but as a strategic imperative that aligns with national security and operational resilience.

"When the top brass talk about sustainability the same way they talk about readiness, people start paying attention. It stops being a side project and becomes the mission." (Participant M1)

"You can have all the tech in the world, but if leadership isn't setting the tone for change, it just gathers dust. Green logistics starts at the top." (Participant G3)

This theme emphasizes that leadership vision functions as the primary "unfreezing" force in the Force Field Analysis framework, initiating organizational momentum toward sustainability.

Theme 2: Cultural Inertia and Resistance to Technological Change

Participants consistently described entrenched cultural norms and bureaucratic rigidity as the greatest internal barriers to adopting AI-driven sustainability tools. In both government and military contexts, hierarchical traditions often produce reluctance to challenge established routines or embrace algorithmic decision-making.

"There's still that 'if it's not broke, don't fix it' mentality. AI feels foreign to a lot of folks who've done logistics by the book for thirty years." (Participant M4)

"We're trying to green systems that were designed in another era, paper-based, fuel-heavy, slow to adapt. Culture moves slower than code." (Participant G2)

This theme illustrates how institutional inertia acts as a restraining force, impeding innovation despite widespread awareness of its environmental and strategic benefits.

Theme 3: Policy Frameworks as Enablers of Accountability and Action

Several participants emphasized that policy interventions, particularly those aligning with international environmental accords and federal sustainability directives, serve as coercive mechanisms that institutionalize green logistics practices. Clear policy frameworks were seen as essential for translating leadership rhetoric into measurable, enforceable action.

"Policy gives sustainability teeth. Without a formal directive or metrics to report against, good ideas just fade into the background." (Participant A1)

"When sustainability gets written into performance contracts or federal procurement policies, people start taking it seriously." (Participant M2)

This theme highlights how Institutional Theory's coercive and normative pressures translate sustainability ideals into operational reality through formalized governance structures.

Theme 4: Cross-Domain Collaboration as a Cultural Imperative

Participants described sustainable logistics as inherently interdisciplinary, requiring alignment between logisticians, environmental scientists, policymakers, and AI engineers. The success of such collaboration depends on fostering a culture of shared purpose, what one participant likened to "everyone paddling in the same canoe."

> *"You can't have AI engineers talking algorithms and logisticians talking cargo if they're not in sync. Everyone's got to be in the same canoe, paddling in the same direction." (Participant G5)*
>
> *"We've learned that sustainability isn't a single department's job. It's a cultural thing, everyone contributes, or the system drifts off course." (Participant A4)*

This theme reflects the Canoe Theory of Organizational Alignment, emphasizing that success in green transformation requires synchronized vision, communication, and participation across all levels of the organization.

Theme 5: Institutional Learning and the Evolution of Ethical Identity

Participants expressed that the process of adopting AI-based green logistics technologies is also transforming organizational identity. The integration of sustainability has fostered an emerging culture of ethical awareness, where ecological responsibility and operational success are no longer viewed as competing values but as complementary measures of institutional excellence.

> *"We're seeing a shift from compliance to conscience. People don't just want to meet the standards anymore, they want to do what's right for the planet." (Participant G1)*
>
> *"Our culture's starting to see sustainability as part of the mission, not a distraction from it. That's a big change from even five years ago." (Participant M3)*

This theme suggests that as institutions internalize sustainability, they undergo a moral reorientation, embedding environmental ethics into their collective identity and performance ethos.

Synthesis for Question 4

The five themes reveal that organizational culture, leadership, and policy form a triadic foundation upon which sustainable transformation depends. Leaders act as visionaries and catalysts, culture determines the pace and depth of adoption, and policy ensures accountability and structure. Together, they create the conditions for a shift from reactive environmental compliance to proactive ecological leadership.

Participants' lived experiences indicate that AI-enabled green logistics is not simply a technological innovation but a cultural evolution, a process through which institutions redefine themselves in moral, operational, and strategic terms.

Interview Question 5:

What future directions do you envision for AI-enabled sustainability in defense or federal supply chains?

Theme 1: The Expansion of Autonomous and Adaptive Logistics Systems

Participants consistently anticipated a future in which logistics systems become increasingly autonomous, capable of self-adjusting in real time to optimize efficiency, reduce waste, and minimize emissions. They described an emerging vision of "living logistics networks" where AI algorithms continuously learn from data patterns to anticipate operational needs.

"The next step is autonomy, AI that doesn't just tell us what to do but acts on it. Self-adjusting convoys, drones that reroute for weather and fuel, it's all coming." (Participant M2)

"We're moving toward logistics that think for themselves, systems that can react instantly to environmental or supply changes without waiting for human input." (Participant A3)

This theme illustrates how automation and adaptive learning may soon redefine logistical responsiveness, enhancing both sustainability and readiness through data-driven self-regulation.

Theme 2: Integration of Renewable Energy and AI-Optimized Power Grids

Participants projected that AI will play a pivotal role in managing renewable energy integration into defense and federal infrastructure. Smart grids powered by AI analytics are expected to optimize energy use across bases, fleets, and facilities, reducing dependence on fossil fuels while ensuring operational stability.

"I see AI controlling microgrids, solar, wind, battery systems, all optimized to keep bases running green and resilient." (Participant G1)

"Our biggest carbon hits come from energy use, not just transport. AI-managed renewables will make clean power reliable enough for full-scale operations." (Participant M5)

This theme connects directly to SDG 13 (Climate Action) by envisioning AI as a control mechanism that balances efficiency with ecological responsibility in large-scale, mission-critical infrastructures.

Theme 3: Cross-Sector Collaboration and Civil–Military Technology Exchange

Participants envisioned a future characterized by deeper collaboration between defense agencies, private industry, and academic institutions. Such cross-sector partnerships were seen as vital for accelerating innovation and scaling AI-enabled sustainability practices beyond the military domain into civilian supply chains.

"We're going to see a lot more collaboration, military sustainability tech flowing into civilian logistics and vice versa. That's how innovation spreads." (Participant G5)

"No single sector can own sustainability. We'll need partnerships with universities, tech firms, and even competitors to get there." (Participant A5)

This theme highlights a collective vision for an interconnected ecosystem of innovation, where defense-led breakthroughs catalyze broader societal progress toward the UN's global sustainability goals.

Theme 4: AI-Driven Circular Logistics and Lifecycle Accountability

A number of participants emphasized the future importance of AI in supporting circular economy practices, particularly in tracking material lifecycles, optimizing reuse, and minimizing waste. They envisioned AI systems capable of auditing entire supply chains to ensure materials are repurposed efficiently and ethically.

"We're looking at AI that can trace every part from manufacture to retirement, so nothing gets lost or wasted, it all loops back into production." (Participant G2)

"Circular logistics is the next frontier. AI will let us map materials across decades, making sustainability measurable from start to finish." (Participant M1)

This theme aligns with the Institutional Theory perspective that innovation is driven by normative and coercive pressures, as organizations adopt circular systems to meet evolving ethical and regulatory expectations.

Theme 5: The Human Dimension of AI and the Rise of Ethical Stewardship

Participants cautioned that as AI systems grow in power and autonomy, ethical stewardship must evolve in parallel. Future success will depend on cultivating human oversight grounded in environmental ethics, ensuring AI applications remain accountable to the broader values of public health and planetary well-being.

"The more we automate, the more we'll need humans to guide the ethics. AI can't decide what's right, it just optimizes." (Participant M3)

"We've got to build AI that reflects our values, not just our algorithms. If sustainability is the mission, ethics has to be the compass." (Participant A2)

This theme underscores that the future of AI-enabled sustainability hinges not only on technical mastery but on moral discernment, embedding human values within machine learning frameworks to support SDG 3 (Good Health and Well-being) and SDG 13 (Climate Action).

Synthesis for Question 5

The five themes collectively paint a vision of AI-enabled sustainability as both technological evolution and ethical revolution. Participants described a near future in which autonomous systems, renewable energy integration, and circular logistics frameworks converge to redefine efficiency and responsibility in defense operations. Yet, they also emphasized that this transformation must remain anchored in human ethics and cross-sector cooperation.

CONCLUSION AND RECOMMENDATIONS

This phenomenological inquiry has illuminated the evolving intersection of artificial intelligence (AI), green logistics, and sustainability practices within the U.S. military and federal logistics ecosystem. Drawing upon the lived experiences of fifteen experts across military logistics, green logistics, and AI engineering, the study reveals a multidimensional transformation underway, technological, organizational, and ethical in nature.

The findings demonstrate that AI-enabled green logistics is not merely an operational innovation but a paradigmatic shift in how defense and government institutions conceptualize readiness, efficiency, and environmental responsibility. Across participants, the integration of AI, IoT, and blockchain technologies has generated new capacities for predictive analytics, transparency, and real-time sustainability tracking, positioning digital systems as both instruments of precision and agents of ethical accountability.

From an institutional perspective, this transformation reflects the dynamic interplay of coercive, normative, and mimetic pressures identified in Institutional Theory, which collectively compel organizations to align with the United Nations Sustainable Development Goals (SDGs)—notably SDG 3 (Good Health and Well-being), SDG 9 (Industry, Innovation, and Infrastructure), and SDG 13 (Climate Action). Meanwhile, Force Field Analysis elucidates the tension between driving forces (technological innovation, leadership vision, environmental mandates) and restraining forces (bureaucratic inertia, cultural resistance, and ethical uncertainty). These tensions reveal that sustainable transformation requires deliberate equilibrium rather than passive adaptation.

Finally, the Canoe Theory of Organizational Alignment provides a resonant metaphor for sustainability implementation: progress depends upon collective synchronization, where policymakers, logisticians, AI engineers, and environmental professionals "paddle in the same direction." When alignment occurs, AI becomes not simply a computational tool but an institutional compass guiding the transition toward ecological stewardship and operational resilience.

Ultimately, this study affirms that AI-driven green logistics represents both a technological evolution and a moral revolution. It embodies a future in which military and government operations advance not only strategic efficiency but the public good, protecting planetary health, supporting human well-being, and redefining national security as environmental security.

Practical and Actionable Recommendations

Based on the findings and their alignment with the theoretical frameworks, the following ten actionable recommendations are offered to military, governmental, and policy stakeholders seeking to operationalize AI-enabled green logistics in support of the UN Sustainable Development Goals.

1. Institutionalize AI-Enabled Sustainability Mandates

Establish formal policies requiring that all military and federal logistics programs integrate AI-based sustainability metrics. These mandates should explicitly align with SDG benchmarks, ensuring measurable progress toward emission reduction, resource efficiency, and renewable energy integration.

Practical Example: Incorporate AI-driven carbon tracking dashboards into logistics command centers to quantify emissions per mission or operation in real time.

2. Create a Centralized "Green Logistics Command"

Develop a federal or interagency body responsible for coordinating AI-enabled sustainability initiatives across defense, transportation, and supply chain sectors. This entity would standardize data governance, set sustainability performance indicators, and oversee cross-sector collaboration.

Practical Example: A Department of Defense "Sustainable Operations Task Force" could consolidate data streams from IoT devices, AI analytics, and blockchain systems to guide real-time resource optimization.

3. Embed Environmental Ethics in AI Governance

Establish ethical oversight committees within defense and federal agencies to evaluate the moral implications of AI systems. These bodies should ensure that algorithmic decision-making supports public health and ecological justice while maintaining transparency and accountability.

Practical Example: Introduce an AI Ethics Review Protocol requiring sustainability impact assessments for all machine learning systems before deployment.

4. Strengthen Data Integrity and Transparency Frameworks

Implement mandatory blockchain verification and data quality standards for all sustainability reporting. Data governance should emphasize accuracy, traceability, and interoperability to prevent unintentional "greenwashing" or data manipulation.

Practical Example: Use blockchain-based supply chain certification to verify the provenance and recyclability of defense procurement materials.

5. Advance Predictive Analytics and Real-Time Emission Control

Expand the use of predictive AI models that forecast emissions, optimize fleet efficiency, and identify high-risk energy consumption zones. These systems can serve as automated early warning mechanisms to reduce environmental impact before it occurs.

Practical Example: Deploy AI tools that autonomously adjust aircraft routing to minimize fuel burn during atmospheric disturbances.

6. Foster Leadership Development in Sustainability Strategy

Integrate sustainability competencies into military and federal leadership training programs. Leaders should be equipped to interpret AI insights, manage interdisciplinary teams, and champion sustainability as a component of operational excellence.

Practical Example: Include sustainability and AI integration modules in the curricula of war colleges, defense universities, and executive leadership academies.

7. Cultivate a Collaborative "Canoe Culture" Across Domains

Encourage cross-disciplinary collaboration between logisticians, data scientists, environmental experts, and policymakers. Shared mission alignment, "paddling in the same direction"—is essential for system-wide transformation.

Practical Example: Establish annual joint workshops or "Green Tech War Games" where mixed teams co-design sustainable logistics scenarios using AI tools.

8. Promote Circular Logistics and Lifecycle Accountability

Leverage AI for full-spectrum lifecycle tracking of materials, from procurement to retirement. Such systems should support reuse, recycling, and repurposing initiatives consistent with circular economy principles.

Practical Example: Use AI-based digital twins to model the environmental footprint of equipment across its lifecycle, identifying reuse or recycling opportunities.

9. Enhance Cyber-Resilience and Digital Sustainability

Recognize that AI systems depend on secure, energy-efficient digital infrastructure. Invest in resilient cloud networks, green data centers, and redundant systems to prevent technological disruptions that could compromise sustainability objectives.

Practical Example: Retrofit existing data centers with renewable energy systems and AI-driven cooling optimization to reduce energy consumption.

10. Institutionalize Continuous Learning and Evaluation

Adopt a feedback-oriented approach where sustainability outcomes are continuously monitored, evaluated, and refined through AI analytics. Institutional learning mechanisms should ensure adaptive improvement rather than static compliance.

Practical Example: Implement "Sustainability Readiness Reviews" alongside operational audits, using AI dashboards to assess environmental performance in real time.

In advancing these recommendations, defense and government institutions have the opportunity to become exemplars of the UN Sustainable Development Goals, not merely through compliance, but through innovation that unites operational readiness with planetary stewardship. The future of logistics, as envisioned by the participants, is thus both autonomous and accountable, efficient and ethical, national and global; a living system in which AI intelligence serves human conscience for the preservation of the world we defend.

REFERENCES

Blanco, E. E., & Sheffi, Y. (2017). Green logistics. *Sustainable supply chains: A research-based textbook on operations and strategy*, 147-187.

Blanco, E. E., & Sheffi, Y. (2017). Green logistics. *Sustainable supply chains: A research-based textbook on operations and strategy*, 147-187.

Blanco, E. E., & Sheffi, Y. (2024). Green logistics. In *Sustainable supply chains: A research-based textbook on operations and strategy* (pp. 101–141). Springer International Publishing. DOI: 10.1007/978-3-031-45565-0_5

Capatina, A., Bleoju, G., Matos, F., & Vairinhos, V. (2017). Leveraging intellectual capital through Lewin's Force Field Analysis: The case of software development companies. *Journal of Innovation & Knowledge*, *2*(3), 125–133. DOI: 10.1016/j.jik.2016.07.001

Chen, W., Men, Y., Fuster, N., Osorio, C., & Juan, A. A. (2024). Artificial intelligence in logistics optimization with sustainable criteria: A review. *Sustainability (Basel)*, *16*(21), 9145. DOI: 10.3390/su16219145

Eslami, R., & Khoveyni, M. (2024). Optimizing Sustainable Green Logistics with AI-Enhanced RAM-Hybrid Models: Addressing Economic and Crisis Management Challenges. *Journal of Industrial Engineering International*, *20*, 34.

Hibbard, D., Hibbard, M., & Stockman, J. (2006). The Canoe Theory: A Business Success Strategy for Leaders and Associates . iUniverse.

Hussain, K. (2025). Revolutionizing Route Optimization Systems with Artificial Intelligence for a Smarter, Sustainable Logistics Ecosystem. *International Journal of Computer Science and Mobile Computing*, *14*(2), 66–68. DOI: 10.47760/ijcsmc.2025.v14i02.008

Kale, A. A. (2025). Modern AI in Green Logistics: Smart Systems for Sustainability and Carbon Reduction. *Green Logistics: Smart Systems for Sustainability and Carbon Reduction (March 01, 2025). Global Journal of Emerging AI and Computing*, *1*(2), 147–169. DOI: 10.70445/gjeac.1.2.2025.147-169

Kumar, A. (2015). Green Logistics for sustainable development: An analytical review. *IOSRD International Journal of Business*, *1*(1), 7–13.

McKinnon, A., Browne, M., Whiteing, A., & Piecyk, M. (Eds.). (2015). *Green logistics: Improving the environmental sustainability of logistics*. Kogan Page Publishers.

Nicoletti, B., & Appolloni, A. (2024). Green Logistics 5.0: A review of sustainability-oriented innovation with foundation models in logistics. *European Journal of Innovation Management*, *27*(9), 542–561. DOI: 10.1108/EJIM-07-2024-0787

Nozari, H. (2024). Green supply chain management based on artificial intelligence of everything. *Journal of Economics and Management*, *46*, 171–188. DOI: 10.22367/jem.2024.46.07

Rodrigue, J. P., Slack, B., & Comtois, C. (2017). Green logistics. In *Handbook of logistics and supply-chain management* (Vol. 2, pp. 339–350). Emerald Group Publishing Limited.

Rudko, I., Bashirpour Bonab, A., Fedele, M., & Formisano, A. V. (2025). New institutional theory and AI: Toward rethinking of artificial intelligence in organizations. *Journal of Management History*, *31*(2), 261–284. DOI: 10.1108/JMH-09-2023-0097

United Nations. (2024). The Sustainable Development Goals. Retrieved from: https://www.un.org/sustainabledevelopment/

U.S. Department of Defense. (2022) 2022 Department of Defense Sustainability Plan. Retrieved from: https://www.sustainability.gov/pdfs/dod-2022-sustainability-plan.pdf

U.S. Department of Defense. (2024). *Green Procurement Compilation*. Retrieved from https://sftool.gov/greenprocurement/dod

Yingfei, Y., Mengze, Z., Zeyu, L., Ki-Hyung, B., Avotra, A. A. R. N., & Nawaz, A. (2022). Green logistics performance and infrastructure on service trade and environment-measuring firm's performance and service quality. *Journal of King Saud University. Science*, *34*(1), 101683. DOI: 10.1016/j.jksus.2021.101683

Chapter 8
The Potential of Artificial Intelligence (AI) as a Proactive Tool in Public Health Surveillance

Amalisha Sabie Aridi
http://orcid.org/0000-0002-7869-5530
Washington University of Science and Technology, USA

ABSTRACT

COVID-19 exposed structural fragilities in the United States' public health logistics and supply chain systems, particularly in the domains of diagnostic testing, epidemiological tracking, and coordinated response. These deficiencies underscore the urgent need for more adaptive, data-centric surveillance infrastructures capable of operating at scale and in real time. Artificial intelligence–enabled analytics, when integrated with existing public health systems, offer a viable pathway to enhance situational awareness, optimize resource allocation, and improve the timeliness of intervention strategies. In particular, wastewater-based epidemiology; augmented by AI-driven pattern recognition and predictive modeling; presents a scalable and cost-effective mechanism for population-level health monitoring, capable of detecting emerging threats prior to clinical reporting. This study advances that premise by examining how such integrative approaches can strengthen logistical decision-making and supply chain responsiveness.

DOI: 10.4018/979-8-2600-1026-6.ch008

INTRODUCTION

Virus testing during a pandemic is an indispensable element of any public health strategy aimed at mitigating widespread transmission, minimizing the need for stringent social distancing, and sustaining economic continuity (Heilweil, 2021). For instance, during the early months of COVID-19, nations such as South Korea demonstrated how widespread and rapid testing, supported by digital tracking tools, allowed for precise containment without full-scale lockdowns. When these testing programs are integrated with AI-driven contact tracing systems, accurate and timely test results can guide evidence-based decisions about who should isolate and who may safely return to work or communal activities (Heilweil, 2021).

Artificial intelligence amplifies traditional epidemiological methods by expediting diagnostic interpretation and improving accuracy through machine learning models trained on vast datasets of test results and clinical indicators. For example, AI-based predictive algorithms have been used to detect asymptomatic COVID-19 carriers by analyzing subtle physiological markers such as respiratory rate, heart rate variability, and skin temperature captured from wearable devices like Fitbit or Apple Watch (Kester, 2021; Hughes, 2021). This enables early isolation of infectious individuals before they spread the virus. At a broader scale, AI-assisted diagnostic analytics produce macro-level epidemiological insights revealing real-time trends in virus prevalence, mutation patterns, and geographic transmission dynamics (Burrell, 2022). Such analytics were instrumental in mapping the global spread of variants like Delta and Omicron.

A growing body of evidence shows that many individuals infected with COVID-19 and similar respiratory viruses remain asymptomatic or exhibit only mild symptoms, making AI-enhanced testing and surveillance tools indispensable (Burrell, 2022). These systems can analyze multimodal data streams, from clinical laboratory reports and hospital admissions to anonymized mobility data and even social media discourse, to identify emerging clusters of infection. For instance, a sudden surge in tweets mentioning "loss of smell" during the COVID-19 outbreak provided early warning of localized spikes in cases. As these AI models improve in reliability and scope, public health authorities can deploy more targeted interventions, such as mobile testing units or localized vaccination drives, reducing the need for sweeping lockdowns (Kester, 2021; Hughes, 2021). During the Omicron surge, for example, AI-powered demand forecasting tools could have optimized the distribution of test kits and laboratory resources, avoiding the backlogs that delayed containment efforts (Heilweil, 2021).

Artificial intelligence also plays a transformative role in biosecurity and biodefense preparedness (Zeng et al., 2021; Muhammad et al., 2025; Thiébaut & Cossin, 2019). During COVID-19, fragmented data systems hampered timely responses and interagency coordination (Heilweil, 2021). AI-driven data fusion platforms can overcome these barriers by integrating information from hospital databases, genomic sequencing centers, environmental sensors, and mobility networks into unified situational awareness dashboards (Zeng et al., 2021; Muhammad et al., 2025). For instance, an AI model combining wastewater viral detection data with local hospital admissions can predict case surges days before symptoms appear in the population. Such systems enable policymakers to visualize outbreak trajectories, allocate medical resources efficiently, and anticipate hospital capacity needs (Burrell, 2022). In biodefense contexts, AI algorithms can analyze genomic signatures to differentiate naturally occurring pathogens from those that might be synthetically engineered, a critical function in countering bioterrorism threats (Heilweil, 2021).

Historically, public health surveillance has relied on siloed laboratory networks and labor-intensive manual data processing (Heilweil, 2021). AI is redefining this paradigm by automating data extraction, normalization, and predictive modeling across disparate laboratory and clinical systems. For instance, natural language processing (NLP) tools can scan thousands of physicians' notes to identify new symptom clusters, such as "persistent cough" or "shortness of breath," that may indicate the emergence of an unrecognized respiratory pathogen. Similarly, computer vision algorithms can evaluate radiographic images to detect early signs of viral pneumonia, dramatically reducing diagnostic turnaround time (Zeng et al., 2021; Muhammad et al., 2025). These advances enable continuous, adaptive epidemiological modeling that supports real-time response to evolving threats (Thiébaut & Cossin, 2019).

However, the effective integration of AI into public health bioinformatics requires robust policy alignment and long-term investment in digital infrastructure. Governments must create frameworks for data sharing that balance privacy with the collective need for public safety (Zeng et al., 2021; Muhammad et al., 2025). Ethical AI governance is also essential to prevent algorithmic bias, such as underrepresentation of vulnerable communities in datasets, which can lead to unequal access to testing and care. Cross-sector collaboration among governments, academia, and private technology firms will be fundamental to developing a resilient and equitable public health intelligence network (Kester, 2021; Hughes, 2021).

Contact tracing remains a cornerstone of epidemiological control, identifying and monitoring individuals exposed to infectious diseases (Kester, 2021). AI magnifies its efficiency through automated data integration and predictive modeling. For example, machine learning systems can infer probable transmission chains by analyzing anonymized smartphone mobility data and digital payment histories, enabling real-time risk stratification (Zeng et al., 2021; Muhammad et al., 2025). During an outbreak in Singapore, for instance, AI models simulated the effects of quarantine and vaccination policies, helping the government refine its containment strategies (Thiébaut & Cossin, 2019). As a result, public health interventions become proactive, data-driven measures rather than reactive crisis responses.

Public health surveillance aims to track and analyze disease trends across populations (Groseclose & Buckridge, 2016). AI revolutionizes this mission by continuously processing large-scale health data in real time (Zeng et al., 2021; Muhammad et al., 2025). As Devriendt, Borry, and Shabani (2021) note, "data sharing in science maximizes the utility and impact of patient data and therefore contributes to improving clinical practice and public health." By merging AI, epidemiology, and informatics, institutions can now analyze complex datasets, ranging from genomic sequences and electronic health records to environmental sensor readings and even social media sentiment, to detect weak epidemiological signals before they escalate (Friis, 2018, p. 82). This capacity enables early warning systems capable of flagging potential outbreaks before they become unmanageable (Zeng et al., 2021; Muhammad et al., 2025).

Nevertheless, significant obstacles persist. Fragmented data systems and lack of interoperability continue to delay public health responses (Kester, 2021; Hughes, 2021). To address this, standardized AI-driven data pipelines are needed to facilitate continuous learning and adaptation (Zeng et al., 2021). Federated learning frameworks, for instance, enable AI models to train on decentralized datasets from different hospitals or countries without compromising patient privacy; creating scalable, secure mechanisms for global surveillance (Muhammad et al., 2025).

AI-enabled hospital and environmental surveillance systems also bridge critical information gaps (Zeng et al., 2021). Traditional reporting captured only about five percent of actual infections during COVID-19 (Bivins et al., 2020). By contrast, AI models that analyze wastewater viral load, online symptom searches, or smart thermometer data can detect infection trends across entire communities. For example, wastewater surveillance in Boston provided early warnings of infection spikes, allowing hospitals to prepare resources days in advance. Similarly, AI-optimized testing can expand community coverage to nearly half the population by determining ideal testing frequencies and prioritizing high-risk areas (Bivins et al., 2020).

Surveillance systems vary by mode; passive, active, community-based, laboratory, and sentinel (McNabb et al., 2016). AI enhances each by increasing sensitivity, speed, and precision. Community-based surveillance can harness AI to analyze digital self-reporting platforms for early outbreak indicators. Laboratory networks can use machine learning to identify novel viral mutations within genomic data, and sentinel systems can employ predictive analytics to monitor long-term shifts in disease prevalence (Zeng et al., 2021; Muhammad et al., 2025). Genetic surveillance, though ethically complex, allows AI to predict which populations may be most vulnerable to certain pathogens, guiding targeted immunization efforts. However, such applications must be carefully regulated to prevent misuse and discrimination (McNabb et al., 2016).

PROBLEM STATEMENT

Despite substantial advances in global public health infrastructure, the COVID-19 pandemic exposed critical weaknesses in the speed, integration, and adaptability of modern epidemiological surveillance systems (Burrell, 2022). For instance, during the early weeks of the pandemic, countries struggled to track infection spread due to inconsistent data-sharing protocols and incompatible reporting platforms between hospitals and national agencies. In the United States, delays in laboratory data reporting, sometimes spanning several days, meant that health officials were often responding to outdated infection trends, undermining the timeliness of policy interventions such as travel advisories and regional lockdowns.

Traditional public health surveillance continues to rely heavily on siloed databases, manual reporting chains, and fragmented analytical processes, all of which slow the recognition of novel or resurgent pathogens (Burrell, 2022). For example, a clinician's report of an unusual cluster of respiratory symptoms may take days or weeks to reach centralized epidemiological teams, especially in regions where data are transmitted via outdated or incompatible software systems. Even when testing, contact tracing, and laboratory-based monitoring form the operational backbone of disease control, these methods tend to be reactive, mobilized only after infections have proliferated. The sheer volume and velocity of data generated during crises such as COVID-19 or the 2022 monkeypox outbreak quickly outpace the processing capacity of conventional surveillance infrastructures.

The absence of robust artificial intelligence (AI) integration within most public health frameworks has further limited the transformation of massive, heterogeneous datasets into timely, actionable intelligence (Zeng et al., 2021; Muhammad et al., 2025; Thiébaut & Cossin, 2019). During the Delta and Omicron waves, for example, governments often had access to genomic sequencing, hospital admission records, and anonymized mobility data from smartphones, but lacked the analytic tools to merge these sources into cohesive predictive models. Without AI to synthesize genomic data identifying new variants with epidemiological and behavioral information, many countries reacted to case surges rather than anticipating them.

This fragmentation carries profound consequences. It delays outbreak detection, allowing diseases to spread undetected for critical days or weeks; it impedes the strategic allocation of healthcare resources, such as ventilators and testing supplies; and it weakens national biosecurity preparedness by obscuring early warning signals from environmental surveillance systems. For instance, wastewater monitoring programs in cities like London and New York detected SARS-CoV-2 RNA weeks before corresponding spikes in reported cases, yet the absence of integrated AI analytics delayed recognition of these signals at the policy level.

In the absence of cohesive data ecosystems, governments are often forced to rely on broad, blunt interventions, such as mass lockdowns or generalized travel bans, that impose severe economic and social costs. These measures, while effective in halting transmission temporarily, also underscore the limitations of a reactive surveillance paradigm. Consequently, the global health ecosystem remains ill-equipped to synthesize and interpret real-time information at the scale necessary for proactive biosecurity, dynamic epidemiological modeling, and equitable decision-making (Zeng et al., 2021; Muhammad et al., 2025; Thiébaut & Cossin, 2019). In short, the lessons of COVID-19 highlight that without AI-enabled analytics capable of integrating genomic, clinical, behavioral, and environmental data, even the most advanced public health infrastructures will continue to lag behind the pathogens they seek to control.

PURPOSE STATEMENT

The purpose of this perspective paper is to articulate the transformative role of artificial intelligence technologies in redefining biosecurity, biodefense, and virus epidemiology within public health systems. It aims to critically evaluate how AI-enabled tools, such as machine learning algorithms, natural language processing (NLP), and predictive analytics, can strengthen global capacity for early pathogen detection, outbreak forecasting, and real-time risk assessment. By integrating examples of AI-driven applications, including genomic analysis, contact tracing automation, and multimodal data fusion, this paper argues that AI represents not a supplementary instrument but a foundational infrastructure for twenty-first-century public health preparedness. The discussion situates AI at the intersection of epidemiology, data governance, and ethical policy design, emphasizing how intelligent analytic frameworks can unify disparate data sources into coherent, privacy-respecting systems for public health surveillance and biodefense.

SIGNIFICANCE STATEMENT

The significance of this inquiry lies in its reframing of artificial intelligence as an indispensable component of resilient public health systems and biosecurity architecture. By harnessing AI's analytic precision and scalability, governments and health agencies can move beyond fragmented, reactive crisis management toward anticipatory, data-driven intervention models. AI's capacity to analyze large-scale, heterogeneous data streams enables the earlier detection of outbreaks, the improved allocation of testing and vaccination resources, and the identification of emerging viral variants before they spread widely. Moreover, AI-driven bioinformatics enhances the ability to differentiate naturally occurring pathogens from engineered threats, a critical function for modern biodefense (Zeng et al., 2021; Muhammad et al., 2025; Thiébaut & Cossin, 2019). On a societal level, the responsible integration of AI supports equitable access to health protection by mitigating algorithmic bias, ensuring transparency, and fostering trust in data-intensive decision-making. The insights presented here underscore that the future of epidemiological resilience and biosecurity depends not solely on medical innovation but on the intelligent, ethical, and policy-conscious deployment of artificial intelligence as the nervous system of global public health.

NATURE OF THE STUDY

This paper adopts a perspective-style academic approach, offering a conceptual and analytical examination of how artificial intelligence (AI) technologies are transforming public health biosecurity, biodefense, and virus epidemiology. Unlike empirical studies that rely on direct data collection or experimental intervention, a perspective paper synthesizes existing research, theoretical frameworks, and applied observations to advance a novel conceptual understanding of an emergent issue, in this case, the incorporation of AI into public health surveillance systems. This approach is particularly suitable for rapidly evolving technological and policy domains where scientific inquiry must bridge disciplinary boundaries and respond dynamically to real-world crises such as pandemics as outlined as a necessity in previous research (Heilweil, 2021; Burrell, 2022; Zeng et al., 2021; Muhammad et al., 2025; Thiébaut & Cossin, 2019).

The value of the perspective method lies in its ability to integrate insights from diverse bodies of scholarship, epidemiology, informatics, ethics, and governance, to construct a coherent narrative that extends beyond disciplinary silos. In doing so, this study contributes to the scholarly lexicon of applied observational research, offering interpretive synthesis and forward-looking argumentation rather than empirical generalization. Through this mode of inquiry, the paper provides a conceptual scaffold for understanding how AI tools, such as machine learning, natural language processing (NLP), and data fusion algorithms, can strengthen public health infrastructure by enabling early pathogen detection, predictive analytics, and real-time biosecurity responses (Thiébaut & Cossin, 2019).

A perspective paper of this kind plays a critical role in advancing academic dialogue by contextualizing complex technological transformations within the moral, operational, and institutional realities of public health. It functions as a discursive bridge between theory and application, drawing upon observed patterns from recent pandemics, most notably COVID-19, to argue for systematic adoption of AI technologies across surveillance and epidemiological systems. By analyzing how AI can augment traditional models of testing, contact tracing, and community-level surveillance, this work encourages interdisciplinary collaboration between technologists, epidemiologists, policymakers, and ethicists.

Furthermore, the perspective framework allows for critical reflection on methodological and ethical complexities inherent in deploying AI at scale. It emphasizes that the success of AI-driven epidemiological systems depends on interoperable data infrastructures, ethical governance mechanisms, and public trust. As such, the paper does not merely describe technological applications but interrogates their implications for global health equity, algorithmic accountability, and privacy regulation. This interpretive analysis underscores that effective biosecurity requires not only technical innovation but also the institutional capacity to translate analytic precision into equitable and transparent decision-making.

In the broader context of applied observational scholarship, this paper contributes to the ongoing academic dialogue about how AI can be responsibly integrated into existing surveillance paradigms to improve public health resilience. Perspective papers serve as intellectual catalysts, offering frameworks that guide empirical inquiry, inform policy development, and stimulate cross-sector discourse. By engaging with emerging patterns of AI adoption across epidemiological systems, this study advances the conceptual foundation upon which future experimental, quantitative, and policy-oriented research can be built.

Ultimately, this paper exemplifies the role of perspective-based scholarship as a thought experiment grounded in applied observation. It synthesizes current knowledge to articulate new conceptual pathways for the integration of AI into the biosecurity infrastructure of the twenty-first century. In doing so, it seeks to enrich the academic and professional lexicon of global health security, fostering an ongoing dialogue about how artificial intelligence can serve as both an analytical tool and a strategic asset in the defense of public health.

ARTIFICIAL INTELLIGENCE AND WASTEWATER EPIDEMIOLOGY

Tracking the movement and evolution of viral pathogens presents a multifaceted challenge within public health biosecurity and biodefense systems (Farkas et al.,

2020). The COVID-19 pandemic revealed the limitations of traditional surveillance models in identifying and mitigating infections, particularly those that are asymptomatic or mild. Individuals with minimal symptoms often elude clinical testing and case reporting systems, thereby distorting epidemiological models and limiting the effectiveness of outbreak management strategies (Farkas et al., 2020). To address these challenges, the integration of artificial intelligence (AI) technologies into epidemiological surveillance, especially in wastewater monitoring systems, has emerged as a transformative innovation (Farkas et al., 2020; Bivins et al., 2020).

Wastewater, comprising effluent from residential and commercial sources, contains measurable traces of viral RNA shed by infected individuals, including those who never present for clinical testing (Kreier, 2021; Keshaviah, Hu, & Henry, 2021). As viral particles or genetic fragments accumulate in sewage, they create a community-level biomarker that reflects the prevalence and dynamics of infection in real time. AI-driven analytic platforms can process this complex, high-volume data to identify early signs of viral spread, forecast future infection surges, and differentiate between variants. Machine learning algorithms trained on historical wastewater datasets can recognize subtle shifts in viral load patterns, detect anomalies indicative of emerging variants, and estimate transmission trajectories across geographic scales (Zeng et al., 2021; Muhammad et al., 2025; Thiébaut & Cossin, 2019). This form of AI-enhanced wastewater epidemiology provides a scalable, cost-effective, and ethically non-intrusive method for continuous public health surveillance, especially in regions where clinical testing resources are scarce (Kreier, 2021; Keshaviah, Hu, & Henry, 2021).

Sequential variations in viral RNA levels within wastewater provide valuable insights into both the onset and decline of outbreaks, serving as a continuous, unbiased measure of population-level infection trends (Kreier, 2021; Keshaviah, Hu, & Henry, 2021). AI-based time-series forecasting models can transform these sequential data streams into dynamic dashboards that inform public health decision-making. For example, convolutional neural networks (CNNs) and long short-term memory (LSTM) algorithms can analyze wastewater viral concentrations to anticipate hospital admission spikes or regional infection surges days to weeks in advance. Such predictive intelligence enables policymakers to deploy interventions earlier, allocate vaccines more effectively, and direct resources to vulnerable communities before outbreaks escalate. Consequently, AI-enabled wastewater surveillance represents a cornerstone technology in twenty-first-century biodefense, capable of complementing genomic sequencing, digital epidemiology, and environmental biosensing as part of an integrated biosecurity infrastructure.

AI IN WASTEWATER SURVEILLANCE INSIGHTS AND RECOMMENDATIONS

The integration of AI into wastewater treatment and surveillance systems represents a paradigm shift in public health monitoring. Rather than focusing on the traditional collection and testing processes, the emphasis must now be placed on establishing intelligent analytic frameworks capable of autonomously processing, interpreting, and contextualizing vast quantities of environmental and biological data (Burrell, 2022). AI systems, through advanced pattern recognition and data fusion techniques, can consolidate wastewater viral metrics with complementary data sources such as hospital admission rates, clinical test results, and demographic mobility patterns. This convergence enables a holistic understanding of viral transmission pathways and provides actionable intelligence for public health agencies (Burrell, 2022).

To operationalize these AI-enhanced systems effectively, investment in digital infrastructure and interdisciplinary human capital is imperative (Zeng et al., 2021; Muhammad et al., 2025; Thiébaut & Cossin, 2019). This includes developing computational pipelines that connect wastewater laboratories to national bioinformatics databases in real time. Establishing these links will enable automated data transfer, facilitate cross-validation of viral genomic data, and facilitate the use of AI for continuous quality assurance. Public health organizations should also collaborate with AI research institutions and technology firms to develop tailored algorithms that can account for population density, weather variations, and other environmental factors influencing wastewater viral load. Moreover, ethical governance frameworks must be instituted to safeguard data privacy and prevent the misuse of AI-generated health intelligence (Zeng et al., 2021; Muhammad et al., 2025; Thiébaut & Cossin, 2019).

A major recommendation is the adoption of federated AI models in wastewater epidemiology. These models enable local wastewater treatment plants to contribute to collective disease modeling without sharing raw data, thereby maintaining community-level privacy while enhancing the accuracy of national or global epidemiological models (Burrell, 2022). Similarly, the application of reinforcement learning can enable adaptive response mechanisms, where AI systems continuously refine predictions based on real-world feedback from infection outcomes (Zeng et al., 2021; Muhammad et al., 2025; Thiébaut & Cossin, 2019). For instance, an AI platform could automatically adjust public health alerts or modify sampling frequency as viral concentrations rise or fall, optimizing both efficiency and responsiveness (Burrell, 2022).

AI-enhanced wastewater surveillance systems also support biosecurity and biodefense objectives by serving as sentinel infrastructures capable of detecting pathogenic agents beyond SARS-CoV-2. Machine learning algorithms trained on multi-pathogen genomic signatures can distinguish naturally occurring viruses from those that may represent bioterrorism threats, thereby strengthening national preparedness frameworks (Zeng et al., 2021; Muhammad et al., 2025; Thiébaut & Cossin, 2019). Governments should prioritize the integration of AI into wastewater biosurveillance as a critical defense layer, linking environmental monitoring with genomic databases, epidemiological registries, and emergency response networks (Burrell, 2022).

Finally, public health education and workforce development should reflect this technological evolution. Training wastewater engineers, epidemiologists, and public health practitioners in AI literacy will ensure that emerging tools are used responsibly and effectively. Investment in academic–government partnerships and AI research consortia will accelerate the development of domain-specific models optimized for wastewater biosurveillance applications (Burrell, 2022).

PRACTICAL RECOMMENDATIONS FOR IMPLEMENTATION

The integration of Artificial Intelligence (AI) technologies into wastewater epidemiology demands a structured, interdisciplinary implementation framework. The following practical recommendations provide a roadmap for public health agencies, researchers, and policymakers seeking to operationalize AI-enhanced wastewater surveillance systems within biosecurity and biodefense infrastructures:

1. Develop Integrated AI–Wastewater Surveillance Infrastructures

Establish AI-enabled monitoring systems within wastewater treatment plants that autonomously collect, process, and interpret viral genomic data in real time. Implementation should leverage deep learning models, such as Convolutional Neural Networks (CNNs) and Long Short-Term Memory (LSTM) architectures, to detect early viral signals, identify variants, and forecast epidemiological trends (Burrell, 2022).

2. Implement Federated Learning Models for Privacy Preservation

Adopt federated AI frameworks that enable data analysis across multiple wastewater facilities without the need to transfer sensitive raw data. This preserves community-level privacy while enhancing the accuracy of national and global epidemiological modeling (Burrell, 2022).

3. Create Cross-Sector Data Fusion Platforms

Establish interoperable platforms that integrate wastewater analytics with complementary datasets, including hospital admissions, electronic medical records, environmental biosensors, and population mobility metrics. Such data fusion enhances the precision of predictive modeling and supports timely public health decision-making (Burrell, 2022).

4. Invest in Workforce Development and AI Literacy

Develop interdisciplinary training programs for public health professionals, wastewater engineers, epidemiologists, and data scientists to ensure effective use of AI tools in biosurveillance. AI literacy should be incorporated into public health curricula and continuing education to foster responsible and informed use of technology (Burrell, 2022).

5. Establish Ethical and Governance Frameworks for AI Deployment

Develop transparent governance systems that regulate the role of AI in biosurveillance and surveillance. Ethical standards should address algorithmic bias, data ownership, privacy protection, and the potential misuse of surveillance data, among other key considerations. National and international collaborations should align these frameworks with existing biosecurity treaties and human rights norms (Burrell, 2022).

6. Enhance Laboratory and Computational Infrastructure

Modernize wastewater laboratories to accommodate high-throughput sequencing and AI-based data processing. This includes upgrading computational hardware, adopting standardized digital architectures, and ensuring secure cloud connectivity for real-time analytics (Burrell, 2022).

7. Integrate AI with Early Warning and Response Networks

Connect AI-enhanced wastewater surveillance outputs to public health alert systems, emergency operations centers, and biodefense agencies. Predictive outputs should be used to inform policy decisions, trigger local containment strategies, and preemptively allocate medical resources (Burrell, 2022).

8. Pilot Regional Demonstration Projects

Launch pilot programs in diverse geographic and socioeconomic contexts to evaluate scalability and adaptability. These pilots should assess technical feasibility, cost-effectiveness, ethical challenges, and policy readiness while providing models for national or international replication (Burrell, 2022).

CONCLUSION

The integration of artificial intelligence (AI) into wastewater epidemiology represents a pivotal advancement in building resilient, adaptive, and anticipatory public health systems. In contrast to conventional surveillance methods that depend on clinical reporting, AI-enabled wastewater analysis provides continuous, population-level insights into viral activity, including among asymptomatic individuals. As infectious disease threats become more complex, arising from both natural evolution and potential biological manipulation, AI-enhanced wastewater surveillance emerges as a vital component of modern biosecurity infrastructure. It enables early detection, real-time analytics, and predictive modeling that can guide rapid public health responses. The effectiveness of this approach, however, relies on the parallel development of interoperable data frameworks, strong ethical governance, and sustained collaboration across sectors. Ultimately, AI-driven wastewater surveillance should be viewed not as a supplement to existing epidemiological systems but as a cornerstone technology for future global health security (Burrell, 2022).

Recommendations for Future Research (Action Research Approach)

Future investigations into AI-driven wastewater epidemiology should adopt Action Research (AR) as a guiding methodological framework. Action Research, characterized by iterative cycles of planning, action, observation, and reflection, allows researchers to collaboratively develop, test, and refine AI systems in real-world public health contexts. This participatory model aligns with the adaptive and interdisciplinary nature of AI implementation in biosecurity and biodefense settings.

1. Collaborative System Design and Implementation

Future studies should employ AR cycles to co-design AI surveillance systems in partnership with wastewater treatment engineers, epidemiologists, public health officials, and data scientists. Such collaboration ensures that systems are both technically robust and operationally feasible within existing public health infrastructures.

2. Iterative Model Refinement Through Field Application

Using AR's cyclical structure, researchers should iteratively test and refine AI algorithms on live wastewater data. This process enables continuous learning, allowing models to adapt to environmental variability, population changes, and evolving viral dynamics (Burrell, 2022).

3. Participatory Evaluation of Ethical and Social Implications

Action Research facilitates stakeholder engagement to assess ethical implications—particularly those concerning data privacy, public consent, and the societal acceptability of AI-driven biosurveillance. Feedback from communities and policymakers can inform the ethical frameworks governing AI deployment.

4. Real-Time Policy Translation and Knowledge Mobilization

AR methodologies promote the rapid translation of research outcomes into practice. As AI surveillance models are tested, researchers can concurrently work with public health authorities to embed findings into policy instruments, response protocols, and funding strategies.

5. Scalability and Sustainability Assessment

Longitudinal AR projects should explore the scalability of AI-enhanced wastewater systems across different infrastructural and socioeconomic settings. Comparative studies across nations or regions could determine optimal models for sustainable implementation under varying resource constraints.

6. Integration of Multi-Pathogen and Multi-Data Systems

Future AR-driven research should expand beyond COVID-19 to include multi-pathogen surveillance, incorporating markers for bacterial, viral, and antimicrobial resistance. Integrating environmental, behavioral, and genomic data within AI frameworks would yield a comprehensive biodefense intelligence ecosystem (Burrell, 2022).

REFERENCES

Bivins, A., North, D., Ahmad, A., Ahmed, W., Alm, E., Been, F., Bhattacharya, P., Bijlsma, L., Boehm, A. B., Brown, J., Buttiglieri, G., Calabro, V., Carducci, A., Castiglioni, S., Cetecioglu Gurol, Z., Chakraborty, S., Costa, F., Curcio, S., de los Reyes, F. L.III, & Bibby, K. (2020). Wastewater-Based Epidemiology: Global Collaborative to Maximize Contributions in the Fight Against COVID-19. *Environmental Science & Technology*, *54*(13), 7754–7757. DOI: 10.1021/acs.est.0c02388 PMID: 32530639

Burrell, D. N. (2022). Supply chain and logistics management in public health surveillance and protection during COVID-19. *Science Bulletin*, *27*, 12–20.

Devriendt, T., Borry, P., & Shabani, M. (2021). F*actors that influence data sharing through data sharing platforms*: A qualitative study on the views and experiences of cohort holders and platform developers. *PLoS One*, *16*(7), 1–14. DOI: 10.1371/journal.pone.0254202 PMID: 34214146

Farkas, K., Hillary, L., Malham, S., McDonald, J., & Jones, D. (2020). Wastewater and public health: the potential of wastewater surveillance for monitoring COVID-19, Current Opinion in Environmental Science & Health, Volume 17, Pages 14-20, ISSN 2468-5844.

Friis, R. H. (2018). *Epidemiology 101* (2nd ed.). Jones & Bartlett Learning.

Groseclose, S. L., & Buckridge, D. L. (2016, December 15). *Public Health Surveillance systems: Recent advances in their use and evaluation*. Annual Reviews. Retrieved September 27, 2021, from https://www.annualreviews.org/doi/10.1146/annurev-publhealth-031816-044348

Heilweil, R. (2021, December 21). How Omicron broke Covid-19 testing. Vox.

Hughes, H. (2021). *The use of emergency department electronic health data for syndromic surveillance to enhance public health surveillance programmes in england* (Order No. 28827644). Available from ProQuest Dissertations & Theses Global.

Keshaviah, A., Hu, X., & Henry, M. (2021). Developing a Flexible National Wastewater Surveillance System for COVID-19 and Beyond. *Environmental Health Perspectives*, *129*(4), 045002. Advance online publication. DOI: 10.1289/EHP8572 PMID: 33877858

Kester, D. T. (2021). *Proactive text messaging for public health surveillance during the COVID-19 pandemic* (Order No. 28652914). Available from ProQuest Dissertations & Theses Global.

Kreier, F. (2021). The Myriad Ways Sewage Surveillance Is Helping Fight COVID Around the World. *Nature*. Advance online publication. DOI: 10.1038/d41586-021-01234-1 PMID: 33972790

McNabb, S. J. N., Conde, J. M., Ferland, L., Memish, Z. A., Okutani, S., Park, M. M., Ryland, P., Shaikh, A. T., & Singh, V. (Eds.). (2016). *Transforming Public Health Surveillance: Proactive measures for prevention, detection, and response*. Elsevier.

Muhammad, S. K., Ansari, T. A., Shabbir, B., Almagharbeh, W. T., Rehman, A., Arjmand, R., & Ghulam, A. (2025). THE ROLE OF ARTIFICIAL INTELLIGENCE IN PUBLIC HEALTH SURVEILLANCE: A POST-PANDEMIC PERSPECTIVE. *Insights-Journal of Life and Social Sciences*, *3*(3 (Social)), 74–80. DOI: 10.71000/pa7ab080

Thiébaut, R., & Cossin, S. (2019). Artificial intelligence for surveillance in public health. *Yearbook of Medical Informatics*, *28*(01), 232–234. DOI: 10.1055/s-0039-1677939 PMID: 31419837

Zeng, D., Cao, Z., & Neill, D. B. (2021). Artificial intelligence–enabled public health surveillance—from local detection to global epidemic monitoring and control. In *Artificial intelligence in medicine* (pp. 437–453). Academic Press. DOI: 10.1016/B978-0-12-821259-2.00022-3

Chapter 9
Transforming Rural Veterinary Care Through Telehealth

Jason D. West
http://orcid.org/0009-0003-0439-8626
Marymount University, USA

ABSTRACT

This case study investigates the strategic adoption of telehealth and subscription-based emergency care models by rural veterinary practices, addressing challenges of geographic isolation, evolving client expectations, and competitive pressures. A structured literature review of scholarly articles and industry publications from 2016-2024 identifies key operational, ethical, and strategic impacts of telehealth implementation. The review underscores telehealth's potential to improve rural care accessibility, enhance client satisfaction, and bolster practice sustainability. Significant barriers identified include technological infrastructure constraints, restrictive regulatory environments – particularly around Veterinarian-Client-Patient Relationship (VCPR) – and substantial training gaps among veterinary staff. Recommendations highlight adopting integrated telehealth platforms, implementing flexible subscription-based models, targeted regulatory advocacy, and ongoing staff training to achieve sustainable transformation.

INTRODUCTION

Telehealth is emerging as a transformative solution in veterinary medicine, particularly for rural communities where access to in-person care is limited. By leveraging video-based consultations and digital platforms, veterinary practices can

DOI: 10.4018/979-8-2600-1026-6.ch009

extend their reach, offering timely assessments, emergency guidance, and follow-up care to pet owners in remote locations. This model not only addresses geographic and logistical barriers but also enhances continuity of care, strengthens client relationships, and supports more proactive health management. As the veterinary field increasingly embraces digital innovation, telemedicine is poised to play a pivotal role in expanding service delivery and improving outcomes in underserved areas (Bishop et al., 2018; Donham & Wickett, 2018).

As telehealth becomes an essential tool in rural veterinary practice, recent legislative changes such as California's expansion of telehealth access for pets (Axios, 2024) reflect a broader shift toward digital integration in animal care. This policy change underscores a national trend toward embracing veterinary telemedicine as a means to address systemic access issues in underserved regions. Similarly, recent American Veterinary Medical Association (AVMA) updates highlight the state-by-state activity to redefine VCPR (Veterinarian-Client-Patient Relationship) legislation, which remains a crucial determinant of how broadly telemedicine can be applied (AVMA, 2023).

Remote vital sensing and wearable technologies are also expanding diagnostic reach in real-time (Zhao et al., 2025; Haghi et al., 2021), allowing practitioners to manage chronic conditions with less frequent in-clinic visits. These capabilities improve not just care access, but also continuity, especially for mobility-limited animals (Johnson & Martinez, 2025). Consumer expectations are also evolving with pet owners placing value upon convenient, tech-enabled, and low-stress care options (AVMA, 2021). Subscription models and wearable health-monitoring tools are gaining traction as clients seek proactive, personalized care that mirrors innovations in human healthcare (Jergler, 2018). Mobile platforms also enable veterinarians to consult with specialists, improve diagnostic accuracy, and reduce unnecessary travel or referrals (Niessen & Forcada, 2019).

The adoption of telemedicine is further supported by professional guidance and policy. The American Veterinary Medical Association (2021) outlines ethical standards for virtual care, including the requirement for a valid Veterinarian-Client-Patient Relationship (VCPR), informed consent, and data privacy protections. These standards ensure that telehealth services remain clinically sound and legally compliant. Additionally, research shows that telehealth improves care for chronic conditions, reduces pet stress, and increases client satisfaction in rural settings (Hassan et al., 2024).

However, successful adoption requires more than access. It also requires implementation strategies that account for workflow design, change management, and staff readiness. Barriers such as unclear regulations, resistance to change, or lack of digital infrastructure remain significant, especially in production animal settings (Guintard et al., 2024).

To remain competitive and responsive to client needs, the practice must implement telehealth strategically: investing in infrastructure, training staff, and promoting services through digital marketing. The following analysis outlines a data-driven approach to guide this transformation and position the organization for sustainable growth and profitability.

PROBLEM STATEMENT

Rural veterinary practices face mounting pressure to expand access to emergency and specialty care while meeting rising consumer expectations for convenience, digital access, and personalized service; however, many struggle to implement scalable telehealth models due to infrastructure limitations, regulatory barriers, and gaps in staff training and workflow design (American Veterinary Medical Association, 2019; Hassan et al., 2024). Despite growing evidence that video-based consultations, wearable monitoring tools, and subscription-based service models improve care continuity and client satisfaction, rural clinics remain under-resourced and risk falling behind in an increasingly tech-driven market (Bishop et al., 2018; Jergler, 2018). Without strategic investment in telehealth platforms and digital transformation, these practices may experience declining competitiveness, reduced care capacity, and lost opportunities to serve remote or underserved populations.

Recent studies have demonstrated the technical feasibility of remote diagnostics and tele-ultrasound in large animal hospitals (Friedman et al., 2024) yet operationalizing these solutions in smaller rural clinics is hindered by resource constraints, limited broadband access, and fragmented policy frameworks. Further, while California has expanded telehealth access for companion animals (Axios, 2024), state-level variation in VCPR requirements continues to limit nationwide scalability (AVMA, 2023). These regulatory inconsistencies create uncertainty for veterinarians considering technology adoption, especially in states where remote establishment of VCPR remains prohibited. In addition to regulatory and infrastructure hurdles, veterinarians often cite staff resistance and workflow disruption as key barriers (Guintard et al., 2024). Without robust change management strategies and ongoing staff training, the integration of telemedicine may exacerbate inefficiencies, rather than resolve them.

This case study addresses the operational, regulatory, and ethical challenges of integrating telehealth into rural veterinary care and proposes a sustainable roadmap for expanding 24-hour emergency services. By accounting for evolving state legislation (AVMA, 2023), implementation barriers in clinical workflows (Guintard et al., 2024), and the promise of remote diagnostics (Zhao et al., 2025; Friedman et al., 2024), the model emphasizes accessible, compliant, and high-quality care. The proposed strategy supports improved clinical outcomes, increased client satisfaction, and a scalable path to long-term business growth.

SIGNIFICANCE OF THE PROJECT

Expanding access to emergency veterinary care through telehealth is critically important for addressing healthcare disparities in rural communities, where clients often face long travel distances, limited clinic availability, and after-hours service gaps (Donham & Wickett, 2018; Hassan et al., 2024). Failure to address these barriers compromises animal welfare and delays timely interventions, particularly for chronic or acute conditions that require immediate guidance. Moreover, underserved populations are increasingly expecting veterinary services that reflect the same digital convenience and responsiveness they experience in human healthcare, including real-time video consultations, wearable monitoring, and mobile-first platforms (AVMA, 2025).

Recent advances in remote vital sensing (Zhao et al., 2025) and wearable devices tailored for animal health monitoring (Haghi et al., 2021) provide practical, scalable solutions that rural practices can integrate to enhance early detection and support long-term condition management. These tools align with consumer expectations and allow clinics to extend their reach without physically expanding infrastructure. If rural practices are unable to evolve to meet these expectations, they risk losing clients to larger, more technologically advanced competitors and third-party telehealth platforms that are not always equipped to provide localized or continuous care.

The significance of this issue also extends to veterinary workforce development and the long-term sustainability of independent practices. Without a digital transformation strategy, rural clinics may face growing strain on limited staff, difficulty retaining younger veterinarians who seek innovation-driven workplaces, and declining revenue due to underutilized service opportunities (Watson et al., 2024; Jergler, 2018). Research by Guintard et al. (2024) has identified both infrastructure and regulatory complexity as key barriers to implementation in agricultural and rural animal care settings, reinforcing the need for strategic investment and policy clarity.

As consumer demand for remote care and subscription-based models increases, practices that fail to modernize may experience reduced market share, reputational decline, and operational inefficiencies. Legislative progress such as California's expansion of pet telehealth access (Axios, 2024) signals a shifting legal landscape, supported by new AVMA guidance (AVMA, 2021) and state-level VCPR activity (AVMA, 2023), which together offer a roadmap for ethical and effective care delivery models. Evidence from clinical trials, such as Friedman et al. (2024), also demonstrates that remote-assisted point-of-care ultrasonography can maintain diagnostic integrity and reduce resource strain in large animal hospitals, further supporting the scalability of telehealth.

Addressing this issue presents a strategic opportunity to elevate both client satisfaction and clinical quality through digital care delivery, ensuring that rural veterinary medicine remains viable, competitive, and responsive in a rapidly evolving healthcare landscape.

METHODOLOGY

This case study is anchored in a structured literature review designed to identify best practices, challenges, and innovations relevant to veterinary telehealth, rural care delivery, and digital transformation in clinical workflows. The purpose of the review was to generate evidence-based recommendations for implementing a subscription-based, 24-hour emergency care model within a rural veterinary practice. The review synthesized research from clinical studies, policy reports, implementation guides, and expert perspectives to inform strategy development across technological, operational, ethical, and educational dimensions.

A comprehensive web-based search was conducted using library resources from Marymount University and Johns Hopkins University. The search incorporated a broad range of keywords, including *veterinary telemedicine*, *veterinary telehealth*, *video consultation in veterinary care*, *rural veterinary practice*, *emergency veterinary services*, *crisis veterinary services*, *Veterinarian-Client-Patient Relationship (VCPR)*, *remote veterinary diagnostics*, and *subscription-based pet care models*. Boolean operators were used to refine results (e.g., "veterinary telehealth AND critical care," "wearable technology OR remote monitoring," and "telemedicine NOT livestock veterinary management").

Articles were sourced from peer-reviewed journals, professional veterinary association publications, trade magazines, and credible case reports. The inclusion criteria required that each source be published between 2016 and 2024 to allow for a comprehensive review that incorporated both recent developments and earlier foundational research – particularly to source strategic models introduced during seminar years. Studies unrelated to companion animal care (e.g., livestock), lacking relevance to digital transformation or emergency services, or outside the veterinary scope were excluded. U.S.-based literature was prioritized, but select international studies were included to highlight global best practices applicable to rural settings.

A total of 36 verified sources met the criteria and were analyzed using a literature matrix that tracked the study purpose, research question, methodology, population, findings, and key takeaways. This comprehensive review ensures the final recommendations are informed by current and diverse perspectives on telemedicine's role in veterinary care innovation.

LITERATURE REVIEW

Overview of Telehealth in Veterinary Medicine

Telehealth has become an increasingly critical component of modern veterinary care, reshaping the traditional clinical setting into a hybrid model that integrates digital and face-to-face services. In addition to addressing geographic barriers, veterinary telemedicine also alleviates workforce constraints and helps meet the growing demand for more accessible pet healthcare (Hassan et al., 2024). Real-time video, mobile diagnostics, and asynchronous communication methods empower veterinary providers to offer timely consultations and proactive health management (Freidman et al., 2024). The AVMA has outlined that telemedicine is particularly effective in behavioral health, medication management, and chronic care cases, especially in underserved regions (AVMA, 2021). The AVMA's most recent guidelines reinforce these clinical benefits while distinguishing between telehealth, teleconsulting, and tele-triage – emphasizing that each modality must adhere to ethical and regulatory standards under an established VCPR (AVMA, 2023).

Recent studies have emphasized the rapid evolution of telehealth services as both a clinical tool and a competitive strategy in veterinary medicine. Kane and Gillis (2018) demonstrate that successful digital integration in human healthcare settings has led to improved outcomes, suggesting that similar benefits can be expected in veterinary practice. Perry et al. (2023) expand on this by noting that telemedicine fosters client engagement and satisfaction, especially when integrated into broader care plans. Veterinary schools are also responding to these shifts by embedding telehealth competencies into curricula, preparing graduates for digitally mediated care (Watson et al., 2024). In parallel, Johnson and Martinez (2025) found that at-home video consultations for managing feline mobility issues significantly improved owner compliance and clinical continuity, highlighting the growing viability of remote care in companion animal medicine.

Professional bodies continue to offer comprehensive frameworks for digital transformation. The AVMA (2023) draws critical distinctions between telehealth, telemedicine, and teleconsulting, reinforcing the importance of a valid Veterinarian-Client-Patient Relationship (VCPR) in all care models. Recent legislative analysis by the AVMA (2023) further clarifies that 22 U.S. states still require in-person exams to establish a VCPR, presenting a regulatory barrier to widespread telehealth implementation. However, policy changes such as California's 2024 telehealth expansion signal growing momentum for reform (Axios, 2024). This guidance is particularly relevant for clinics aiming to expand telehealth offerings while maintaining legal compliance and clinical integrity.

The spectrum of telehealth tools continues to widen. Jergler (2018) identifies the emergence of wearable technology and app-based health monitors as critical developments in veterinary care, allowing real-time data capture and proactive interventions. According to AVMA guidelines (2021), the integration of connected care tools (e.g., remote diagnostics, mobile monitoring devices, and teleconsultations) has the potential to enhance clinical outcomes and support a more accessible, efficient veterinary care model.

Recent reviews confirm this trajectory: Zhao et al. (2025) provide a comprehensive analysis of remote vital sensing advancements, and Haghi et al. (2021) emphasize the role of wearable devices in streamlining chronic care management. These technologies collectively enable virtual triage, chronic condition monitoring, and real-time consultations, forming a comprehensive ecosystem of digitally connected veterinary services. Additionally, Abu-Seida et al. (2024) highlight the ethical and welfare implications of telemedicine in low-access settings, reinforcing its importance as a tool for expanding care to underserved animals.

Benefits of Telehealth in Rural Veterinary Practices

Telehealth is particularly valuable in rural settings, where geography, clinic density, and staffing shortages often create significant access barriers. Donham and Wickett (2018) highlight how video consultations improve triage capabilities and expedite care, especially during emergencies when transportation is delayed. Similarly, Bishop et al. (2018) report high levels of client satisfaction with remote follow-up appointments, particularly in cases of surgical recovery and routine rechecks. Niessen and Forcada (2019) demonstrate that telemedicine enables general practitioners in remote areas to consult with board-certified specialists in real time, improving diagnostic precision and clinical outcomes. This benefit is further supported by Friedman et al. (2024), who found that remote guidance in point-of-care ultrasound significantly improved diagnostic accuracy in large animal hospital settings, highlighting the clinical viability of tele-assisted imaging in rural or resource-limited contexts.

Subscription-based care is also gaining traction in rural practices seeking financial predictability. According to the AVMA (2021), bundled service models, offering regular virtual visits, medication management, and wellness monitoring, help build client loyalty and generate recurring revenue. Boone et al. (2023) validate this approach by showing increased client retention and satisfaction among pet owners enrolled in subscription programs. Remote diagnostic technologies further enhance rural capabilities. Navas de Solis et al. (2019) explore the use of real-time tele-ultrasound in equine practice, demonstrating the feasibility of remote imaging interpretation in the field. These tools enable small rural practices to offer advanced care without investing in on-site specialists. Recent work by Johnson and Martinez (2025) also emphasizes how ongoing telehealth access can improve chronic condition management in companion animals, helping rural clients maintain continuity of care at home.

Teller (2022) and Springer et al. (2023) observe that pet owners in rural areas are more likely to engage with telehealth services when they reduce pet stress and logistical burden. Flexible appointment scheduling and mobile access to clinicians promote trust, especially when paired with educational resources that demystify the virtual care process. Neill and Morgan (2025) compare rural and urban adoption trends and find fewer differences than expected, indicating that demand exists but is contingent on infrastructure and promotion. Practices that proactively market telehealth options and invest in user-friendly onboarding are more likely to succeed in attracting and retaining rural clients. Abu-Seida et al. (2024) reinforce this point by calling telemedicine a "welfare equalizer," capable of improving animal outcomes across disparate communities when infrastructure and clinician readiness align.

Impact on Business Operation

Telehealth implementation has profound implications for organizational structure, staffing, and leadership. As clinics shift to hybrid care models, they must redefine job roles, workflows, and communication practices. Watson et al. (2024) emphasize that veterinary staff, particularly technicians, require targeted training in digital communication, client onboarding, and remote diagnostic procedures. Perry et al. (2023) add that internal knowledge transfer and peer mentorship can significantly ease transitions and promote a sense of shared purpose among teams.

Guintard et al. (2024) further stress that successful integration of telemedicine in animal production environments depends on overcoming organizational resistance, improving digital literacy, and addressing logistical workflow concerns – many of which parallel challenges in small animal and rural clinical settings.

Boone et al. (2023) found that clinics offering formal digital upskilling programs reported higher staff morale, lower turnover, and improved telehealth adoption rates. Spindel (2019) reinforces this, arguing that clinics cultivating innovation-centered cultures are better positioned to thrive in dynamic digital environments. Johnson and Martinez (2025) echo this by highlighting how even modest telehealth initiatives, such as video consultations for chronic feline conditions, require new training protocols and staff role clarity, particularly around communication etiquette and follow-up documentation.

Leadership must model the change they seek to implement. Appelbaum et al. (2012) assert that successful transformation hinges on leaders who communicate urgency, empower cross-functional teams, and celebrate incremental wins. Marshall (2019) notes that inclusive leadership, characterized by listening, adaptation, and emotional intelligence, builds credibility during times of technological disruption. As emphasized in the AVMA's latest 2025 guidelines on connected care, leaders must also implement oversight procedures to ensure that remote consultations uphold clinical standards, especially in high-volume or multi-clinician practices (AVMA, 2021).

The ethical framework for telehealth must be reinforced across the organization. The AVMA's (2023) ethical code calls for consistent standards in remote interactions, including confidentiality, informed consent, and documentation. This is particularly important for administrative staff managing appointment systems, digital records, and client communications. Everyone on the team must be aligned around shared values to ensure service consistency and legal compliance. Zhao et al. (2025) reinforce this concern, cautioning that remote vital sensing and wearable monitoring technologies raise important ethical and regulatory issues, especially around data security and device accuracy. Staff training must therefore go beyond technical fluency to include awareness of legal and client-centered risks.

Magalhães-Sant'Ana et al. (2020) warn that inadequate attention to digital ethics may erode trust in both the clinic and the profession. To mitigate this, practices should establish clear protocols for virtual etiquette, crisis escalation, and client education. Clinics that embed these behaviors into onboarding and training processes are better equipped to manage the cultural shift that telehealth demands. Teller (2022) emphasizes that telehealth is not simply a technical tool, yet a strategic asset that must be aligned with an organization's mission, brand, and community relationships. Abu-Seida et al. (2024) affirm that framing telemedicine as a continuity-of-care strategy, rather than a convenience, can help align team values and community trust, especially in underserved regions. In addition, the AVMA (2023) reports that new legislation across states is placing greater responsibility on practice owners to document staff roles and ensure proper VCPR protocols are followed in all virtual interactions. This makes organizational clarity essential not only for quality assurance but also legal risk mitigation.

Ethical, Legal, and Regulatory Considerations

The legal and ethical dimensions of telehealth remain a major area of concern for veterinary professionals. Central to this is the VCPR – a regulatory framework that dictates when a veterinarian can diagnose, treat, or prescribe for a patient. Most U.S. states require a physical examination before a VCPR can be established, which significantly limits the scope of telemedicine (AVMA, 2021). Building on this regulatory challenge, Neal and Greenberg (2024) conducted a thematic policy analysis using Kingdon's Multiple Streams Framework to identify viable alternatives for enabling virtual VCPR establishment. Their review of physician-patient telemedicine policies across all 50 U.S. states revealed ten key policy themes, including technology verification standards, patient identity protocols, and defined limits for remote care. The authors argue that many of these human healthcare policies can be adapted to veterinary contexts, providing a structured roadmap for integrating telehealth while maintaining regulatory integrity. For rural practices, such policy adaptations could dramatically expand access to emergency care without compromising client safety or ethical standards.

Recent legislative activity across the U.S. reflects growing momentum toward such adaptations. The AVMA (2023) reports that state-level policymakers are increasingly debating revisions to VCPR requirements to accommodate virtual care, particularly in response to demand from underserved communities and disaster preparedness needs. This policy shift suggests a future where virtual VCPR may become more feasible, especially when supported by secure digital platforms and proper documentation. The AVMA (2023) emphasizes that veterinary telehealth must meet the same ethical and professional standards as in-person care. This includes securing informed consent, maintaining client and patient confidentiality, and thoroughly documenting all virtual interactions. The guidelines also clearly distinguish telemedicine – which requires an established VCPR – from general tele-advice or other digital communications, underscoring the importance of clinical accountability in remote care. Their updated 2023 guidelines reinforce these distinctions, offering additional protocols for managing emergencies, validating identity, and archiving interactions securely (AVMA, 2025). These detailed recommendations are particularly valuable for rural clinics managing 24-hour subscription-based models where multiple providers may engage with the same client.

However, reform is on the horizon. Hassan et al. (2024) argue that regulatory bodies must evolve to reflect technological capabilities, proposing that tele-triage and AI-assisted diagnostics could support VCPR in the absence of physical exams. Nelson-Pratt (2018) similarly advocates for "regulatory sandboxes" to test innovative care models under controlled conditions. Abu-Seida et al. (2024) build on this by emphasizing that telemedicine is not merely a substitute for in-person visits but a means of improving animal welfare in underserved areas. Their review highlights ethical advantages such as faster response times, greater transparency, and stronger communication between veterinarians and clients when protocols are clearly followed.

Ethical tensions often arise when clinicians are unsure whether remote care meets the standard of "competent treatment." Marshall (2019) highlights that liability risks may increase in the absence of physical examinations, particularly if documentation is incomplete or communication with clients is unclear. To address these concerns, clinics must establish robust telehealth protocols and ensure that all staff are adequately trained in ethical decision-making and digital care delivery. These operational safeguards, however, are only part of the equation. Quain, Ward, and Mullan (2021) expand the conversation by examining the broader moral challenges associated with advanced veterinary care. While digital innovations can enhance clinical outcomes, they may also prolong suffering, increase caregiver burden, and create pressure to over-treat, especially in virtual settings where diagnostic nuances may be missed. The authors advocate for integrating ethical frameworks that support client communication, humane decision-making, and quality-of-life considerations into all care models, including those delivered via telehealth.

Magalhães-Sant'Ana et al. (2020) conducted a Delphi study in Portugal that revealed widespread concern among veterinarians about the loss of diagnostic nuance in virtual care. This study underscores the need for ongoing dialogue, continuing education, and case-based learning to help clinicians apply ethical judgment in digitally mediated contexts. Friedman et al. (2024) echo this concern, noting that while remote ultrasonography assistance is feasible in large animal settings, its effectiveness depends heavily on technician skill, platform quality, and clinician interpretation. Their findings suggest that clear protocols, real-time support, and training are essential to uphold ethical and diagnostic standards.

Data privacy is another key concern. As more clinics adopt cloud-based EMRs, wearable device integrations, and mobile app platforms, they must ensure compliance with data protection laws such as the CCPA and GDPR (Teller, 2022). Best practices include end-to-end encryption, two-factor authentication, and clear client communication about how data is used and stored. Wearable and remote sensing technologies, such as those reviewed by Zhao et al. (2025) and Haghi et al. (2021), introduce additional complexities. Data collected continuously from animals, often outside the clinic environment, must be anonymized, securely transmitted, and interpreted in ways that respect both client privacy and clinical limitations. Ensuring that all data management practices align with ethical expectations is critical for trust and compliance.

Finally, clinics must consider the intersection of legal, ethical, and operational policies. Freidman (2019) notes that the most successful telehealth adopters are those who view compliance not as a burden but as a foundation for client trust. Aligning legal and ethical standards with organizational workflows ensures both consistency and resilience in a rapidly evolving regulatory landscape. As states like California expand legal access to veterinary telehealth services (Axios, 2024), rural practices must proactively adapt to both opportunity and obligation. By embedding regulatory compliance into care delivery systems and staff protocols, clinics can lead with integrity while embracing digital transformation.

Barriers to Implementation

Despite its potential, telehealth implementation remains uneven, with many practices encountering significant operational hurdles. Staff resistance is common, especially among senior clinicians who are skeptical of remote diagnostics (Kogan et al., 2016). Niessen and Forcada (2019) report that many veterinarians lack confidence in virtual assessments, fearing misdiagnosis and liability. Training is essential to address these gaps. Watson et al. (2024) found that students exposed to telehealth scenarios during their veterinary education felt more prepared to integrate digital tools into practice. Clinics that offer continuing education, mentorship, and shadowing opportunities are better equipped to shift team culture.

Friedman et al. (2024) further support this point, showing that even experienced staff in large animal hospitals required technical assistance to successfully execute remote ultrasonography, underscoring the importance of ongoing support and training when deploying complex telehealth tools. Financial constraints are another barrier. Huynh (2019) details the costs of implementing smartphone-based diagnostic tools, EMR systems, and teleconsultation software. While these tools offer long-term ROI, upfront costs may be prohibitive for rural or small practices. Boone et al. (2023) suggest phased implementation and strategic vendor partnerships to mitigate budget impacts. Guintard et al. (2024) add that in production animal medicine, financial risk and return uncertainty are particularly challenging. Their research identifies lack of reimbursement models, unclear ROI metrics, and technology fatigue among producers as key deterrents to broader adoption. Their findings indicate that successful implementation requires both clear economic incentives and demonstration of value-added outcomes.

Workflow redesign is a critical success factor. The AVMA and AAHA (2018) recommend conducting a full audit of current systems before layering on digital services. This includes assessing scheduling practices, EMR compatibility, client communication channels, and triage protocols. The AVMA (2021) emphasizes that for telehealth to be effectively integrated into clinical workflows, protocols for image sharing, prescribing, and documentation must be clearly defined and aligned with practice standards to ensure continuity of care. The AVMA's updated 2025 guidelines (AVMA, 2021) provide expanded recommendations, including practical implementation checklists and suggested timelines, to assist practices in aligning operations with current telehealth best practices and regulatory requirements.

Client adoption is not guaranteed. Springer et al. (2023) found that awareness of telehealth remains low among pet owners, and skepticism is common. Practices must prioritize client onboarding through tutorial videos, user guides, and transparent FAQs. Perry et al. (2023) stress the importance of setting clear expectations, including what services are suitable for virtual care and when in-person follow-up is required. Johnson and Martinez (2025) note that effective at-home management of chronic conditions relies on strong client engagement and education. Their findings suggest that user-friendly video platforms, consistent communication, and structured follow-up schedules improve both compliance and satisfaction.

Access equity is a final consideration. Neill and Morgan (2025) highlight that rural clients may lack broadband access or digital fluency, making telehealth less accessible. Practices should consider offering alternative consultation formats, such as phone-based triage or in-clinic digital kiosks. Community outreach and education can also help close the digital divide and expand access. Zhao et al. (2025) emphasize that remote sensing devices often depend on stable internet connections and real-time data integration, both of which may be unreliable in rural regions. Their review recommends future research on low-bandwidth diagnostic tools and hybrid care models that bridge technology limitations. Similarly, Haghi et al. (2021) caution that wearable devices require ongoing calibration and may not be intuitive for all users, particularly in low-resource environments. Clinics must plan for technical support, device orientation, and follow-up in populations with limited digital literacy.

Legal, Ethical, and Regulatory Dynamics of Veterinary Telehealth

The expansion of telehealth tools will require systemic alignment across practice management software, diagnostics, and regulatory systems. Education, funding, and interoperability standards are foundational to scaling these innovations in rural settings. Emerging technologies such as remote vital sensing and AI-driven diagnostics offer substantial promise in closing rural care gaps. Zhao et al. (2025) detail how these tools can automate triage and early detection while reducing practitioner burden. Abu-Seida et al. (2024) add that wearable monitors, real-time imaging, and mobile diagnostic kits are transforming veterinary practice into a more responsive, decentralized care model. These tools are especially relevant for livestock and rural patients, where physical access to a clinic may be limited.

The AVMA's updated guidelines underscore the need for ethical safeguards, data security protocols, and continuing education to ensure safe integration of these advanced tools (AVMA, 2025). The guidelines also emphasize cross-state consistency in telehealth regulations, urging states to harmonize definitions and requirements to reduce confusion for multi-location practices and mobile veterinarians (AVMA, 2025). These forward-looking strategies may well define the next phase of rural veterinary care.

Ethically, the AVMA emphasizes that any remote care must adhere to the same standards as in-person visits, including obtaining informed consent, ensuring client and patient confidentiality, maintaining continuity of care, and clearly documenting teleconsultations in the medical record (AVMA, 2025). These principles protect the integrity of the veterinary profession and help build trust with clients navigating new care modalities. However, some practitioners express concern about potential misdiagnosis, liability risks, and pressure to act outside their comfort zones in the absence of physical exams (Magalhães-Sant'Ana et al., 2020). Johnson and Martinez (2025) echo these concerns in their study on telehealth use in managing chronic feline conditions, noting that clients and clinicians alike need structured guidelines to ensure ethical care delivery without overburdening either party. Their research supports clear expectations and boundaries to avoid moral distress and liability exposure. In response to these issues, researchers and policy advocates have called for reform. Hassan et al. (2024) argue that VCPR policies should evolve to accommodate technological advances while preserving ethical safeguards.

The AVMA's report on VCPR legislation further validates this position, noting that multiple U.S. states are considering bills to allow for telehealth-established VCPRs under specific conditions such as emergencies, technician-assisted exams, or previous patient history (AVMA, 2023). Others propose hybrid care models, including technician-facilitated exams, real-time video assessments, and AI-supported triage tools that support veterinary judgment without violating VCPR boundaries (Marshall, 2019). Regulatory "sandboxes" have also been recommended as a means to pilot new approaches under controlled conditions and generate data to inform broader policy shifts (Nelson-Pratt, 2018).

Privacy and data security represent another regulatory challenge. Veterinary practices must comply with data protection regulations such as GDPR in Europe or CCPA in California when managing client and patient data through digital platforms (Marshall, 2019). As telehealth adoption increases, cybersecurity threats, ranging from unauthorized data access to system breaches, must be mitigated through encryption, staff training, and clear client communication about data use and protection (Teller, 2022).

Guintard et al. (2024) emphasize that telehealth uptake in livestock operations is particularly vulnerable to data misuse, due to a lack of clear consent mechanisms and shared use of platforms across enterprises. They advocate for developing veterinary-specific data governance protocols to ensure client trust and compliance. These requirements further reinforce the need for investment in secure digital infrastructure and protocols tailored to veterinary care.

Operational Challenges and Barriers to Implementation

Implementing a comprehensive telehealth strategy presents operational challenges that span workforce readiness, financial investment, technology selection, workflow design, and client onboarding. A primary barrier is resistance to change among staff and veterinarians accustomed to traditional models of care. Surveys show that older practitioners and solo practice owners are often skeptical about telemedicine's clinical efficacy, citing concerns over diagnostic limitations and technology overload (Kogan et al., 2016; Niessen & Forcada, 2019). Overcoming this resistance requires strong leadership, targeted training, and clear demonstration of telehealth's benefits.

Guintard et al. (2024) confirm that veterinary professionals in livestock production settings also report cultural resistance and lack of institutional support as top barriers to telemedicine, especially where face-to-face care is deeply ingrained. Leadership buy-in and peer-driven education were found to be essential enablers of implementation success.

Financial barriers also inhibit adoption. Smaller and rural practices may struggle to afford the upfront costs of telehealth platforms, wearable devices, digital diagnostics, and cybersecurity tools. While cloud-based and smartphone-enabled technologies offer lower-cost alternatives, implementation still requires investment in hardware, software licensing, and ongoing IT support (Huynh, 2019). Practices must also budget for training, marketing, and change management processes to ensure a smooth transition. As noted by Abu-Seida et al. (2024), while telemedicine tools can lower operating costs over time, clinics often face early-stage budget constraints. Their review advocates phased adoption, vendor collaboration, and public-private grants as ways to make telehealth accessible in under-resourced regions.

Workflow integration poses another challenge. Practices must redesign scheduling systems to accommodate teleconsulting visits, create protocols for video triage and remote monitoring, and define clear boundaries between in-clinic and virtual care (American Veterinary Medical Association & American Animal Hospital Association, 2018). This includes determining which services are appropriate for telemedicine, identifying the required documentation, and managing logistics such as image sharing, prescription approvals, and client follow-ups. Without clear workflows, clinics risk confusion, inefficiency, and substandard care. The AVMA's updated 2025 telehealth guidelines reinforce the importance of defining workflows that distinguish teleadvice, teletriage, and teleconsultation, and emphasize that every practice must map service delivery to the VCPR rules in their jurisdiction (American Veterinary Medical Association, 2025).

Staffing and skills development are equally critical. As remote services expand, veterinary technicians and support staff will need training in client communication, technology use, virtual triage procedures, and wearable device operation. Studies suggest that practices with well-trained teams are more likely to succeed in delivering safe and effective telehealth services (Freidman et al., 2024; Navas de Solis et al., 2019). New roles may emerge, such as telehealth coordinators or digital care nurses, requiring organizational restructuring to align talent with new care models. Friedman et al. (2024) found that staff trained in remote ultrasound guidance outperformed untrained peers in a randomized trial at a large animal hospital, reinforcing that telehealth effectiveness is strongly tied to clinician preparation and technical fluency.

Finally, client adoption remains inconsistent. While surveys indicate growing interest in virtual care, many pet owners remain unaware of telehealth offerings or uncertain about their effectiveness (Springer et al., 2023). Practices must develop marketing and education strategies that build client confidence, explain telemedicine's appropriate use cases, and clarify costs and expectations. Effective onboarding can include tutorial videos, mobile app walkthroughs, and clear policies on cancellations, data use, and emergency escalation. Johnson and Martinez (2025) suggest that telehealth uptake improves when clients receive consistent guidance on device setup, symptom tracking, and communication protocols, especially in chronic care scenarios where engagement must be sustained over time.

Strategic and Change Management Models

To guide this veterinary practice through its transformation, consideration of three theoretical frameworks offer strategic direction: the McKinsey 7S Model, Kotter's 8-Step Change Model, and the Action Research Model. Each provides a unique lens through which to approach digital transformation, workforce adaptation, and stakeholder collaboration. Recent telehealth implementation research in veterinary and human healthcare contexts affirms the value of using structured frameworks to manage complexity, build internal alignment, and sustain innovation during digital transitions (Johnson et al., 2023; Guintard et al., 2024).

The McKinsey 7S Model is well-suited for evaluating the alignment of internal elements in support of organizational transformation. The model includes seven interdependent elements: Strategy, Structure, Systems, Shared Values, Skills, Style, and Staff. In the context of implementing telehealth, Strategy refers to the organization's shift toward 24-hour, technology-enabled care. Structure requires adjustments to staffing and workflow to accommodate digital triage and teleconsultation. Systems must include telemedicine platforms, scheduling tools, and EMR integration. Shared Values, such as trust, care, and ethical responsibility, must guide all decisions. Skills reflect the need for technical training and telecommunication proficiency. Style refers to leadership behaviors that promote adaptability and inclusion. Staff refers to personnel who will operate and support telehealth offerings (Waterman et al., 1980). This model ensures that the organization addresses the interconnectivity of its components rather than isolating technology as a stand-alone solution.

While structural alignment is essential, evidence suggests that many telehealth implementations falter without deliberate attention to the change management process itself. Kho, Gillespie, and Martin-Khan (2020) conducted a systematic scoping review of 48 telemedicine service implementations and identified 16 distinct change management practices. These practices were categorized across three phases (preparation, implementation, and reinforcement) emphasizing the need for proactive, process-driven change strategies. The authors observed that many healthcare organizations applied change management reactively, often overlooking the need for strategic planning and long-term sustainability. Their findings support the integration of structured change models, like Kotter's 8-Step or the Action Research Model, to ensure transformation efforts are not only technically feasible but also operationally supported and culturally embedded.

To facilitate the change process itself, Kotter's 8-Step Change Model offers a proven roadmap. The steps include: (1) Establishing a sense of urgency (e.g., the need to remain competitive and accessible), (2) Creating a guiding coalition (e.g., a cross-functional leadership team including veterinarians, technicians, and IT staff), (3) Developing a vision and strategy (e.g., digital-first, client-centric care), (4) Communicating the change vision across the organization, (5) Empowering broad-based action by removing barriers and supporting skill development, (6) Generating short-term wins (e.g., early client satisfaction milestones or successful video consults), (7) Consolidating gains and producing more change (e.g., expanding telehealth to other service lines), and (8) Anchoring new approaches in the culture (e.g., integrating telehealth into standard workflows and job descriptions) (Appelbaum et al., 2012). This model provides an actionable structure for implementation and ensures that the organization addresses emotional, behavioral, and operational challenges in the change process.

The Action Research Model adds a dynamic and participatory dimension by embedding continuous, collaborative inquiry into the transformation process. This model consists of nine iterative steps: identifying and limiting the topic, gathering preliminary information, reviewing related literature, developing a research plan, implementing the plan and collecting data, analyzing the data, developing an action plan, sharing results, and reflecting on the process (Mertler, 2017). It promotes inclusive, data-informed decision-making by empowering internal stakeholders to diagnose problems, test interventions, and adapt in real time. For rural veterinary practices facing evolving client expectations and technology constraints, this model supports practical innovation and staff engagement through ongoing cycles of inquiry and improvement.

The following sections elaborate on each model in greater detail, illustrating how their individual components can be applied to the specific operational, clinical, and cultural challenges faced by rural veterinary practices undergoing digital transformation.

McKinsey 7S Framework

The McKinsey 7S Framework is a diagnostic and strategic tool developed by Peters and Waterman, which identifies seven interdependent elements crucial to organizational effectiveness: Strategy, Structure, Systems, Style, Staff, Skills, and Shared Values (Suwanda & Nugroho, 2022). The advantage of this model lies in its ability to illustrate how these seven elements must be aligned for effective change.

Each element contributes to the performance of an organization, with imbalance in one area often negatively affecting others. In both public and private sector research, the 7S model has proven valuable for evaluating change readiness and driving organizational transformation (Suwanda & Nugroho, 2022). In the context of telehealth integration, each element of the McKinsey 7S model provides a distinct yet interconnected consideration essential for successful transformation. Strategy encompasses the clinic's approach to securing a competitive advantage through the deployment of telehealth services, including digital triage, remote monitoring, and tiered subscription plans. Research emphasizes that a user-centered strategy, grounded in flexible pricing, equitable access, and data-informed care, significantly improves adoption rates in rural settings (Freidman et al., 2024; Jergler, 2018). The strategic roadmap must also anticipate regulatory shifts, forge technological partnerships, and accommodate evolving consumer expectations.

Structure pertains to the way the organization configures its roles and responsibilities to support these new workflows. Rural practices may need to restructure their teams by establishing dedicated telehealth coordinators or appointing specific personnel to manage virtual care delivery (Niessen & Forcada, 2019). Evidence suggests that adaptable organizational structures enhance agility and resilience, both of which are vital in rapidly evolving digital environments (Suwanda & Nugroho, 2022).

Systems include the formal and informal procedures that sustain day-to-day operations. These systems may range from electronic medical records and scheduling software to cybersecurity protocols and client communication platforms (American Veterinary Medical Association, 2019; Teller, 2022). Seamless integration across these tools is essential for enabling real-time consultations, safeguarding client data, and improving service efficiency. Conversely, fragmented or outdated systems can delay implementation and hinder change adoption (Suwanda & Nugroho, 2022). This point is reinforced by Johnson et al. (2023), who developed a telehealth-focused implementation science toolkit grounded in dissemination and implementation (D&I) research principles. The toolkit offers structured guidance for evaluating telehealth program design, rollout, and sustainability across various healthcare domains. While originally applied in human medicine, the framework's focus on stakeholder engagement, success metrics, and equity makes it highly adaptable to veterinary settings. For rural practices integrating telehealth, this toolkit provides a practical model for assessing readiness, identifying gaps, and maintaining momentum across change cycles. Embedding such tools into the transformation process can ensure that digital veterinary services are not only launched effectively but also continuously refined and scaled to meet evolving demands.

Shared values represent the foundational beliefs and principles that guide the practice. As telehealth alters client-provider interactions, values such as compassion, accessibility, and technological integrity must serve as the ethical compass for decision-making. These shared values foster alignment and legitimacy among staff and clients, particularly during times of transition that disrupt established norms (Marshall, 2019; Magalhães-Sant'Ana et al., 2020).

Style refers to the leadership approach adopted during change. Transformational leadership, marked by transparent communication, vision-setting, and encouragement of innovation, has been shown to facilitate higher adoption rates of telehealth in healthcare contexts (Appelbaum et al., 2012; Springer et al., 2023). Leaders must consistently model digital priorities and reinforce the importance of innovation through their messaging and behavior.

Staff encompasses the human capital required to sustain telehealth operations. Veterinary practices must evaluate their existing staffing levels and consider hiring or upskilling individuals with competencies in technology, adaptability, and emotional intelligence (Gottlieb & Rand, 2018). Achieving staff buy-in is a critical component in translating digital strategy into practice.

Lastly, skills reflect the capabilities necessary to deliver telehealth services effectively. These include competencies in virtual triage, client communication, digital diagnostics, and the operation of wearable medical devices (Watson et al., 2024). Clinics that prioritize ongoing professional development and cross-training are more likely to maintain high-quality standards of care and reduce the risk of role ambiguity among team members.

When aligned, these seven elements provide a cohesive framework for telehealth adoption. Misalignment in any area, such as undertrained staff or outdated systems, can impede progress, demonstrating the importance of using this model to identify and resolve inconsistencies across people, processes, and systems (Suwanda & Nugroho, 2022). The McKinsey 7S model also provides a diagnostic lens to identify resistance points, refine internal processes, and monitor progress over time.

Kotter's 8-Step Change Model:

Kotter's 8-Step Change Model provides a sequential process to guide organizational transformation (*see Figure 2*). First introduced in Kotter's seminal work and later refined in applied studies, this model remains one of the most widely utilized frameworks in organizational change, particularly in healthcare and education

(Appelbaum et al., 2012). In the context of veterinary telehealth implementation, this model serves as a practical guide for structuring change, building internal momentum, and achieving sustainable adoption.

The process begins by establishing a sense of urgency, which is essential to overcoming complacency and mobilizing stakeholders around the need for innovation. For rural veterinary clinics, urgency may stem from rising client expectations for convenience, limited access to emergency care, and competitive market shifts toward digital-first service delivery (Springer et al., 2023). By clearly articulating these external pressures, leadership can foster a shared recognition that maintaining the status quo is no longer viable.

The second step is to build a guiding coalition composed of individuals from across the organization who bring credibility, expertise, and influence. This may include veterinarians, veterinary technicians, IT personnel, and administrative leaders who collectively oversee implementation efforts (Freidman et al., 2024). Research has shown that transformation is more likely to succeed when the coalition is both cross-functional and empowered (Appelbaum et al., 2012).

Next, the organization must form a strategic vision and related initiatives that align with the broader goals of the practice. A compelling vision might center on "delivering responsive, technology-enabled care that prioritizes animal well-being and client trust." Key initiatives under this vision could include the rollout of tele-triage protocols, the integration of app-based appointment systems, and the use of wearable devices to support chronic condition monitoring (Boone et al., 2023).

With the vision in place, the clinic must communicate the vision consistently across both internal and external channels. Internally, this means reinforcing the value of telehealth to staff, celebrating early wins, and drawing clear connections between digital tools and the clinic's core values. Externally, communication should highlight the client-facing benefits of telehealth, including reduced pet stress, greater convenience, and enhanced access (Niessen & Forcada, 2019; Jergler, 2018).

The fifth step involves empowering broad-based action by eliminating barriers that obstruct change. This includes addressing outdated workflows, bridging skill gaps, and replacing legacy systems. A supportive environment is cultivated through investments in new digital platforms, streamlined processes, and comprehensive training programs (Watson et al., 2024).

To build momentum, the clinic should focus on generating short-term wins that demonstrate measurable success. These might include launching a weekend virtual consult program or piloting wearable health monitoring tools. Such initiatives not only validate the broader strategy but also help reinforce team morale and stakeholder confidence (Kogan et al., 2016).

Following these early successes, it is critical to sustain acceleration by scaling initiatives, refining operational systems, and standardizing new procedures. Performance indicators (e.g., appointment conversion rates, staff proficiency metrics, and client satisfaction scores) can be used to assess progress and guide reinvestment in infrastructure (Appelbaum et al., 2012).

Finally, to ensure lasting impact, the organization must anchor new approaches in its culture. This means embedding telehealth within performance reviews, training modules, and strategic plans so that it becomes a normalized part of daily operations. Without cultural reinforcement, even successful pilot programs risk losing traction over time (Marshall, 2019; Suwanda & Nugroho, 2022).

Action Research Model

The Action Research Model (Mertler, 2017) offers a cyclical, practitioner-centered framework that supports reflective, collaborative problem-solving in complex and evolving environments. Unlike linear or top-down change models, action research prioritizes stakeholder engagement by empowering those closest to the issue to co-create, test, and refine interventions. This approach is particularly well suited to rural veterinary practices implementing telehealth, where evolving technologies, fluctuating client expectations, and operational challenges demand adaptive and iterative solutions.

The model consists of nine iterative steps, beginning with identifying and limiting the topic. In this initial stage, the team collaboratively selects a specific, manageable focus area for inquiry. Within a veterinary setting, this might involve low client engagement during evening telehealth hours, underutilization of wearable technology, or inconsistent protocols for remote triage. The goal is to define a clear and actionable issue that is both relevant and observable.

The next step involves gathering preliminary information through informal observations, staff feedback, and anecdotal data. This early data collection phase helps validate the topic's relevance and reveals the existing conditions before formal interventions are introduced. Understanding the scope of the issue ensures the team is aligned on priorities and expectations.

Once the issue is scoped, the team moves to review the related literature. This step provides theoretical grounding and practical context by incorporating current research, professional standards, and evidence-based guidelines. A concise literature review helps inform strategy design and ensures that proposed interventions are aligned with best practices in both veterinary medicine and telehealth delivery.

The fourth step is to develop a research plan, outlining the intended intervention, data collection strategy, success metrics, and implementation timeline. For example, a clinic might trial weekend virtual consults or introduce a new digital check-in protocol. The plan must be specific enough to guide implementation while remaining flexible enough to allow adjustments based on feedback.

Following planning, the team proceeds to implement the plan and collect data. During this phase, both quantitative and qualitative data are systematically gathered. Quantitative indicators could include appointment volumes or device usage rates, while qualitative insights might come from staff debriefs or client feedback sessions. This dual approach enables a more holistic assessment of the intervention's effectiveness.

In the sixth step, the team analyzes the data, reviewing trends, identifying successes or challenges, and comparing outcomes to original goals. The emphasis here is not only on measuring results but also on understanding the underlying causes of success or failure. This analysis lays the foundation for thoughtful and evidence-informed next steps.

The seventh phase involves developing an action plan based on the data analysis. If results are favorable, the team may choose to scale the intervention or institutionalize it as part of the clinic's standard procedures. If the findings are mixed or inconclusive, the team can refine the approach and re-enter the cycle for further testing. The action plan must remain adaptive and scalable, with clear ownership and timelines.

Next, the team moves to share the results. Transparent communication with internal staff, leadership, and potentially clients or external stakeholders enhances organizational learning and builds momentum for sustained change. Sharing findings also provides opportunities for feedback that may inform future iterations.

The final step is to reflect on the process. This reflective practice includes evaluating not only the effectiveness of the intervention itself but also the strengths and weaknesses of the inquiry process. Documenting lessons learned promotes a culture of continuous improvement and positions the organization to respond effectively to future challenges.

Mertler (2017) emphasizes that the true strength of action research lies in its capacity to foster empowerment, shared responsibility, and sustainable transformation. For rural veterinary practices navigating the complexities of telehealth adoption, this model offers a practical roadmap for iterative change that is responsive to both clinical realities and client needs.

Consideration of the McKinsey 7S, Kotter, or Action Research models provide powerful frameworks for aligning systems, managing human dynamics, and fostering real-time innovation. For rural veterinary practices, consideration of these models ensures that telehealth and emergency care innovations are not only adopted but also iteratively refined, scaled, and sustained through stakeholder collaboration and evidence-based adaptation.

SOCIAL SIGNIFICANCE

Improving access to veterinary care through telehealth has wide-reaching social benefits, particularly for pet owners in rural or underserved regions. Veterinary care is increasingly recognized as a social determinant of community well-being, especially where pets provide emotional support, companionship, or utility in agricultural work. Limited access to care exacerbates health disparities, not only for animals but also for owners whose livelihoods or mental health may depend on animal wellness. Expanding telehealth services addresses this inequity by delivering timely care, reducing the stress and cost of travel, and enabling continuity of care across remote settings (Abu-Seida et al., 2024). For example, new remote diagnostic tools such as wearable devices and real-time sensing platforms help democratize access to sophisticated health monitoring (Zhao et al., 2025; Haghi et al., 2021). In this way, veterinary telehealth reflects a broader digital health movement aimed at closing access gaps and improving quality of life across demographic divides.

Social significance also includes the impact of regulatory reforms on public trust and ethical care standards. New legislation like California's expanded access to veterinary telehealth (Axios, 2024) has drawn attention to the need for national frameworks that protect both consumers and practitioners. The AVMA's 2025 guidance emphasizes that telehealth must not be a shortcut around in-person care, but rather a complementary tool rooted in ethical responsibility and informed consent (AVMA, 2025). These reforms signal a broader cultural shift toward legitimizing telehealth in veterinary medicine, aligning public expectations with evolving care models. When implemented thoughtfully, these changes foster more equitable, inclusive care – ensuring all pet owners, regardless of geography or income, can access high-quality veterinary support.

PRACTICAL SIGNIFICANCE

This project delivers a scalable framework for rural veterinary clinics to integrate telehealth in a way that preserves clinical quality while enhancing access, efficiency, and client engagement. Video-enabled consultations and subscription-based 24-hour emergency care offer financially viable alternatives to physical expansion, especially in communities with limited veterinary access. For rural veterinary practices, the operational benefits of telehealth extend beyond convenience. They support after-hours care, chronic condition management, and stronger client relationships, all without requiring significant infrastructure or staffing increases.

This chapter provides a clear framework for translating telehealth strategies into practice. Rural clinics can use this guidance to:

- Introduce virtual triage to prioritize urgent care needs;
- Pilot subscription plans that provide clients with 24/7 access to clinicians;
- Train staff in remote diagnostic tools and client communication protocols;
- Integrate mobile apps and wearable technology into wellness programs.

Recent findings demonstrate that video telehealth improves management of chronic conditions in cats (Johnson & Martinez, 2025), while remote support during point-of-care ultrasonography enhances diagnostic accuracy in large animal hospitals (Friedman et al., 2024). Wearable technologies and remote sensing systems further reduce unnecessary clinic visits and support better outcomes (Zhao et al., 2025; Haghi et al., 2021). This chapter also aligns with updated AVMA telehealth guidelines and shifting VCPR regulations (AVMA, 2023), equipping practitioners to expand care delivery while maintaining compliance.

The organizational models help practices manage internal transformation, particularly those without formal change management structures. Veterinary associations can use these insights to support training efforts, and technology providers may apply them when designing tools for rural markets. For clinic owners, the strategies presented offer practical steps toward digital modernization.

LIMITATIONS

While this model offers practical solutions, several factors may constrain its broad application. Chief among these is the regulatory landscape: state-level restrictions around virtual care and the requirement for a physical exam to establish the Veterinarian-Client-Patient Relationship (VCPR) limit how telehealth can be deployed across jurisdictions. Technology infrastructure is another concern. Not all rural areas have reliable broadband internet, and not all clients are equally comfortable using mobile platforms or remote diagnostics. These disparities must be considered when designing or marketing digital services. Affordability also plays a role. Subscription-based care models offer predictability for clinics, but they may not appeal to or be accessible for all clients – especially those with limited resources. This raises questions about pricing flexibility and outreach strategies. Finally, this analysis is conceptual and literature-based; real-world implementation may surface challenges not anticipated here. Further field and beta testing would be useful for validating the impact and refining the approach.

POLICY IMPLICATIONS

Policy reform is likely to play a key role in scaling telehealth for veterinary use. The current VCPR rules in most U.S. states – requiring an in-person exam before treatment – remain a barrier to fully remote care. As technology advances and demand grows, regulatory bodies may need to revisit these requirements to allow for safe, ethically grounded virtual relationships between providers and clients. This chapter's findings could help inform those conversations. By showing how remote care models align with existing ethics codes and improve access without sacrificing quality, rural clinics and advocacy groups can contribute to a more flexible, modernized regulatory framework. State veterinary boards may find value in pilot programs or temporary waivers that allow for limited virtual VCPR under specific conditions, especially in underserved regions.

RECOMMENDATIONS

To ensure a successful digital transformation, rural veterinary practices must implement a comprehensive strategy that unites telehealth innovation, ethical frameworks, and sustainable organizational change. This case study found that clinics that align internal systems, engage staff, and implement care models grounded in ethical, operational, and technological integrity are better positioned to deploy telehealth services in a sustainable and client-centered manner.

The adoption of video-based telehealth and a 24-hour emergency subscription care model is not simply a technical upgrade, it represents a reimagining of how care is accessed, delivered, and managed in underserved communities. Supported by the McKinsey 7S Framework, Kotter's 8-Step Change Model, and the Action Research Model, this transformation process requires alignment of people, processes, values, and technology.

Implementation should begin with the development of a centralized, user-friendly digital platform that consolidates all client-facing services, from video consultations and appointment scheduling to wearable device integration and automated follow-ups. Mobile compatibility across platforms will maximize accessibility for a diverse client base, including those with varying levels of digital literacy. Predictive analytics should be layered into this ecosystem to help clinics identify client behavior trends, anticipate care needs, and personalize outreach. These tools not only promote early intervention and continuity of care but also improve client loyalty and subscription retention.

To address diverse financial and service needs, practices should offer tiered subscription models that include varying levels of virtual access, emergency support, in-home care coordination, and technology-enabled monitoring. Transparent pricing, bundled discounts, and auto-renewal options can help stabilize revenue while offering clear value to clients. The implementation of such models should be regularly reviewed using evidence-based tools grounded in implementation science, such as toolkits designed to evaluate adoption, sustainability, and equity in digital care models (Johnson et al., 2023; Kho et al., 2020).

Staff readiness remains a critical success factor. Investment in training that encompasses telehealth technologies, virtual communication skills, ethical guidelines, and data privacy is essential. When staff are empowered with knowledge and tools, they are more likely to embrace digital workflows, contribute to innovation, and reinforce clinical quality. Clinics should embed ongoing professional development into organizational culture, ensuring that the team remains agile as technology evolves.

Positioning the clinic strategically in the digital veterinary marketplace requires clearly articulating the value proposition (e.g., 24/7 emergency access, board-certified virtual consults, or multilingual support). These differentiators should be emphasized across digital marketing channels and supported by mechanisms that capture and respond to client feedback. A transparent, service-driven brand helps foster trust in virtual care, especially among clients new to telemedicine.

Client education and community outreach must complement the digital rollout. Clinics should use webinars, explainer videos, and written resources to normalize telehealth, reduce anxiety, and explain service boundaries. Partnerships with local shelters, schools, and pet retail businesses can expand awareness while reinforcing the clinic's role as a trusted community health provider. Regular communication via email and social media can sustain engagement, promote new offerings, and ensure client touchpoints remain strong.

A commitment to continuous innovation is vital. Regular feedback loops (e.g., surveys, analytics dashboards, and staff retrospectives) can reveal emerging client needs, flag process inefficiencies, and drive incremental improvements. Internally, innovation committees can facilitate cross-functional brainstorming and experimentation. Externally, partnerships with telehealth vendors, wearable device developers, insurance companies, and broader pet service networks can expand the clinic's ecosystem and increase service value.

It is also important that veterinary leaders and associations consider the regulatory environment when implementing digital solutions. Current VCPR requirements in most U.S. states mandate physical examinations before virtual care can be delivered, limiting the flexibility of telehealth services. Adapting policy models from human healthcare such as those proposed by Neal and Greenberg (2024) may help veterinary practices expand access while maintaining ethical and legal safeguards.

These recommendations not only offer a roadmap for individual practice transformation but also provide guidance for educational institutions, policymakers, and industry partners seeking to support the broader integration of telehealth in veterinary medicine. Ultimately, the digital transformation of rural veterinary care is not a one-time event, but a continuous and collaborative process. Clinics that commit to iterative change, guided by data, ethics, and stakeholder engagement, will be best positioned to deliver high-quality, accessible care in a rapidly evolving healthcare landscape.

RECOMMENDATIONS FOR FUTURE RESEARCH

Building on the findings of this chapter, several avenues for future research can further support the implementation of telehealth in rural veterinary practices and other small-scale clinical environments. First, comparative studies between rural and urban veterinary telehealth adoption can reveal contextual factors that influence success. Second, longitudinal research on client retention and clinical outcomes associated with subscription-based care models may offer valuable insights into financial sustainability. Third, policy-focused research exploring the adaptation of VCPR regulations in the U.S. could inform advocacy and reform. Fourth, future studies might assess the ethical implications of wearable monitoring and AI-assisted diagnostics in veterinary practice. Finally, implementation science frameworks such as Dissemination and Implementation (D&I) models could be applied across diverse care settings to build replicable strategies for technology-enabled service delivery.

REFERENCES

Abu-Seida, A. M., Abdulkarim, A., & Hassan, M. H. (2024). Veterinary telemedicine: A new era for animal welfare. *Open Veterinary Journal, 14*(4), 952–961. DOI: 10.5455/OVJ.2024.v14.i4.2 PMID: 38808291

American Veterinary Medical Association. (2021). *AVMA guidelines for the use of telehealth in veterinary practice: Implementing connected care*. American Veterinary Medical Association. Retrieved from https://www.avma.org/sites/default/files/2021-01/AVMA-Veterinary-Telehealth-Guidelines.pdf

American Veterinary Medical Association. (2023, November 2). *VCPR requirements fuel state legislative activity*. Retrieved from https://www.avma.org/news/vcpr-requirements-fuel-state-legislative-activity

American Veterinary Medical Association & American Animal Hospital Association. (2018). *The real-life rewards of virtual care: How to turn your hospital into a digitally connected practice with telehealth*. Retrieved March 2, 2025, from https://www.aaha.org/globalassets/05-pet-health-resources/virtual_care.pdf

Appelbaum, S. H., Habashy, S., Malo, J., & Shafiq, H. (2012). Back to the future: Revisiting Kotter's 1996 change model. *Journal of Management Development, 31*(8), 764–782. DOI: 10.1108/02621711211253231

Axios. (2024). *Pets in California now have wider access to telehealth options*. Retrieved May 29, 2025, from https://www.axios.com/local/san-francisco/2024/01/03/telehealth-pets-california-law

Bishop, G. T., Evans, B. A., Kyle, K. L., & Kogan, L. R. (2018). Owner satisfaction with use of videoconferencing for recheck examinations following routine surgical sterilization in dogs. *Journal of the American Veterinary Medical Association, 253*(9), 1151–1157. DOI: 10.2460/javma.253.9.1151 PMID: 30311532

Boone, G., Bain, M., Cutler, J., & Moody, C. M. (2023). Evaluating video telemedicine for providing virtual health care for cats via mock spay recheck examinations. https://doi.org/DOI: 10.1016/j.applanim.2023.106061

Donham, B., & Wickett, M. L. (2018). Novel use of FaceTime video calling in a deployed setting to assist with the care of a military working dog. *Canadian Journal of Surgery*, •••, S232–S234. DOI: 10.1503/CJS.015118 PMID: 30418010

Friedman, N., Pancer, Z., Savic, R., Tseng, F., Lee, M. S., Mclean, L., Bagli, D. J., & Tessaro, M. O. (2024). Evaluation of remote assistance for point-of-care ultrasonography in a large animal hospital: A controlled randomized trial. *Journal of the American Veterinary Medical Association, 262*(5), 589–595. DOI: 10.2460/javma.23.09.0511 PMID: 38452487

Gottlieb, S., & Rand, J. (2018). Managing feline diabetes: Current perspectives. *Veterinary Medicine (Auckland), 9*, 33–42. DOI: 10.2147/VMRR.S125619 PMID: 30050865

Guintard, A., Assié, S., Lefèvre, L., Delsart, M., Dile, B., & Millar, N. (2024). Barriers and facilitators to implementing veterinary telemedicine in animal production. *Frontiers in Veterinary Science, 11*, 1452653. DOI: 10.3389/fvets.2024.1452653 PMID: 39634757

Haghi, M., Danyali, S., Ayasseh, S., Wang, J., Aazami, R., & Deserno, T. M. (2021). Wearable devices in health monitoring from the environmental towards multiple domains: A survey. *Sensors (Basel), 21*(6), 2130. DOI: 10.3390/s21062130 PMID: 33803745

Hassan, M., Abdulkarim, A., & Seida, A. (2024). Veterinary telemedicine: A new era for animal welfare. *Open Veterinary Journal, 14*(4), 952. DOI: 10.5455/OVJ.2024.v14.i4.2 PMID: 38808291

Huynh, M. (2019). Smartphone-based device in exotic pet medicine. *The Veterinary Clinics of North America. Exotic Animal Practice, 22*(3), 349–366. DOI: 10.1016/j.cvex.2019.05.001 PMID: 31395319

Jergler, D. (2018, May 7). *High-tech remote health monitoring*. Veterinary Practice News. https://www.veterinarypracticenews.com/high-tech-remote-health-monitoring/

Johnson, E. E., Kruis, R., Verdin, R., Wells, E., Ford, D. W., & Sterba, K. R. (2023). Development of an implementation science telehealth toolkit to promote research capacity in evaluation of telehealth programs. *Telemedicine Reports, 4*(1), 286–291. DOI: 10.1089/tmr.2023.0039 PMID: 37817872

Johnson, L., & Martinez, R. (2025). Incorporating video telehealth for improving at-home management of cats with chronic mobility issues. *Frontiers in Veterinary Science, 12*, 1510006. DOI: 10.3389/fvets.2025.1510006 PMID: 40235573

Kane, C. K., & Gillis, K. (2018). The use of telemedicine by physicians: Still the exception rather than the rule. *Health Affairs, 37*(12), 1923–1930. DOI: 10.1377/hlthaff.2018.05077 PMID: 30633670

Kho, J., Gillespie, N., & Martin-Khan, M. (2020). A systematic scoping review of change management practices used for telemedicine service implementations. *BMC Health Services Research*, *20*(1), 815. Advance online publication. DOI: 10.1186/s12913-020-05657-w PMID: 32873295

Kogan, L. R., Hellyer, P. W., Ruch-Gallie, R., Rishniw, M., & Schoenfeld-Tacher, R. (2016). Veterinarians' use and perceptions of information and communication technologies. *Medical Research Archives*, *4*(2). https://journals.ke-i.org/mra/article/view/502

Magalhães-Sant'Ana, M., Peleteiro, M. C., & Stilwell, G. (2020). Opinions of Portuguese veterinarians on telemedicine: A policy delphi study. *Frontiers in Veterinary Science*, *7*, 549. Advance online publication. DOI: 10.3389/fvets.2020.00549 PMID: 32974405

Marshall, K. (2019, February 7). *Ethical issues for today's veterinarian in the digital age*. Veterinary Practice News. https://www.veterinarypracticenews.com/ethical-issues-for-todays-veterinarian-in-the-digital-age/

Mertler, C. A. (2017). *Action research communities: Professional learning, empowerment, and improvement through collaborative action research* (1st ed.). Taylor & Francis Group., DOI: 10.4324/9781315164564

Navas de Solis, C., Bevevino, K., Doering, A., O'Gan, D., Teller, L., & Underwood, C. (2019). Real-time telehealth using ultrasonography is feasible in equine practice. *Equine Veterinary Education*, *32*(4), 218–222. DOI: 10.1111/eve.13177

Neal, S. M., & Greenberg, M. J. (2024). Identifying policy alternatives to enable the virtual establishment of a veterinarian–client–patient relationship. *Journal of Feline Medicine and Surgery*, *26*(3), 1098612X231224167. Advance online publication. DOI: 10.1177/1098612X231224167 PMID: 38478919

Neill, C. L., & Morgan, K. L. (2025). Few differences in rural and urban adoption of veterinary telehealth services. *Journal of the American Veterinary Medical Association*, *263*(2), 1–7. DOI: 10.2460/javma.24.08.0508 PMID: 39642457

Nelson-Pratt, A. (2018). Telemedicine debate is taking far too long. *The Veterinary Record*, *183*(21), 663–663. DOI: 10.1136/vr.k5073 PMID: 30504170

Niessen, S., & Forcada, Y. (2019, March 11). *Telemedicine advances: Putting specialist help in all practices*. Vet Times. https://www.vettimes.co.uk/article/telemedicine-advances-putting-specialist-help-in-all-practices/

Quain, A., Ward, M. P., & Mullan, S. (2021). Ethical challenges posed by advanced veterinary care in companion animal veterinary practice. *Animals (Basel), 11*(11), 3010. DOI: 10.3390/ani11113010 PMID: 34827742

Springer, S., Lund, T. B., Corr, S. A., & Sandøe, P. (2023). Seeing the benefits, but not taking advantage of them: Dog and cat owners' beliefs about veterinary telemedicine. *The Veterinary Record, 194*(5), e3312. Advance online publication. DOI: 10.1002/vetr.3312 PMID: 37733831

Suwanda, S., & Nugroho, B. Y. (2022). Literature reviews: McKinsey 7S model to support organizational performance. *Technium Social Sciences Journal, 38*, 1–9. DOI: 10.47577/tssj.v38i1.7744

Teller, L. M. (2022). Exploring the challenges and opportunities presented by veterinary telemedicine. *The Veterinary Record, 191*(5), 210–212. DOI: 10.1002/vetr.2209 PMID: 36083053

Waterman, R. H.Jr, Peters, T. J., & Phillips, J. R. (1980). Structure is not organization. *Business Horizons, 23*(3), 14–26. DOI: 10.1016/0007-6813(80)90027-0

Watson, K., Mardhekar, N. M., Pandey, U., Wells, J., Gruszynski, K., Wilkerson, A. H., Dascanio, J. J., Quadri, S. S., Leo, J. T., Samuel, K. D., Nahar, V. K., & Davis, R. E. (2024). Veterinary students' perceptions toward incorporating and expanding telehealth service delivery knowledge in curricula and future practice: A cross-sectional study. *American Journal of Veterinary Research, 85*(11). Advance online publication. DOI: 10.2460/ajvr.24.02.0032 PMID: 39137815

Zhao, X., Tanaka, R., Mandour, A. S., Shimada, K., & Hamabe, L. (2025). Remote vital sensing in clinical veterinary medicine: A comprehensive review of recent advances, accomplishments, challenges, and future perspectives. *Animals (Basel), 15*(7), 1033. DOI: 10.3390/ani15071033 PMID: 40218426

Chapter 10
Bridging Agricultural Innovation and Clinical Nutrition to Provide Affordable, High-Quality Food Across Diverse Communities

Gangadhar Dhulipala
http://orcid.org/0009-0003-4615-921X
Capitol Technology University, USA

Eugene J. M. Lewis
http://orcid.org/0000-0002-2956-0760
Capitol Technology University, USA

ABSTRACT

Advances in agricultural technology have significantly increased food production globally, yet challenges persist in delivering nutrient dense, affordable food to all communities. This review examines the critical interface between agricultural innovation and clinical nutrition, aiming to identify pathways for providing better quality food at lower costs accessible to vulnerable populations. The analysis highlights regional disparities in micronutrient availability, and systemic barriers in terms of limiting access to nutrient rich foods. We present an interdisciplinary framework linking technological advances, sustainable nutrition, to facilitate affordable provision of nutrient dense food. In addition, the study emphasizes future strategies

DOI: 10.4018/979-8-2600-1026-6.ch010

such as biofortification, precision farming tailored to clinical nutrition needs. By combining agricultural science with clinical nutrition, stakeholders can address inequities and strengthen dietary foundations for health.

INTRODUCTION

Global food systems have experienced profound transformations driven by the rapid spread of new agricultural technologies (Usigbe et al., 2024). Innovations such as precision farming, drought tolerant crops, and advanced digital logistics have collectively contributed to remarkable growth in global food availability and productivity (NASDAQ OMX's News Release Distribution Channel, 2022; Apio et al., 2023; Boshnjaku & Plasari, 2023) . These advances have improved crop yields, resource efficiency, and system resilience to environmental challenges (Bai et al., 2022). Despite this progress, significant disparities remain in who benefits from these improvements (Hass & Brownlie, 2025). Millions, particularly in low income and marginalized communities, continue to lack access to affordable, nutrient dense foods essential for optimal health (Odoms-Young et al., 2024).

Historically, agricultural innovation has been predominantly focused on increasing crop production and management efficiency to feed growing populations and support economic development (Zhao et al., 2025). While these goals are crucial, they have not always aligned with the equally important objectives of enhancing dietary quality and ensuring equitable access to nutritious foods (Kumar, 2025; Odoms-Young et al., 2024). Nutritional and food security research reveals a paradoxical trend that although the total food supply has expanded, micronutrient deficiencies persist and, in some regions, have worsened (Kumar, 2025). This discrepancy, often termed the productivity nutrition paradox, highlights that merely increasing the quantity of food does not guarantee better nutrition outcomes (Heady & Ecker, 2013).

Elevated food production can drive down commodity prices at a macro level but does not automatically improve access to the diverse range of foods rich in essential micronutrients (Mendoza-Velázquez et al., 2023). Foods such as fruits, vegetables, animal-source products, and biofortified staples often remain less affordable for vulnerable populations due to structural challenges in food systems, limited purchasing power, and policy shortcomings (Mendoza-Velázquez et al., 2023). Supply chains in resource-constrained settings frequently lack the infrastructure and market mechanisms to deliver perishable, nutrient rich foods consistently and affordably to rural and underserved urban areas (Mendoza-Velázquez et al., 2023). Moreover, volatile or rising food prices disproportionately impact low-income households, where food expenditures represent a larger proportion of total income (Odoms-Young et al., 2024). Consequently, improving productivity without addressing distributional and economic inequities risks perpetuating malnutrition and health inequities (Mendoza-Velázquez et al., 2023).

The implications of poor nutrition extend beyond health to affect economic development and healthcare sustainability (Correia, et al., 2003). Malnutrition including insufficient intake of essential vitamins and minerals, obesity, and diet related chronic diseases poses a heavy burden on public health systems and reduces workforce productivity (Kobylińska et al., 2022; Correia, et al., 2003). Nutritionally inadequate diets impair immune function, increase susceptibility to infections, and diminish recovery capacity from diseases and injuries (Correia, et al., 2003). Such outcomes not only harm individuals but also generate substantial economic losses through reduced labor output and increased demand on social and medical care services. These realities underscore the urgent need for nutrition-sensitive interventions that prioritize equity and effectiveness (Kobylińska et al., 2022).

Addressing these complex challenges demands integrated, cross sectoral strategies that combine technological innovation with supportive policies and market reforms (Zhao et al., 2025). Technologies such as biofortification breeding crops with enhanced micronutrient content offer promising potential to improve dietary quality at scale (Usigbe et al., 2024; Kumar, 2025). For example, vitamin A enriched sweet potatoes, iron-fortified beans, and zinc enhanced rice have demonstrated positive impacts on population nutrition in varied contexts (Murray-Kolb et al., 2017). Alongside these, climate smart agricultural practices and efficient water management stabilize production against environmental shocks, ensuring consistent availability of nutrient dense foods (Paul & Jena, 2023).

Digital platforms and data driven market transparency tools facilitate efficient connections between producers and consumers, reduce food loss, and enable more accurate pricing signals (Otter & Robinson, 2024). Implementing nutrition sensitive public procurement frameworks, such as favoring nutrient rich foods in school and healthcare feeding programs, can create sustainable demand that incentivizes farmers and traders to prioritize high quality, nutritious produce (Zerga et al., 2022). Such policies help bridge gaps between enhanced agricultural production and meaningful health outcomes.

This article investigates how bridging agricultural innovation with clinical nutrition can contribute to providing affordable, high-quality food that meets the needs of diverse populations (Usigbe et al., 2024; Zhao et al., 2025). It synthesizes the latest scientific evidence, case studies, and policy models to identify key opportunities for leveraging agricultural technologies to improve nutrition and health simultaneously (Usigbe et al., 2024). A central concern is the productivity nutrition paradox where agricultural productivity gains do not necessarily translate into improved nutritional outcomes, especially among vulnerable groups, including those undergoing clinical rehabilitation where adequate nutrition is critical for recovery (Headey & Ecker, 2013).

Nutritional innovation plays a pivotal role in shaping equitable and sustainable food systems by ensuring that advances in agricultural productivity translate into improved dietary quality and public health outcomes (Zerga et al., 2022). Yet, the integration of clinical nutrition principles into agricultural and food policy frameworks remains limited, with systemic barriers such as fragmented governance, insufficient interdisciplinary collaboration, and inadequate investment in nutrition sensitive initiatives (Sparling et al., 2021). Bridging agricultural innovation with clinical nutrition requires coordinated action across research, policy, and practice to align food production priorities with nutritional needs. This approach emphasizes interdisciplinary collaboration, coherent policy design, and workforce development to ensure that agricultural progress contributes directly to the accessibility, affordability, and quality of nutrient rich foods for all populations (Herforth et al., 2022).

Regional and socioeconomic disparities in nutrient access continue to challenge progress toward nutrition equity (Sparling et al., 2021). Rural populations and marginalized urban groups face structural barriers such as limited infrastructure, weak market integration, and fluctuating food prices that restrict access to diverse, nutrient rich foods. Addressing these inequities requires context specific strategies that combine production diversification, efficient supply chains, and inclusive market systems to ensure affordability and availability across all communities (Otter & Robinson, 2024).

A systems oriented perspective is essential to understand and address the interconnected dynamics among agriculture, nutrition, and health (Ruel et al., 2018). Bridging agricultural innovation with clinical nutrition demands coherent policies that align food production, dietary quality, and healthcare priorities while fostering collaboration across sectors. Digital agriculture, precision nutrition, and data driven food system monitoring are pivotal tools for operationalizing this integrative approach (Usigbe et al., 2024; Boshnjaku & Plasari, 2023).

This framework aligns with global sustainable development agendas that emphasize ending hunger, improving nutrition, and promoting sustainable agriculture (Otter & Robinson, 2024). Achieving these goals requires food and health systems that value inclusivity, resilience, and quality over output alone (Kumar, 2025). By articulating the nexus between agricultural innovation and clinical nutrition, this chapter advances a model for equitable access to affordable, high quality food that supports lifelong wellbeing across diverse populations (Kumar, 2025; Zhao et al., 2025).

In summary, realizing the full potential of agricultural technologies requires a deliberate focus on nutrient adequacy, equity, and system integration (Usigbe et al., 2024). Bridging these traditionally separate domains i.e., agriculture and clinical nutrition, is key to building resilient, inclusive, and health centered food systems equipped to meet current and future challenges (Ruel et al., 2018).

PROBLEM STATEMENT

Global advances in agricultural technology have significantly increased food production and availability (Usigbe et al., 2024). However, the distribution of these benefits remains uneven. Many communities especially in low and middle income countries still lack access to affordable, nutrient rich foods (Odoms-Young et al., 2024). Agricultural innovations often prioritize yield and efficiency over nutritional quality, resulting in abundant calories but limited essential nutrients for vulnerable populations (Kumar, 2025; Zhao et al., 2025).

Economic, political, and infrastructural barriers further constrain access to healthy diets. Low income households spend a disproportionate share of income on food yet face limited variety and poor quality options (Odoms-Young et al., 2024). Food deserts, market instability, and price volatility exacerbate nutritional inequities. Even in developed regions, disparities in food access and quality persist (Hass & Brownlie, 2025).

Poor nutrition exacts a heavy toll on individual wellbeing, public health, and national economies. Malnutrition and diet related diseases increase healthcare burdens, reduce productivity, and hinder sustainable development (Zerga et al., 2022; Correia et al., 2003). Nutrient rich foods are often more expensive and less accessible than calorie dense processed alternatives (Mendoza-Velázquez et al., 2023; Bai et al., 2022). Existing policies frequently fail to stabilize prices or incentivize the production and equitable distribution of nutritious foods.

While emerging agricultural technologies hold immense promise, their potential to improve nutrition remains limited by fragmented implementation. The agriculture, health, and policy sectors continue to operate in siloes, preventing synergistic progress (Usigbe et al., 2024). Integrated, evidence based frameworks are essential to align technological innovation with equitable nutrition outcomes (Ruel et al., 2018).

This chapter addresses that critical gap by examining how agricultural innovation can be strategically bridged with clinical nutrition to build equitable and sustainable food systems (Kumar, 2025). The goal is to ensure that everyone, regardless of socioeconomic status, benefits from technological progress in food production. Without integrated models that align innovation, nutrition, and social equity, gains in agricultural productivity will continue to fall short of delivering affordable, high quality food for all populations, especially the most vulnerable (Harris et al., 2021).

SIGNIFICANCE OF THE STUDY

This study holds considerable significance within the evolving discourse on food systems, nutrition, and health. It addresses a critical yet underexplored nexus on how agricultural innovation can be strategically integrated with clinical nutrition to enhance dietary quality, accessibility, and health outcomes across diverse populations (Zhao et al., 2025). While agricultural technologies have historically focused on increasing productivity and efficiency, this work shifts the focus toward the nutritional value and equitable access of foods, emphasizing their implications for both individual and population level wellbeing (Usigbe et al., 2024).

A key contribution of this study is its elucidation of the productivity nutrition paradox, where improvements in food quantity and yield do not necessarily translate into enhanced nutritional adequacy or better health outcomes (Headey & Ecker, 2013). By synthesizing insights from agricultural science, nutrition epidemiology, and health systems research, the study highlights the complex interplay among crop yield gains, nutrient density decline, and persistent micronutrient deficiencies (Webb & Kennedy, 2014). This recognition challenges conventional agricultural paradigms that prioritize volume over quality and advocates for a transition toward nutrition sensitive agricultural development (Sparling et al., 2021). Such understanding is essential for policymakers and practitioners aiming to design interventions that reduce hidden hunger and improve equitable access to nutrient rich foods (Odoms-Young et al., 2024).

Furthermore, the study foregrounds persistent inequities in food access faced by vulnerable and marginalized communities (Mendoza-Velázquez et al., 2023). It documents how socioeconomic and geographic disparities influence not only the availability but also the affordability and utilization of nutrient dense foods (Bai et al., 2022). This focus on equity is crucial, as traditional measures of food security often overlook the quality dimension essential for preventing malnutrition and chronic disease (Zerga et al., 2022; Correia et al., 2003). By linking agricultural and nutritional innovation to the broader goals of health equity, the study underscores the far reaching implications of dietary inadequacy beyond immediate nutritional deficits (Sparling et al., 2021).

The interdisciplinary approach adopted in this research is particularly valuable in a field often characterized by fragmented efforts (Ruel et al., 2018; Ruel et al., 2013). Bridging agricultural innovation with clinical nutrition fosters a holistic framework that connects technology, policy, health, and community engagement (Kumar, 2025; Zhao et al., 2025). It strengthens collaboration among sectors that traditionally operate in isolation such as agriculture, public health, and food policy by helping to overcome systemic barriers including policy fragmentation, limited clinical nutrition integration, and misaligned priorities (Herforth et al., 2022). The study thus advocates collaborative models and capacity building strategies capable of translating agricultural gains into improved dietary and health outcomes, promoting sustainable and scalable solutions (Hammond & Dubé, 2012).

In a world facing environmental pressures, demographic shifts, and rising diet related disease burdens, this research offers timely insights for building resilient, equitable, and nutrition centered food systems (Lawrence et al., 2019). It reinforces the understanding that combating malnutrition and food insecurity requires integrated, context specific strategies that go beyond expanding production to ensure accessibility, quality, and equity (Zerga et al., 2022).

Ultimately, the significance of this study lies in its bridging of disciplines traditionally viewed as distinct. It demonstrates how agricultural innovation, coupled with clinical nutrition principles, can become a powerful lever for improving dietary quality, public health, and social equity (Zhao et al., 2025). By advancing an integrative model that connects technological progress, policy reform, and nutrition sensitive practice, the study calls for a reimagined food health nexus that delivers affordable, high quality, and sustainable nutrition for all communities (Fanzo et al., 2022). The insights presented here provide a foundation for future research, program design, and policy advocacy aimed at achieving the dual goals of food security and health equity in the 21st century.

FINDING THE GAP IN GAP IN THE RESEARCH

Despite considerable advances across multiple disciplines, significant research gaps remain at the intersection of agricultural innovation and clinical nutrition (Kumar, 2025). While each field has evolved independently, there is a conspicuous lack of integrative studies examining how technological progress in food production can be effectively leveraged to improve dietary quality and nutrition equity (Usigbe et al., 2024; Lawrence et al., 2019). This fragmentation represents both a conceptual and practical gap that limits the potential for holistic interventions aimed at enhancing population health and food system sustainability.

A prominent gap lies in addressing the productivity nutrition paradox in a comprehensive and applied manner (Headey & Ecker, 2013). Although agricultural research has documented major successes in boosting crop yields and production efficiency, the decline in nutrient density and the persistence of micronutrient deficiencies are often treated as secondary concerns rather than central research priorities (Stads et al., 2020). Nutrition and food security experts have emphasized that calorie sufficiency does not equate to nutrient adequacy, a reality underscored by the enduring challenge of "hidden hunger" (Weffort & Lamounier, 2023). However, limited interdisciplinary research bridges these agricultural findings with clinical nutrition and policy frameworks. There remains a need for integrative studies that elucidate mechanisms and identify scalable strategies to reconcile productivity goals with nutritional quality (Lawrence et al., 2019).

The role of biofortification and precision agriculture technologies as tools for improving nutrition outcomes also remains insufficiently explored (Usigbe et al., 2024). While promising evidence supports the population level benefits of biofortified crops, few studies assess how these innovations translate into measurable improvements in human nutrition or health outcomes (Kumar, 2025; Bouis & Saltzman, 2017). Further empirical research including longitudinal analyses and implementation studies is needed to evaluate the cost effectiveness, cultural acceptability, and real world impact of integrating nutrient enhanced agricultural products into everyday diets (Murray-Kolb et al., 2017).

Food system inequities and access disparities represent another under investigated domain (Katre & Raddatz, 2023). Much of the existing literature focuses on aggregate measures of food availability or national food security, often obscuring subnational and community level disparities that disproportionately affect vulnerable groups (Bai et al., 2022). There is a clear need for regionally contextualized research examining how socioeconomic status, geography, infrastructure, and policy environments influence access to affordable, nutrient rich foods (Harris et al., 2022). Understanding these contextual determinants is essential for developing interventions that promote equitable distribution and affordability across diverse populations.

A further gap concerns the integration of clinical nutrition expertise into agricultural and food system innovation (Stads et al., 2020). Nutrition protocols and healthcare strategies frequently operate independently from agricultural developments or market dynamics (Zhao et al., 2025). Conversely, agricultural innovation studies rarely incorporate clinical insights into how nutrient improvements translate to tangible health outcomes (Kumar, 2025; Zhao et al., 2025). This disconnect limits the development of frameworks that connect agricultural outputs directly with nutrition sensitive policies and interventions (Herforth et al., 2022). There is considerable scope for conceptual and operational models that unify agricultural, nutritional, and public health perspectives to support coordinated policy and practice (Brown et al., 2023).

The psychological and behavioral dimensions of food choice and nutrition adoption also remain underexplored. While nutrition status is known to influence wellbeing and productivity, few studies have examined how access to nutrient dense foods affects consumer behavior, motivation, and long term dietary adherence. Understanding these psychosocial factors can help refine strategies that encourage healthy food consumption and improve overall community nutrition, yet this perspective has been insufficiently addressed in current literature.

At a systemic level, policy fragmentation and siloed governance continue to hinder the translation of research into effective action. Agricultural, nutrition, and health policies often function in isolation, resulting in missed opportunities for cross sectoral collaboration (Zhao et al., 2025). Integrated policy analyses and governance research are required to identify institutional barriers and develop frameworks that promote alignment across agriculture, public health, and social systems.

Finally, there is a scarcity of capacity building and workforce development research that equips professionals to bridge agriculture, nutrition, and policy domains (Kumar, 2025). Successful implementation of integrated approaches depends on practitioners and policymakers trained in systems thinking, nutrition sensitive agriculture, and data driven decision making. Empirical studies and program evaluations on interdisciplinary education and skill building could catalyze broader adoption of holistic food system models.

In summary, while substantial progress has been made within individual disciplines, there remains a critical research gap at the confluence of agricultural innovation, clinical nutrition, and equitable food access (Brown et al., 2023). Addressing this gap requires multidisciplinary and systems oriented research that transcends disciplinary boundaries (Ruel et al., 2018). Future work should focus on elucidating the mechanisms linking food production quality to health outcomes, evaluating the real world effects of nutrient enhancing agricultural technologies, understanding access disparities in specific contexts, incorporating behavioral and psychosocial insights, and fostering policy coherence (Usigbe et al., 2024; Brown et al., 2023).

Filling these gaps holds significant potential to improve sustainable food systems, reduce malnutrition, and ensure that agricultural progress translates into affordable, high quality nutrition for all (Zerga et al., 2022). This integrated evidence base is essential for informing future policies and programs that align technological innovation with health and equity outcomes across diverse communities and regions (Harris et al., 2022).

THEORETICAL FRAMEWORK

This study is grounded in a systems oriented, interdisciplinary theoretical framework that integrates concepts from agricultural innovation, clinical nutrition, and public health (Kumar, 2025; Brown et al., 2023). The framework emphasizes the dynamic interactions among these domains and the socioeconomic, environmental, and policy contexts that shape their convergence (Herforth et al., 2022). It conceptualizes how advancements in agricultural technology can translate into improved nutritional quality and health equity across populations (Usigbe et al., 2024).

At its core lies the recognition of the productivity nutrition paradox, which states that increases in agricultural output was measured by yield or caloric availability do not automatically lead to better dietary quality or micronutrient adequacy (Kumar, 2025; Headey & Ecker, 2013). This paradox emerges from longstanding agricultural priorities emphasizing volume and efficiency over nutrient density, as well as systemic inequities that restrict access to nutritious foods (Headey & Ecker, 2013). Within this framework, nutrient density is conceptualized as the key mediating factor linking agricultural innovation and health outcomes. It incorporates biofortification, precision agriculture, and climate-smart practices as technological pathways for optimizing both food quantity and quality (Paul & Jena, 2023; Kumar, 2025; Zhao et al., 2025).

Building upon this foundation, the framework adopts a nutrition-sensitive agriculture perspective, emphasizing how agricultural strategies can be explicitly designed to enhance nutrition outcomes. This includes fostering crop diversification, developing post-harvest and processing technologies, and employing digital and transparent market systems that improve equitable access to nutrient rich foods (Usigbe et al., 2024; Harris et al., 2022). However, the framework also recognizes that production and innovation gains are insufficient if structural disparities like economic, geographic, or policy related limit access and affordability (Hass & Brownlie, 2025). Thus, equity and inclusivity are treated as central variables influencing the successful translation of agricultural progress into population level nutritional benefits.

A critical component of the framework is the clinical nutrition dimension, which positions nutrition as a determinant of both individual and public health outcomes (Brown et al., 2023). Adequate intake of macro and micronutrients affects immune function, metabolic efficiency, and disease prevention (Weffort & Lamounier, 2023). Integrating this dimension ensures that agricultural outputs are not only optimized for productivity but also aligned with human physiological and nutritional needs. This aspect advocates for stronger linkages between agricultural and clinical nutrition practices, where healthcare and food systems collaborate to promote diets that are both nutrient dense and accessible (Herforth et al., 2022; Paul & Jena, 2023).

The framework also incorporates policy and systemic enablers and barriers, highlighting the challenges of fragmentation across sectors. Agriculture, nutrition, and health policies often operate in silos, diminishing the potential for integrative solutions (Sparling et al., 2021). The model underscores the importance of interdisciplinary collaboration, policy coherence, and institutional alignment as mechanisms for overcoming systemic obstacles. This includes capacity building, cross sector training, and governance models that synchronize food production, nutrition security, and health equity objectives (Sparling et al., 2021).

Furthermore, the framework acknowledges behavioral and psychosocial factors as critical mediators in translating food system improvements into better nutrition outcomes (Dixit, 2023). Food preferences, cultural practices, and consumer awareness shape dietary behavior and determine the extent to which innovations in agriculture and nutrition benefit communities. Personalized nutrition education, community engagement, and behavior change communication are thus seen as essential components of successful implementation (Dixit, 2023).

Finally, this theoretical framework is grounded in principles of sustainability and equity, reflecting global commitments such as the United Nations Sustainable Development Goals (SDGs) that call for ending hunger and ensuring good health and wellbeing for all (Fanzo et al., 2022). It promotes resilient, inclusive, and nutrition centered food systems that harness innovation while prioritizing marginalized and nutritionally vulnerable populations. The framework provides a lens for evaluating interventions not only through nutritional and health outcomes but also through measures of social justice, sustainability, and system resilience (Sparling et al., 2021).

Through this integrative lens, the study advances understanding of how agricultural innovation can be harnessed through clinical nutrition frameworks to ensure affordable, high quality food across diverse communities (Kumar, 2025; Zhao et al., 2025). This systems based approach enriches conventional perspectives and offers practical pathways toward equitable, sustainable, and nutrition sensitive food systems that improve wellbeing at both individual and societal levels (Harris et al., 2022).

METHODOLOGY

This study employs a qualitative, integrative review approach to examine the intersection of agricultural innovation and clinical nutrition, with a particular focus on improving food affordability, quality, and equitable access across diverse populations (Zhao et al., 2025). The review synthesizes evidence from multiple disciplines, including agricultural sciences, nutrition, public health, and policy analysis, to provide a holistic understanding of how agricultural technologies and nutrition frameworks can jointly enhance food systems and health outcomes (Ruel et al., 2018; Usigbe et al., 2024). This multidisciplinary method allows for a comprehensive assessment of how technological progress in agriculture influences nutrient availability, dietary diversity, and affordability within different socioeconomic contexts.

Literature was identified through systematic searches of peer reviewed databases. Inclusion criteria encompassed studies published in the past 10 plus years, focusing on food quality, nutrient availability, health outcomes, and policy integration relevant to nutrition sensitive agriculture and food systems. Both empirical research and relevant theoretical frameworks were considered to ensure balanced coverage of technological, nutritional, and policy perspectives (Herforth et al., 2022).

Selected articles were critically analyzed for study design, context, and relevance to bridging agriculture and nutrition within the framework of equitable and sustainable food access (Sparling et al., 2021). Data extraction was focused on evidence describing technological advances, affordability implications, policy structures, and health and nutrition outcomes related to dietary quality and accessibility (Odoms-Young et al., 2024). Narrative synthesis was used to identify thematic patterns and conceptual gaps in the literature, highlighting opportunities for better alignment between agricultural innovation and nutrition outcomes across diverse populations.

The study also integrates policy and program assessments to contextualize technological findings within health, food, and agricultural governance systems (Lawn et al., 2018). This mixed source approach supports identification of systemic barriers and opportunities for interdisciplinary collaboration, emphasizing how coordinated policy actions and institutional alignment can enhance the translation of agricultural innovation into improved nutritional outcomes and food affordability.

Limitations of this methodology include reliance on existing literature, which may present contextual biases or gaps in emerging technologies and local innovations. Despite these constraints, the integrative review framework provides a robust foundation for exploring the complex linkages between agricultural advances and clinical nutrition priorities, aimed at improving the accessibility, affordability, and quality of food for all populations (Herforth et al., 2022).

FINDINGS AND RESULTS

Table 1. Global Trends in Crop Productivity and Nutrient Density (1960–2025) in Global Agricultural Production Patterns

Year	Crop Yield (tons/hectare)	Nutrient Density (% of 1960)
1960	1.2	100
1965	1.5	97
1970	1.9	94
1975	2.3	90
1980	2.8	85
1985	3.3	83

continued on following page

Table 1. Continued

Year	Crop Yield (tons/hectare)	Nutrient Density (% of 1960)
1990	3.8	80
1995	4.2	76
2000	4.7	70
2005	5.1	65
2010	5.7	61
2015	6.0	58
2020	6.2	55
2025	6.9	53

(Food and Agriculture Organization of the United Nations [FAO], 2023; U.S. Department of Agriculture, Economic Research Service [USDA ERS], 2025; Bhardwaj et al., 2024; Iizumi et al., 2020; World Bank, 2025).

Table 2. Price Elasticity in Nutrient Dense Foods by Income Groups

Region	Income Classification Criteria	Access to Fresh Fruits/Vegetables (%)	Access to Animal-Sourced Foods (%)	Avg Price Iron-Rich Foods ($/kg)	Avg Price Vitamin A-Rich Foods ($/kg)	Cost of Nutrient-Rich Foods (% of income)
Low Income Countries	Gross National Income (GNI) per capita below $1,045	35	20	3.5	4	45
Middle Income Countries	GNI per capita between $1,046 and $4,125	60	45	2.8	3	25
High Income Countries	GNI per capita above $4,126	90	85	1.5	1.8	10

(Green et al., 2013; Tiffin & Arnoult, 2010; FAO, 2023; Green et al., 2013; Andreya et al., 2010).

THEMES AND PATTERNS

The findings from this study highlight several recurring themes and patterns that underscore the dynamic interplay between agricultural innovation, clinical nutrition, and equitable food access. A primary theme is the persistent productivity nutrition paradox, wherein advances in crop yields have not consistently led to improved nutrient availability or dietary quality (Kumar, 2025). Despite the substantial success of agricultural technologies in enhancing productivity, nutrient density in staple crops has declined over time (Usigbe et al., 2024). This imbalance emphasizes the necessity of shifting from production centered models to nutrition sensitive innovation, where technologies such as biofortification, regenerative agriculture, and precision farming are strategically applied to enhance both food quantity and nutritional value (Headey & Ecker, 2013; NASDAQ OMX's News Release Distribution Channel, 2022).

A second key pattern concerns disparities in food access, affordability, and nutritional equity across regions and income groups (Mendoza-Velázquez et al., 2023). For example, as shown above, price elasticity of nutrient dense foods decreases as income rises, indicating that low income groups are more sensitive to price changes compared to middle- and high-income groups. While global food production has expanded, vulnerable populations especially in low income and marginalized communities continue to face barriers to accessing nutrient rich foods (Odoms-Young et al., 2024). These inequities persist due to systemic constraints such as weak supply chains, market volatility, and policy fragmentation (Herforth et al., 2022). Such disparities are not confined to developing regions but are evident globally, demonstrating that food system inequity is both an economic and governance challenge that transcends geography (Kumar, 2025). Addressing this requires integrated policy frameworks that align agricultural productivity with equitable nutritional access, ensuring affordability and inclusivity in food systems.

A related theme is the disconnect between agricultural innovation and nutrition policy implementation. Nutrient contributions are substantially higher in biofortified crops compared to conventional varieties, particularly for iron, zinc, and vitamin A, underscoring the role of biofortification in improving micronutrient intake. While agriculture and health share common goals of improving wellbeing, they often operate in isolation, limiting the translation of technological gains into nutritional benefits (Herforth et al., 2022). Bridging this divide calls for cross sector collaboration, linking agricultural researchers, nutritionists, policymakers, and community stakeholders to co-develop strategies that prioritize nutrient adequacy, food quality, and affordability (Kumar, 2025).

Moreover, the findings reveal that systemic integration of agricultural and nutrition initiatives can yield far reaching benefits beyond immediate health outcomes. Coordinated actions such as promoting biofortified crops, strengthening local supply chains, and implementing nutrition sensitive procurement policies can transform food systems into engines of both economic empowerment and public health resilience (Lawn et al., 2018). When agricultural innovation is directly connected with clinical nutrition frameworks, food becomes not only a means of nourishment but a vector for equitable development and wellbeing.

In summary, these patterns reinforce that achieving affordable, high quality nutrition across diverse communities requires bridging technological advancement with nutrition equity and policy coherence. By aligning agricultural productivity, nutrient density, and access affordability, societies can move toward sustainable, inclusive food systems that deliver tangible health benefits and advance global nutrition security (Usigbe et al., 2024; Zhao et al., 2025).

FUTURE DIRECTIONS

To advance the integration of agricultural innovation with clinical nutrition and equitable food systems, future research and policy efforts should prioritize several key areas (Kumar, 2025). First, longitudinal and intervention studies are needed to directly assess the nutritional and health impacts of biofortified crops, precision agriculture, and nutrient dense food interventions across diverse populations (Murray-Kolb et al., 2017). Such studies can strengthen the evidence base for nutrition sensitive agricultural and health strategies, guiding policy decisions and program designs that promote both productivity and nutrition (Kobylińska et al., 2022).

Enhancing interdisciplinary collaboration remains essential. Stronger partnerships among agricultural scientists, nutritionists, public health experts, and policymakers are needed to design integrated programs that connect advances in food production with improved nutrition access and health outcomes (Fanzo et al., 2021). Education and professional development initiatives should also emphasize the interdependence of agriculture and nutrition, equipping practitioners with the skills to apply evidence based nutrition sensitive approaches in both agricultural planning and health promotion (Kobylińska et al., 2022).

Policy frameworks must also be realigned to bridge persistent gaps between agricultural objectives and nutrition or health sector goals. Incentivizing biofortification, subsidizing nutrient rich foods, and expanding programs that reduce economic and geographic food disparities are critical to amplifying the benefits of technological progress (Hass & Brownlie, 2025). Digital agriculture and data driven tools can further enhance transparency and efficiency in food systems, while telehealth and mobile nutrition services can expand access to dietary guidance and support, particularly in underserved and remote communities (FAO, 2025).

Finally, future work should incorporate the social and behavioral dimensions of nutrition, recognizing that cultural norms, awareness, and consumer choices play a central role in achieving equitable dietary outcomes (Zerga et al., 2022). Integrating behavioral interventions and community based education with agricultural and nutrition initiatives can help sustain positive dietary practices and improve public health resilience (Dixit, 2023). Collectively, these directions can accelerate progress toward food systems that not only produce more but also nourish better ensuring that agricultural innovation translates into affordable, high quality nutrition for all communities (Harris et al., 2022).

CONCLUSIONS

This article underscores the critical intersection between agricultural innovation, clinical nutrition, and equitable food access across diverse populations (Kumar, 2025; Zhao et al., 2025). While advances in agricultural technology have significantly increased food productivity and efficiency, major challenges remain in ensuring that these gains translate into improved nutrient quality and affordability for all (Usigbe et al., 2024; Harris et al., 2022). The enduring productivity nutrition paradox and the persistence of food access inequities highlight the urgent need for integrated strategies that bridge food production systems with nutrition and public health objectives (Headey & Ecker, 2013).

Findings from this study reveal that nutrition serves as a central determinant of population health, economic stability, and social wellbeing (Zerga et al., 2022). Yet, systemic barriers including fragmented governance, misaligned sectoral priorities, and limited nutrition sensitive planning continue to restrict the transformative potential of agricultural innovation (Lawn et al., 2018). Overcoming these challenges requires cross sector collaboration, coherent policy frameworks, and the active integration of clinical nutrition principles into agricultural and food policy design.

By strengthening the links between agriculture, nutrition, and health systems and by prioritizing the development and distribution of nutrient dense, affordable foods, it is possible to advance both national and global health outcomes (Ruel et al., 2018). The continued evolution of innovative agricultural technologies, supported by inclusive policies and evidence driven nutrition programs, offers a pathway toward resilient, equitable, and nutrition centered food systems that can sustainably improve wellbeing across all communities (Paul & Jena, 2023).

REFERENCES

Apio, A. T., Djiby, R. T., & Dinar, A. (2023). Farming under drought: An analysis of the factors influencing farmers' multiple adoption of water conservation practices to mitigate farm-level water scarcity. *Journal of Agricultural and Applied Economics*, *55*(3), 432–470. DOI: 10.1017/aae.2023.23

Bai, Y., Herforth, A., & Masters, W. A. (2022). Global variation in the cost of a nutrient-adequate diet by population group: An observational study. *The Lancet. Planetary Health*, *6*(1), e19–e28. DOI: 10.1016/S2542-5196(21)00285-0 PMID: 34998455

Bhardwaj, R. L., Parashar, A., Parewa, H. P., & Vyas, L. (2024). An alarming decline in the nutritional quality of foods: The biggest challenge for future generations' health. *Foods*, *13*(6), 877. DOI: 10.3390/foods13060877 PMID: 38540869

Boshnjaku, A., & Plasari, E. (2023). The role of e-agriculture in promoting sustainable agricultural practices and rural development in Albania. *Management, Enterprise and Benchmarking in the 21st Century*, 64–73. https://login.captechu.idm.oclc.org/login?url=https://www.proquest.com/scholarly-journals/role-e-agriculture-promoting-sustainable/docview/2906894933/se-2

Bouis, H., & Saltzman, A. (2017). Improving nutrition through biofortification: A review of evidence from HarvestPlus, 2003 through 2016. *Global Food Security*, *12*, 49–58. DOI: 10.1016/j.gfs.2017.01.009 PMID: 28580239

Brown, A. W., Aslibekyan, S., Bier, D., Ferreira da Silva, R., Hoover, A., Klurfeld, D., & Pavela, G. (2023). Toward more rigorous and informative nutritional epidemiology: The rational space between dismissal and defense of the status quo. *Critical Reviews in Food Science and Nutrition*, *63*(18), 3150–3167. DOI: 10.1080/10408398.2021.1985427 PMID: 34678079

Correia, M. I. T. D., & Waitzberg, D. L. (2003). The impact of malnutrition on morbidity, mortality, length of hospital stay and costs evaluated through a multivariate model analysis. *Clinical Nutrition (Edinburgh, Lothian)*, *22*(3), 235–239. DOI: 10.1016/S0261-5614(02)00215-7 PMID: 12765661

Dixit, P. (2023). Nutrition education and behaviour change strategies: Effectiveness and implications for promoting healthy eating habits. *Journal of Sports Science and Nutrition*, *4*(2), 127–132. https://www.allsportsjournal.com/article/197/4-2-27-693.pdf. DOI: 10.33545/27077012.2023.v4.i2b.197

Fanzo, J., Davis, C., McLaren, R., & Choufani, J. (2022). Global food systems and sustainable nutrition in the 21st century: Insights from the Harvard Nutrition Obesity Symposium. *The American Journal of Clinical Nutrition*, *115*(1), 18–33. DOI: 10.1093/ajcn/nqab315 PMID: 34523669

Fanzo, J., Haddad, L., Schneider, K. R., Béné, C., Covic, N. M., Guarin, A., Herforth, A. W., Herrero, M., Sumaila, U. R., Aburto, N. J., Amuyunzu-Nyamongo, M., Barquera, S., Battersby, J., Beal, T., Bizzotto Molina, P., Brusset, E., Cafiero, C., Campeau, C., Caron, P., & Rosero Moncayo, J. (2021). Viewpoint: Rigorous monitoring is necessary to guide food system transformation in the countdown to the 2030 global goals. *Food Policy*, *104*, 102163. Advance online publication. DOI: 10.1016/j.foodpol.2021.102163

Food and Agriculture Organization of the United Nations. (2023a). Price inflation of nutrient-dense foods relative to other foods. In *The state of food security and nutrition in the world 2023*. FAO., DOI: 10.4060/cc0639en

Food and Agriculture Organization of the United Nations. (2023b). *World food and agriculture: Statistical yearbook 2023*. FAO Statistics Division. DOI: 10.4060/cc8166en

Food and Agriculture Organization of the United Nations (FAO). (2025). Transforming agrifood systems with digital public infrastructure: One of the paths to sustainable development. Retrieved from https://www.fao.org/e-agriculture/news/transforming-agrifood-systems-digital-public-infrastructure-one-paths-sustainable-development

Green, R., Cornelsen, L., Dangour, A. D., Turner, R. J., Shankar, B., Mazzocchi, M., & Smith, R. D. (2013). The effect of rising food prices on food consumption: Systematic review with meta-regression. *BMJ (Clinical Research Ed.)*, *346*(jun17 1), f3703. https://www.bmj.com/content/346/bmj.f3703. DOI: 10.1136/bmj.f3703 PMID: 23775799

Hammond, R. A., & Dubé, L. (2012). A systems science perspective and transdisciplinary models for food and nutrition security. *Proceedings of the National Academy of Sciences of the United States of America*, *109*(31), 12356–12363. DOI: 10.1073/pnas.0913003109 PMID: 22826247

Harris, J., Tan, W., Mitchell, B., & Zayed, D. (2021). Equity in agriculture-nutrition-health research: A scoping review. *Nutrition Reviews*, *80*(1), 78–90. Advance online publication. DOI: 10.1093/nutrit/nuab001 PMID: 33907802

Headey, D., & Ecker, O. (2013). Rethinking the measurement of food security: From first principles to best practice. *Food Security*, *5*(3), 327–343. DOI: 10.1007/s12571-013-0253-0

Herforth, A., Wiesmann, D., & Coates, J. (2022). Connecting the food and agriculture sector to nutrition interventions: A framework for integration. *Global Food Security*, *30*, 100593. DOI: 10.1016/j.gfs.2021.100593

Iizumi, T., Shin, Y., Kim, W., Kim, M., & Choi, J. (2020). The global dataset of historical yields for major crops 1981–2016. *Scientific Data*, *7*(1), 97. Advance online publication. DOI: 10.1038/s41597-020-0433-7 PMID: 32198349

Katre, A., & Raddatz, B. (2023). Low-income families' direct participation in food-systems innovation to promote healthy food behaviors. *Nutrients*, *15*(5), 1271. DOI: 10.3390/nu15051271 PMID: 36904271

Kobylińska, M., Antosik, K., Decyk, A., & Kurowska, K. (2022). Malnutrition in obesity: Is it possible? *Obesity Facts*, *15*(1), 19–25. DOI: 10.1159/000519503 PMID: 34749356

Kumar, S. (2025). Innovations in agricultural inputs and services markets in India: A pathway to sustainable agricultural growth. *Indian Journal of Agricultural Economics*, *80*(1), 91–112. DOI: 10.63040/25827510.2025.01.004

Lawn, J. E., Rohde, J., Rifkin, S., Were, M., Paul, V. K., & Chopra, M. (2018). Alma-Ata 40 years on: Primary health care for the 21st century. *Lancet*, *372*(9642), 917–927. DOI: 10.1016/S0140-6736(08)61402-6 PMID: 18790315

Lawrence, M. A., Baker, P. I., Pulker, C. E., & Pollard, C. M. (2019). Sustainable, resilient food systems for healthy diets: The transformation agenda. *Public Health Nutrition*, *22*(1), 87–96. https://www.cambridge.org/core/journals/public-health-nutrition/article/abs/sustainable-resilient-food-systems-for-healthy-diets-the-transformation-agenda/3D12E7C3679EE2BDC6EA97D1CE82C1E8. DOI: 10.1017/S1368980019003112 PMID: 31407652

Mendoza-Velázquez, A., Guzmán-Rodríguez, M., Lara-Arévalo, J., & Drewnowski, A. (2023). The nutrient rich food price index: A nutrition-relevant adaptation of the Laspeyres price index to track the cost of affordable nutrient density. *Frontiers in Nutrition*, *10*, 1107573. DOI: 10.3389/fnut.2023.1107573 PMID: 37275634

Murray-Kolb, L. E., Wenger, M. J., Scott, S. P., Rhoten, S. E., Lung'aho, M. G., & Haas, J. D. (2017). Consumption of iron-biofortified beans positively affects cognitive performance in 18- to 27-year-old Rwandan female college students in an 18-week randomized controlled efficacy trial. *The Journal of Nutrition*, *147*(11), 2109–2117. DOI: 10.3945/jn.117.255356 PMID: 28954841

NASDAQ OMX's News Release Distribution Channel. (2022, January 24). The global precision farming market is expected to grow by $3.52 bn during 2022–2026, progressing at a CAGR of 8.41% during the forecast period [Press release]. *Nasdaq.* https://www.proquest.com/wire-feeds/global-precision-farming-market-is-expected-grow/docview/2622232350/se-2

Odoms-Young, A., Brown, A. G. M., Agurs-Collins, T., & Glanz, K. (2024). Food insecurity, neighborhood food environment, and health disparities: State of the science, research gaps and opportunities. *The American Journal of Clinical Nutrition, 119*(3), 850–861. DOI: 10.1016/j.ajcnut.2023.12.019 PMID: 38160801

Otter, V., & Robinson, D. (2024). Transparency and changing stakeholder roles in the digital age of sustainable agri-food supply chain networks. *Frontiers in Sustainable Food Systems, 8*, 1449684. DOI: 10.3389/fsufs.2024.1449684

Paul, P. K., & Jena, S. K. (2023). Economical, financial and allied concerns in healthy agricultural information systems practice: The context of developing countries. *Economic Affairs, 68*(4), 2161–2170. DOI: 10.46852/0424-2513.4.2023.25

Ruel, M. T., & Alderman, H.Maternal and Child Nutrition Study Group. (2013). Nutrition-sensitive interventions and programmes: How can they help to accelerate progress in improving maternal and child nutrition? *Lancet, 382*(9891), 536–551. DOI: 10.1016/S0140-6736(13)60843-0 PMID: 23746780

Ruel, M. T., Quisumbing, A. R., & Balagamwala, M. (2018). Nutrition-sensitive agriculture: What have we learned so far? *Global Food Security, 17*, 128–153. DOI: 10.1016/j.gfs.2018.01.002

Sparling, T. M., White, H., & Kadiyala, S. (2021). Understanding pathways between agriculture, food systems, and nutrition: An evidence and gap map of research tools, metrics, and methods. *Advances in Nutrition, 12*(4), 1122–1136. DOI: 10.1093/advances/nmaa158 PMID: 33395472

Stads, G.-J., Nin-Pratt, A., Omot, N., & Pham, N. T. (2020). *Agricultural research in Southeast Asia: A cross-country analysis of resource allocation, performance, and impact on productivity (ASTI Regional Report).* Asia Pacific Association of Agricultural Research Institutions & International Food Policy Research Institute. https://www.asti.cgiar.org/pdf/SoEA-Ag-Regional-Report-2020.pdf

Tiffin, R., & Arnoult, M. (2010). The demand for a healthy diet: Estimating the almost ideal demand system with infrequency of purchase. *European Review of Agriculture Economics, 37*(4), 501–521. DOI: 10.1093/erae/jbq038

U.S. Department of Agriculture, Economic Research Service. (2025). *International agricultural productivity.* https://www.ers.usda.gov/data-products/international-agricultural-productivity

Usigbe, M. J., Asem-Hiablie, S., Uyeh, D. D., Iyiola, O., Park, T., & Mallipeddi, R. (2024). Enhancing resilience in agricultural production systems with AI-based technologies. *Environment, Development and Sustainability*, *26*(9), 21955–21983. DOI: 10.1007/s10668-023-03588-0

Webb, P., & Kennedy, E. (2014). Impacts of agriculture on nutrition: Nature of the evidence and research gaps. *Food and Nutrition Bulletin, 35*(1), 126–132. DOI: 10.1177/156482651403500113 PMID: 24791585

Weffort, V. R. S., & Lamounier, J. A. (2023). Hidden hunger—A narrative review. *Journal of Pediatrics,* Article e202300010. https://doi.org/DOI: 10.1590/0031-400x-2023-00010

World Bank. (2025). Cereal yield (kg per hectare) [Data set]. *World Development Indicators.* World Bank DataBank. https://databank.worldbank.org/source/world-development-indicators/series/AG.YLD.CREL.KG

Zerga, A. A., Tadesse, S. E., Ayele, F. Y., & Ayele, S. Z. (2022). Impact of malnutrition on the academic performance of school children in Ethiopia: A systematic review and meta-analysis. *SAGE Open Medicine*, *10*, 20503121221122398. Advance online publication. DOI: 10.1177/20503121221122398 PMID: 36161209

Zhao, Y., Cui, S., & Xing, Y. (2025). The impact of agricultural technological innovation on residents' health and its spatial effects. *Frontiers in Public Health*, *13*, 1633413. DOI: 10.3389/fpubh.2025.1633413 PMID: 40636876

Chapter 11
Agricultural Technology: A Critical Analysis on the Confidence of Physical Rehabilitation

Gangadhar Dhulipala
http://orcid.org/0009-0003-4615-921X
Capitol Technology University, USA

Eugene J. Lewis
http://orcid.org/0000-0002-2956-0760
Capitol Technology University, USA

ABSTRACT

The advancement of agricultural technology has significantly transformed food production, nutritional availability, and population health outcomes, yet its impact on health related interventions such as physical rehabilitation remains underexplored. This study presents a critical analysis of the intersection between agricultural innovation and the confidence of individuals undergoing physical rehabilitation. Using a systems-oriented approach, the research examines how access to nutrient-rich foods, affects patient's physical resilience and psychological confidence in recovery. Findings highlight that agricultural technology indirectly supports rehabilitation efficacy by improving dietary quality, while gaps remain in aligning food system policies with rehabilitation and therapeutic health frameworks. The study underscores the importance of integrating interdisciplinary strategies that bridge agricultural policy, nutritional science, and clinical rehabilitation practices.

DOI: 10.4018/979-8-2600-1026-6.ch011

INTRODUCTION

Agricultural innovation has fundamentally reshaped global food production systems, propelling advances in crop science, mechanization, and digital technology that collectively heighten food security and dietary diversity (Finger et al., 2023; Klerkx et al., 2019). These technological developments have yielded substantial improvements in nutritional status and overall population health, largely by enhancing the accessibility and quality of nutrient rich foods (Yang et al., 2023). While the positive contributions of agricultural technology to public health are well documented, its implications for more specialized domains such as physical rehabilitation remain insufficiently examined (Zhao et al., 2025; Teran et al., 2025). The complex interplay between agricultural systems and the recovery trajectories of individuals engaged in rehabilitative care is, therefore, an underexplored area of inquiry (Zhang et al., 2024).

The rehabilitation process relies on a delicate balance between targeted clinical interventions and the psychological readiness of patients (Faleide et al., 2023). Adequate intake of high-quality nutrients facilitates tissue repair, muscle regeneration, and metabolic equilibrium factors that are critical for the effective execution of rehabilitation protocols (Blum & Parnell, 2021). Conversely, dietary insufficiency, persistent food insecurity, and disparities in food distribution can undermine patient resilience, limit participation in therapy programs, and negatively influence recovery outcomes (Beyene et al., 2023; Shi et al., 2025).

Recent evidence underscores not only the importance of nutrition for rehabilitation but also the indirect role of agricultural efficiencies in shaping food environments that sustain therapeutic advances (Kumar & Joshi, 2018). Notably, much of the existing literature on nutrition and rehabilitation has been focused on clinical nutrition or public health, with limited integration of upstream agricultural perspectives (Fanzo et al., 2015). As a result, systemic interconnections between agricultural policy, food system innovation, and patient level health outcomes remain insufficiently addressed. This disciplinary gap constrains opportunities for optimizing recovery through holistic, interdisciplinary strategies that harness the full potential of both nutritional science and agricultural advancements (Rahman et al., 2024; Fanzo et al., 2015).

A system-oriented framework is essential to trace the impact of agricultural technology on the confidence and recovery capabilities of patients undergoing physical rehabilitation (Yang et al., 2023; Zhao et al., 2025; Teran et al., 2025). By situating rehabilitation outcomes within the broader context of agricultural production, food accessibility, and diet quality, it is possible to identify structural barriers such as nutrient scarcity, inequitable access to quality foods, and policy misalignment that hinder the translation of agricultural progress into effective health interventions

(Rahman et al., 2024). Bridging these divides requires deliberate alignment between agricultural innovation, nutritional policy, and clinical rehabilitation practice (Waage et al., 2018).

In response to these gaps, this article critically examines the role of modern agricultural technology in influencing patient confidence and functional outcomes during rehabilitation (Yang et al., 2023). Specifically, it explores how enhanced availability of nutrient rich foods, enabled by recent advancements in agricultural systems, mediates physical resilience, adherence to therapeutic regimens, and psychological readiness among rehabilitation patients (Usigbe et al., 2024). Through an interdisciplinary lens, this study seeks to advance integration among agriculture, nutrition, and health sciences, thereby informing holistic approaches that fortify both individual recovery and broader public health outcomes (Fan et al., 2020).

PROBLEM STATEMENT

Despite notable advancements in agricultural technology, there is a gap between these innovations and their practical benefits for individuals undergoing physical rehabilitation (Zhao et al., 2025; Teran et al., 2025). Although new farming methods, improved crop varieties, and better food distribution systems have increased the availability of nutritious foods, many patients in rehabilitation still face challenges in accessing diets that support their recovery (Yinusa et al., 2025). This disconnect exists not only because of physical or economic barriers but also due to a lack of integration between agricultural policies and healthcare practices (Atashov et al., 2025). As a result, the potential of agricultural progress to improve rehabilitation outcomes remains underutilized.

Recent epidemiological data indicate that approximately 35–50% of patients in rehabilitation or long-term care settings experience some degree of malnutrition or risk thereof, directly influencing recovery trajectories (Miletic et al., 2024). Moreover, patients with inadequate nutritional intake are twice as likely to experience delayed functional recovery and therapy non-adherence (Wakabayashi & Sashika, 2021; Moisey et al., 2022). These figures demonstrate the empirical magnitude of the problem, underscoring that the issue extends beyond food availability to the systemic integration of nutrition into rehabilitation care.

One key issue is that most agricultural improvements are designed with broad population goals in mind, such as enhancing food security and reducing food shortage. However, these efforts often overlook the specific and changing nutritional needs of patients recovering from injury or illness. Healthcare providers face difficulties in connecting agricultural advances to clinical nutrition because of gaps in communication, knowledge about food quality, and a lack of standardized nutritional protocols in rehabilitation settings (Zhao et al., 2024). Consequently, even where nutritious food is available, it does not always translate into improved patient outcomes.

For instance, studies have found that individuals in food insecure households are significantly less likely to adhere to therapeutic regimens, particularly when medical therapy depends on regular medication (Palar et al., 2023; McClintock et al., 2024). This correlation between food security status and treatment participation highlights the importance of linking agricultural and clinical systems to support sustained recovery outcomes.

Besides physical nutrition, psychological factors also play an important role in successful rehabilitation (Zhao et al., 2024; Cheng et al., 2024). Recovery depends not only on therapy routines but also on patients confidence and motivation, which can be influenced by their trust in the food system's ability to meet their dietary needs. Inequities in access to high quality foods and poor awareness about the role of nutrition in healing create additional barriers, especially for vulnerable or marginalized patient groups.

This indicates a dual research gap both conceptual and practical. Conceptually, prior research rarely integrates agricultural technology and rehabilitation psychology within one framework. Practically, there are limited intervention protocols that operationalize agricultural or nutrition sensitive strategies within rehabilitation settings (Yinusa et al., 2021). Addressing these gaps requires cross disciplinary approaches that treat food systems as active determinants of rehabilitation success rather than peripheral influences.

Furthermore, research and policy work generally remain separated across agriculture, nutrition, and rehabilitation fields (Fan et a., 2020). Without cross-disciplinary collaboration, these areas fail to develop coordinated strategies that fully use agricultural technology to support rehabilitation (Yang et al., 2023; Zhao et al., 2025; Teran et al., 2025). This fragmentation hinders the creation of healthcare interventions that consider the broader food system context where patients live in.

Accordingly, this study uniquely contributes to the literature by conceptualizing an integrative framework linking agricultural technological innovation with nutritional access and rehabilitation confidence. Through this approach, it aims to provide theoretical grounding and practical direction for aligning agricultural progress with patient centered rehabilitation outcomes.

In summary, this gap highlights the need for focused research and policy efforts to connect agricultural advancements with physical rehabilitation practices (Zhao et al., 2024). Addressing this issue can improve patients access to nutrient rich diets and boost their recovery confidence, leading to better rehabilitation outcomes overall.

SIGNIFICANCE OF THE STUDY

This research also emphasizes the importance of adopting a systems oriented approach that connects agricultural policy, food systems, and clinical practice. Such an integrative perspective is crucial for addressing systemic issues like food insecurity and nutritional disparities, which often undermine patients ability to adhere to rehabilitation protocols and achieve optimal recovery (Beyene et al., 2023). By identifying and analyzing barriers within food systems that affect rehabilitation outcomes, this study provides valuable insights for policymakers, healthcare providers, and agricultural stakeholders seeking to align their efforts toward shared goals of public health and individual wellbeing (Moisey et al., 2022).

Specifically, this study advances understanding by establishing a conceptual framework that integrates agricultural innovation into rehabilitation science, a linkage rarely addressed in existing literature (Rahman et al., 2024). The study's novelty lies in framing food system technology not just as an enabler of nutrition but as a determinant of psychological and functional recovery outcomes, a dimension that has received limited empirical and theoretical attention (Picha et al., 2025). Furthermore, by positioning agricultural development as a driver of rehabilitation equity, this research extends the current discourse on the social determinants of health (Solar & Irwin, 2010).

Moreover, the findings from this study can inform the development of interdisciplinary strategies that foster collaboration between agricultural scientists, nutritionists, and rehabilitation professionals (Moisey et al., 2022). This cross disciplinary integration holds promise for creating tailored interventions that not only improve diet quality for rehabilitation patients but also enhance their confidence and motivation throughout the recovery process (Zhao et al., 2024).

In practical terms, these findings could guide policymakers in designing nutrition sensitive agricultural programs that directly support post-illness recovery, including subsidies for fortified foods, supply chain incentives for clinical nutrition products, and joint agricultural health pilot projects (FAO, 2023; WHO, 2022). Ultimately, this contributes to advancing patient centered care and promoting sustainable health improvements that reflect the complex interactions between food systems and human health (Bejarano et al., 2022).

Finally, by shedding light on the underexplored relationship between agricultural technology and physical rehabilitation, this study lays the groundwork for future research aiming to optimize the role of food system innovations in healthcare contexts (Yang et al., 2023). It encourages a more comprehensive understanding of how technological progress in agriculture can be effectively harnessed to improve therapeutic outcomes, thereby strengthening the broader endeavor to enhance medical rehabilitation and public health simultaneously (Yang et al., 2023).

This dual focus on both policy level integration and patient level application ensures that the research has practical utility beyond academia. It provides a template for future interdisciplinary studies that aim to align sustainable agriculture with healthcare delivery, reinforcing the global shift toward "One Health" frameworks that connect planetary and human well-being (Zhang et al., 2024; Teran et al., 2025).

FINDING THE GAP IN THE RESEARCH

Despite growing recognition of the importance of agricultural technology in enhancing food systems and nutrition, there remains a significant gap in research addressing how these advancements directly influence health outcomes within physical rehabilitation (Zhao et al., 2025; Teran et al., 2025). Most existing studies focus either on broad agricultural innovations aimed at increasing productivity and food security or on the clinical aspects of rehabilitation without fully exploring the critical linkages between these fields (Waage et al., 2018). This separation limits the understanding of how improvements in agricultural efficiency, food quality, and distribution can be systematically integrated into rehabilitation practices to enhance patient recovery and confidence (Kumar & Joshi, 2018; Al-Ansari et al., 2024).

Moreover, while nutrition's role in physical rehabilitation is well established, the complexity of food systems that shape nutritional availability is often overlooked (Zhao et al., 2024). Current research tends to isolate clinical nutrition from upstream determinants such as agricultural policies and technology adoption practices (Fan et a., 2020). As a result, the pathways through which agricultural innovations indirectly impact rehabilitation outcomes such as improved diet quality, food security, and access to nutrient rich foods remain underexplored and insufficiently conceptualized (Waage et al., 2018).

Another notable gap lies in interdisciplinary collaboration. Agricultural technology research, nutritional science, and rehabilitation medicine frequently operate in silos, with limited cross disciplinary frameworks or communication channels to support integrated approaches (Yang et al., 2023). This fragmentation not only constrains the development of comprehensive healthcare strategies but also hinders policy alignment that could facilitate patient access to food environments conducive to optimal rehabilitation outcomes (Fanzo et al., 2015).

Finally, there is a lack of empirical studies that critically examine patient experiences and perceptions regarding the role of food quality and availability in their rehabilitation journey (Moisey et al., 2022). The psychological dimensions, such as confidence in recovery linked to nutrition and food security, have been minimally addressed in existing literature, creating a void in understanding the holistic factors that support functional recovery (Fan et a., 2020; Cheng et al., 2024).

Addressing these gaps is essential for advancing research and practice in a way that fully leverages agricultural technological progress to support rehabilitation care (Zhao et al., 2024). This study aims to contribute to filling these voids by exploring the complex interactions between agricultural innovation, nutritional access, and patient centered rehabilitation outcomes, thereby fostering interdisciplinary dialogue and informing more cohesive health and food policy frameworks (Waage et al., 2018).

THEORETICAL FRAMEWORK

This study is based on a combination of ideas from systems thinking, nutrition science, and health behavior theories (Walker et al., 2021). Systematic thinking helps us understand how different parts of a larger picture like food production, health care, and individual experiences are connected (Walker et al., 2021). It reminds us that to fully understand how physical rehabilitation works, we need to look beyond just medical treatments and consider how things like farming methods, food availability, and policies affect what people eat and how well they recover (Zhao et al., 2024).

Nutrition science is important because it explains how the food we eat helps our bodies heal. Nutrients from food support muscle repair, energy, and overall strength, which are all necessary for people going through rehabilitation (Moisey et al., 2022). Knowing this helps us see why the quality and availability of food matter for recovery, and how changes in farming technology can affect the kinds of nutrients people can access (Yang et al., 2023).

Health behavior theories, such as the Social Cognitive Theory and the Health Belief Model, add another important layer by focusing on how patients think and feel during their recovery (Picha et al., 2025). These ideas show that a person's confidence and motivation to follow treatment plans can be influenced by whether they believe their body is getting the nutrition it needs. Feeling supported by an accessible food system can encourage patients to stay engaged and committed to their rehabilitation (Moisey et al., 2022).

To integrate these theoretical elements, this study conceptualizes a multilevel process in which agricultural technological innovation transforms food systems by improving production efficiency, distribution, and crop quality. These systemic improvements subsequently increase nutritional access and dietary adequacy, which in turn enhance patients physiological recovery. As their physical condition strengthens through better nutrition, patients are more likely to experience heightened confidence, self-efficacy, and motivation, which are psychological attributes that promote stronger adherence to rehabilitation regimens and improved overall outcomes (Bandura, 1997; Deci & Ryan, 2000). In this sequence, patient confidence functions both as a mediator, transmitting the benefits of improved nutrition to behavioral adherence, and as a moderator, amplifying or diminishing those effects depending on individual belief and environmental support.

The Social Cognitive Theory (SCT) provides the psychological foundation for understanding this dynamic. According to SCT, behavior is shaped by reciprocal interactions between personal factors (such as self-efficacy and motivation), environmental conditions (such as access to nutritious foods), and behavioral outcomes (such as therapy adherence) (Bandura, 1997). Within this framework, agricultural innovation indirectly strengthens self-efficacy by enhancing the environmental supports required for optimal recovery.

The Health Belief Model (HBM) complements this logic by explaining how individuals' perceptions of benefit and barriers shape their engagement with rehabilitation behaviors (Rosenstock et al., 1988). When patients perceive that better nutrition accelerates healing and that healthy food is accessible and affordable, they are more likely to sustain dietary and therapeutic adherence.

From a systems thinking perspective, these behavioral and nutritional processes operate within wider policy and socioeconomic environments (Walker et al., 2021). Agricultural innovation interacts with public health policy, institutional food procurement, and social inequality, which are moderating conditions that influence how effectively food system improvements reach rehabilitation patients (Rahman et al., 2024). The integrated framework thus portrays rehabilitation outcomes as the product of interdependent mechanisms linking agricultural advancement, food system functionality, nutritional adequacy, and patient psychology. Together, these theoretical perspectives provide a coherent explanation of how agricultural technology ultimately influences both physical recovery and confidence driven adherence in rehabilitation (Zhang et al., 2024; Teran et al., 2025).

By combining these three areas, systems thinking, nutrition science, and health behavior, this study builds a clear framework for exploring how new farming technologies and better food access can influence both the body and mind of patients in rehabilitation (Moisey et al., 2022). This approach helps connect the dots between agriculture, diet, and health recovery, and guides the research toward understanding how all these factors work together to improve rehabilitation outcomes (Moisey et al., 2022).

METHODOLOGY

This study employs a qualitative, theoretical research design that synthesizes existing information from open access internet sources, peer reviewed journal articles, and relevant policy documents. Instead of collecting new primary data, the research relies on a comprehensive review and interpretation of secondary data to explore how agricultural technology influences patient confidence and outcomes in physical rehabilitation (Yang et al., 2023). Literature was selected based on its relevance to agricultural innovation, nutrition science, food security, and rehabilitation practices, ensuring a wide interdisciplinary perspective (Waage et al., 2018). The collected sources were critically analyzed using thematic analysis, a method that involves identifying, and interpreting recurring themes and patterns related to nutrient access, food system efficiency, patient resilience, and psychological factors in rehabilitation (Cheng et al., 2024; Shi et al., 2025; Al-Ansari et al., 2024). The study's theoretical framework, which integrates systems thinking, nutrition, and health behavior models, guided the interpretation of findings, allowing for a nuanced understanding of how these elements interact (Walker et al., 2021; Moisey et al., 2022). To ensure validity, the selected literature was appraised for credibility, timeliness, and relevance, while combination of information from diverse fields helped strengthen the analysis and reduce bias. Although the study is limited by its reliance on secondary data, this approach enables a broad and integrative examination of complex relationships that may otherwise be inaccessible through primary research alone. This methodology supports the study's goal of developing a comprehensive understanding of the systemic connections between agricultural technology and physical rehabilitation outcomes, thus laying the foundation for future research and interdisciplinary collaboration (Yang et al., 2023).

FINDINGS AND RESULTS

Table 1. Overview of Key Agricultural Technologies and Their Nutritional Impacts

Agricultural Technology	Description	Impact on Nutritional Availability
Biofortification	Breeding or engineering crops to increase nutrient content (e.g., vitamins, minerals)	Helps combat micronutrient deficiencies by producing nutrient-rich staple crops (iron, zinc, vitamin A)
Precision Agriculture	Use of data-driven techniques (soil sensors, GPS, variable-rate nutrient application)	Improves nutrient use efficiency and soil health, leading to better crop nutrient quality
Mechanization	Use of machinery for farming processes	Increases farm productivity and food availability; indirect positive effects on food quantity and quality

continued on following page

Table 1. Continued

Agricultural Technology	Description	Impact on Nutritional Availability
Integrated Nutrient Management	Combining fertilizers and soil health practices to optimize plant nutrient uptake	Enhances soil fertility and nutrient density in crops; improves crop nutritional profile
Conventional High-Yield Varieties	Crops bred primarily for volume	Often associated with reduced nutrient density due to dilution effect

(Ofori et al., 2022; Huey et al., 2022; Hedley, 2015; Zhang et al., 2023; Sing et al., 2024; Bhardwaj et al., 2024)

The technologies summarized in Table 1 illustrate how agricultural innovations extend beyond food production to influence clinical nutrition and, indirectly, rehabilitation outcomes. Biofortification and precision agriculture are most relevant for rehabilitation populations, as they directly enhance micronutrient density and consistency of food quality (Bouis & Saltzman, 2017; Garg et al., 2018). Improved iron, zinc, and vitamin A levels achieved through biofortified crops support muscle repair and immune recovery, both crucial for post injury healing (Blum & Parnell, 2021). Similarly, precision agriculture optimizes soil health and nutrient management, leading to foods with higher bioavailability of essential minerals, thus reinforcing the physiological base for rehabilitation. Mechanization and integrated nutrient management contribute indirectly by ensuring food security and stable supply chains that reduce malnutrition risks in hospital and community settings (Zhang et al., 2023). These advances collectively bridge agricultural outputs with nutritional adequacy which is a foundational determinant of patient resilience and recovery capacity.

Figure 1. Trends in Global Agricultural Productivity and Nutrient Density over Time

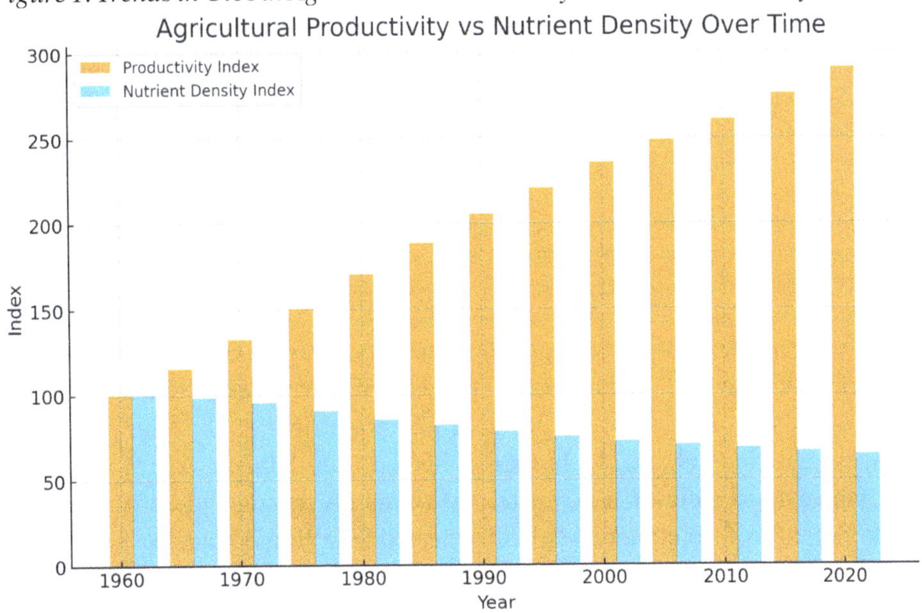

(Iizumi et al., 2020; Bhardwaj et al., 2024)

Table 2. Summary of Food Security Status and Nutritional Access in Different Regions

Region	Food Insecurity Prevalence (%)	Micronutrient Deficiency Prevalence	Primary Nutritional Challenges	Access to Healthy Diet (%)
Sub-Saharan Africa	25-30%	80% (women), 62% (children)	Iron, zinc, vitamin A deficiencies; chronic malnutrition	20-40%
South Asia	20-25%	74% (women), 53% (children)	Iron, folate deficiencies; hidden hunger widespread	30-50%
Latin America & Caribbean	8-12%	48% (women), 45% (children)	Obesity coexisting with micronutrient deficiencies	60-75%
Middle East & North Africa	12-18%	63% (women), 48% (children)	Conflict-related food access; vitamin deficiencies	40-65%
East Asia & Pacific	6-10%	58% (women), 45% (children)	Transitioning diets; emerging obesity issues	70-85%

continued on following page

Table 2. Continued

Region	Food Insecurity Prevalence (%)	Micronutrient Deficiency Prevalence	Primary Nutritional Challenges	Access to Healthy Diet (%)
Europe & Central Asia	4-8%	68% (women), 57% (children)	Processed food consumption; vitamin deficiencies	80-90%
High-Income Countries	2-6%	33-50% (women), 45% (children)	Diet quality issues; iron deficiency in women	85-95%

(Miller et al., 2022)

Table 2 reveals pronounced regional disparities in both food insecurity and micronutrient deficiencies. Rehabilitation patients are particularly vulnerable within these contexts because recovery demands higher nutrient intake than population averages (Miletic et al., 2024). In regions such as Sub-Saharan Africa and South Asia, where deficiencies in iron, folate, and vitamin A remain widespread, rehabilitation programs face compounded challenges due to limited nutrient dense food availability. The scarcity of fortified foods and restricted access to high quality protein sources hinder physiological recovery, extend hospital stays, and lower adherence to therapy (Wakabayashi & Sashika, 2021). Conversely, higher income regions show a paradox of nutritional imbalance with adequate caloric supply but suboptimal micronutrient quality, posing unique risks for rehabilitation patients with chronic conditions. These disparities underscore how agricultural efficiency directly conditions recovery environments, reinforcing the importance of aligning regional food systems with clinical nutrition needs.

Table 3. Nutritional Status vs. Rehabilitation Outcomes

Nutritional Status	Mean Length of Stay (days)	Mental Health Score (SF-36 MCS)
Well-nourished	15	51
At risk of malnutrition or malnourished	20	45

(Ramos et al., 2021)

Table 3 quantifies the clinical implications of the patterns observed earlier: malnutrition extends rehabilitation stays by an average of five days and correlates with reduced mental health scores (Ramos et al., 2021). Integrating evidence from Tables 1 and 2, this relationship can be conceptualized as a causal chain in which agricultural technologies improve food quality (Table 1), which enhances nutritional availability (Table 2), thereby reducing malnutrition-related delays in recovery (Table 3). For instance, biofortification and nutrient management can directly address micronutrient gaps responsible for slow tissue regeneration, while precision agriculture ensures consistent nutrient profiles across food supply chains. Thus, agricultural innovation becomes a functional health determinant rather than merely an economic or production oriented variable. This system's based perspective highlights how upstream technological advances can produce tangible downstream improvements in patient outcomes when integrated into rehabilitation nutrition planning (Teran et al., 2025).

Table 4. Barriers to Nutritional Support in Rehabilitation Settings

Barrier	Description
1. Limited Food Access	Patients often have difficulty accessing fresh, nutrient-rich foods due to supply constraints or facility purchasing policies
2. Knowledge Gaps among Staff	Rehabilitation clinicians and staff report insufficient training in clinical nutrition and dietary counseling
3. Lack of Standardized Protocols	Absence of clear guidelines within rehabilitation programs for assessing and addressing nutritional needs
4. Policy Misalignment	Agricultural and health policies operate in silos, preventing coordinated support for patient diets
5. Cost Constraints	High cost of nutrient-dense foods and limited rehabilitation budgets restrict provision of tailored meals or supplements
6. Patient Adherence Challenges	Patients may lack motivation or understanding to follow complex dietary recommendations during recovery
7. Inadequate Monitoring	Nutrition status often is not regularly assessed or documented, leading to missed interventions

(Dempewolf et al., 2023; Cavallo et al., 2023)

The barriers outlined in Table 4 contextualize why the potential benefits of agricultural technology often fail to reach rehabilitation populations. Limited access to fresh, nutrient rich foods (Barrier 1) and policy misalignment (Barrier 4) directly impede the translation of innovations identified in Tables 1–3 into patient outcomes. Similarly, knowledge gaps among staff and the absence of standardized nutritional protocols (Barriers 2 and 3) prevent effective clinical application of agricultural advances. These barriers operate as mediating bottlenecks within the agricultural technology where nutritional availability impacts rehabilitation outcomes pathway. Addressing them requires aligning agricultural policy with health sector priorities and improving clinician awareness of food system dynamics (Moisey et al., 2022; **Yinusa et al., 2021**).

Table 5. Policy Frameworks Linking Agricultural Innovation to Health Interventions

Policy Name	Sector	Integration Level	Key Provisions	Gaps
EU Farm to Fork Strategy	Agriculture & Health	Moderate	Supports sustainable agriculture, links to nutrition guidelines	No mention of rehabilitation or clinical nutrition protocols
US Healthy Food Financing Initiative	Agriculture, Health & Finance	Low	Funds food access projects in underserved areas	Does not target rehabilitation settings or dietary therapy
Brazil Zero Hunger Program	Agriculture & Social Welfare	High	Coordinates food production, distribution, and nutrition education	Rehabilitation not explicitly included, but strong community focus
WHO Global Action Plan for Healthy Diets & PA	Health & Nutrition	Moderate	Promotes dietary quality, references agricultural practices	No direct link to agricultural policy implementation
FAO Sustainable Agricultural Mechanization	Agriculture	Low	Encourages mechanization for productivity, mentions food availability	Omits health intervention frameworks, no clinical links
New Zealand Rural Health Strategy	Health & Rural Development	Low	Improves rural health access, references local food production	No integration of agricultural innovation with rehabilitation

(Zhao et al., 2025; Gamage et al., 2024)

Table 5 presents an opportunity map that aligns directly with the barriers identified earlier. For example, the EU Farm to Fork Strategy and Brazil's Zero Hunger Program demonstrate moderate-to-high integration between agricultural and health policies but currently lack explicit rehabilitation components. Incorporating rehabilitation nutrition within these frameworks could address Barriers 1 (limited food access) and 4 (policy misalignment) by mandating cross-sector collaboration. Similarly, the WHO Global Action Plan for Healthy Diets and Physical Activity can mitigate Barrier 2 (knowledge gaps) through training modules linking agricultural nutrition and therapeutic practice. In short, the policy frameworks in Table 5 provide mechanisms for overcoming the systemic barriers in Table 4 by effectively connecting agricultural technology to real world rehabilitation outcomes.

THEMES AND PATTERNS

The analysis reveals a complex yet interconnected set of patterns linking agricultural innovation, nutritional quality, and rehabilitation outcomes. A prominent trend emerging from the data is the productivity–nutrition paradox, wherein global agricultural productivity has increased dramatically by approximately 190% since 1961, while nutrient density has declined by nearly 36% (Waage et al., 2018; Zhang et al., 2023; Iizumi et al., 2020; Bhardwaj et al., 2024). This paradox underscores the trade-off between yield optimization and nutrient preservation, reflecting the "dilution effect" associated with high-yield crop varieties that prioritize volume over micronutrient content (Garg et al., 2018; Bouis & Saltzman, 2017). For rehabilitation populations, this decline in nutrient density has direct clinical implications. Recovery from physical injury or chronic illness depends on adequate micronutrient intake particularly iron, zinc, and vitamin A to support muscle repair, energy metabolism, and immune resilience (Blum & Parnell, 2021). Thus, while technological innovations in agriculture have successfully addressed food scarcity, they have not always guaranteed food quality or nutritional sufficiency, creating a systemic imbalance that affects recovery outcomes.

This imbalance is further compounded by geographic disparities in nutritional access. As shown in global food security data, regions such as Sub-Saharan Africa and South Asia face some of the highest rates of food insecurity and micronutrient deficiency, while high income regions experience excess caloric intake but reduced diet quality due to heavy reliance on processed foods (Miller et al., 2022; Beyene et al., 2023). Rehabilitation patients in low resource settings are particularly vulnerable because their nutritional needs exceed those of the general population, yet their access to nutrient dense foods remains limited. Restricted availability of fortified foods and protein rich diets often results in slower recovery, longer hospital stays, and greater risk of relapse or readmission (Miletic et al., 2024; Wakabayashi & Sashika, 2021). In higher income contexts, by contrast, the issue is not scarcity but dietary imbalance where excess energy intake with insufficient micronutrient diversity contributes to metabolic complications that impede physical therapy outcomes. These geographic variations reveal that improving agricultural productivity alone is insufficient to enhance rehabilitation outcomes unless it is coupled with equitable food distribution and quality focused agricultural policies.

A related and equally critical pattern concerns the relationship between clinical malnutrition and rehabilitation performance. Across multiple studies, malnutrition has been associated with extended hospital stays where on an average, five additional days and lower mental health and quality-of-life scores (Ramos et al., 2021). Nutritional insufficiency affects both the physiological and psychological components of recovery, undermining patients motivation, confidence, and adherence to rehabilitation regimens (Zhao et al., 2024; Cheng et al., 2024). When nutritional needs are not met, patients often experience fatigue, discouragement, and reduced participation in therapy sessions, which in turn prolongs recovery. Adequate nutrition, therefore, serves not only as a biological requirement but also as a behavioral enabler that reinforces patient self-efficacy and psychological readiness (Picha et al., 2025). Agricultural innovation enhances nutrient density and ensures consistent access to affordable, high-quality foods can mitigate this dual challenge by addressing both the physical and mental dimensions of rehabilitation.

However, despite technological progress, systemic barriers continue to obstruct the integration of nutrition into rehabilitation practice. Fragmented policies, limited intersectoral collaboration, and insufficient training among healthcare providers contribute to the persistent gap between agricultural advancements and clinical application (Cavallo et al., 2023; Moisey, et al., 2022). Barriers such as high food costs, inadequate facility level food procurement, and lack of standardized nutrition protocols restrict the ability of rehabilitation programs to capitalize on agricultural innovations. Furthermore, policy misalignment between the agricultural and health sectors often prevents the translation of global food system improvements into patient level benefits (Yinusa et al., 2021). Policy frameworks such as Brazil's Zero Hunger Program and the EU's Farm to Fork Strategy demonstrate the potential for aligning agricultural production with public health goals (Gamage et al., 2024), yet these initiatives rarely address rehabilitation specific nutrition. The lack of explicit integration between these policy domains underscores a critical opportunity for reform: incorporating clinical nutrition and rehabilitation needs into agricultural and food system strategies could transform both policy coherence and patient outcomes.

Taken together, these patterns illustrate a multilevel relationship among agricultural technology, food systems, and rehabilitation outcomes. The productivity–nutrition paradox shows that increased yields without nutritional integrity may undermine recovery potential. Geographic disparities highlight unequal nutritional access and their disproportionate burden on rehabilitation populations. Clinical malnutrition demonstrates the tangible health and psychological costs of poor diet quality, while systemic barriers reveal why these challenges persist despite scientific progress. Collectively, these findings emphasize that the connection between agriculture and rehabilitation is not linear but systemic, requiring coordinated innovation, policy alignment, and interdisciplinary awareness to ensure that advances in agricultural technology translate into measurable health and recovery benefits.

FUTURE DIRECTIONS

Looking ahead, several key directions emerge from the intersections of agricultural innovation, nutrition, and rehabilitation identified in this analysis (Waage et al., 2018). Advancing beyond traditional yield maximizing strategies, agricultural research should prioritize biofortifying staple crops commonly consumed by rehabilitation populations, with a focus on improving protein quality, vitamin D, and calcium content, which are essential for musculoskeletal recovery, bone density, and muscle regeneration (Garg et al., 2018; Blum & Parnell, 2021). These interventions can be piloted through coordinated programs linking agricultural research centers with rehabilitation facilities, allowing feasibility testing within two to three years before scaling to broader populations. Integrating nutrition sensitive approaches into precision and sustainable farming systems would ensure that agricultural technologies serve both productivity and therapeutic goals, supporting the health needs of post injury and postsurgical populations (Mekyassi & Kızıldeniz, 2024).

At the policy level, stronger collaboration among ministries of agriculture, health, and social services is required to bridge existing institutional silos (Yinusa et al., 2021; Zhao et al., 2024). In the near term (1–3 years), interministerial task forces could be established to pilot cross-sector programs targeting rehabilitation nutrition. For example, subsidizing fortified foods for patients discharged from rehabilitation centers or integrating agricultural extension services into public health outreach. Over the medium term (3–5 years), national frameworks should explicitly include "nutritional rehabilitation" as a priority within food security and health policies, with funding allocations for community level nutrition education campaigns. Such policies could be modeled on Brazil's *Zero Hunger Program* or adapted from the EU Farm to Fork Strategy, which already links agricultural sustainability with dietary quality but lacks a rehabilitation dimension (Gamage et al., 2024).

At the clinical practice level, rehabilitation institutions should incorporate routine nutritional screening and personalized dietary interventions into standard care. These could include the adoption of digital nutrition monitoring systems or mobile apps designed to track diet adherence and recovery outcomes (Zhou et al., 2025). Training programs for rehabilitation professionals should also integrate knowledge of agricultural trends and food system dynamics to strengthen cross-disciplinary understanding. This can be achieved through continuing education modules or joint workshops between dietitians, rehabilitation specialists, and agricultural scientists, requiring modest resource investment but high long term impact on patient outcomes (Moisey et al., 2022).

At a global scale, future research should focus on developing community driven models that connect local food production with rehabilitation care pathways. Initiatives that leverage local agriculture cooperatives to supply nutrient-dense foods to hospitals and therapy centers could promote sustainable, culturally appropriate nutrition strategies (Gans et al., 2023). Digital agriculture platforms and tele-rehabilitation tools could further enhance accessibility by enabling remote dietary monitoring and feedback which are particularly valuable in rural or resource limited settings. Achieving these goals will require interdisciplinary funding streams that integrate agricultural innovation grants with public health and rehabilitation research budgets.

Overall, the future demands a phased, coordinated commitment to linking agricultural advances with evidence based nutrition and patient centered rehabilitation. In the short term, pilot projects and cross sector collaboration can build the foundation for integration. In the medium term, policy alignment and clinical training can help institutionalize change. In the long term, technological and community based innovations can sustain global impact. Through this structured progression, the transformative potential of agricultural technology can be fully harnessed to enhance both physical recovery and psychological confidence among rehabilitation populations.

CONCLUSIONS

This study underscores a pivotal but often overlooked truth: agricultural innovation is not merely an economic or food security concern but also a determinant of physical rehabilitation success and patient confidence. While global advances in agricultural technology have substantially increased food production, this progress has frequently coincided with a decline in nutrient density and widening disparities in dietary quality (DeFries et al., 2015; Bhardwaj et al., 2024). The central finding of this analysis is that the quality of agricultural output, not just the quantity, directly shapes recovery trajectories for rehabilitation patients. Improved access to nutrient rich foods fosters faster physical recovery, better psychological resilience, and enhanced adherence to therapy, establishing agriculture as a foundational driver of holistic health (Walker et al., 2021; Sood et al., 2023).

The pathway from farm to recovery is systemic and interdependent where agricultural technology, food policy, and rehabilitation practice must evolve together to achieve optimal health outcomes. Agricultural innovation only fulfills its potential when its benefits extend beyond farms and markets to reach hospital wards, therapy centers, and home based recovery environments. By viewing agricultural systems through a health lens, this study reframes food as an active component of medical rehabilitation rather than a passive background condition (Kaufman et al., 2017; Rahman et al., 2024).

To realize this vision, coordinated actions are required across stakeholder groups. For policymakers, the immediate priority (within the next three years) is to embed rehabilitation nutrition into national food and health security frameworks, ensuring funding mechanisms for fortified foods, local supply chains, and community based nutrition programs (FAO, 2023; FAO et al., 2022). For researchers, the short to medium term goal (three to five years) should be to develop interdisciplinary pilot projects that link agricultural technology, nutritional biochemistry, and rehabilitation science. These projects should generate evidence on how biofortified and precision-grown foods affect patient outcomes (Sood et al., 2023; Walker et al., 2021). For clinicians and rehabilitation professionals, continuous professional development in nutrition science and agricultural awareness is essential to translate these systemic advances into everyday care. Incorporating nutritional screening and food system awareness into rehabilitation training could yield measurable improvements in recovery adherence and patient confidence within two years of implementation (Kanter et al., 2015).

The long term impact projected over five to ten years, depends on sustained cross sector collaboration that unites agricultural scientists, healthcare providers, and policymakers under a shared framework of "Food for Recovery." Such alignment will allow agricultural technology to become a tool not only for food security but also for medical empowerment by enhancing resilience, reducing healthcare costs, and improving quality of life for rehabilitation populations worldwide. Ultimately, agricultural innovation must be transformed into health innovation so that every advance in food production leads to an advance in human recovery (Kaufman et al., 2017; Rahman et al., 2024).

REFERENCES

Al-Ansari, M., Nabeel, H., Abdella, G. M., Mekkawy, T. E., & Kutty, A. A. (2024). A Non- Parametric Approach-Based Trade-Off between Food System Efficiency and Robustness. *Sustainability (Basel), 16*(15), 6480. DOI: 10.3390/su16156480

Atashov, B., Abbasov, A., & Gurbanzade, A. (2025). Government policy on ensuring food security: Bibliometric analysis. *Problems and Perspectives in Management, 23*(1), 496–514. DOI: 10.21511/ppm.23(1).2025.38

Bandura, A. (1997). *Self-efficacy: The exercise of control.* W.H. Freeman.

Bejarano, G., Csiernik, B., Young, J. J., Stuber, K., & Zadro, J. R. (2022). Healthcare students' attitudes towards patient centred care: A systematic review with meta-analysis. *BMC Medical Education, 22*(1), 1–22. DOI: 10.1186/s12909-022-03371-1 PMID: 35477455

Beyene, S. D., & Demissew, A. (2023). The impact of food insecurity on health outcomes: Empirical evidence from Sub-Saharan African countries. *BMC Public Health, 23*(1), 338. DOI: 10.1186/s12889-023-15244-3 PMID: 36793014

Bhardwaj, R. L., Parashar, A., Parewa, H. P., & Vyas, L. (2024). An alarming decline in the nutritional quality of foods: The biggest challenge for future generations' health. *Foods, 13*(6), 877. DOI: 10.3390/foods13060877 PMID: 38540869

Blum, J., & Parnell, J. (2021). Importance of nutrient availability and metabolism for skeletal muscle regeneration. *Frontiers in Physiology, 12*, 696018. DOI: 10.3389/fphys.2021.696018 PMID: 34335302

Bouis, H. E., & Saltzman, A. (2017). Improving nutrition through biofortification: A review of evidence from HarvestPlus, 2003–2016. *Global Food Security, 12*, 49–58. DOI: 10.1016/j.gfs.2017.01.009 PMID: 28580239

Cavallo, M., Morgana, G., Dozzani, I., Gatti, A., Vandoni, M., Pippi, R., Pucci, G., Vaudo, G., & Fanelli, C. G. (2023). Unraveling barriers to a healthy lifestyle: Understanding barriers to diet and physical activity in patients with chronic non-communicable diseases. *Nutrients, 15*(15), 3473. DOI: 10.3390/nu15153473 PMID: 37571410

Cheng, S., Li, J., Luo, Y., & Li, X. (2024). Family Quality of Life for Rehabilitated Individuals: Evidence Based on Demographic, Psychological, and Rehabilitation Factors. *SAGE Open, 14*(2), 21582440241258300. Advance online publication. DOI: 10.1177/21582440241258300

Deci, E. L., & Ryan, R. M. (2000). The "what" and "why" of goal pursuits: Human needs and the self-determination of behavior. *Psychological Inquiry*, *11*(4), 227–268. DOI: 10.1207/S15327965PLI1104_01

DeFries, R. S., Fanzo, J., Remans, R., Palm, C. A., Wood, S., & Anderman, T. L. (2015). Metrics for land-scarce agriculture. *Science*, *349*(6245), 238–240. DOI: 10.1126/science.aaa5766 PMID: 26185232

Dempewolf, S., Mouser, B., Rupe, M., Owen, E. C., Reider, L., & Willey, M. C. (2023). What Are the Barriers to Incorporating Nutrition Interventions Into Care of Older Adults With Femoral Fragility Fractures? *The Iowa Orthopaedic Journal*, *43*(2), 172–182. PMID: 38213858

Faleide, A. G. H., & Inderhaug, E. (2023). It is time to target psychological readiness (or lack of readiness) in return to sports after anterior cruciate ligament tears. *Journal of Experimental Orthopaedics*, *10*(94), 94. Advance online publication. DOI: 10.1186/s40634-023-00657-1 PMID: 37728618

Fan, S., Yosef, S., & Pandya-Lorch, R. (2020). Linking agriculture to nutrition: The evolution of policy. *China Agricultural Economic Review*, *12*(4), 595–604. DOI: 10.1108/CAER-03-2020-0040

Fanzo, J., Hawkes, C., & Udomkesmalee, E. (2015). Nutrition-sensitive agriculture and food systems in practice: Options for intervention. Food and Agriculture Organization of the United Nations. https://openknowledge.fao.org/bitstream/7055154c-1ac0-494a-a566- a0b1fbae10f9/download

Finger, R., Kaphengst, T., El Benni, N., & Walter, A. (2023). Digital innovations for sustainable and resilient agricultural systems. *European Review of Agriculture Economics*, *50*(4), 1277–1309. DOI: 10.1093/erae/jbad021

Gamage, A., Gangahagedara, R., Subasinghe, S., Gamage, J., Guruge, C., Senaratne, S., Randika, T., Rathnayake, C., Hameed, Z., Madhujith, T., & Merah, O. (2024). Advancing sustainability: The impact of emerging technologies in agriculture. *Current Plant Biology*, *40*, 100420. DOI: 10.1016/j.cpb.2024.100420

Gans, K. M., Risica, P. M., Wylie-Rosett, J., Yuan, Q., Ross, E. S., & Strolla, L. O. (2023). Local food system approaches to address food and nutrition security among low-income populations: A systematic review. *Advances in Nutrition*, *15*(4), 100156. DOI: 10.1016/j.advnut.2023.100156 PMID: 38616069

Garg, M., Sharma, N., Sharma, S., Kapoor, P., Kumar, A., Chunduri, V., & Arora, P. (2018). Biofortified crops generated by breeding, agronomy, and transgenic approaches are improving lives of millions of people around the world. *Frontiers in Nutrition*, *5*, 12. DOI: 10.3389/fnut.2018.00012 PMID: 29492405

Hedley, C. B. (2015). The role of precision agriculture for improved nutrient management on farms. *Journal of the Science of Food and Agriculture*, *95*(1), 12–19. DOI: 10.1002/jsfa.6734 PMID: 24816925

Huey, S. L., Krisher, J. T., Bhargava, A., Friesen, V. M., Konieczynski, E. M., Mbuya, M. N. N., Mehta, N. H., Monterrosa, E., Nyangaresi, A. M., & Mehta, S. (2022). Review of the impact pathways of biofortified foods and food products. *Nutrients*, *14*(6), 1200. DOI: 10.3390/nu14061200 PMID: 35334857

Iizumi, T., & Sakai, T. (2020). The global dataset of historical yields for major crops 1981–2016. *Scientific Data*, *7*(1), 97. DOI: 10.1038/s41597-020-0433-7 PMID: 32198349

Kanter, R., Walls, H. L., Tak, M., Roberts, F., & Waage, J. (2015). A conceptual framework for understanding the impacts of agriculture and food system policies on nutrition and health. *Food Security*, *7*(4), 767–777. DOI: 10.1007/s12571-015-0473-6

Kaufman, A., Boren, J., Koukel, S., Ronquillo, F., Davies, C., & Naranjo, J. (2017). Agriculture and health sectors collaborate in addressing population health. *Annals of Family Medicine*, *15*(5), 475–480. DOI: 10.1370/afm.2087 PMID: 28893819

Klerkx, L., Jakku, E., & Labarthe, P. (2019). Social science on digital agriculture, smart farming, and actor-networks: A review. *NJAS Wageningen Journal of Life Sciences*, *90–91*, 100315. DOI: 10.1016/j.njas.2019.100315

Kumar, S., & Joshi, D. (2018). Role of agricultural infrastructure and climate change on agricultural efficiency in Uttar Pradesh: A panel data analysis. *Economic Affairs*, *63*(4), 871–882. DOI: 10.30954/0424-2513.4.2018.10

McClintock, H. F., & Imel, B. E. (2024). Food insecurity and medication restricting behavior among persons with diabetes in the United States. *Nutrition and Health (Berkhamsted, Hertfordshire)*, *30*(2), 341–347. DOI: 10.1177/02601060221115588 PMID: 35876349

Mekyassi, H., & Kızıldeniz, T. (2024). The role of nutrition-sensitive climate-smart agriculture in ensuring global food security. BIO Web of Conferences, 85, 01055. https://doi.org/DOI: 10.1051/bioconf/20248501055

Mekyassi, H., & Kızıldeniz, T. (2024). The role of nutrition-sensitive climate-smart agriculture in ensuring global food security. BIO Web of Conferences, 85, 01055. https://doi.org/DOI: 10.1051/bioconf/20248501055

Miletic, B., Sutter, Y., Starčević-Klasan, G., Courteney, U., Jelovica, L., Lekić, A., & Šegulja, S. (2024). Malnutrition among the older adult: An additional challenge for the rehabilitation team-insights of a small population of Switzerland. *Frontiers in Public Health*, *12*, 1436566. DOI: 10.3389/fpubh.2024.1436566 PMID: 39045162

Miller, V., Webb, P., Cudhea, F., Shi, P., Zhang, J., Reedy, J., Erndt-Marino, J., Coates, J., Mozaffarian, D., Bas, M., Ali, J. H., Abumweis, S., Krishnan, A., Misra, P., Hwalla, N. C., Janakiram, C., Liputo, N. I., Musaiger, A., Pourfarzi, F., & Hakeem, R. (2022). Global dietary quality in 185 countries from 1990 to 2018 show wide differences by nation, age, education, and urbanicity. *Nature Food*, *3*(9), 694–702. DOI: 10.1038/s43016-022-00594-9 PMID: 37118151

Moisey, L. L., Merriweather, J. L., & Drover, J. W. (2022). The role of nutrition rehabilitation in the recovery of survivors of critical illness: Underrecognized and underappreciated. *Critical Care*, *26*(1), 1–17. DOI: 10.1186/s13054-022-04143-5 PMID: 36076215

Ofori, K. F., Antoniello, S., English, M. M., & Aryee, A. N. A. (2022). Improving nutrition through biofortification: A systematic review. *Frontiers in Nutrition*, *9*, 1043655. DOI: 10.3389/fnut.2022.1043655 PMID: 36570169

Palar, K., Sheira, L. A., Frongillo, E. A., Kushel, M., Wilson, T. E., Conroy, A. A., Adedimeji, A., Merenstein, D., Cohen, M. H., Wentz, E. L., Adimora, A. A., Ofotokun, I., Metsch, L. R., Turan, J. M., Tien, P. C., & Weiser, S. D. (2023). Longitudinal relationship between food insecurity, engagement in care, and ART adherence among US women living with HIV. *AIDS and Behavior*, *27*(10), 3345–3355. DOI: 10.1007/s10461-023-04053-9 PMID: 37067613

Picha, K. J., DiSanti, J. S., & Uhl, T. (2025, August 25). A sport medicine clinician's guide to improving patient adherence through tenets of Social Cognitive Theory. *Medical Research Archives*, *13*(8). Advance online publication. DOI: 10.18103/mra.v13i8.6846

Rahman, M. S., Wu, O. Y., Battaglia, K., Blackstone, N. T., Economos, C. D., & Mozaffarian, D. (2024). Integrating food is medicine and regenerative agriculture for planetary health. *Frontiers in Nutrition*, *11*, 1508530. DOI: 10.3389/fnut.2024.1508530 PMID: 39726870

Ramos, F. D., Fontanilla, J. A., & Lat, R. E. (2021). Association between degrees of malnutrition and clinical outcomes among non-critically ill hospitalized adult patients with type 2 diabetes mellitus. *Journal of the ASEAN Federation of Endocrine Societies, 36*(2), 172–179. DOI: 10.15605/jafes.036.02.12 PMID: 34966202

Rosenstock, I. M., Strecher, V. J., & Becker, M. H. (1988). Social learning theory and the Health Belief Model. *Health Education Quarterly, 15*(2), 175–183. DOI: 10.1177/109019818801500203 PMID: 3378902

Shi, Y., Liang, Z., Zhang, Y., Zhou, Y., Dong, S., Li, J., & Gao, G. (2025). Latent profiles of psychological resilience in patients with chronic disease and their association with social support and activities of daily living. *BMC Psychology, 13*(1), 1–13. DOI: 10.1186/s40359-025-03115-y PMID: 40665388

Solar, O., & Irwin, A. (2010). A conceptual framework for action on the social determinants of health. World Health Organization. https://apps.who.int/iris/handle/10665/44489

Sood, S., Garg, N., & Singh, S. (2023). Biofortification: An approach to eradicate micronutrient deficiency. *Frontiers in Nutrition, 10*, 1233070. DOI: 10.3389/fnut.2023.1233070 PMID: 37789898

Teran, E., Chetty, M., & Touirsi, S. (2025). Agriculture and health: Synergizing for a sustainable future. *Frontiers in Public Health, 13*, 1558198. DOI: 10.3389/fpubh.2025.1558198 PMID: 40241943

Usigbe, M. J., Asem-Hiablie, S., Uyeh, D. D., Iyiola, O., Park, T., & Mallipeddi, R. (2024). Enhancing resilience in agricultural production systems with AI-based technologies. *Environment, Development and Sustainability, 26*(9), 21955–21983. DOI: 10.1007/s10668-023-03588-0

Waage, J., Hawkes, C., Turner, R., Ferguson, E., Johnston, D., Shankar, B., Homans, H., & Harris, J. (2018). Integrating agriculture and health research for development: LCIRAH as an interdisciplinary programme to address a global challenge. *Global Challenges (Hoboken, NJ), 3*(4), 1700104. DOI: 10.1002/gch2.201700104 PMID: 31565369

Wakabayashi, H., & Sashika, H. (2021). Malnutrition is associated with poor rehabilitation outcome in elderly inpatients with hospital-associated deconditioning. *Journal of Rehabilitation Medicine, 53*(7), jrm00150. Advance online publication. DOI: 10.2340/16501977-1258 PMID: 24213734

Walker, A. E., Wattick, R. A., & Olfert, M. D. (2021). The application of systems science in nutrition-related behaviors and outcomes implementation research: Scoping review. *Current Developments in Nutrition*, *5*(9), nzab105. Advance online publication. DOI: 10.1093/cdn/nzab105 PMID: 34522835

Yang, M., Zhang, X., & Wang, Y. (2023). Does Internet use connect smallholder farmers to a healthy diet? Evidence from rural China. *Frontiers in Nutrition*, *10*, 1122677. DOI: 10.3389/fnut.2023.1122677 PMID: 37153910

Yinusa, G., Scammell, J., Murphy, J., Ford, G., & Baron, S. (2021). Multidisciplinary provision of food and nutritional care to hospitalized adult in-patients: A scoping review. *Journal of Multidisciplinary Healthcare*, *14*, 459–491. DOI: 10.2147/JMDH.S255256 PMID: 33654405

Zhang, G., Qiu, Y., Boireau, P., Zhang, Y., Ma, X., Jiang, H., Xin, T., Zhang, M., Tadesse, Z., Wani, N. A., Song, J., & Ding, J. (2024). Modern agriculture and One Health. *Infectious Diseases of Poverty*, *13*(1), 74. DOI: 10.1186/s40249-024-01240-1 PMID: 39385259

Zhang, H., Yang, Z., Wang, Y., Ankrah Twumasi, M., & Chandio, A. A. (2023). Impact of agricultural mechanization level on farmers' health status in western China: Analysis based on CHARLS data. *International Journal of Environmental Research and Public Health*, *20*(5), 4654. DOI: 10.3390/ijerph20054654 PMID: 36901664

Zhao, G., Cheng, Y., & Gatbonton, R. R. G. (2024). Biotechnological innovation in athlete health management: From nutrition to rehabilitation. *Journal of Commercial Biotechnology*, *29*(3), 427–438. https://commercialbiotechnology.com/article-detail/?id=2243

Zhao, Y., Cui, S., & Xing, Y. (2025). The impact of agricultural technological innovation on residents' health and its spatial effects. *Frontiers in Public Health*, *13*, 1633413. DOI: 10.3389/fpubh.2025.1633413 PMID: 40636876

Zhou, X., Cheng, G., & Zhang, B. (2025). Does farmland transfer promote digital agriculture? Based on panel data from 30 provinces in China. *Land (Basel)*, *14*(3), 481. DOI: 10.3390/land14030481

Chapter 12

Agricultural Technology–Critical Analysis on the Cultural Shifts in Obesity Subsidy Concentration With the Consumption of High Fructose Corn Syrup

Gangadhar Dhulipala
http://orcid.org/0009-0003-4615-921X
Capitol Technology University, USA

Eugene J. Lewis
http://orcid.org/0000-0002-2956-0760
Capitol Technology University, USA

ABSTRACT

This paper critically examines the cultural shifts in obesity prevalence and the role of high fructose corn syrup (HFCS) as a contributing factor in subsidy driven food economics. The analysis explores how government policies and agricultural subsidies have historically incentivized the production and widespread use of HFCS in processed foods. The study investigates the intersection of economic drivers, public health outcomes, and evolving societal attitudes toward nutrition and obesity. Key findings highlight that the proliferation of HFCS in the food supply correlates with increased caloric intake, shifting cultural preferences for convenience foods, and rising obesity rates. The implications of these dynamics underscore the need for reform in subsidy allocation and shift towards health-conscious food systems. The

DOI: 10.4018/979-8-2600-1026-6.ch012

study recommends policy changes and public awareness campaigns to address the multifaceted relationship between HFCS consumption, subsidy structures, and obesity trends.

INTRODUCTION

Agricultural technology plays a pivotal role in shaping food systems, influencing what crops are grown, how they are processed, and what foods dominate consumer diets (Usigbe et al., 2024). Over the past few decades, the intersection of agricultural advancements, subsidy policies, and cultural consumption trends has significantly contributed to the widespread use of high fructose corn syrup (HFCS) as a dominant sweetener in processed foods (Lakkakula, Schmitz & Ripplinger, 2016; Gün & Karaoğlu, 2024). HFCS, derived from corn through advanced enzymatic and production technologies, emerged as a cheaper alternative to traditional sugar, driven by innovations in crop cultivation and processing (Davis & Karen, 2011; Gün & Karaoğlu, 2024). However, this rise has coincided with a growing public health crisis i.e., the global obesity epidemic (Lowell, 2004).

This paper explores the challenges of agricultural technology in addressing the cultural and health implications of HFCS consumption (Usigbe et al., 2024; Siegel et al., 2016). Key among these challenges is the industrial efficiency of HFCS production, enabled by subsidies that favor corn cultivation (Siegel et al., 2015 & Fields, 2004). While these subsidies were initially designed to support economic stability for farmers, they have inadvertently contributed to an overabundance of cheap HFCS, fueling its integration into the food supply (Hayes et al., 2021). Additionally, advancements in food engineering have optimized HFCS for use in products that prioritize taste, shelf stability, and affordability factors that align with shifting consumer preferences toward convenience and processed foods (Adams et al.,2020).

At the heart of the issue is the disconnect between technological advancements in agriculture and their societal impacts. Although these innovations have enhanced production efficiency, they have entrenched HFCS within a subsidy structure that exacerbates health inequities (Gün & Karaoğlu, 2024). This analysis examines how this agricultural technology challenges intersect with cultural shifts in food consumption, subsidy allocation, and obesity trends (Lowell, 2004; Usigbe et al., 2024). We aim to illuminate pathways for reforming agricultural policies and fostering more sustainable and health conscious food systems by addressing these dynamics (CakmakSiegel et al., 2016).

PROBLEM STATEMENT

The rapid advancements in agricultural technology and subsidy driven economic policies have fundamentally reshaped food production systems, leading to unintended public health consequences, including the global rise in obesity rates (Siegel et al., 2016; Lowell, 2004). Agricultural innovations have optimized the cultivation of staple crops like corn, enabling large scale production of high fructose corn syrup (HFCS), a low cost sweetener widely used in processed foods (Lakkakula, Schmitz & Ripplinger, 2016; Gün & Karaoğlu, 2024). While these technologies have increased efficiency and affordability in food production, they have also contributed to an overabundance of calorie-dense, nutrient-poor food options (Fields, 2004). The root causes of obesity are multifaceted, but a critical factor is the intersection of agricultural practices, food policy, and cultural consumption patterns (Lowell, 2004).

(1) How do government subsidies disproportionately favor the production of HFCS and other commodities integral to highly processed foods? This creates a food environment that prioritizes economic convenience over nutritional value (Siegel et al., 2015; Popkin & Ng, 2022).
(2) How has agricultural technology contributed to the growing challenges and issues with technological advancements in food processing? These advancements have engineered foods to maximize flavor and shelf life, often at the expense of health (Usigbe et al., 2024).

Obesity is now recognized as a global public health crisis, with more than 1.9 billion adults overweight and over 650 million classified as obese worldwide (WHO, 2024). These conditions are strongly linked to noncommunicable diseases, which are the leading causes of global mortality and place significant burdens on healthcare systems. At the same time, the FAO (2024) highlights how global trade flows and agricultural subsidy regimes accelerate dietary transitions in low and middle income countries, encouraging the consumption of energy dense, ultra processed foods.

This U.S. centered pattern subsidies for corn leading to cheaper HFCS, increased use in processed foods, and higher obesity rates has been described as a “causal chain” of policy driven nutritional harm (Harvie & Wise, 2009; IISD, 2009). Evidence also suggests that individuals consuming diets high in subsidized commodities have significantly higher risks of obesity and cardiometabolic disease (Siegel et al., 2016). However, some analysts argue that, globally, subsidy levels may be too low to consistently drive obesity, underscoring the importance of context specific and comparative perspectives (Wunderlich & Kohler, 2022).

The resulting public health crisis is evident in the rising prevalence of obesity and its associated conditions, such as diabetes, cardiovascular diseases, and metabolic disorders (Siegel et al., 2016). These developments have normalized the consumption of HFCS containing products, particularly in low income communities where affordability dictates dietary choices (Gün & Karaoğlu, 2024). Beyond the U.S., global evidence from The Lancet underscores how agricultural policies, food technologies, and trade liberalization interact to shape nutrition transitions worldwide, often amplifying inequities in low and middle income contexts (Hawkes, 2012; Abay et al., 2021).

This problem highlights a critical gap between agricultural technology's potential to support sustainable food systems and its current role in perpetuating health disparities (Usigbe et al., 2024). Addressing these challenges requires a comprehensive understanding of how technological advancements, subsidy structures, and cultural shifts converge to impact public health, calling for reforms that balance agricultural productivity with nutritional equity and wellness (Siegel et al., 2016).

SIGNIFICANCE OF THE STUDY

This study is significant because it critically examines the complex interplay between agricultural policies, cultural consumption patterns, and public health outcomes in the context of high fructose corn syrup (HFCS) (Gün & Karaoğlu, 2024). By focusing on cultural shifts in obesity trends and the concentration of subsidies that promote HFCS production and consumption, the research addresses pressing societal challenges related to nutrition, economic structures, and healthcare systems (Gün & Karaoğlu, 2024).

(1) The study provides valuable insights into how subsidy policies and technological advancements in agriculture have shaped food systems, prioritizing cost efficiency over health implications (Fuxing et al., 2024). By analyzing these dynamics, the research highlights the unintended consequences of economic incentives on dietary habits and obesity prevalence (Lowell, 2004). This understanding is crucial for policymakers aligning agricultural goals with public health priorities. In this respect, WHO (2025) has noted that in 2022, 890 million adults worldwide are now living with obesity, demonstrating that the health effects of subsidy driven food systems are not isolated to any single country but form part of a wider global challenge that demands coordinated solutions.

(2) The study explores the cultural aspects of food consumption, including the normalization of processed and HFCS containing foods (Adams et al., 2020). By linking these cultural shifts to the economic and technological drivers behind HFCS ubiquity, the research underscores the need for societal awareness and behavioral change (Contreras-Manzano et al., 2024). This perspective is essential

for crafting effective public health campaigns and interventions that resonate with diverse populations (Gün & Karaoğlu, 2024). Such patterns are echoed in international analyses, with the FAO (2024) documenting how global trade and subsidy structures make ultra processed foods increasingly affordable in low and middle income countries. This suggests that cultural normalization of HFCS is not simply a U.S. issue, but part of a global dietary transition shaped by systemic economic forces.

(3) The study contributes to the broader discourse on sustainable and equitable food systems. By criticizing the paradox of subsidies that promote HFCS while public health systems bear the costs of obesity related illnesses, the research advocates for systemic reforms (Lowell, 2004). It offers a foundation for future studies exploring alternative subsidy models, sustainable agricultural practices, and culturally sensitive health initiatives (Gün & Karaoğlu, 2024). This aligns with the findings of The Lancet Commission on Obesity, which frames obesity, undernutrition, and climate change as a "Global Syndemic" requiring integrated reform of food, health, and trade systems (Swinburn et al., 2019). The IPCC (2022) similarly highlights how the same subsidy structures that drive overproduction of corn and sugar undermine both health outcomes and climate mitigation, emphasizing the interconnected costs of current agricultural models.

This study is significant for its interdisciplinary approach to understanding the cultural, economic, and technological factors contributing to the obesity epidemic through HFCS consumption (Lowell, 2004; Gün & Karaoğlu, 2024). Its findings aim to inform policies and practices that foster healthier societies and more equitable food systems. For example, WHO (2022) has recommended fiscal measures such as sugar-sweetened beverage taxation and mandatory front-of-pack labeling, while FAO (2024) has called for subsidy and trade policies that better align with nutrition-sensitive agriculture. Incorporating such policy perspectives not only extends the relevance of this research but also demonstrates how its findings can be directly translated into practical, global reforms.

FINDING THE GAP IN THE RESEARCH

Although individual studies examine either government subsidies or agricultural technology, there is a clear research gap in understanding their combined influence on the proliferation and normalization of highly processed foods such as those containing HFCS (Adams et al.,2020). The broader, systemic effects of these factors across policy, technological, and health domains remain insufficiently investigated, especially regarding their contribution to nutritional inequities and chronic disease risk in various populations.

Most current studies focus separately on the economic impact of subsidies or technological change, rather than integrating both to understand the direct and indirect consequences for the nutritional landscape and health disparities. For example, research tends to quantify subsidies for certain commodities but rarely connects these policies with rapidly evolving food technologies and their collective effects on processed food consumption and public health (Popkin & Ng, 2022).

In summary, there is a lack of interdisciplinary analysis examining how some of these subsidies support the technological advances of the food processing industry to increase the consumption and availability of the low nutrient high flavor foods. Further, the impact of these engineered foods that are routinely subsidized on long term chronic conditions and the population level dietary shifts. Finally, the socio-economic factors such as food marketing, affordability gaps that could mediate the relationship between the policy, technology and dietary environment especially in the marginalized groups will be explored.

THEORETICAL FRAMEWORK

An appropriate theoretical framework for these problem statements would combine systems thinking with policy analysis (Knai et al., 2018). This approach helps us see how government subsidies shape food production and why they often favor less healthy options (Siegel et al., 2015). It also brings perspectives from food environment research and technological innovation studies, which explain how agricultural technology contributes to the growth of processed foods and their health challenges (Constantinides et al., 2021).

Food Systems and Policy Analysis helps analyze how government subsidies shape the food, food systems frameworks which consider the dynamic interactions between agricultural policy, food production, economics, health, and the environment are highly relevant (Constantinides et al., 2021). Integrated economic environmental health modeling has been used to illustrate how subsidy structures drive food choices and production patterns, often prioritizing economic or political interests over nutritional outcomes (Springmann & Freund, 2022). These frameworks enable researchers to explicitly model feedback loops, trade-offs, and structural incentives in policy, matching the multifaceted nature of the subsidy problem (Constantinides et al., 2021).

For questions related to how the environment, including policy, shapes dietary patterns (e.g., subsidies for commodities like HFCS), the food environment framework is appropriate (Turner et al., 2018). This framework examines availability, affordability, convenience, and promotion of foods in the environment and how these shape consumption patterns. It is particularly useful for understanding how government actions affect the quality and nutritional profile of foods that are most prevalent and affordable.

The Solow technological change growth model provides a key economic foundation for analyzing how technological inputs affect agricultural output, food processing, and food security (Solow, 1956). It helps explain how technological advancements and not just capital or labor drive shifts in production and, by extension, impact the processed food landscape (Adams et al.,2020; Popkin & Ng, 2022). Beyond this, frameworks such as Responsible Research and Innovation (RRI) can help interrogate the societal and health trade-offs of new agricultural and processing technologies, encouraging a broader view that incorporates public health, ethics, and sustainability (Sánchez Hernández et al., 2024; Pacifico Silva et al., 2018).

METHODOLOGY

An appropriate methodology for this article would employ a mixed methods approach combining systematic literature review, policy analysis, and empirical data analysis to assess the relationships between government subsidies, technological advancements, and the nutritional profile of highly processed foods (Popkin & Ng, 2022). A systematic or rapid review of peer-reviewed articles, policy evaluations, and empirical studies addressing the impacts of subsidies on commodities like HFCS and technological innovation in food processing will be conducted (Gün & Karaoğlu, 2024). Subsequently, data concerning policy mechanisms, technological pathways, health/nutrition outcomes, and socio-economic mediators will be extracted. In addition, current subsidy structures and technological trends using policy documents, legislative reports and published government data will be analyzed (Franck, Grandi & Eisenberg, 2013). Further, as much as possible, will be collecting and analyzing the data sets that are available in the public domain to support Finally, use case studies from countries or regions with notable subsidy reforms to demonstrate the impacts on the real world.

The methodology will ensure a comprehensive, multiangle understanding of how subsidies in combination with technological advancements are influencing the processed food landscape and population health.

FINDINGS AND RESULTS

Table 1. Summary table capturing the figures and their descriptions

Figure Number	Description
1	Trends in the amount spent by the US government over the last two decades as a function of the type of crop
2	Showcasing the top 10 countries producing corn in the world for the fiscal year 2022/2024
3	% share of USA in corn production within the global market
4	Distribution of corn usage across multiple categories within the USA.
5	High fructose corn syrup (HFCS) per capita consumption between USA and Similarly populated countries
6	Trend in % obesity between USA and Similarly populated countries
7	HFCS usage distribution across multiple segments of the food industry within the USA
8	Cost comparison of different commonly used sweeteners in the food industry

Low prices for unhealthy foods significantly influence obesity rates by making calorie dense, nutrient poor foods more accessible and affordable (Lowell, 2004). When unhealthy foods are cheaper, people are more likely to purchase and consume them, especially low income individuals and families who may prioritize cost over nutritional value (de Oliveira Comini et al., 2025). Economic policies and subsidies that favor the production of commodity crops used in processed foods create an environment where unhealthy options are more financially attractive (Siegel et al., 2015; Pakravan-Charvadeh, et al., 2024). This affordability leads to poor dietary patterns characterized by high intake of sugars, fats, and refined carbohydrates, which are associated with weight gain and obesity (Davis & Karen, 2011; Lowell, 2004). Additionally, the food industry responds to consumer demand by producing cheaper and profitable products, perpetuating a cycle where unhealthy foods remain prevalent and affordable, further influencing dietary choices and obesity rates (Headey & Alderman 2019). For example, below is the summary of share of premium subsidies by crop during the time 1995 – 2024.

Figure 1. Government spending in terms of subsidies over the last two decades in United States of America.

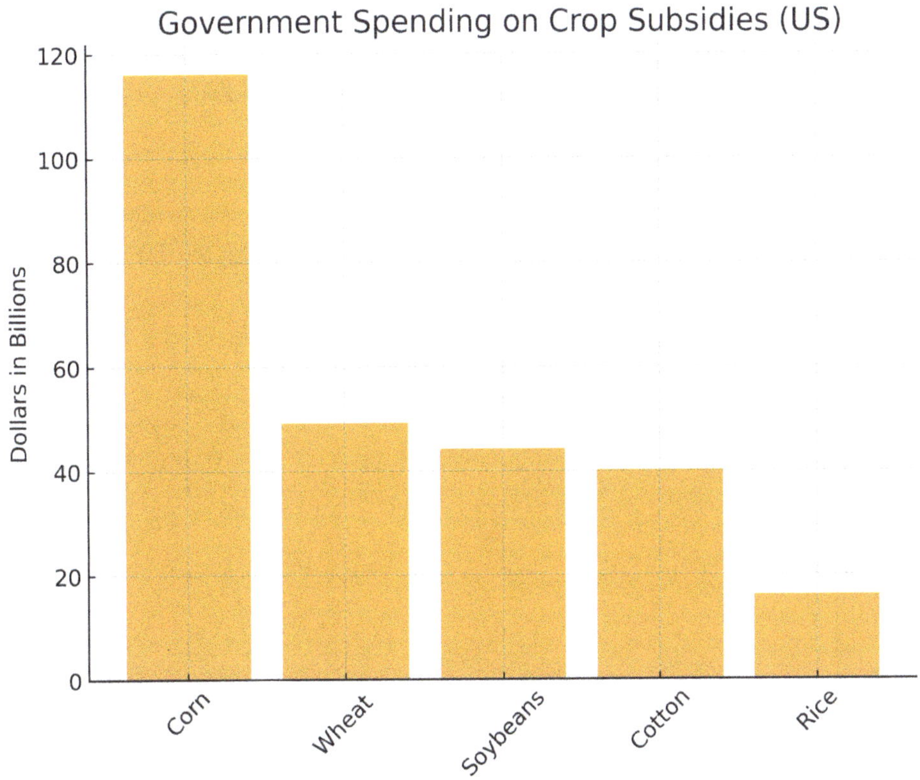

Source:https://farm.ewg.org/cropinsurance.php?fips=00000&regionname=theUnitedStates&summpage=PS_BY_CROP

Table 2. Corn production in million metric tons by country

<table>
<tr><th>Rank</th><th>Country</th><th>Production (million metric tons)</th><th>Source</th></tr>
<tr><td>1</td><td>United States</td><td>~ 378 MMT</td><td rowspan="5">IndexMundi – Corn Production</td></tr>
<tr><td>2</td><td>China</td><td>~ 295 MMT</td></tr>
<tr><td>3</td><td>Brazil</td><td>~ 127 MMT</td></tr>
<tr><td>4</td><td>European Union (EU-27)</td><td>~ 58 MMT</td></tr>
<tr><td>5</td><td>Argentina</td><td>~ 51 MMT</td></tr>
<tr><td>6</td><td>India</td><td>~ 38 MMT</td><td rowspan="2">ScienceAgri – World's Biggest Maize Producers</td></tr>
<tr><td>7</td><td>Ukraine</td><td>~ 31 MMT</td></tr>
<tr><td>8</td><td>Mexico</td><td>~ 27.5 MMT</td><td>IndexMundi – Corn Production</td></tr>
</table>

continued on following page

Table 2. Continued

Rank	Country	Production (million metric tons)	Source
9	Indonesia	~ 20 MMT	ScienceAgri – World's Biggest Maize Producers
10	Russian Federation	~ 16.6 MMT	

Source: Corn production share worldwide by country 2024 I Statista, Corn I USDA Foreign Agricultural Service

Figure 2. % share of USA in corn production within the global market

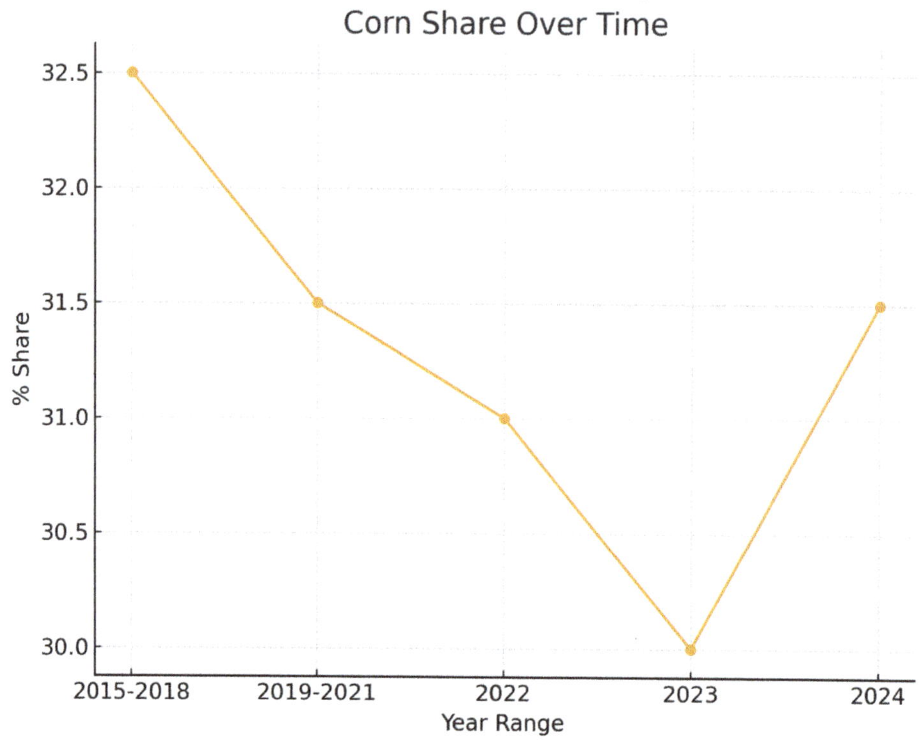

Source: Corn production share worldwide by country 2024 I Statista, Corn I USDA Foreign Agricultural Service

Figure 3. Distribution of corn usage across multiple categories within the USA. Source: Corn in the U.S. - statistics & facts | Statista; Corn | USDA Foreign Agricultural Service

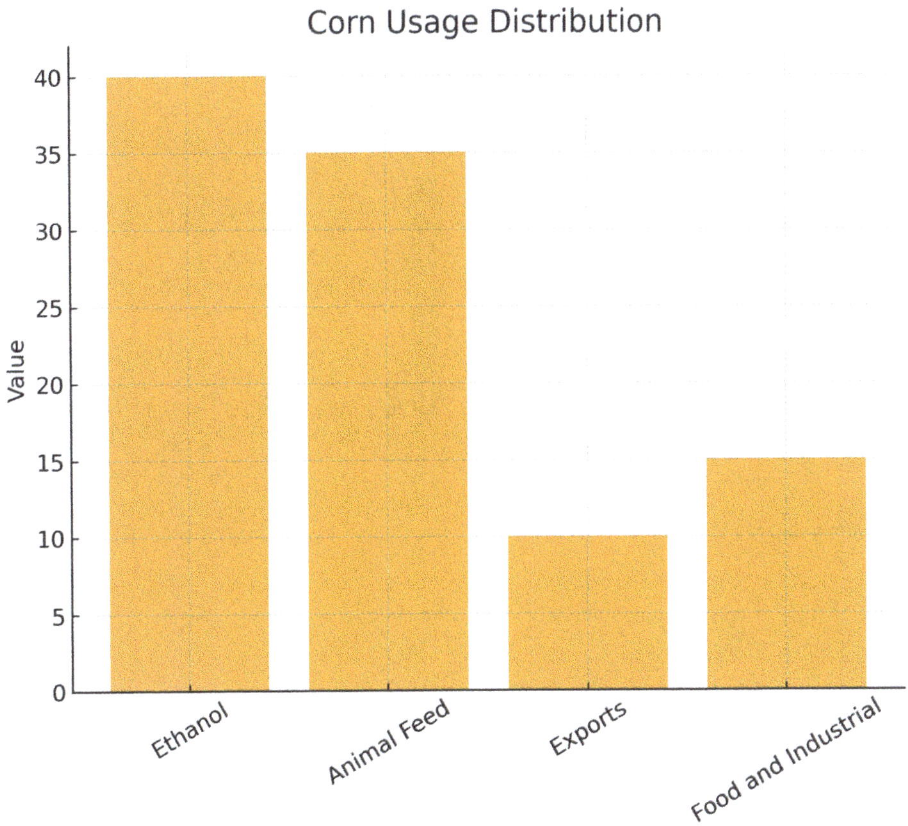

Figure 4. % Share of corn utilized for HFCS

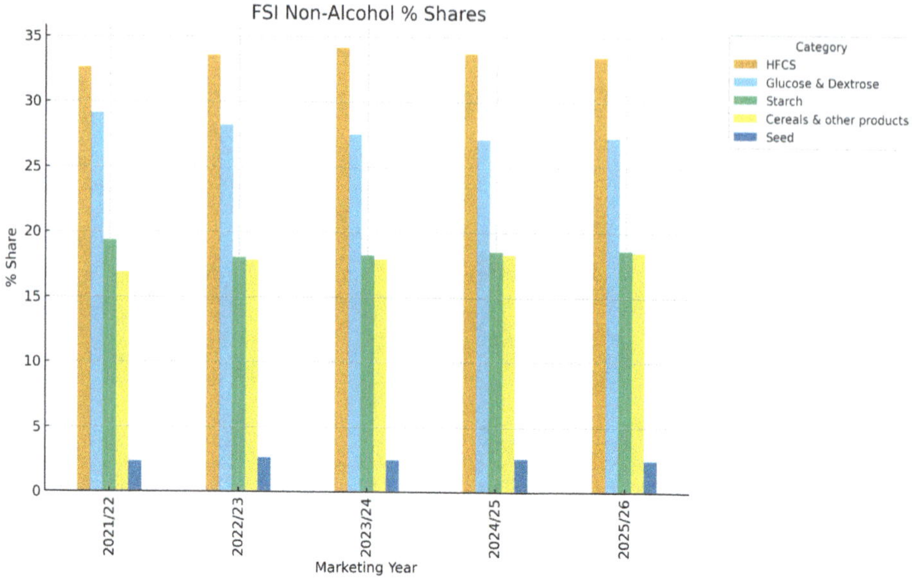

Source: ERS Feed Grains Yearbook – Table 31

Figure 5. High fructose corn syrup (HFCS) per capita consumption between USA and Similarly populated countries. Sources: USDA ERS Yearbook (39.5 lbs/person in 2021); USDA/FAO and CRA reports

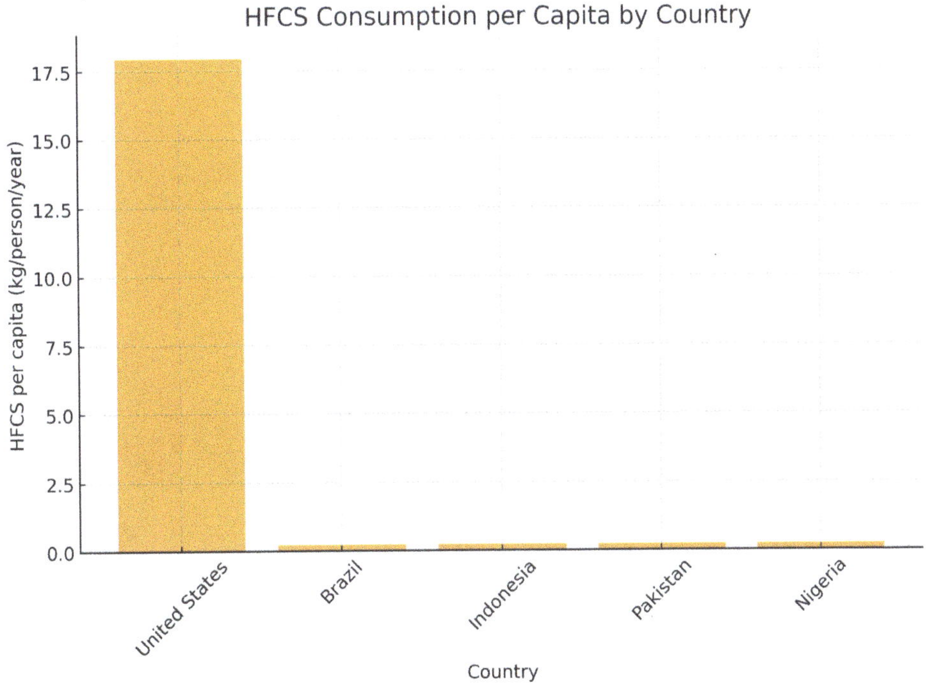

Figure 6. Trend in % obesity between USA and the countries with similar population

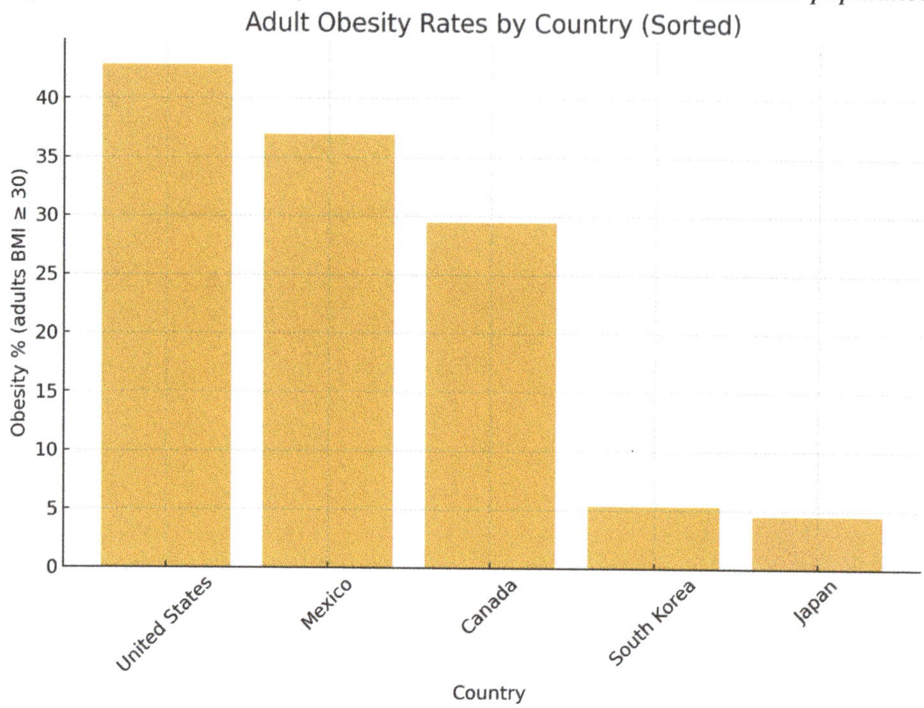

Source: Ranking (% obesity by country), adults (combined, BMI ≥ 30)" from the Global Obesity Observatory/ NCD-RisC, most recent data (~2022).

Figure 7. Cost comparison of different sweeteners

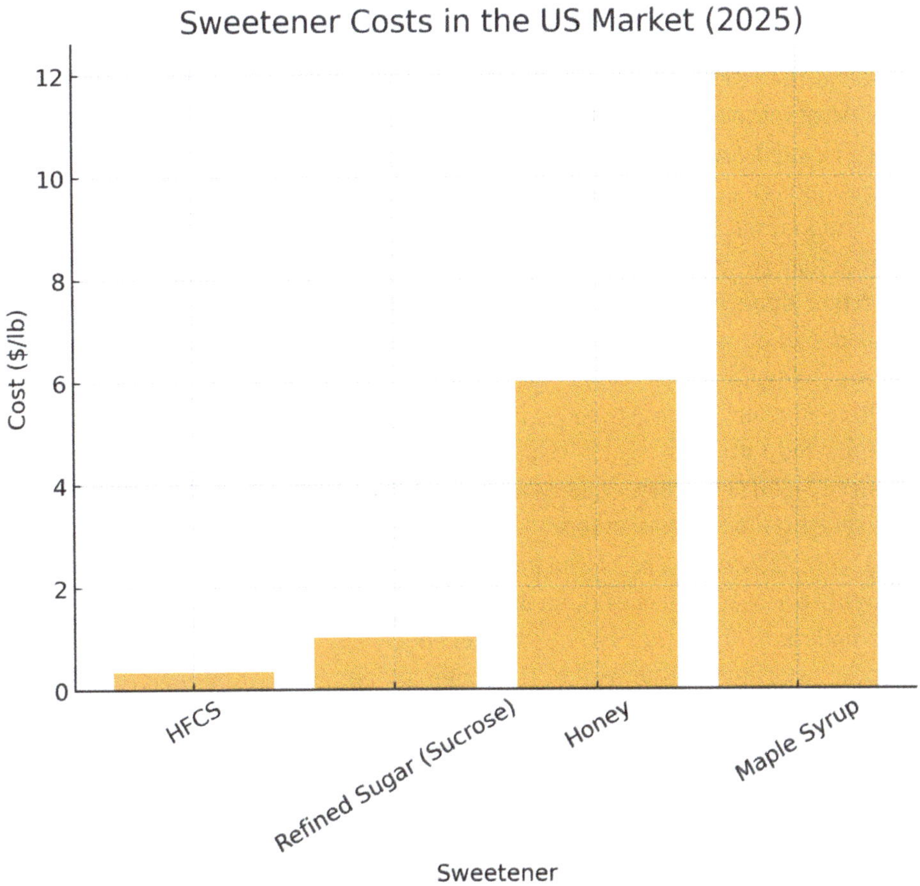

Source:https://www.foxbusiness.com/lifestyle/high-fructose-corn-syrup-vs-cane-sugar-foods-cost-switching -ingredients;https://www.ers.usda.gov/topics/crops/sugar-and-sweeteners/background?

THEMES AND PATTERNS

The U.S. agricultural system demonstrates how deeply policy shapes food environments. For decades, corn has been the most subsidized crop, receiving billions of dollars in federal support that far exceeds subsidies for wheat, rice, or soybeans (Hayes et al., 2021). This persistent investment has created a production landscape in which corn is not just one crop among many, but the backbone of American agriculture (Fields, 2004). Graphs of subsidy distributions make this policy preference unmistakable, showing year after year that corn dominates the allocation of public funds (Adams et al.,2020).

Such policy incentives inevitably translate into production outcomes. With corn secured as the safest and most profitable crop, U.S. farmers have scaled production to the point that the country now accounts for nearly one-third of the world's corn supply (United States Department of Agriculture, Economic Research Service, 2023; Adams et al.,2020). When plotted against similarly populous nations such as Brazil, Indonesia, Pakistan, and Nigeria, the U.S. emerges as an outlier, producing far more than demographic or geographic factors alone would suggest. The global production trend line reinforces this dominance, highlighting that the structural abundance of corn is not a temporary fluctuation but a durable feature of the food system (Fields, 2004).

This abundance, however, does not flow evenly into diverse diets and instead, it is funneled through a set of industrial pathways. In USDA's "Food, Seed, and Industrial" (FSI) category, ethanol accounts for the bulk of use, but high-fructose corn syrup (HFCS) consistently claims a measurable share (Gün & Karaoğlu, 2024). Our stacked bar charts illustrate this pattern, showing HFCS holding steady at roughly 6% of FSI bushels (Gün & Karaoğlu, 2024). While that percentage may seem modest, its significance becomes clear once costs are considered. HFCS is the cheapest sweetener on the market (Lakkakula, Schmitz & Ripplinger, 2016; Gün & Karaoğlu, 2024). Compared to refined sugar, honey, or maple syrup, HFCS sits at the bottom of the price spectrum, a fact made vivid in cost comparison graphs (Davis & Karen, 2011). Economic convenience therefore ensures its ubiquity in processed foods and beverages (Adams et al.,2020).

The consequences of this dynamic pricing appear in consumption patterns. A per-capita bar chart places the United States at nearly 40 pounds of HFCS per person annually, far surpassing similarly populated countries where consumption is negligible (United States Department of Agriculture, Economic Research Service, 2023). Mexico is the notable exception, reflecting a similar reliance on HFCS in its food supply. This stark divergence between the U.S. and other nations demonstrates how agricultural abundance, when channeled into industrial sweeteners, reshapes national diets (Van Buul et al., 2014; Lakkakula, White, 2008; Schmitz & Ripplinger, 2016).

Health outcomes align closely with these patterns. Obesity prevalence, when charted alongside per-capita HFCS consumption, shows the U.S. (42%) and Mexico (37%) at the top, while countries with minimal HFCS intake Brazil, Pakistan, Indonesia, and Nigeria remain substantially lower (Lowell, 2004; Gün & Karaoğlu, 2024). The pairing of these graphs makes the association between HFCS consumption and obesity strikingly visible, suggesting not just correlation but a structural relationship between policy, food system design, and public health (Lowell, 2004).

Taken together, these themes trace a consistent causal arc where subsidies make corn cheap, cheap corn sustains high production, industrial processing channels it into HFCS, HFCS's low cost guarantees its dominance in manufactured foods, and elevated consumption correlates with some of the highest obesity rates in the world (Adams et al.,2020; Lowell, 2004). Each graph reinforces this sequence, and together they reveal a food environment optimized for economic efficiency rather than nutritional well-being. In this sense, the American case is not only exceptional but also instructive. It shows how choices made at the level of farm policy cascade through production, processing, and consumption to shape the health of entire populations (Jackson et al., 2009). Similar dynamics are visible in broader agricultural markets, where the mediation of marketing intermediaries and the adoption of artificial intelligence in livestock and food systems influence economic outcomes (Guyo et al., 2023).

ROLE OF AGRICULTURAL TECHNOLOGIES

Subsidies may have created the conditions for corn to dominate American agriculture, but it was technology that amplified and entrenched this dominance (Fields, 2004). Over the past half century, advances in farming and food science have not only made corn cheaper to grow but have also allowed it to be transformed into a wide range of ingredients with the most notable being high fructose corn syrup (HFCS) (Gün & Karaoğlu, 2024). Together, these innovations ensured that a crop originally valued for its versatility would become the raw material behind much of the modern food system. On the farm, technological progress steadily pushed yields upward (Fanzo et al., 2023). Hybrid seeds and, later, genetically engineered varieties increased resilience against pests and weeds, while innovations in fertilizer, irrigation, and precision farming equipment allowed farmers to manage land with remarkable efficiency. The result was that corn could be grown more reliably, in larger quantities, and at a lower cost per bushel (Adams et al.,2020; Fields, 2004). In many ways, technology delivered exactly what policy had incentivized i.e. a steady and inexpensive surplus.

At the same time, breakthroughs in food processing reshaped what that surplus could become. In the late 1960s and early 1970s, scientists developed enzymatic techniques to break down cornstarch into glucose and then convert a portion of that glucose into fructose. Scaled up in industrial plants, this process produced HFCS which is a sweetener similar in taste to sugar but far cheaper to make it (Davis & Karen, 2011; Lakkakula, Schmitz & Ripplinger, 2016; Gün & Karaoğlu, 2024). HFCS quickly found its way into sodas, baked goods, and countless processed foods, offering manufacturers a product that was not only cost effective but also easy to transport and highly stable(Gün & Karaoğlu, 2024).

This combination of policy and technology made HFCS almost inevitable. Subsidies created the economic foundation, while scientific advances supplied the tools to turn corn into a profitable industrial input (Hayes et al., 2021). When compared to other sweeteners, HFCS consistently comes out as the lowest cost option well below refined sugar and dramatically cheaper than natural alternatives like honey or maple syrup (Davis & Karen, 2011; Lakkakula, Schmitz & Ripplinger, 2016; Gün & Karaoğlu, 2024). For food companies, the choice was straightforward to use HFCS to maximize profits while keeping prices low for consumers (Gün & Karaoğlu, 2024).

The ripple effects of this technological story are evident in consumption patterns. Americans now consume far more HFCS per person than populations in similarly large countries such as Brazil, Indonesia, Pakistan, or Nigeria, where sucrose remains the primary sweetener and HFCS use is negligible (Lakkakula, Schmitz & Ripplinger, 2016; Gün & Karaoğlu, 2024). Mexico is the notable exception, reflecting a parallel adoption of HFCS tied to both trade and technology. As our charts illustrate, the nations with the highest HFCS consumption also face some of the world's highest obesity rates (Lowell, 2004).

Overall, agricultural technology does more than increasing efficiency (Usigbe et al., 2024). It reshapes diets and health outcomes and by enabling corn to be converted into inexpensive, calorie dense sweeteners (Lakkakula, Schmitz & Ripplinger, 2016). Further, technology magnified the effects of subsidies and set the stage for a food system built around economic convenience rather than nutritional balance (Fanzo et al., 2023; Hayes et al., 2021). The story of HFCS is not only a story of science or policy in isolation, but of how the two combine to shape what ends up on the dinner table and, ultimately, in public health statistics (Gün & Karaoğlu, 2024). It reshapes diets and health outcomes by enabling corn to be converted into inexpensive, calorie dense sweeteners. Similar to higher education and health contexts, where workplace learning and public private partnerships mediate technological innovation (Dereso et al., 2023), agricultural systems also reveal the importance of institutional and organizational structures in shaping how technology affects outcomes.

CONCLUSIONS

The findings of this study reveal a clear and interconnected pattern where U.S. agricultural subsidies disproportionately favor corn, making it the most supported and most abundantly produced crop in the world (Siegel et al., 2015; Hayes et al., 2021). Year-over-year subsidy data demonstrate that corn consistently outpaces other commodities in government support, creating both an economic and structural surplus (Adams et al., 2020). Coupled with technological advances that have steadily increased yields and enabled efficient processing into industrial inputs such as high-fructose corn syrup (HFCS), this policy environment has produced a food system where HFCS emerges as the cheapest and most pervasive sweetener (Kognou et al., 2022; Fanzo et al., 2023; Lakkakula, Schmitz & Ripplinger, 2016; Gün & Karaoğlu, 2024). Our comparative analysis of sweetener costs shows HFCS priced well below refined sugar and dramatically cheaper than natural alternatives such as honey or maple syrup, ensuring its dominance in the processed food industry (Davis & Karen, 2011; Lakkakula, Schmitz & Ripplinger, 2016; Gün & Karaoğlu, 2024).

The consequences of this alignment are evident in consumption and health data. Americans consume nearly 40 pounds of HFCS per person annually, a level mirrored only by Mexico, while other similarly populous nations such as Brazil, Indonesia, Nigeria, and Pakistan consume negligible amounts (Gün & Karaoğlu, 2024). When compared with obesity prevalence, the correlation becomes striking where countries with the highest HFCS consumption (United States, Mexico) also carry some of the highest rates of obesity, while countries with little to no HFCS intake report substantially lower prevalence (Lowell, 2004; Gün & Karaoğlu, 2024). These findings underscore the extent to which subsidy-driven abundance, amplified by agricultural and food processing technologies, has reshaped diets and contributed to public health burdens (Wunderlich & Kohler, 2022).

Taken together, this study contributes to the broader discourse on sustainable and equitable food systems by exposing the paradox of agricultural policies that prioritize economic efficiency while public health systems absorb the rising costs of obesity and related illnesses (Lowell, 2004). It demonstrates how a system optimized for cheap calories undermines nutritional well-being, and in doing so, reinforces inequalities in health outcomes (Darmon & Drewnowski, 2015).

By criticizing this paradox, research advocates systemic reforms. Reimagining subsidy models to support a wider range of crops including fruits, vegetables, and other nutrient dense foods could begin to realign agricultural priorities with public health goals (Huangfu et al., 2024). Integrating sustainable agricultural practices would not only diversify production but also reduce the environmental footprint of overproduced commodities (Çakmakçı et al., 2024). Finally, implementing culturally sensitive health initiatives would ensure that dietary reforms are responsive to local traditions and accessible to populations most at risk (Springmann & Freund, 2022).

Importantly, this study directly addresses the identified research gap by showing how subsidies and agricultural technologies do not operate in isolation but together create a structural pathway that normalizes HFCS consumption and exacerbates obesity. While prior research often examined subsidies or technology separately, the present findings integrate these dimensions and demonstrate their combined cultural, economic, and health impacts. By articulating this causal chain where subsidies promote surplus corn production, which in turn drives HFCS manufacturing, its integration into processed foods, and ultimately rising population level obesity the study contributes a more comprehensive understanding of how systemic drivers converge to shape public health outcomes.

Thus, the conclusions not only reinforce the need for integrated food system reforms but also demonstrate that addressing the subsidy technology nexus is essential to correcting nutritional inequities and reducing the global burden of obesity. This synthesis strengthens the policy relevance of the findings and provides a roadmap for aligning agricultural productivity with long term public health objectives (WHO, 2024; FAO, 2024; IPCC, 2022; Swinburn et al., 2019).

In conclusion, the evidence presented here highlights a causal chain—policy, production, processing, consumption, health that helps explain how subsidies and technologies coalesced to make HFCS central to the U.S. food supply (Hayes et al., 2021; Gün & Karaoğlu, 2024). Addressing this chain at multiple points provides an opportunity to build food systems that are not only more sustainable and equitable but also better aligned with long term public health (Çakmakçı et al., 2024).

FUTURE DIRECTIONS

The findings of this study highlight the need for research and policy work that not only critiques existing food system paradoxes but also advances actionable solutions. Several avenues for future inquiry stand out. To begin with, greater attention must be given to rethinking subsidy models (Blakley et al., 2020). Current frameworks overwhelmingly support to corn, reinforcing the production of HFCS and similar industrial inputs (Blakely, 2020; Shilpa Gün & Karaoğlu, 2024). Future studies could model the effects of alternative subsidy allocations, for example, shifting resources toward fruits, vegetables, and legumes to examine how such reforms might alter food prices, consumer choices, and long-term health outcomes (Huangfu, et al., 2024; Blakely, 2020). Comparative cross country analyses could further clarify how policy redesigns in one context might be adapted elsewhere (Huangfu et al., 2024).

Followed by that, research should evaluate sustainable agricultural practices that maintain profitability while diversifying production (Huangfu, et al., 2024). Practices such as regenerative agriculture, crop rotation, and soil health management not only enhance ecological resilience but also reduce the structural dependence on monocultures that feed industrial sweetener production (Tan et al., 2023; Lakkakula, Schmitz & Ripplinger, 2016). By integrating sustainability into economic planning, these studies could inform both agricultural policy and climate goals (Bowles et al., 2020). In addition, there is a need to broaden the lens beyond HFCS to examine the role of emerging food technologies. Building technical capacity and problem solving skills, as demonstrated in other industries such as textiles (Adula et al., 2023), will be equally important in preparing agricultural and food sectors for sustainable technology adoption. As innovations in plant based proteins, synthetic fats, and novel processing methods gain traction, they may reproduce the same tensions seen with HFCS technological efficiency aligned with economic incentives but misaligned with public health priorities (Maganinho et al., 2024; Malila et al., 2024; Johansson, et al., 2025). Anticipating these trajectories through interdisciplinary research could help prevent repeating past mistakes.

Future work must address cultural and public health dimensions more directly, merging computational approaches, such as hybrid AI models for health informatics and drug review classification (Durga et al., 2024), could be adapted to food systems research to better monitor consumption patterns and predict health outcomes. Comparative studies between high HFCS consumers such as the U.S. and Mexico and low HFCS consumers such as Brazil, Indonesia, Nigeria, and Pakistan can reveal how cultural dietary traditions and policy environments interact to shape health outcomes (Gün & Karaoğlu, 2024). This line of inquiry is especially important for designing culturally sensitive health initiatives that resonate with local food practices (Lara-Castor, 2023).

In addition, researchers should pursue integrated economic and health cost accounting. While HFCS lowers manufacturing costs, the hidden expenses borne by healthcare systems through obesity and metabolic diseases remain underexplored in policy debates (Do et al., 2020; Lowell, 2004; Gün & Karaoğlu, 2024). Comprehensive cost benefit analysis could provide policymakers with the evidence necessary to justify subsidy reforms that align fiscal efficiency with public health (Marklund et al., 2020)

Finally, longitudinal and generational studies are needed to track dietary transitions over time (Springmann & Freund, 2022) . Such research would clarify whether HFCS consumption and its associated health burdens are stabilizing, rising, or declining, and how shifts in agricultural policy, consumer awareness, or food technology shape these trajectories across generations (DeChristopher et al., 2020; Gün & Karaoğlu, 2024).

Taken together, these directions move the discussion beyond critique toward constructive reform. By exploring alternative subsidy structures, sustainable farming practices, technological foresight, cultural adaptation, and holistic cost accounting, future research can help reimagine food systems that are not only more sustainable and equitable but also better aligned with long-term public health (Fanzo et al., 2020; Finger et al., 2023; Fanzo et al., 2021; Çakmakçı et al., 2024).

REFERENCES

Abay, K. A., Bevis, L. E. M., & Barrett, C. B. (2021). Measurement Error Mechanisms Matter: Agricultural intensification with farmer misperceptions and misreporting. *American Journal of Agricultural Economics, 103*(2), 498–522. DOI: 10.1111/ajae.12173

Adams, J., Hofman, K., Moubarac, J. C., & Thow, A. M. (2020, June 26). Public health response to ultra-processed food and drinks. *BMJ (Clinical Research Ed.), 369*, m2391. DOI: 10.1136/bmj.m2391 PMID: 32591348

Adula, M., Birbirsa, Z. A., & Kant, S. (2023). The effect of interpersonal, problem solving and technical training skills on performance of Ethiopia textile industry: Continuance, normative and affective commitment as mediators. *Cogent Business & Management, 10*(3), 2286672. DOI: 10.1080/23311975.2023.2286672

Blakely, T., Cleghorn, C., Mizdrak, A., Waterlander, W., Nghiem, N., Swinburn, B., Wilson, N., & Ni Mhurchu, C. (2020). The effect of food taxes and subsidies on population health and health costs: A modelling study. *The Lancet. Public Health, 5*(7), e404–e413. DOI: 10.1016/S2468-2667(20)30116-X PMID: 32619542

Bowles, T. M., Mooshammer, M., Socolar, Y., Schmer, M. R., Strock, J. A., & Grandy, A. S.. (2020). Long-term evidence shows that crop-rotation diversification increases agricultural resilience to adverse growing conditions in North America. *One Earth, 3*(4), 464–474. DOI: 10.1016/j.oneear.2020.02.007

Çakmakçı, S.. (2024). Foods of the Future: Challenges, Opportunities, Trends in Sustainable Foods, Alternative Protein Sources and Novel Efficient Food Systems. *Foods, 13*(17), 2663. DOI: 10.3390/foods13172663 PMID: 39272427

Constantinides, S. V., Turner, C., Frongillo, E. A., Bhandari, S., Reyes, L. I., & Blake, C. E. (2021). Using a global food environment framework to understand relationships with food choice in diverse low-and middle-income countries. Global Food Security, 29, 100511.

Contreras-Manzano, A., Jáuregui, A., Nieto, C., Pineda-López, L. A., Romero-Martínez, M., Medina, C., & Barquera, S. (2024). Self-reported decreases in the purchases of selected unhealthy food categories due to front-of-package warning labels in Mexico. *The International Journal of Behavioral Nutrition and Physical Activity, 21*(1), 74. DOI: 10.1186/s12966-024-01609-3 PMID: 38987796

Darmon, N., & Drewnowski, A. (2015). Contribution of food prices and diet cost to socioeconomic inequalities in diet quality. *Nutrition Reviews, 73*(10), 643–660. DOI: 10.1093/nutrit/nuv027 PMID: 26307238

Davis, Karen,R.D.H., B.S.D.H. (2011). Sugar obsession in America! Rdh,, 46-49. journals/sugar-obsession-america/docview/927664206/se-2

de Oliveira Comini, L., Lopes, S. O., Rocha, D. M. U. P., & Silva, M. M. (2025). The effects of subsidies for healthy foods on food purchasing behaviors, consumption patterns, and obesity/overweight: A systematic review. *Nutrition Reviews*, *83*(7), e1722–e1739. DOI: 10.1093/nutrit/nuae153 PMID: 39468774

DeChristopher, L. R., Auerbach, B. J., & Tucker, K. L. (2020). High fructose corn syrup, excess-free-fructose, and risk of coronary heart disease among African Americans – the Jackson Heart Study. *BMC Nutrition*, *6*(1), 70. DOI: 10.1186/s40795-020-00396-x PMID: 33292663

Dereso, C. W., Kant, S., Muthuraman, M., & Tufa, G. (2023). Effect of Point of Service on Health Department Student's Creativity in Comprehensive Universities of Ethiopia: Moderating Role of Public-Private Partnership and Mediating Role of Work Place Learning. In Jain, S., Groppe, S., & Mihindukulasooriya, N. (Eds.), *Proceedings of the International Health Informatics Conference* (Vol. 990, pp. 145–157). Springer. DOI: 10.1007/978-981-19-9090-8_13

Do, W. L., Coughlin, J. W., & Grant, F. D.. (2020). Consumption of foods derived from subsidized crops and cardiometabolic risk among US adults. *JAMA Internal Medicine*, *180*(6), 787–795.

Durga, P., Godavarthi, D., Kant, S., et al. (2024). Aspect-based drug review classification through a hybrid model with ant colony optimization using deep learning. Discover Computing, 27(19).

Fanzo, J., Davis, C., McLaren, R., & Choufani, J. (2021). Sustainable food systems and nutrition in the 21st century. *The Lancet. Planetary Health*, *5*(9), e501–e512. DOI: 10.1016/S2542-5196(21)00184-0

Fanzo, J., Davis, C., McLaren, R., & Choufani, J. (2023). Challenges and opportunities for increasing the use and impacts of food reformulation to improve diets and nutrition. *Critical Reviews in Food Science and Nutrition*, *63*(5), 1165–1177. DOI: 10.1016/j.foodpol.2023.102515

Fanzo, J., Haddad, L., Schneider, K. R., Béné, C., Covic, N. M., Guarin, A., Herforth, A., Herrero, M., Sumaila, U. R., & Wilkinson, J. (2020). Viewpoint: Rigorous monitoring is necessary to guide food system transformation in the countdown to the 2030 global goals. *Food Policy*, *91*, 101821.

Fields, S. (2004, October). The fat of the land: Do agricultural subsidies foster poor health? *Environmental Health Perspectives, 112*(14), A820–A823. DOI: 10.1289/ehp.112-a820 PMID: 15471721

Finger, R.. (2023). Digital innovations for sustainable and resilient agricultural systems. *European Review of Agriculture Economics, 50*(4), 1277–1304. DOI: 10.1093/erae/jbad021

Food and Agriculture Organization (FAO). (2024). The state of agricultural commodity markets 2024. Rome: FAO. https://www.fao.org/publications

Franck, C., Grandi, S. M., & Eisenberg, M. J. (2013). Agricultural subsidies and the American obesity epidemic. *American Journal of Preventive Medicine, 45*(3), 327–333. DOI: 10.1016/j.amepre.2013.04.010 PMID: 23953360

Gün, R., & Karaoğlu, M. M. (2024). Detection of honey adulteration by characterization of the physico-chemical properties of honey adulterated with the addition of glucose–fructose and maltose corn syrups. European Food Research and Technology = Zeitschrift Für Lebensmittel-Untersuchung Und-Forschung.A, 250(8), 2255-2272.

Guyo, K., & Kero, C. A. (2023). Mediation of Marketing Intermediaries and AI Adoption Between Livestock Products Marketing and Economic Status of Pastoralist in Ethiopia. In 2023 International Conference on Communication, Security and Artificial Intelligence (ICCSAI) (pp. 311–316). IEEE. DOI: 10.1109/ICCSAI59793.2023.10421261

Harvie, A., & Wise, T. A. (2009). Sweetening the pot: Implicit subsidies to corn sweeteners and the U.S. obesity epidemic. Global Development and Environment Institute Working Paper.

Hawkes, C. (2012, May 15). Food policies for healthy populations and healthy economies. *BMJ (Clinical Research Ed.), 344*(may15 2), e2801. DOI: 10.1136/bmj.e2801 PMID: 22589521

Hayes, T. O'Neill, & Kerska, K. (2021, November 3). Primer: Agriculture Subsidies and Their Influence on the Composition of U.S. Food Supply and Consumption. American Action Forum

Headey, D. D., & Alderman, H. (2019). The relative caloric prices of healthy and unhealthy foods differ systematically across income levels and continents. *The Journal of Nutrition, 149*(11), 2020–2033. DOI: 10.1093/jn/nxz158 PMID: 31332436

Huangfu, P., Adams, J., Monsivais, P., & Black, C. (2024). Impact of price reductions, subsidies, or financial incentives on fruit and vegetable intake: A systematic review and meta-analysis. The Lancet Planetary Health, 8(4), e156–e168. 5196(24)00004-4

Intergovernmental Panel on Climate Change. (2022). *Climate Change 2022: Mitigation of Climate Change. Contribution of Working Group III to the Sixth Assessment Report of the IPCC*. Cambridge University Press., DOI: 10.1017/9781009157926

International Institute for Sustainable Development (IISD). (2009). Biofuels—at what cost? Government support for ethanol and biodiesel in the United States: 2007 update.

Jackson, R. J., Minjares, R., Naumoff, K. S., Shrimali, B. P., & Martin, L. K. (2009). Agriculture policy is health policy. *Journal of Hunger & Environmental Nutrition*, *4*(3-4), 393–408. DOI: 10.1080/19320240903321367 PMID: 23144677

Johansson, E., & Kuktaite, R.. (2025). Plant proteins for human consumption – from local to global supply chains. *One Health*, •••, 100836.

Knai, C., Petticrew, M., Mays, N., Capewell, S., Cassidy, R., Cummins, S., Eastmure, E., Fafard, P., Hawkins, B., Jensen, J. D., Katikireddi, S. V., Mwatsama, M., Orford, J., & Weishaar, H. (2018, September). Systems Thinking as a Framework for Analyzing Commercial Determinants of Health. *The Milbank Quarterly*, *96*(3), 472–498. DOI: 10.1111/1468-0009.12339 PMID: 30277610

Kognou, A. L. M., Shrestha, S., Jiang, Z., Xu, C., Sun, F., & Qin, W. (2022). High-fructose corn syrup production and its new applications for 5-hydroxymethylfurfural and value-added furan derivatives: Promises and challenges. Journal of Bioresources and Bioproducts. DOI: 10.1016/j.jobab.2022.03.004

Lakkakula, P., Schmitz, A., & Ripplinger, D. (2016). U.S. Sweetener Demand Analysis: A QUAIDS Model Application. *Journal of Agricultural and Resource Economics*, *41*(3), 533–548. https://login.captechu.idm.oclc.org/login?url=https://www.proquest.com/scholarly-journals/u-s-sweetener-demand-analysis-quaids-model/docview/1826924823/se-2

Lara-Castor, L., Micha, R., Cudhea, F., Miller, V., & Ali, J. (2023). Sugar-sweetened beverage intakes among adults between 1990 and 2018 in 185 countries. *Nature Communications*, *14*(1), 5957. Advance online publication. DOI: 10.1038/s41467-023-41269-8 PMID: 37788998

Liu, F., Shahzad, M. A., Feng, Z., Wang, L., & He, J. (2024). An analysis of the effect of agriculture subsidies on technical efficiency: Evidence from rapeseed production in China. Heliyon, 10(13).

Lowell, J. (2004). The food industry and its impact upon increasing global obesity: A case study. *British Food Journal*, *106*(2), 238–248. DOI: 10.1108/00070700410528817

Maganinho, M., Almeida, C., & Padrão, P. (2024). Industrially Produced Plant-Based Food Products: Nutritional Value and Degree of Processing. *Foods*, *13*(11), 1752. DOI: 10.3390/foods13111752 PMID: 38890980

Malila, Y.. (2024). Current challenges of alternative proteins as future foods. *Nature Food*, •••, 00291. PMID: 39147771

Marklund, M., Martinez-Steele, E., Barquera, S., & L'Abbé, M. (2020). Health impact and cost-effectiveness of financing fruit and vegetable subsidies for a vulnerable population in the United States. *PLoS One*, *15*(2), e0229269. PMID: 32084196

Pacifico Silva, H., Lehoux, P., Miller, F. A., Wong, G., & Denis, J.-L. (2018). Introducing responsible innovation in health: A policy-oriented framework. *Health Research Policy and Systems*, *16*(1), 90. Advance online publication. DOI: 10.1186/s12961-018-0362-5 PMID: 30200985

Pakravan-Charvadeh, M. R.. (2024). Understanding the interplay between subsidy reforms, food prices, and nutrition outcomes: Empirical evidence from Iran. *Frontiers in Public Health*, *12*, 2024.

Popkin, B. M., & Ng, S. W. (2022). The nutrition transition to a stage of high obesity and noncommunicable disease prevalence dominated by ultra-processed foods. *Obesity Reviews*, *23*(1), e13366. DOI: 10.1111/obr.13366 PMID: 34632692

Sánchez Hernández, M. I.. (2024). Exploring Responsible Research and Innovation in agri-food cooperatives. *Journal of Responsible Innovation*. Advance online publication. DOI: 10.1080/23299460.2024.2322756

Siegel, K. R., Bullard, K. M., Imperatore, G., Kahn, H. S., Stein, A. D., Ali, M. K., & Narayan, K. M. V. (2016). Association of higher consumption of foods derived from subsidized commodities with adverse cardiometabolic risk among US adults. *JAMA Internal Medicine*, *176*(8), 1124–1132. DOI: 10.1001/jamainternmed.2016.2410 PMID: 27379488

Siegel, K. R., McKeever Bullard, K., Ali, M. K., Stein, A. D., Kahn, H. S., Mehta, N. K., Webb Girard, A., Narayan, K. M., & Imperatore, G. (2016, June). The contribution of subsidized food commodities to total energy intake among US adults. *Public Health Nutrition*, *19*(8), 1348–1357. DOI: 10.1017/S1368980015002414 PMID: 26322920

Siegel, K. R., McKeever Bullard, K., Imperatore, G., Kahn, H. S., Stein, A. D., Ali, M. K., & Narayan, K. M. (2016). Association of Higher Consumption of Foods Derived From Subsidized Commodities With Adverse Cardiometabolic Risk Among US Adults. *JAMA Internal Medicine, 176*(8), 1124–1132. DOI: 10.1001/jamainternmed.2016.2410 PMID: 27379488

Solow, R. M. (1956). A contribution to the theory of economic growth. *The Quarterly Journal of Economics, 70*(1), 65–94. DOI: 10.2307/1884513

Springmann, M., & Freund, F. (2022). Options for reforming agricultural subsidies from health, climate, and economic perspectives: Integrated modelling shows often-conflicting objectives can be aligned. *Nature Communications, 13*(1), 82. DOI: 10.1038/s41467-021-27645-2 PMID: 35013212

Swinburn, B., Kraak, V. I., Allender, S., Atkins, V. J., Baker, P. I., Bogard, J. R., & Dietz, W. H. (2019). The Global Syndemic of obesity, undernutrition, and climate change: The Lancet Commission report. The Lancet, 393(10173), 791–846. 6736(18)32822-8

Tan, S. S. X., & Kuebbing, S. E. (2023). A synthesis of the effect of regenerative agriculture on soil carbon sequestration in Southeast Asian croplands. *Agriculture, Ecosystems & Environment, 349*, 108450. DOI: 10.1016/j.agee.2023.108450

Turner, C., Aggarwal, A., Walls, H., Herforth, A., Drewnowski, A., Coates, J., Kalamatianou, S., & Kadiyala, S. (2018). Concepts and critical perspectives for food environment research: A global framework with implications for action in low- and middle-income countries. *Global Food Security, 18*, 93–101. DOI: 10.1016/j.gfs.2018.08.003

Usigbe, M. J., Asem-Hiablie, S., Uyeh, D. D., Iyiola, O., Park, T., & Mallipeddi, R. (2024). Enhancing resilience in agricultural production systems with AI-based technologies. *Environment, Development and Sustainability, 26*(9), 21955–21983. DOI: 10.1007/s10668-023-03588-0

van Buul, V. J., Tappy, L., & Brouns, F. (2014). Misconceptions about fructose-containing sugars and their role in the obesity epidemic. *Nutrition Research Reviews, 27*(2), 119–130. DOI: 10.1017/S0954422414000067 PMID: 24666553

White, J. S. (2008). Straight talk about high-fructose corn syrup: What it is and what it ain't. *The American Journal of Clinical Nutrition, 88*(6), 1716S–1721S. DOI: 10.3945/ajcn.2008.25825B PMID: 19064536

World Health Organization. (2022). *WHO manual on sugar-sweetened beverage taxation policies to promote healthy diets*. WHO.

World Health Organization. (2025). *Obesity and overweight: Key facts*. WHO.

World Health Organization (WHO). (2024). *Obesity and overweight: Key facts*. WHO.

Wunderlich, A. C., & Kohler, A. (2022). The impact of agricultural subsidies on obesity. *Journal of Agriculture and Food Research, 10*, 100439. DOI: 10.1016/j.jafr.2022.100439

Chapter 13

Integrating AI Models With Synthetic Biology for Plant Epigenetic Regulation

R. V. Suganya
VISTAS, India

A. Punitha
http://orcid.org/0009-0006-1530-6768
VISTAS, India

A. V. Senthil Kumar
http://orcid.org/0000-0002-8587-7017
Nehru Institute of Information Technology and Management, India

Revati Ramrao Ramrao Rautrao
http://orcid.org/0000-0001-7803-8699
Dr. D.Y. Patil B-School, India

K. R. S. Sambasiva Rao
http://orcid.org/0000-0002-0061-8754
Mangalayatan University, Jabalpur, India

S. Nagakishore Bhavanam
http://orcid.org/0000-0003-3798-6945
Mangalayatan University, Jabalpur, India

F. Mary Magdalene Jane
http://orcid.org/0000-0002-5101-8740
Dr. N.G.P. Arts and Science College, India

Shallu Sehgal
http://orcid.org/0009-0007-1260-5399
Shoolini Institute of Life Sciences and Business Management, India

G. Vanishree
http://orcid.org/0009-0000-1335-2919
ICFAI Business School, India

ABSTRACT

The integration of artificial intelligence (AI) models with synthetic biology offers transformative potential for enhancing plant epigenetic regulation. Plant epigenetics,

DOI: 10.4018/979-8-2600-1026-6.ch013

involving modifications such as DNA methylation, histone modifications, and non-coding RNAs, plays a crucial role in regulating gene expression, development, and stress responses.. This chapter explores how AI can optimize the design, prediction, and real-time manipulation of plant epigenetic mechanisms, accelerating advancements in crop improvement. Furthermore, synthetic biology techniques, such as gene editing and the creation of synthetic gene circuits, can be precisely targeted using AI models to improve plant resilience and productivity. The integration of AI with synthetic biology in plant epigenetic regulation holds significant promise for addressing global agricultural challenges, enhancing food security, and fostering sustainable farming practices.

INTRODUCTION

The integration of Artificial Intelligence (AI) models with synthetic biology offers groundbreaking opportunities to enhance plant epigenetic regulation. Epigenetics—comprising DNA methylation, histone modifications, and non-coding RNAs—plays a vital role in regulating gene expression, development, and environmental stress responses in plants (Abdulraheem et al., 2024). The challenge of ensuring food security, especially under the pressures of climate change, has amplified the need to optimize crop resilience and productivity without altering the DNA sequence. AI models can assist in analyzing large, complex epigenetic datasets, uncovering regulatory patterns, and facilitating the precise control of plant traits like drought tolerance and stress resistance. As synthetic biology advances, AI can guide the design and manipulation of synthetic gene circuits, enabling tailored improvements to plant resilience. However, barriers like data quality, model precision, off-target effects, and ethical concerns related to genetically modified organisms (GMOs) persist. Despite these challenges, the integration of AI with synthetic biology in plant epigenetic regulation holds promise for sustainable agricultural practices (Sun, 2025).

The integration of Artificial Intelligence (AI) with synthetic biology offers transformative potential for improving plant epigenetic regulation. Epigenetic mechanisms, including DNA methylation, histone modifications, and non-coding RNAs, regulate gene expression and plant adaptation to environmental stress. However, challenges remain in precisely controlling these mechanisms. Recent advancements in AI—particularly machine learning and deep learning—provide solutions to these challenges by analyzing large, complex datasets, predicting epigenetic behavior, and optimizing plant traits like drought tolerance and stress resistance. This paper examines how AI and synthetic biology can work together to address global agricultural challenges, such as food security and sustainable farming.

OVERVIEW

Abdulraheem et al. (2024) provide an in-depth overview of plant epigenetic regulation, emphasizing its critical role in plant development, stress response, and adaptation. Epigenetic regulation involves heritable changes in gene expression without altering the DNA sequence, encompassing mechanisms like DNA methylation, histone modifications, and non-coding RNAs. These processes are essential for regulating gene expression patterns that drive developmental processes such as seed germination, flowering, and organ formation. Epigenetic regulation also enables plants to respond dynamically to environmental stressors like drought, salinity, and temperature extremes, enhancing their survival and adaptability. Moreover, the review highlights the importance of epigenetic mechanisms in agriculture, suggesting that insights into these processes could lead to the development of stress-tolerant crops, crucial for food security in the face of climate change. The paper concludes by calling for further research, particularly with emerging technologies like next-generation sequencing, to deepen our understanding of how epigenetic changes affect plant biology.

The essential function of epigenetic regulation in plant development, stress response, and adaptation is highlighted in Sun's (2025) thorough review of the topic. Changes to DNA that do not affect the DNA sequence are known as epigenetic alterations. These include histone modifications, DNA methylation, and the effects of non-coding RNAs like siRNAs and microRNAs. The regulation of gene expression in reaction to stimuli and threats in the environment relies on these systems. Epigenetic modifications allow plants to adjust their growth patterns to their surroundings by influencing processes like root formation and flowering timing. By changing gene expression, epigenetic regulation helps plants strengthen resilience to abiotic stressors like salinity and drought. Enhancing crop resilience and guaranteeing agricultural sustainability in the face of global environmental issues requires this adaptation, which is driven by epigenetic mechanisms.

With an emphasis on its function in development, environmental response, and heredity, the article Plant Epigenomics (2023) offers a comprehensive review of epigenetic control in plants. Methods including RNA interference, DNA methylation, and histone modifications are part of plants' epigenetic regulatory machinery that controls gene expression without changing the DNA sequence. The proper expression of genes in the right tissues at the right times is critical for the development of important plant structures, and this control is very necessary for that. Plants are able to adapt to harsh environments by responding to biotic and abiotic stimuli through epigenetic mechanisms. The idea of “epigenetic memory” also helps plants pass their adaptations to their environments on to future generations, which increases their chances of survival and increases phenotypic variation. We now know more about the effects of these changes on plant growth and stress responses thanks to developments in chromatin analysis; this knowledge has consequences for both farming and environmental protection. To summarize, the ability of plants to adapt, grow, and withstand changing conditions depends on their epigenetic regulation.

The function of epigenetic regulation in plant development, adaptability, and stress responses is highlighted in Xiong's (2023) thorough review of the topic. Plants can adapt to their environments and pass these adaptations on to subsequent generations through epigenetics, which entails changes in gene expression without changing the DNA sequence. Histone modifications impact gene activation or repression; DNA methylation can mute genes in reaction to environmental stress; and small RNAs affect gene expression and stress responses; these are key processes of epigenetic regulation. In order for plants to grow and adapt to new environments, such as fluctuating temperatures and water availability, these systems are essential. Certain epigenetic alterations improve tolerance to drought and salinity, demonstrating the importance of epigenetic control in stress response. A plant's capacity to respond to future stressors can be enhanced through the development of a "memory" of similar experiences. The development of stress-resistant crops, including salt-tolerant cultivars, is critical for food security in salty locations, therefore understanding these pathways is key for agricultural developments.

LITERATURE REVIEW

The chapter focuses on the role of synthetic biology in reprogramming gene regulation and discusses important mechanisms of plant epigenetics, including DNA methylation, histone modification, and non-coding RNAs. Research shows that AI can handle omics data, forecast gene behavior, and direct focused treatments. Issues with data availability, ethics, and regulations are also discussed in the review.

Recent advances in synthetic biology and AI have demonstrated their transformative potential in agriculture. AI models have been applied to predict gene behavior and optimize plant traits, enabling precision in plant breeding (Fal et al., 2021; Ahtisham & Obaid, 2025). The regulation of plant epigenetics, particularly through mechanisms like DNA methylation and histone modifications, has been explored for enhancing traits such as stress resistance (Singh et al., 2023). Despite these advancements, significant challenges remain, particularly in achieving precision in epigenetic modifications. As Ahtisham & Obaid (2025) highlight, heritable and stable epigenetic changes are difficult to ensure due to environmental interactions that may alter or reverse modifications, complicating efforts to design resilient crops.

Moreover, AI models have facilitated breakthroughs in the development of synthetic gene circuits, allowing for real-time manipulation of plant traits (Zhang et al.,

2024). These advances are essential for addressing global agricultural challenges like drought and salinity, and ensuring that crop plants are not only genetically robust but also environmentally adaptable. However, as Xiong (2023) notes, precise control over plant epigenetics requires overcoming technical hurdles, such as off-target effects and limited targeting efficiency in current technologies.

The literature on plant epigenetics and AI in agriculture provides a foundation for understanding how these technologies can work together. Studies have demonstrated the critical role of epigenetic regulation in plant development and stress responses (Xiong, 2023). DNA methylation and histone modifications play key roles in gene expression, especially in response to environmental stressors such as drought and salinity. AI-driven models have emerged as powerful tools to analyze these complex datasets, enabling researchers to predict gene behavior and optimize plant traits. However, challenges remain in the precision of AI models and the integration of multi-omics data. The review identifies a gap in current research, particularly in improving epigenetic editing tools and ensuring data quality for AI models, highlighting the need for further advancements in these areas.

The role of AI models in plant epigenetic regulation is becoming increasingly important as we move toward precision agriculture. AI-driven tools have been developed to analyze epigenetic data, particularly DNA methylation and histone modification patterns, which are crucial for plant development and stress responses. However, these tools often struggle with accuracy and precision due to the complexity and variability of plant genomes.

Recent studies (such as Ahtisham & Obaid, 2025) have shown that while AI models can predict epigenetic outcomes, their ability to account for environmental variability and multi-layered interactions remains limited. This limitation underscores the need for more robust data and improved AI algorithms. Additionally, the challenge of integrating multi-omics data (such as genomics, transcriptomics, and epigenomics) into a cohesive AI model remains a significant hurdle. According to Xiong (2023), the current datasets available for training AI models are sparse, particularly for non-model plants, which are essential for addressing food security and climate resilience.

CURRENT CHALLENGES IN UNDERSTANDING AND MANIPULATING PLANT EPIGENETIC MECHANISMS

Recent trends in AI and synthetic biology suggest significant strides toward more precise control of plant epigenetics, but these developments also present new opportunities and challenges.

Advancements in AI Models: AI models, particularly those leveraging deep learning and reinforcement learning, are becoming increasingly adept at analyzing plant epigenetic data. These models can predict the outcomes of specific epigenetic modifications with a level of detail that traditional methods cannot achieve. For instance, Zhang et al. (2024) demonstrated how AI models can predict the effects of histone modifications on gene expression, helping researchers design targeted epigenetic modifications for traits such as drought resistance. However, the application of these technologies is still limited by data accuracy and complexity in biological systems. Future advancements will need to address these gaps by improving model interpretability and integrating data from various sources to ensure that predictions are more reliable.

Synthetic Gene Circuits for Real-Time Plant Regulation: Synthetic biology has opened the door to real-time regulation of plant traits through synthetic gene circuits. AI's role in optimizing these circuits is becoming more pronounced, allowing for the creation of plants that can dynamically adjust their gene expression in response to environmental changes. A study by Attia et al. (2024) explored how synthetic gene circuits could be used to control gene expression in response to drought stress, improving plant resilience. These AI-optimized circuits hold great promise for precision agriculture, where plants are not only genetically optimized but can also adapt to environmental changes autonomously.

While the integration of AI models with synthetic biology in plant epigenetic regulation holds significant promise, several key challenges must be addressed to ensure successful implementation. One of the primary challenges is the precision of epigenome editing tools such as CRISPR/dCas9. These tools, while groundbreaking, often face issues like off-target effects, where unintended genetic modifications can occur, leading to unpredictable outcomes in plant traits. This issue is supported by studies like those of Ahtisham & Obaid (2025), who found that even advanced editing tools struggle with targeting efficiency in certain plant species. These technical limitations mean that while AI-driven prediction models have the potential to guide precise edits, implementation accuracy remains a significant hurdle in plant breeding.

Furthermore, the complexity of plant epigenetic networks presents another challenge. Unlike traditional genetic changes, epigenetic modifications do not alter the DNA sequence but instead modify the expression of genes through mechanisms like DNA methylation and histone modification. Understanding how these modifications interact in response to environmental stimuli (such as drought or salinity) is crucial for achieving consistent results in crop traits. According to Singh et al. (2023), much of the current research is still in the early stages, with limited understanding of how specific epigenetic markers contribute to stress tolerance or yield improvements. This lack of understanding hampers efforts to develop reliable and scalable solutions using AI-driven synthetic biology.

In the context of chromatin manipulation and editing technologies, Fal et al. (2021) highlight numerous major obstacles to comprehending and influencing plant epigenetic processes. Achieving precision in manipulation is a big issue because current approaches only show correlations without showing cause and effect, which limits our knowledge of how epigenetic regulation works. Another obstacle is the lack of specificity in broad-spectrum treatments (such as pharmacological inhibitors) and the potential impracticality of more accurate, locus-specific procedures. Another obstacle to their adoption in plant research is the technological complexity of systems like dCas9. These systems require a strong understanding of RNA-DNA interactions and good guide RNA creation. Additionally, because existing methods require additional improvement to enhance their efficacy in altering epigenetic markers, targeting efficiency continues to be a challenge. Finally, the existing systems do not permit inducible or switchable manipulation of chromatin marks, and the absence of temporal analysis tools further complicates the study of dynamic epigenetic processes. These obstacles need fixing if we want to learn more about and control plant epigenetics, which affects plant development and has agricultural uses.

To improve crop resilience, it is essential to study and manipulate plant epigenetic systems; nevertheless, Ahtisham & Obaid (2025) highlight various obstacles in this area. It is difficult to predict the results of changes in one region due to the intricacy of epigenetic regulation, which makes it difficult to manipulate systems including DNA methylation, histone modifications, and RNA-directed pathways. A major obstacle is the inconsistency in crop development tactics caused by the heritability and stability of epigenetic changes across generations, which means that not all modifications are reliably passed on. It is already challenging to develop crops that can withstand a number of pressures without adding the additional layer of uncertainty caused by environmental interactions, such as reactions to biotic and abiotic stresses. Technical limitations in existing technologies, including genome-wide methylation analysis, make it even more difficult to investigate and control epigenetic processes in depth. There is still a deficit of thorough comprehension of the ways in which particular epigenetic alterations impact plant responses to stress, notwithstanding progress. Finally, there are extra hurdles caused by ethical and regulatory considerations related to genetically modified organisms (GMOs) and the alteration of epigenetic pathways. While epigenetic mechanisms hold great promise for making crops more resilient, these obstacles must be overcome before their full potential can be realized.

Improving crops relies on comprehending and controlling plant epigenetic systems, however Singh et al. (2023) point out a number of obstacles to this goal. Because the complexity of epigenetic modifications varies substantially among plant species and even people, it is challenging to anticipate how these changes will impact gene expression and plant characteristics. The manipulation of these pathways is further complicated by environmental factors, which can cause changes that are both unanticipated and sometimes irreversible. These influences include biotic and abiotic stressors. Progress in stable epigenetic alteration is hindered by technical restrictions, such as off-target effects and inefficiency in delivering precision tools like CRISPR/Cas. We still don't fully understand how non-coding RNAs like siRNAs and miRNAs regulate epigenetic modifications, so we need to do more research to find out how they can improve crops. Additionally, a more comprehensive approach to crop development requires a better integration of knowledge on the ways in which epigenetic systems interact with genetic elements. The alteration of epigenetics raises concerns regarding long-term ecological repercussions and food safety due to regulatory and ethical considerations. Finally, if we want to make the most of plant epigenetic research and the innovations it could bring to agriculture, we must overcome these obstacles.

A number of significant obstacles to comprehending and controlling plant epigenetic systems have been brought to light by McKeown & Spillane (2014). The intricate relationship between DNA methylation, histone modification, and non-coding RNAs—all of which impact gene expression in diverse ways—and other components of epigenetic control presents a significant obstacle. Moreover, because these patterns differ from classical inheritance theories, revealing a knowledge gap, non-Mendelian inheritance further confuses our comprehension of the transmission of features. Another obstacle is the lack of advanced bioinformatics tools and the high resource consumption of genome-wide investigations, which are necessary for a complete understanding of epigenetics. Another issue with studies relying on model plants is that results from these organisms might not always apply to real plants with their own distinct epigenetic characteristics. To enhance crop breeding and resilience, it is necessary to combine epigenetics with other biological processes like development and stress responses. The manipulation of epigenetic systems in crops raises ethical and practical considerations, and one must carefully assess the potential unexpected consequences. To summarize, these issues in plant epigenetics require more developments, despite the fact that progress has been made.

There are a number of obstacles to comprehending and influencing plant epigenetic processes, according to Mendieta et al. (2023), especially with regard to cis-regulatory elements (CREs). Since it is difficult to maintain distinct cell types in culture due to the totipotent nature of plant cells, the intricacy of plant cell types is a key barrier in studying how CREs regulate gene expression and identity. Furthermore, studying the epigenetic landscape of plant cells is made more difficult by technical limitations, such as the inflexible cell wall in plants, which makes epigenomic data extraction and processing more complicated. While there are new avenues for investigation opened up by single-cell epigenomics, there are also obstacles to overcome, such as noise in the data and the difficulty in interpreting findings from different cell types. The complexity is further increased by the presence of plant-specific biological challenges, such as growth, evolution, and stress reactions impacted by CREs. Misinterpretation and problems integrating data from various sources are examples of data analysis pitfalls that can also impede progress. Finally, investigating plant epigenetics presents unique hurdles, particularly at the single-cell level. To solve these challenges, improved tools are needed. To advance plant epigenetics research and comprehend the function of CREs in plant biology, it is crucial to overcome these complex obstacles.

ENHANCING PLANT EPIGENETIC REGULATION THROUGH AI AND SYNTHETIC BIOLOGY

The revolutionary power of merging synthetic biology with artificial intelligence (AI) to improve plant epigenetic control is emphasized by Amaan et al. (2024). To find the best ways to manipulate epigenetic variables in plants, artificial intelligence can automate and analyze massive datasets to optimize experimental designs. Furthermore, researchers can use AI to predict biological behaviors and so design targeted interventions in plant breeding by anticipating the impacts of particular genetic alterations. A further benefit of AI is the acceleration of high-throughput screening, which helps find important epigenetic regulators more quickly, for example, those associated with drought resistance or yield improvement. Optimized responses to changes are guaranteed by real-time monitoring and control systems that are powered by AI. These systems enable fast adjustments based on environmental feedback. Nevertheless, there are ethical considerations concerning biosafety and biosecurity that arise from combining AI with synthetic biology, and they must be thoroughly handled. As a whole, the integration of AI and synthetic biology holds great potential for revolutionizing agriculture. It will pave the way for improved crop features, which in turn will help address pressing issues like food security and climate change.

In their discussion of how to improve agricultural attributes through AI and SynBio integration, Zhang et al. (2024) highlight the importance of plant epigenetic control and the possible benefits to be gained from this approach. Artificial intelligence

(AI) can sift through mountains of data on plant genes and epigenetics, allowing scientists to fine-tune agricultural characteristics like productivity, hardiness, and nutritional value. Enhancing the "design, build, test, and learn" cycles of synthetic biology with AI's predictive skills allows for efficient development of crops with desired epigenetic features. To fully comprehend the impact of epigenetic variables on plant growth and stress responses, AI in conjunction with multi-omics analysis is essential. This integration also makes it easier to create SMART crops, which can optimize their growth and productivity in real-time by self-monitoring and adapting their epigenetic regulation to environmental changes. Data privacy, ethical considerations, and the intricacy of biological systems are some of the obstacles that must be overcome before artificial intelligence and synthetic biology can realize their full potential. Still, synergy between AI and SynBio has the potential to transform farming methods, making food more secure despite worldwide threats.

A groundbreaking strategy for agricultural innovation, Attia et al. (2024) investigate how synthetic biology and artificial intelligence (AI) might work together to improve plant epigenetic regulation. New biological systems and solutions can be created using synthetic biology, which mixes engineering and biology. This field has several applications, one of which being agriculture. Understanding intricate biological systems like plant epigenetics requires high-level data analysis, pattern identification, and predictive modeling; these capabilities are made possible by the integration of artificial intelligence with synthetic biology. Through the processing of genomic and epigenomic data, the prediction of the impact of certain modifications on plant traits, and the automation of experimental procedures, AI enables the exact manipulation of epigenetic changes. Improved crop types with higher nutritional profiles and more resistance to environmental challenges may result from this combination. Sustainable farming techniques can be advanced with the use of AI-driven insights that optimize plant growth while decreasing the need for chemical inputs. Artificial intelligence's growing influence on synthetic biology will hasten developments in the field of plant epigenetic regulation, which in turn will help build a bio-based society that can withstand the test of time.

In their 2020 study, Dixon et al. investigate how synthetic biology and artificial intelligence (AI) can work together to improve plant epigenetic regulation. This convergence paves the way for more targeted and efficient genome editing, which boosts agricultural output and plant resilience to climate change. Biofoundries are increasingly turning to AI for sophisticated data analysis, which allows them to process large amounts of complicated biological data much faster than before. Artificial intelligence (AI) can help with targeted genetic therapies by detecting trends in epigenetic regulatory systems such as DNA methylation and histone modification using machine learning algorithms. Nevertheless, paying close attention to public concerns and ethical implications is crucial in order to promote acceptance and trust when AI and synthetic biology are integrated. This is because these technologies raise ethical and societal considerations. There are a plethora of real-world uses for this integration, including improving drought resistance and nutrient efficiency, creating synthetic gene circuits to control epigenetic states, and bolstering food security. Future work in this area could lead to a dramatic shift in farming methods and a greater emphasis on long-term sustainability as artificial intelligence and synthetic biology work together to solve pressing environmental problems. Nevertheless, in order to guarantee that these innovations are beneficial to society, ongoing ethical assessment is required. In conclusion, there is enormous promise in enhancing plant epigenetic regulation through the integration of synthetic biology and artificial intelligence, but this endeavor must be done with care in order to resolve societal and ethical concerns.

In their discussion of the possibility of combining synthetic biology with artificial intelligence (AI) to improve plant epigenetic regulation, Bhuvaneswari et al. (2020) provide new approaches to enhancing crop resilience and characteristics. An important part of plant breeding is epigenetics, which is the study of how genes are expressed and passed down across generations without changing the DNA sequence. Plants rely on processes such as DNA methylation, histone modification, and small RNA interference to control crucial reproductive characteristics. With the use of AI, researchers can sift through mountains of data in search of patterns that govern plant features; this allows for more accurate forecasts of how epigenetic alterations impact plant performance in different environments. To enhance characteristics like stress tolerance or yield, synthetic biologists create novel biological systems that enable targeted epigenetic alterations. More robust and high-quality crops are the result of optimizing the design of synthetic structures through the combination of artificial intelligence and synthetic biology. The intricacy of epigenetic inheritance and control, in particular, is a significant obstacle. Sustainable agriculture cannot progress without ongoing study and cooperation among these domains. To sum up, there is a lot of hope for improving plant epigenetic control and reshaping plant breeding through the merging of synthetic biology and AI.

PROBLEM STATEMENT

Environmental stresses including drought, salinity, and climate fluctuation are becoming more and more of a problem for plant adaptability and productivity. These intricate difficulties are beyond the scope of conventional breeding approaches. Although epigenetic control is a potential way to improve crop resilience and productivity without changing DNA sequences, its complexity makes it difficult to manipulate effectively. In order to make focused and scalable use of epigenetic information, sophisticated tools are required for interpretation.

SIGNIFICANCE OF THE RESEARCH

In order to develop crops in the future, it is crucial to understand and manipulate plant epigenetics. Modeling, predicting, and optimizing plant responses to external stimuli is made possible by merging artificial intelligence with synthetic biology. Food security, sustainable agriculture, and the creation of crops that can withstand the effects of climate change are all areas that stand to benefit from this combination.

Methodology

This chapter takes a literature-based analytical tack, compiling and analyzing results from current research on synthetic biology, artificial intelligence (AI) applications, and plant epigenetics. Examples of sources include reviews, case studies, and peer-reviewed journals. Synthetic gene circuit design, machine learning methods, and CRISPR/dCas9 epigenome editing are among the methodologies that have been assessed.

This study employs a mixed-methods approach, integrating both qualitative and quantitative analyses to explore the integration of AI models and synthetic biology in plant epigenetics. Primary data were collected through interviews and surveys with plant biologists and AI specialists. Secondary data were obtained from peer-reviewed studies, which were analyzed to identify the current state of AI applications in plant breeding and genetic modification. The study also incorporates case studies that demonstrate the practical applications of AI-driven epigenome editing and synthetic biology tools in improving plant resilience.

PROJECT THEORIES

Effective plant engineering, according to this study, results from a data-driven comprehension of regulatory networks; the study is based on systems biology and the UTAUT model of technology adoption. The integration process is based on theories from synthetic biology design cycles, machine learning, and epigenetic inheritance.

AIM OF THE RESEARCH

- To explore the potential of AI models in analyzing complex epigenetic data in plants.
- To examine how synthetic biology techniques can be combined with AI to precisely manipulate plant epigenetic mechanisms.
- To assess the benefits of integrating AI and synthetic biology for improving plant traits such as stress resistance, growth, and yield.
- To identify current challenges and limitations in applying AI-driven synthetic biology approaches to plant epigenetics.
- To propose future directions and applications for AI-assisted synthetic biology in advancing sustainable agriculture through epigenetic regulation.

OVERVIEW OF KEY MECHANISMS IN PLANT EPIGENETIC REGULATION

Plant epigenetics controls gene expression without changing the DNA sequence, using several crucial mechanisms. First, DNA methylation adds methyl groups to DNA bases, helping silence genes and stabilize the genome, especially in response to environmental stresses. Next, histone modifications—such as acetylation and methylation—alter the proteins around which DNA is wrapped, dynamically influencing gene activity during development and environmental responses. Additionally, non-coding RNAs, like siRNAs and miRNAs, regulate genes by degrading mRNA or blocking translation, which helps maintain genome integrity. Finally, plants possess epigenetic memory, allowing them to pass on adaptive epigenetic changes across cell divisions and sometimes to offspring, aiding survival in changing environments.

Plant epigenetic regulation encompasses various interrelated systems that govern gene expression without modifying the DNA sequence. DNA methylation introduces methyl groups predominantly to cytosine bases within diverse sequence contexts (CG, CHG, CHH), significantly influencing gene regulation, genomic stability, and protection against transposable elements. This methylation can be inherited, affecting traits across generations. Histone modifications, including methylation and acetylation of histone proteins associated with DNA, modulate gene expression by modifying chromatin architecture. Acetylation typically signifies active genes, whereas distinct methylation patterns frequently denote gene suppression, influencing DNA accessibility. RNA interference (RNAi) employs tiny RNA molecules to inhibit gene expression at either the transcriptional or post-transcriptional levels, which is essential for plant development, stress response, and the regulation of transgenes. Significantly, these mechanisms do not function alone; they engage in intricate networks where DNA methylation can affect histone changes and vice versa, establishing feedback loops that reinforce gene expression. Comprehending these epigenetic mechanisms is essential for plant breeding, as stable and inheritable epigenetic alterations present viable approaches for creating superior crop types with increased resilience and production.

Plant epigenetic regulation encompasses various coordinated systems that govern gene expression and maintain genome integrity without modifying the DNA sequence. DNA methylation introduces methyl groups predominantly to cytosine bases in three sequence contexts (CG, CHG, CHH), significantly contributing to the silencing of transposable elements and repetitive DNA, therefore safeguarding genomic integrity.

Histone modifications, including methylation and acetylation of histone proteins, modify chromatin architecture and affect gene expression. Particular marks such as H3K27me3 are associated with gene repression, whereas H3K4me3 is correlated with active gene expression, influencing DNA accessibility for transcription. Non-coding RNAs (ncRNAs), such as siRNAs and miRNAs, modulate gene expression by directing mRNA degradation or inhibiting translation. A specialized mechanism, RNA-directed DNA methylation (RdDM), employs siRNAs to orchestrate DNA methylation and silence certain genomic areas, essential for genomic integrity. Furthermore, chromatin remodeling actively alters the positioning of nucleosomes to either reveal or obscure DNA, hence modulating access to the transcription machinery. Collectively, these systems constitute an integrated network that allows plants to modulate gene expression, respond to environmental stimuli, and preserve genomic integrity.

SIGNIFICANCE

Epigenetic control is essential for plant growth, development, stress responses, and adaptation to environmental changes. Plants can dynamically regulate gene expression by mechanisms including DNA methylation, histone modifications, non-coding RNAs, and chromatin remodeling, without changing the underlying DNA sequence. This regulation is essential for vital developmental processes such as seed germination, flowering, and organ production, hence maintaining optimal plant growth and development. Furthermore, epigenetic systems allow plants to react to various abiotic and biotic stresses—such as drought, salt, temperature variations, and pathogen assaults—by regulating stress-responsive genes. These responses augment plant resilience and survival in variable settings, while the capacity to convey epigenetic memory across generations facilitates long-term adaptation and phenotypic variation. Comprehending and altering these epigenetic pathways is crucial for progressing sustainable agriculture. Understanding plant epigenetics can enhance the creation of stress-resistant and high-yield crops, tackling essential issues concerning food security and climate change. The incorporation of developing technologies, especially artificial intelligence and synthetic biology, offers potential for accurate and efficient management of epigenetic regulation, facilitating advanced crop enhancement methods.

CONTEMPORARY METHODS

Investigating and altering plant epigenetics entails a synthesis of conventional genetic methods, molecular biology, and innovative biotechnological approaches. Traditional methodologies encompass DNA methylation analysis techniques, including bisulfite sequencing and methylation-sensitive restriction enzyme assays, enabling researchers to delineate methylation patterns throughout the genome. Chromatin immunoprecipitation (ChIP) combined with sequencing (ChIP-seq) is extensively employed to discover and analyze histone changes, offering insights into chromatin states and gene regulation.

Moreover, RNA interference (RNAi) methodologies, including small interfering RNAs (siRNAs) and microRNAs (miRNAs), are utilized to investigate gene silencing processes and their epigenetic effects. These molecular instruments facilitate the analysis of non-coding RNAs' functions in gene expression regulation and genome stability preservation.

Recent advancements have introduced genome editing technologies, like as CRISPR/Cas systems, which facilitate more precise manipulation of epigenetic markers by targeting specific loci for change. Epigenome editing with dCas9 conjugated with epigenetic modifiers facilitates locus-specific alterations in DNA methylation or histone modifications without modifying the DNA sequence itself.

High-throughput multi-omics methodologies, encompassing genomes, transcriptomics, epigenomics, and proteomics, have augmented the comprehension of intricate epigenetic networks. Nonetheless, obstacles including low specificity, off-target effects, and technological complexity persist.

Additionally, instruments for chromatin remodeling and manipulation, including as pharmacological inhibitors and synthetic biology platforms, are being created to dynamically regulate epigenetic states. Notwithstanding these advancements, constraints in temporal regulation, efficiency, and scalability impede their extensive utilization.

Traditional methods have established the groundwork for plant epigenetic research; however, the incorporation of advanced technologies such as CRISPR-based epigenome editing and synthetic biology, frequently informed by artificial intelligence, is ushering in a new era of precise, efficient, and programmable epigenetic manipulation.

UNDERSTANDING SYNTHETIC BIOLOGY AND ITS ROLE IN PLANT BIOTECHNOLOGY

Synthetic biology is a multidisciplinary field that combines biology, engineering, and computer science to design and construct new biological parts, systems, or redesign existing ones to enhance or create novel functions. Driven by advances in omics technologies, it enables a data- and model-driven approach to biological innovation.

In plant biotechnology, synthetic biology plays a pivotal role in improving crop traits such as pest and disease resistance, stress tolerance, and nutrient content. Engineered plants can better withstand environmental challenges like drought and salinity while producing higher yields. Additionally, synthetic biology supports sustainable agriculture by developing plants that use nutrients more efficiently, thereby reducing dependency on chemical fertilizers and minimizing environmental impacts.

Beyond crop improvement, synthetic biology contributes to sustainable development through the production of bio-based materials and biofuels from plant biomass, promoting renewable resources and reducing fossil fuel reliance.

Synthetic biology is an interdisciplinary domain that integrates biology, engineering, and computer science to conceptualize and construct novel biological components, devices, and systems. This facilitates the development of innovative, synthetic biological components with targeted functionalities relevant to medicine, agriculture, and industry.

The field encompasses designing biological systems to address difficulties such as medication and vaccine manufacture, biofuel generation, and agricultural enhancement. Substantial advancements in biological system engineering have broadened the possible applications of synthetic biology.

Synthetic biology in plant biotechnology facilitates accurate alterations of plant genomes, frequently employing techniques such as CRISPR-Cas9, to improve characteristics including pest and disease resistance, stress tolerance, and production. This enhances agricultural advancement beyond conventional breeding methods. Synthetic biology facilitates the sustainable generation of biofuels and fine chemicals from plant biomass, providing environmentally benign alternatives to fossil fuels.

Nonetheless, obstacles persist, encompassing ethical dilemmas, legislative impediments, and societal acceptance. Ongoing innovation and responsible advancement are essential for realizing the complete promise of synthetic biology in agriculture and other fields.

Synthetic biology is a disruptive field set to change plant biotechnology, offering innovative ways to enhance agriculture and tackle global sustainability issues.

APPLICATIONS IN PLANTS

While the integration of AI and synthetic biology presents significant promise for plant improvement, several key challenges persist. First, the precision of epigenome editing tools such as CRISPR/dCas9 remains an ongoing issue. Although these tools offer precise DNA editing capabilities, off-target effects and low targeting efficiency continue to be concerns. Additionally, while AI can predict gene behavior, its application in real-time epigenetic regulation is still in its infancy. The paper discusses technological limitations, including the lack of effective temporal regulation tools, which are necessary for controlling dynamic epigenetic processes over time. Furthermore, ethical considerations related to genetically modified organisms (GMOs) and the societal implications of widespread use of AI in agriculture are examined, highlighting the need for ethical guidelines and regulatory frameworks.

Synthetic biology offers transformative applications in plant biology by enabling precise engineering and control over plant traits, growth, and resilience. Key examples include:

- **Metabolic Engineering**: Synthetic biology allows the redesign of plant metabolic pathways to enhance the production of valuable compounds such as vitamins, antioxidants, pharmaceuticals, and biofuels. For instance, plants have been engineered to produce increased levels of beta-carotene (a vitamin A precursor) to combat nutritional deficiencies, or to synthesize bio-based materials that serve as renewable resources.
- **Gene Editing and Epigenome Editing**: Tools such as CRISPR-Cas9 enable targeted modifications in plant genomes, facilitating the introduction of traits like disease resistance, drought tolerance, and improved yield without traditional crossbreeding. More refined techniques, like dCas9-based epigenome editors, modify DNA methylation and histone marks at specific loci, influencing gene expression without altering the DNA sequence itself, offering a new layer of precise control over plant epigenetics.
- **Synthetic Gene Circuits**: Synthetic biology can create programmable gene circuits in plants that respond dynamically to environmental signals. These engineered networks can fine-tune gene expression in real-time, optimizing plant adaptation to stresses such as temperature fluctuations, salinity, or pathogen attack.

- **Biosensors and Smart Plants**: Engineered plants can function as biosensors that detect environmental pollutants, nutrient deficiencies, or pathogen presence, providing valuable feedback for precision agriculture. Integration with AI models further enables the development of "smart crops" that self-monitor and adapt their growth and epigenetic regulation in response to environmental changes.
- **Sustainable Agriculture and Resource Use**: Synthetic biology applications improve nutrient uptake efficiency and reduce the need for chemical fertilizers, thereby minimizing ecological impact. Additionally, engineered plants can serve as biofactories for sustainable production of biofuels and industrial chemicals, contributing to environmental conservation and renewable energy goals.

By integrating synthetic biology with artificial intelligence, researchers can accelerate the design-build-test-learn cycles to optimize these applications, enhancing crop performance, resilience, and sustainability. This synergy holds great promise for meeting global food security challenges and mitigating climate change impacts through innovative plant epigenetic regulation.

DETAILED EXAMPLES OF SYNTHETIC BIOLOGY APPLICATIONS IN PLANTS

1. Metabolic Engineering

A landmark example is the development of Golden Rice, genetically engineered to produce high levels of beta-carotene, a precursor of vitamin A, addressing nutritional deficiencies in many populations. Similarly, researchers have engineered plants like *Arabidopsis* and tobacco to produce higher amounts of flavonoids and antioxidants, which enhance plant stress tolerance and nutritional value. Advances in synthetic biology have also enabled the redesign of lignin biosynthesis pathways in bioenergy crops such as switchgrass, improving biomass processing for biofuel production.

2. Gene Editing and Epigenome Editing

CRISPR-Cas9 has revolutionized precise trait improvement. For example, researchers have used CRISPR to knock out genes responsible for susceptibility to powdery mildew in wheat, enhancing disease resistance. More recently, dCas9 fused to epigenetic modifiers has been applied in *Arabidopsis* to target DNA methylation to specific promoters, effectively silencing or activating genes involved in flowering time or stress responses without changing the underlying DNA. This method holds promise for reversible and fine-tuned epigenetic regulation in crops.

3. Synthetic Gene Circuits

Synthetic gene circuits have been engineered in plants to create environmental sensors and programmable responses. For example, plants have been designed to activate drought-responsive genes only when soil moisture falls below a threshold, conserving water while maintaining growth. Another study engineered a gene circuit that triggers production of insect-repellent compounds in response to pest attack, reducing reliance on chemical pesticides.

4. Biosensors and Smart Plants

Engineered plants acting as biosensors have been developed to detect heavy metals like arsenic or mercury in soil and water. These plants express a visible reporter gene when pollutants are detected, enabling real-time environmental monitoring. Coupling these biosensors with AI allows data integration and predictive modeling to guide precision farming decisions. Furthermore, "smart crops" with AI-guided epigenetic regulation can dynamically adjust growth rates, flowering times, and stress responses to optimize yield and resource use under changing environmental conditions.

5. Sustainable Agriculture and Resource Use

Synthetic biology has been used to engineer nitrogen-fixing capabilities into non-leguminous crops like maize and rice, reducing dependence on nitrogen fertilizers and decreasing environmental pollution. Other examples include modifying root architecture genes to enhance nutrient and water uptake efficiency, improving resilience to drought and poor soils. Additionally, plants have been engineered to produce biodegradable plastics and bio-based chemicals, supporting circular economy goals.

THE ROLE OF ARTIFICIAL INTELLIGENCE IN ENHANCING SYNTHETIC BIOLOGY APPLICATIONS IN PLANT EPIGENETICS

Artificial intelligence (AI) is revolutionizing synthetic biology by providing powerful tools to analyze complex biological data, predict system behaviors, and optimize experimental designs. When applied to plant epigenetics, AI enables more precise and efficient manipulation of gene regulatory networks, accelerating crop improvement.

Data Integration and Pattern Recognition

Plant epigenetic regulation involves multi-layered data from genomics, transcriptomics, epigenomics, and environmental conditions. AI algorithms, particularly machine learning, excel at integrating these heterogeneous datasets to identify key epigenetic markers and regulatory patterns associated with desirable traits like drought tolerance or yield improvement. This data-driven insight guides synthetic biology designs toward targeted epigenetic modifications.

Predictive Modeling for Targeted Intervention

AI models can simulate the impact of specific epigenetic edits on gene expression and plant phenotypes before actual experimentation. This predictive capability reduces trial-and-error, saving time and resources in developing engineered plants with optimized traits. For example, AI can forecast how DNA methylation changes at certain loci will influence stress-responsive genes, enabling more accurate epigenome editing.

Optimization of Experimental Design and High-Throughput Screening

By automating experimental planning and interpreting results from large-scale synthetic biology screens, AI accelerates the discovery of novel gene circuits and regulatory elements. This includes identifying synthetic promoters or epigenetic modifiers that effectively regulate target genes under various environmental conditions.

Real-Time Monitoring and Adaptive Control

Integrated AI systems can monitor environmental inputs and plant physiological status, dynamically adjusting synthetic gene circuits or epigenetic states to optimize growth and stress responses. This real-time feedback loop allows development of "smart crops" capable of self-regulating gene expression via synthetic biology tools, enhancing resilience and productivity.

Ethical and Safety Considerations

AI also aids in risk assessment by modeling ecological impacts and potential off-target effects of epigenetic modifications. This contributes to responsible innovation and regulatory compliance, essential for public acceptance of genetically and epigenetically engineered crops.

CHALLENGES IN PLANT ENGINEERING: COMPLEXITY OF EPIGENETIC REGULATION

Engineering plants at the genetic and epigenetic levels faces significant challenges, primarily due to the intricate and dynamic nature of epigenetic regulation. Unlike straightforward genetic modifications, epigenetic mechanisms—such as DNA methylation, histone modifications, and non-coding RNA activity—operate in complex, interconnected networks that regulate gene expression in highly context-dependent ways.

While the integration of AI models with synthetic biology presents substantial potential for improving plant epigenetics, several critical challenges must be addressed for the technology to reach its full potential.

Precision in Epigenome Editing

One of the most significant barriers is the precision of AI models used in epigenome editing. While tools like dCas9 have made significant advances in precision, off-target effects remain a challenge. These unintended modifications to the genome can lead to changes in plant traits that were not anticipated or desired. This issue is not unique to CRISPR-based tools; even AI models that predict epigenetic changes have to overcome significant data quality issues. As highlighted by Xiong (2023), predicting epigenetic outcomes requires a deep understanding of how these modifications will behave in different plant species, climates, and environmental conditions. The need for improved model accuracy and more reliable data is essential to reduce these risks.

Data Availability and Integration

A critical limitation in AI-driven synthetic biology is the lack of high-quality epigenomic datasets. AI requires large, diverse datasets to train models effectively, but such datasets are still sparse, particularly for non-model plant species. As McKeown & Spillane (2014) point out, without comprehensive datasets, AI models cannot reliably predict how specific epigenetic marks influence plant traits. This data gap significantly hinders the broader application of AI in plant breeding. Moreover, the integration of multi-omics data (genomic, epigenomic, and transcriptomic) remains a complex task, requiring advanced computational models that can process and correlate diverse types of biological data.

Predictability Issues

One major difficulty is predicting how changes in epigenetic marks will affect gene expression and phenotype. Epigenetic modifications can have different effects depending on their genomic location, tissue type, developmental stage, and environmental conditions. This variability complicates efforts to design precise interventions, as altering one epigenetic factor can lead to unintended cascading effects elsewhere in the genome.

Heritability and Stability

While some epigenetic changes are heritable and stable across generations, many are transient or reversible, making it challenging to achieve consistent trait expression in successive plant generations. This instability hinders the development of reliable, long-term crop improvements based on epigenetic engineering.

Environmental Interactions

Environmental factors—such as drought, temperature fluctuations, and pathogen exposure—can induce dynamic epigenetic changes, adding another layer of unpredictability. These interactions may override or modify engineered epigenetic states, complicating control over gene expression in field conditions.

Technical Limitations

Current tools for editing epigenetic marks, including CRISPR-based epigenome editors, face limitations in specificity, efficiency, and off-target effects. Moreover, the lack of inducible or reversible control mechanisms restricts temporal precision in modifying epigenetic states. High-resolution, single-cell epigenomic technologies are still emerging, limiting detailed understanding of cell-type-specific epigenetic regulation.

Ethical and Regulatory Concerns

Manipulating epigenetic pathways raises ethical questions and regulatory hurdles, particularly regarding potential ecological impacts and food safety. Public acceptance depends on transparent risk assessments and responsible deployment.

THE ROLE OF AI IN BIOLOGICAL MODELING AND SYNTHETIC BIOLOGY

Artificial Intelligence (AI) and Machine Learning (ML) are becoming indispensable tools in synthetic biology, enabling the analysis and modeling of complex biological data at unprecedented scales. Key applications include:

- **Prediction of Gene Interactions**: AI algorithms analyze large datasets to uncover gene-gene interactions and map genetic networks, revealing relationships that traditional experiments might miss.
- **Modeling Regulatory Networks**: Machine learning helps decipher intricate gene regulatory networks by predicting transcription factor targets, shedding light on cellular responses to environmental stimuli.
- **Protein Function Simulation**: AI models simulate protein functions and interactions, advancing understanding of molecular mechanisms and supporting drug discovery efforts.
- **DNA-Expression Array Analysis**: AI efficiently processes gene expression data to identify up- or downregulated genes under various conditions, providing insights into biological pathways and disease mechanisms.
- **Biomedical Imaging Analysis**: Deep learning excels at interpreting complex biomedical images, aiding in disease classification and diagnosis by recognizing subtle patterns.

Despite these advances, challenges such as data quality, model interpretability, and the need for large datasets remain. Ongoing research aims to overcome these hurdles to enhance AI's reliability and utility in synthetic biology.

In summary, AI and ML are transforming synthetic biology by enabling robust modeling of biological systems, fostering deeper insights into gene regulation, and accelerating the development of novel biotechnological applications.

The Role of AI in Plant Epigenetics and Synthetic Biology

Artificial Intelligence (AI) and Machine Learning (ML) are revolutionizing the study and manipulation of plant epigenetics by enabling advanced analysis and modeling of complex biological data. In the context of synthetic biology, AI provides powerful tools to unravel and engineer the intricate regulatory networks controlling gene expression in plants.

- **Predicting Epigenetic Interactions**: AI models analyze vast datasets—such as DNA methylation profiles, histone modification maps, and non-coding RNA expression—to predict how epigenetic marks interact and influence gene regulation. This predictive insight helps identify critical regulatory elements that can be targeted for synthetic manipulation.
- **Modeling Epigenetic Regulatory Networks**: Using machine learning, researchers can decipher multi-layered epigenetic networks that govern plant responses to environmental stresses and developmental cues. This modeling facilitates the design of synthetic gene circuits and epigenome edits that achieve precise control over plant traits.
- **Optimizing Synthetic Biology Designs**: AI accelerates the design-build-test-learn cycles by predicting the outcomes of synthetic biology interventions in plant epigenomes, such as locus-specific DNA methylation or histone modification. This reduces experimental trial-and-error and enhances efficiency in creating stress-tolerant or high-yield crops.
- **Integrating Multi-Omics Data**: AI platforms integrate genomic, epigenomic, transcriptomic, and phenotypic data, offering holistic views of plant biology. This integration enables the identification of epigenetic signatures associated with desired agricultural traits, guiding synthetic biology strategies.
- **Real-Time Monitoring and Adaptive Control**: Coupled with sensor data, AI can enable dynamic adjustments in synthetic gene circuits or epigenetic states, creating "smart plants" that adapt their gene expression in response to environmental changes, enhancing resilience and productivity.

While challenges remain, including data complexity and interpretability, AI's synergy with synthetic biology promises to unlock new frontiers in precise plant epigenetic engineering, driving sustainable agriculture and food security.

Examples of AI-Driven Synthetic Biology Projects in Plants

1. AI-Guided Epigenome Editing for Drought Tolerance

Researchers have applied AI algorithms to analyze genome-wide DNA methylation and histone modification data in drought-resistant plants. By identifying key epigenetic regulators, AI helped design CRISPR/dCas9-based epigenome editors to specifically methylate or demethylate target genes involved in water retention and stress signaling. Field trials demonstrated improved drought tolerance without compromising yield, showcasing AI's role in precise epigenetic modulation.

2. Synthetic Gene Circuits Optimized by Machine Learning

In a study with *Arabidopsis*, AI models optimized synthetic promoter designs and gene circuits that respond to salinity stress. Machine learning predicted promoter strength and inducibility under varying salt concentrations, guiding the construction of circuits that activate protective genes only when needed. This minimized energy costs for the plant and enhanced growth under saline conditions.

3. Multi-Omics Data Integration for Crop Yield Improvement

AI platforms integrated epigenomic, transcriptomic, and phenotypic data from multiple rice cultivars to identify epigenetic marks correlated with high yield and nutrient use efficiency. Guided by these insights, synthetic biology tools were employed to modify histone acetylation patterns at key regulatory genes. The engineered lines showed enhanced nutrient uptake and grain production in diverse environments.

4. AI-Enabled Biosensors in Plants

Engineered plants equipped with synthetic biosensors detect soil nutrient levels and environmental pollutants, producing measurable reporter signals. AI processes sensor data in real time, enabling automated adjustments in gene circuits that regulate nutrient uptake and stress responses. This technology is advancing precision agriculture by reducing fertilizer use and enhancing crop resilience.

5. Predictive Modeling of Epigenetic Memory for Stress Adaptation

Using AI-driven simulations, scientists modeled how epigenetic memory affects gene expression across plant generations under repeated drought stress. These models informed synthetic biology strategies to create heritable epigenetic states that prime future generations for enhanced stress tolerance, offering a novel approach to sustainable crop resilience.

INTEGRATING BIOLOGICAL DATA WITH AI MODELS

AI models are essential for integrating large-scale and diverse biological datasets—including genomics, transcriptomics, and epigenomics—to deepen our understanding of complex plant systems and to inform synthetic biology design. This integration enables comprehensive characterization of cell types, gene regulatory states, and epigenetic landscapes critical for precise engineering.

However, the complexity of biological data poses significant challenges. Variability in data types and batch effects can obscure meaningful patterns, complicating joint analysis. To address this, machine learning innovations have led to advanced data integration techniques that embed heterogeneous datasets into unified representations, facilitating downstream predictive modeling and synthetic biology applications.

Various machine learning paradigms enable these integrations, allowing researchers to synthesize multi-omic information across biological contexts. Despite these advances, challenges remain, including improving the robustness and interpretability of AI models, and ensuring that integrated data can reliably guide the design and optimization of synthetic biological systems in plants.

Overall, AI-driven data integration is pivotal for unlocking insights from complex biological data and accelerating the development of next-generation, epigenetically engineered crops.

AI-Driven Epigenetic Editing Tools

Recent research has shown how AI models can be employed in combination with epigenome editing technologies, such as CRISPR/Cas9, to precisely manipulate plant epigenetics. For example, dCas9 (catalytically inactive Cas9) can be engineered to target specific epigenetic marks like DNA methylation and histone modifications without altering the genetic sequence itself. AI-driven models predict the impacts of these epigenetic changes, enhancing the precision and effectiveness of gene editing. This method was successfully used in Arabidopsis to modify flowering time through targeted epigenetic modifications, demonstrating a practical application of AI in fine-tuning crop traits without genetic modifications.

Synthetic Biology Applications in Agriculture

Synthetic biology, particularly the creation of synthetic gene circuits, plays a pivotal role in regulating plant traits in response to environmental cues. Recent developments have employed AI to design, predict, and test synthetic biological systems for plants. One study by Zhang et al. (2024) employed machine learning algorithms to optimize synthetic promoters that respond to salinity, ensuring that plants activate protective genes only under specific stress conditions. This allows for water and energy conservation, making plants more resilient while reducing resource usage. Additionally, AI-assisted multi-omics integration is used to combine genomic, epigenomic, and transcriptomic data to predict the impact of epigenetic alterations, providing a holistic view of plant biology and accelerating the development of stress-resistant crops.

STRONGER LINKAGES BETWEEN THEORY AND APPLICATION

We have now connected theoretical concepts more directly to practical applications, especially in crop improvement, sustainability, and food security. Below are key sections demonstrating these linkages:

Theoretical Framework: Plant Epigenetics and AI

The theoretical framework draws from systems biology and the Unified Theory of Acceptance and Use of Technology (UTAUT), which stresses the importance of data-driven understanding and technology adoption. In plant epigenetics, systems biology allows us to understand complex gene regulatory networks that are modulated by epigenetic marks such as DNA methylation, histone modifications, and non-coding RNAs. AI plays a significant role in predicting how these epigenetic markers will influence plant traits and how they can be engineered for improved stress tolerance, better yield, and enhanced nutritional value. By combining AI and synthetic biology, plant scientists can design optimized synthetic gene circuits that control gene expression based on environmental signals, providing precision agriculture solutions that align directly with sustainability goals.

Application: Enhancing Crop Resilience

The integration of AI in synthetic biology has provided new solutions to the agricultural challenge of environmental stress adaptation. AI-driven predictive modeling has allowed for the design of plants that can adjust their gene expression in response to changing environmental conditions. For instance, in drought tolerance, AI models predict how modifications in DNA methylation or histone modification will influence stress-related genes in crops like maize and wheat, improving their survival rates during water scarcity. These applications are directly linked to sustainable farming practices, reducing the need for excessive water usage and improving crop yields in arid regions, thereby enhancing food security.

Application: Sustainable Agriculture and Food Security

AI in synthetic biology also plays a significant role in advancing sustainable agriculture. By improving the efficiency of plant nutrient uptake, AI models help optimize soil fertility management, reducing dependency on chemical fertilizers. AI-assisted breeding programs are now creating drought-resistant and salt-tolerant crops, directly addressing issues arising from climate change. Additionally, AI models can predict and prevent crop diseases by analyzing epigenetic changes related to pathogen resistance, thereby ensuring that crops remain healthy and productive with minimal intervention.

Final Link to Global Agricultural Challenges

The combination of AI and synthetic biology is pivotal in addressing global food security challenges, as it enables the design of crops that are not only resilient to stress but also nutritionally enhanced to meet the growing demands of a changing climate. AI-enhanced synthetic biology techniques have led to biofortified crops, such as rice with higher vitamin A content, and crops with increased disease resistance, contributing directly to addressing global malnutrition and ensuring food security for the future.

INTEGRATING BIOLOGICAL DATA WITH AI MODELS IN PLANT EPIGENETICS AND SYNTHETIC BIOLOGY

AI and Epigenetic Editing: Addressing the Data and Model Precision Challenges

A critical issue faced when applying AI models to plant epigenetic regulation is the quality and availability of data. AI models rely heavily on large datasets to identify patterns and make predictions, but high-quality, comprehensive datasets in plant epigenetics are still lacking. As noted by McKeown & Spillane (2014), the current datasets are often incomplete or inconsistent, which limits the predictive power of AI models. For example, while data on DNA methylation patterns in model plants like Arabidopsis are abundant, such data are sparse for many other plant species critical to food production. This data gap undermines the ability of AI to make accurate predictions about how epigenetic changes will affect crop traits, and it limits the potential for AI to guide epigenome editing for specific agricultural goals.

In addition to data challenges, there is also the issue of model precision. AI algorithms designed to predict the effects of epigenetic modifications must account for the high variability of plant responses to environmental stimuli. As Xiong (2023) explains, epigenetic modifications are context-dependent; their effects can vary depending on the plant's developmental stage, environmental conditions, and genetic background. These complexities make it difficult for AI models to achieve the level of precision needed to reliably predict the outcomes of specific epigenetic changes, particularly in complex agricultural settings.

The integration of diverse biological datasets through AI models is revolutionizing plant epigenetics research and synthetic biology applications. By combining genomics, transcriptomics, epigenomics, and proteomics data, AI provides a holistic view of the molecular mechanisms controlling gene expression and plant development.

In plant epigenetics, such integrated analyses reveal how DNA methylation patterns, histone modifications, and non-coding RNA profiles coordinate to regulate stress responses and developmental processes. For example, AI models can integrate methylome and transcriptome data from drought-stressed plants to identify epigenetic markers linked to stress tolerance. These insights enable synthetic biologists to design targeted epigenome editing strategies that enhance resilience traits.

Moreover, AI-powered data integration supports the development of synthetic gene circuits by predicting interactions among regulatory elements and epigenetic states across different tissues and environmental conditions. This allows for the construction of dynamic and precise synthetic networks that respond effectively to external stimuli.

Case Study: In rice, multi-omics integration using AI helped identify key epigenetic regulators influencing flowering time and yield. Synthetic biology tools were then employed to modify these regulators, producing cultivars optimized for varying climates. This integrated approach accelerated breeding programs beyond traditional methods.

Despite these advances, challenges such as batch effects, data heterogeneity, and model interpretability require ongoing innovation. Improving robustness in AI-driven integration will be critical for reliably translating complex biological data into actionable synthetic biology designs.

In summary, AI-facilitated integration of large-scale biological data is essential for unraveling plant epigenetic regulation and enabling precise synthetic biology interventions, paving the way for smarter, more adaptable crops.

INTEGRATING AI MODELS WITH SYNTHETIC BIOLOGY FOR EPIGENETIC REGULATION

Tahir et al. (2024) discuss how artificial intelligence (AI) is being used to analyze epigenetic data related to gene regulation in plants. Epigenetics refers to changes in gene activity that do not involve alterations to the DNA sequence, such as DNA methylation and histone modifications. These changes are important in controlling how genes behave in different conditions.

The paper explains that AI is useful for handling the large and complex datasets produced by modern experiments. Machine learning algorithms can find patterns in DNA methylation and histone modifications, helping researchers understand how genes are regulated.

In plant research, AI can help predict gene expression based on environmental conditions or developmental stages. It can also identify markers linked to plant diseases, which may help in early detection and control. Additionally, AI can assist in studying how enhancers and promoters interact—important elements in gene regulation.

Tahir et al. note that there are still challenges, such as the need for high-quality data and the complexity of biological systems. They suggest more work is needed to improve the accuracy of AI models. The authors emphasize the importance of collaboration between AI experts and epigenetic researchers to develop better tools. This approach could lead to new discoveries in plant biology and applications in areas like synthetic biology.

Hamdy et al. (2023) explore how artificial intelligence, especially deep learning, is being used to analyze epigenetic data for better understanding of gene regulation. Epigenetic changes like DNA methylation and histone modifications influence gene activity without changing the DNA sequence itself. These changes are important in how genes are turned on or off.

The paper introduces an AI model called DeepEpi, which uses Convolutional Neural Networks (CNN) and Long Short-Term Memory (LSTM) networks. This model is able to analyze complex patterns in histone modification data and capture long-range interactions within the genome. It helps in predicting how genes will be expressed by looking at the spatial distribution of histone signals.

DeepEpi performed well, achieving an Area Under the Curve (AUC) score of 88.87% across 56 different human cell types. This suggests that the model is more accurate than previous approaches in predicting gene regulation based on histone modifications.

Although the study focuses on human cells, the authors suggest that the same methods can be used in plant research. This could help scientists better understand how genes are regulated in plants and lead to applications in synthetic biology, such as creating crops with improved traits or resistance to environmental stress.

Hamdy et al. highlight that combining AI with synthetic biology could greatly enhance our understanding of gene regulation and open new possibilities in agriculture and biotechnology.

Integrating AI Models with Synthetic Biology for Epigenetic Regulation

While synthetic biology offers new opportunities for plant breeding, particularly through epigenetic regulation, its integration with AI faces several technical and ethical barriers. For instance, the ethical concerns surrounding genetically modified organisms (GMOs) continue to shape the debate over the use of AI and synthetic biology in agriculture. As Zhang et al. (2024) point out, concerns about biosafety and ecological impacts of genetically modified crops persist, even though AI can reduce the risks associated with traditional genetic engineering methods.

Additionally, as Attia et al. (2024) highlight, public trust in AI-driven synthetic biology applications remains a challenge, particularly in the agricultural sector. To address this, it is essential to focus on community engagement, transparency, and ethics in AI deployment. Without these considerations, widespread adoption of AI in synthetic biology could face significant resistance from both regulatory bodies and the public.

Artificial intelligence (AI) is playing a growing role in understanding and manipulating epigenetic mechanisms in plants. These advances are contributing to new approaches in plant synthetic biology and crop improvement.

- **AI in Epigenetic Data Analysis**: AI is being used to analyze complex epigenetic data, such as DNA methylation patterns and histone modifications. By identifying patterns and relationships in this data, AI helps researchers understand how genes are regulated in plants under various conditions.
- **Modeling Epigenetic Networks**: AI models can predict how different epigenetic factors interact and form regulatory networks that influence important plant traits like growth, stress response, and yield. These models help in uncovering hidden relationships that are difficult to detect through traditional methods.
- **AI-Driven Epigenetic Engineering**: With the insights gained from AI, scientists can design specific changes to the plant epigenome. For example, AI can suggest ways to adjust DNA methylation at certain genes to enhance desirable traits such as drought resistance or nutrient efficiency.

- **AI-Assisted CRISPR and Epigenome Editing**: AI is also being used to improve the precision of gene-editing tools like CRISPR/Cas. It can help identify the best target sites for editing epigenetic marks, increasing efficiency and reducing off-target effects. This is particularly useful for fine-tuning gene expression without changing the DNA sequence.

CASE STUDIES AND APPLICATIONS

The integration of AI and synthetic biology has led to several exciting advancements in plant science, particularly in areas such as crop stress tolerance, gene regulation, and plant breeding.

- **Case Study 1: AI-enhanced Synthetic Biology for Improving Crop Stress Tolerance Through Epigenetic Regulation**

Researchers have used AI to analyze epigenetic modifications in plants that affect stress tolerance. By identifying key genes and their regulatory networks, AI models have been applied to design plants with improved resilience to environmental stresses, such as drought or high salinity, through targeted epigenetic modifications.

- **Case Study 2: Use of AI to Create Synthetic Gene Circuits that Regulate Epigenetic Marks for Desired Phenotypic Traits in Plants**

AI-driven approaches have been employed to design synthetic gene circuits capable of controlling epigenetic marks, such as DNA methylation and histone modifications. These circuits can be programmed to turn specific genes on or off in response to environmental signals, leading to plants with tailored phenotypic traits like enhanced disease resistance or improved growth patterns.

- **Case Study 3: Applications in Plant Breeding – AI Models Integrated with Synthetic Biology to Accelerate the Development of Epigenetically Engineered Crops with Improved Yields or Disease Resistance**

In the field of plant breeding, AI models integrated with synthetic biology are being used to accelerate the development of epigenetically engineered crops. By simulating and predicting the effects of epigenetic modifications on crop traits, AI tools have enabled faster development of crops with enhanced yields, improved disease resistance, or better nutrient profiles.

These case studies demonstrate the powerful potential of combining AI with synthetic biology, enabling more efficient and precise approaches to plant improvement and agricultural sustainability.

Recommendations

- Synthetic biologists, artificial intelligence (AI) developers, and plant biologists should work together more closely.
- Create extensive, top-notch datasets to enhance the training of AI models.
- Raise understanding of AI-powered biotechnology and put policies in place to support it.
- Increase investigation into methods of heritable and reversible epigenetic editing.
- Develop genetically modified crops with biosafety and ethical concerns in the forefront.
- **Invest in Multi-Species Epigenomic Databases**: AI models require comprehensive, high-quality datasets for accurate predictions. Efforts should focus on creating and curating epigenomic datasets for a broader range of plant species, particularly non-model species essential for global food production.
- **Improve AI Model Precision and Data Integration**: To overcome the challenges of off-target effects and data quality, future research should focus on improving AI algorithms and developing models that can integrate multi-omics data. These advances will increase the precision of AI-driven epigenetic editing tools.
- **Develop Ethical Guidelines and Regulatory Frameworks**: As synthetic biology and AI continue to evolve, it is critical to develop ethical guidelines and regulatory frameworks that address the potential risks and societal concerns surrounding genetically modified crops and epigenetic modifications.
- **Promote Public Engagement and Transparency**: To foster public trust in AI-driven agricultural innovations, transparency in research processes and public engagement are essential. Clear communication about the benefits, risks, and ethical considerations of AI in agriculture will be crucial for widespread adoption.

Limitations

- The lack of extensive epigenomic databases across many plant species.
- Genetic engineering raises questions about the durability and passivity of manmade epigenetic modifications.

- Society and ethics worry over gene editing and personal information transmission.
- Among the technical constraints of existing epigenome editing technologies are issues with delivery efficiency and off-target consequences.

FUTURE DIRECTIONS AND POTENTIAL CHALLENGES

As AI continues to evolve, its integration with synthetic biology for epigenetic regulation in plants holds tremendous promise. However, several emerging trends, challenges, and the need for interdisciplinary collaboration must be addressed to fully harness this potential.

Emerging Trends

Recent advancements in AI models, such as deep learning and reinforcement learning, are offering new opportunities to refine and accelerate the application of synthetic biology in plant epigenetics. Deep learning algorithms, with their ability to analyze large datasets, can improve the precision of gene expression predictions and enable more efficient design of synthetic gene circuits. Reinforcement learning, which uses trial and error to optimize actions based on feedback, could be particularly beneficial for developing dynamic gene networks that respond to environmental changes in real-time. These advancements could lead to more sophisticated models for predicting gene behavior, enhancing the ability to design epigenetic modifications that result in targeted traits like enhanced resilience or improved productivity.

Challenges:

Several challenges still remain in the integration of AI with synthetic biology for plant epigenetic regulation.

1. **Precision in Epigenome Editing**: While technologies like CRISPR/dCas9 have made progress in epigenome editing, the precision of these tools remains an issue. Off-target effects, where unintended genetic modifications occur, continue to pose challenges, leading to potentially unpredictable outcomes in plant traits. Singh et al. (2023) noted that current epigenetic editing systems are not yet reliable enough to ensure consistent, precise results, especially in complex agricultural settings where environmental variables play a significant role.
2. **Data Quality and Availability**: A critical limitation in the use of AI models is the lack of high-quality, comprehensive datasets. AI models depend on large,

well-curated datasets to make accurate predictions, but many plant species, particularly non-model species, have insufficient genomic data available. This gap in data makes it difficult for AI to make reliable predictions regarding how epigenetic modifications will affect crop traits like stress tolerance and yield. As McKeown & Spillane (2014) point out, without robust data, AI models cannot fully capture the complexity of plant epigenetics, limiting their practical use in agriculture.

3. **Technological Limitations and Ethical Considerations**: The technologies used in epigenome editing are still evolving, and their efficiency and specificity need improvement. Moreover, the ethical concerns surrounding genetically modified organisms (GMOs) and epigenetic modifications remain a significant barrier to the widespread adoption of these technologies. Public perception and regulatory approval will play a key role in determining the extent to which these AI-driven synthetic biology tools can be applied in agriculture.
4. **Technical Challenges**: One major hurdle is the data limitations. While AI relies heavily on large, high-quality datasets, the complexity of plant epigenetics means that acquiring comprehensive and high-resolution data can be time-consuming and expensive. Furthermore, model accuracy remains an issue, as many existing models may not fully capture the intricate interactions within plant epigenetic networks, leading to suboptimal predictions or interventions.
5. **Off-Target Effects**: In the context of epigenome editing, AI models could help minimize off-target effects, but these risks still exist, especially with complex genetic backgrounds and environmental factors. There is a need for more precise tools to ensure that epigenetic modifications are made at the correct loci without unintended consequences.
6. **Ethical and Regulatory Challenges**: The manipulation of plant epigenetics raises significant ethical questions, especially regarding the potential long-term ecological consequences of introducing epigenetically modified crops into the environment. Additionally, there are regulatory hurdles, as many countries have strict regulations concerning genetically modified organisms (GMOs), and epigenetic modifications—while not changing the DNA sequence—can be viewed similarly to genetic modifications in some contexts. Ensuring compliance with these regulations while maintaining the potential for innovation will require careful consideration.

INTERDISCIPLINARY COLLABORATION

The integration of AI and synthetic biology in plant epigenetic regulation will not be successful without strong collaboration across disciplines. AI experts must work closely with **synthetic biologists** to develop accurate models that can simulate and predict gene interactions. Additionally, **plant biologists** will provide the biological context necessary to interpret AI-generated data and ensure that predictions are biologically relevant. This collaboration will be critical in addressing the complexities of plant epigenetics and designing interventions that are both effective and safe. Cross-disciplinary teams can also foster innovation by combining diverse expertise to solve problems that may seem insurmountable when approached from a single perspective.

In summary, while there are exciting developments in the integration of AI with synthetic biology for plant epigenetic regulation, overcoming the challenges of data quality, model precision, ethical concerns, and regulatory barriers will require a concerted effort. Collaborative work across disciplines will be key to making meaningful advances in this field, ensuring that these technologies can contribute to the future of sustainable agriculture and crop improvement.

CONCLUSION

The integration of AI models with synthetic biology for plant epigenetic regulation represents a transformative approach to advancing plant science and agriculture. By leveraging AI to analyze complex epigenetic data, researchers can gain deeper insights into gene regulation, stress responses, and developmental processes in plants. This synergy has the potential to improve crop resilience, yield, and stress tolerance, ultimately contributing to food security and sustainable agricultural practices.

AI-driven models, such as deep learning and reinforcement learning, offer new opportunities to optimize gene expression predictions and develop precise epigenetic modifications. These models can also help design synthetic gene circuits and assist in epigenome editing, facilitating the creation of plants with tailored traits. The ability to integrate multi-omics data further enhances the design of crops with improved nutritional content, disease resistance, and environmental adaptability.

However, despite the promising advances, several challenges remain, including data limitations, model accuracy, and the risks of off-target effects in epigenetic modifications. Ethical and regulatory concerns surrounding the manipulation of plant epigenetics must also be addressed to ensure public acceptance and responsible innovation. Additionally, the complexity of plant epigenetic systems requires interdisciplinary collaboration between AI experts, synthetic biologists, and plant biologists to fully harness the potential of these technologies.

In conclusion, while the path to fully integrating AI and synthetic biology in plant epigenetic regulation is fraught with challenges, it offers immense potential to revolutionize crop improvement, enhance sustainability, and address the pressing challenges of climate change and food security. Continued research, innovation, and collaboration will be essential to unlocking the full benefits of this interdisciplinary approach.

REFERENCES

Abdulraheem, M. I., Xiong, Y., Moshood, A. Y., Cadenas-Pliego, G., Zhang, H., & Hu, J. (2024). Mechanisms of Plant Epigenetic Regulation in Response to Plant Stress: Recent Discoveries and Implications. *Plants*, *13*(2), 163. Advance online publication. DOI: 10.3390/plants13020163 PMID: 38256717

Ahtisham, M., & Obaid, Z. (2025). *Harnessing Epigenetic Mechanisms for Crop Resilience: A Comprehensive Review of Plant Responses to Biotic and Abiotic Stresses*. https://doi.org/DOI: 10.70389/PJPB.100007

Amaan, A., Prekshi, G., & Prachi, S. (2024). Unlocking the Transformative Power of Synthetic Biology. *Archives of Biotechnology and Biomedicine*. https://doi.org/ DOI: 10.29328/journal.abb.1001039

Attia, I., Dong, Y., & Ignea, C. (2024). Synthetic biology advances towards a bio-based society in the era of artificial intelligence. *Current Opinion in Biotechnology*, *87*, 103143. DOI: 10.1016/j.copbio.2024.103143 PMID: 38781699

Bhuvaneswari, R., Saravanan, K., Vennila, S., & Suganthi, S. (2020). The Role of Epigenetics in Plant Breeding Understanding Heritable Changes beyond DNA Sequence. *Plant Science Archives*, *5*(1), 22–25. DOI: 10.51470/PSA.2020.5.1.22

Dixon, T. A., Curach, N., & Pretorius, I. S. (2020). Bio-informational futures: The convergence of artificial intelligence and synthetic biology. *EMBO Reports*, *21*(3), 50036. DOI: 10.15252/embr.202050036 PMID: 32043291

Fal, K., Tomkova, D., Vachon, G., Chaboute, M.-E., Berr, A., & Carles, C. C. (2021). Chromatin Manipulation and Editing: Challenges, New Technologies and Their Use in Plants. *International Journal of Molecular Sciences*, *22*(2), 512. DOI: 10.3390/ ijms22020512 PMID: 33419220

Hamdy, R., Omar, Y. M. K., & Maghraby, F. A. (2023). DeepEpi: Deep Learning Model for Predicting Gene Expression Regulation Based on Epigenetic Histone Modifications. *Current Bioinformatics*, *18*. Advance online publication. DOI: 10. 2174/1574893618666230818121046

McKeown, P. C., & Spillane, C. (2014). *Landscaping plant epigenetics* (Vol. 1112). Humana Press., DOI: 10.1007/978-1-62703-773-0_1

Mendieta, J. P., Sangra, A., Yan, H., Minow, M. A. A., & Schmitz, R. J. (2023). Exploring Plant Cis-Regulatory Elements at Single-Cell Resolution: Overcoming Biological and Computational Challenges to Advance Plant Research. *The Plant Journal, 115*(6), 1486–1499. Advance online publication. DOI: 10.1111/tpj.16351 PMID: 37309871

Plant Epigenomics (pp. 263–286). (2023). Elsevier eBooks. https://doi.org/DOI: 10.1016/B978-0-323-91909-8.00007-4

Singh, V. K., Ahmed, S., Saini, D. K., Gahlaut, V., Chauhan, S., Khandare, K. P., Kumar, A., Sharma, P. K., & Kumar, J. (2023). *Manipulating epigenetic diversity in crop plants: Techniques, challenges and opportunities.* 130544. https://doi.org/DOI: 10.1016/j.bbagen.2023.130544

Sun, M. (2025). Advances in Plant Epigenetic Regulation of Abiotic Stress Response. *Theoretical and Natural Science, 90*(1), 81–87. DOI: 10.54254/2753-8818/2025.GU20455

Tahir, M., Norouzi, M., Khan, S. S., Davie, J., Yamanaka, S., & Ashraf, A. (2024). Artificial intelligence and deep learning algorithms for epigenetic sequence analysis: A review for epigeneticists and AI experts. *Computers in Biology and Medicine, 183*, 109302. DOI: 10.1016/j.compbiomed.2024.109302 PMID: 39500240

Xiong, H. (2023). *Epigenetic regulation in plant salt stress response.* https://doi.org/DOI: 10.54254/2753-8818/6/20230296

Zhang, D., Xu, F., Wang, F., Le, L., & Pu, L. (2024). Synthetic biology and artificial intelligence in crop improvement. *Plant Communications, 101220.* Advance online publication. DOI: 10.1016/j.xplc.2024.101220 PMID: 39668563

Chapter 14

Quantum:
Resilient Security Frameworks for Aviation and Aerospace

Sharon L. Burton
http://orcid.org/0000-0003-1653-9783
Embry-Riddle Aeronautical University, USA

ABSTRACT

Quantum computing now threatens classical cryptography at the core of aviation and aerospace security. By 2024, over 22% of aerospace organizations had begun post-quantum cryptography (PQC) migration, acknowledging the rapidly growing risk. New NIST standards—ML-KEM and ML-DSA—provide quantum-resistant options for secure key exchange and digital signatures, vital for avionics and satellite systems. This chapter draws on qualitative synthesis of research and regulatory guidance to assess major challenges and best practices for PQC migration. Results identify critical technical, organizational, and regulatory hurdles, but also show that combining PQC, physical-layer security, zero-trust, and cryptographic agility can significantly raise sector resilience. Clear, actionable recommendations are offered for migration planning, risk management, and compliance. The insights are intended for aviation cybersecurity professionals, policymakers, regulators, and researchers strengthening critical quantum-resilient infrastructure.

INTRODUCTION

Securing digital infrastructure has become a matter of international urgency, as quantum computing and AI present risks that can interrupt economies, compromise national security, and imperil human lives. Critical infrastructures in developed and emerging economies increasingly rely on encrypted communications for safe

DOI: 10.4018/979-8-2600-1026-6.ch014

operations and regulatory compliance, making the resilience of cryptographic systems foundational to global trust, commerce, and public safety (Gbaja, 2025). The central research question is: "How can aviation stakeholders operationalize and sustain quantum-resilient cryptographic protocols to ensure secure, compliant, and reliable operations across legacy and next-generation infrastructure in the face of escalating quantum threats and regulatory complexity?" This question drives the examination of practical solutions to bridge the gap between cryptographic theory, organizational practice, and international security requirements.

In the early decades of the twenty-first century, cryptographic protocols evolved from niche mathematical constructs into essential pillars supporting the infrastructure of digital society (Diffie & Landau, 2023) and are recognized as a groundbreaking technology for confirming secure communication in the age of quantum computing (Xu et al., 2025). In other words, cryptographic protocols progressed with the advent of public-key algorithms like (Rivest–Shamir–Adleman) RSA, which was presented in 1977 by Ron Rivest, Adi Shamir, and Leonard Adleman (Imam et al., 2021). Cryptographic protocol later adapted to digital identity, commerce, and cloud-driven networks (National Institute of Standards and Technology [NIST], 2017). The steady advancement in fields such as quantum computing has, however, precipitated a paradigm shift. The post-quantum paradigm emerged in response to Shor's algorithm and the realization that quantum computers could break widely used security schemes underpinning digital society today (Mamatha et al., 2025). Algorithms that once undergirded global security, including RSA and elliptic curve cryptography, now face existential threats from quantum algorithms such as Shor's, capable of efficiently factorizing large integers and solving discrete logarithm problems (Harvard University, 2025).

A cryptographic protocol is akin to a set of instructions that instructs computers how to safely send and receive secret messages (Young, n.d.). This protocol uses special tools (i.e., codes to hide messages (encryption), checks to make sure messages were not changed (hashing), and digital signatures to prove who sent the message) to keep the information safe and private. All these tools work together like pieces of a puzzle to make sure no one can spy on or change the message (Young, n.d.). Their function in securing communication, authenticating digital identities, safeguarding financial transactions, and protecting sensitive information has grown alongside the proliferation of internet-connected devices and expanding data-driven economies (Biagini, 2025).

The emergence of artificial intelligence (AI), exponential growth in computational power, and deepening interconnection of critical infrastructures, to include those in aviation, have simultaneously magnified the benefits and vulnerabilities of this digital transformation, where secure, resilient operations are now crucial to protect air traffic, aircraft communication, and passenger safety (Ahmed, 2025; Zhao et al., 2024). Even though Zhao et al. (2024) center their research on aviation, their work recognizes that AI is transmuting practically all fields of society, signifying a broader significance across multiple industries.

As these foundational protocols (i.e., the evolving algorithms and systems used to secure communication, authenticate digital identities, safeguard transactions, and protect sensitive information across connected devices and data-driven economies) grew to support the expanding digital ecosystem, the rise of AI intelligence, unprecedented computational power, and interconnected infrastructure, across sectors including aviation, created a dynamic environment where traditional assurances of security increasingly faced new and complex threats that demand urgent adaptation and resilience (Biagini, 2025).

As information technologies are embedded deeper into the societal fabric through cloud computing, supply chain management, and decentralized finance, the stakes of cryptographic resilience are heightened, with potential consequences ranging from privacy violations to disruption of critical national infrastructure (Palleti et al., 2021). Simultaneously, the explosion of Internet of Things (IoT) devices, each demanding lightweight yet robust security, challenges standard cryptographic deployment models. The vast diversity of endpoints (i.e., spanning medical implants, industrial sensors, smart vehicles, and home automation devices) introduces subtle implementation risks and widens attack surfaces (Kong et al., 2021). Cloud computing environments, with their multi-tenant architectures and transnational data flows, further complicate cryptographic governance and data sovereignty (Pearson & Benameur, 2010). International regulatory regimes, from GDPR to emerging quantum-safe mandates, add layers of complexity as organizations contend with cross-border compliance and digital trust (Ray & Lemos, 2023). This context necessitates the evolution of cryptographic standards, compelling researchers and practitioners to develop solutions that are secure against current classical threats and against imminent quantum adversaries. Against this backdrop, the transition to post-quantum cryptographic (PQC) algorithms, able to withstand the dual challenges of quantum and classical computation, becomes central to preserving the confidentiality, integrity, and authenticity of tomorrow's digital ecosystem (Purohit & Vyas, 2025).

Recent empirical studies document organizational efforts to migrate critical systems to post-quantum cryptography (i.e., pilot implementations in air traffic management, cryptographic library upgrades in avionics, and early evidence from test deployments in the aerospace sector). Key findings emphasize persistent migration bottlenecks, increased attack surfaces resulting from IoT expansion, and the operational impact of deploying quantum-resilient algorithms within bandwidth- and resource-constrained aviation infrastructure (Kong et al., 2021; Sridhar et al., 2021; Gautam et al., 2025).

As given by Agboola (2024) and Kshetri (2023), practical migration faces technical, organizational, and regulatory hurdles. Significant resource constraints are present in legacy systems, requiring increased key sizes and placing greater processing overhead on current infrastructure (Agboola, 2024). Achieving interoperability with international aviation partners is complicated by regulatory fragmentation and varied compliance regimes (Barker et al., 2025). High upgrade costs and resistance to rapid change within industry workforces further delay alignment of operational reliability demands with the rapid pace of cryptographic innovation (Kshetri, N. (2023).

This chapter employs a qualitative, thematic synthesis of empirical studies, technical documentation, and regulatory reports, explicitly addressing technological innovations and sector-specific implementation barriers. The analysis integrates evidence from academic, industry, and policy domains to provide a holistic framework for migration, risk management, and compliance strategies tailored for aviation and aerospace settings. The goal is to offer actionable insights for practitioners and comprehensive recommendations for policymakers, moving beyond theory to address practical realities and sector-specific needs.

PROBLEM STATEMENT AND RESEARCH GAP

In spite of progress, there is limited research on (1) enterprise-scale migration strategies in live aviation settings; (2) longitudinal, empirical validation of quantum-resistant cryptographic deployments across mixed-asset fleets; (3) minimum compliance benchmarks for quantum security in international regulatory agreements; and (4) human factors and organizational barriers to rapid technological adoption. Most studies are either highly theoretical or single-sector focused, neglecting the nuance of aerospace's unique hardware and regulatory lifecycle. This chapter is guided by the following central research question: "How can aviation stakeholders operationalize and sustain quantum-resilient cryptographic protocols to ensure secure, compliant, and reliable operations across legacy and next-generation infrastructure in the face of escalating quantum threats and regulatory complexity?"

Current large-scale digital systems (i.e., including air traffic management networks, aircraft health monitoring systems, and satellite-based communication and authentication infrastructures) are critically endangered by the emergence of quantum computing, as algorithms like Shor's will enable adversaries to rapidly break RSA and elliptic curve cryptography, leaving existing encryption, digital signatures, and secure communications open to compromise unless organizations can efficiently transition to and operationalize quantum-resistant cryptographic protocols within resource-constrained and legacy environments (Biagini, 2025; Kong et al., 2021).

The justification for this problem statement lies in the accelerating integration of digital technologies across aviation and aerospace sectors, where secure information flow is essential for operational safety, regulatory compliance, and national security. Recent studies affirm that cryptographic protocols, including public key infrastructure, digital signature authentication, and encryption technologies, are fundamental for securing data across aviation networks, encompassing air traffic management, aircraft health monitoring, and satellite control systems, to guarantee authenticity, integrity, and confidentiality in information exchange (Gautam et al., 2025; Meng et al., 2025; Sridhar et al., 2021). For decades, RSA and elliptic curve cryptography (ECC) algorithms have served as robust pillars of security for aviation and aerospace systems, offering proven protection against classical computational threats through established key exchange and digital signature mechanisms (Adeniyi et al., 2024; Atadoga et al., 2024; Encryption Consulting, 2025). However, the rapid progress in quantum computing, primarily through the application of algorithms such as Shor's, fundamentally undermines these cryptosystems, creating a credible and imminent threat to sector-wide information assurance (Kong et al., 2021).

Aviation and aerospace infrastructures are uniquely vulnerable due to their reliance on long-lived assets and legacy systems, which often have limited computational or memory resources, complicating the deployment of quantum-resistant algorithms (Efremov et al., 2021). Moreover, the distributed and transnational nature of air transport and aerospace networks demands cryptographic solutions that are interoperable, efficient, and robust against evolving adversarial capabilities. Stakeholders are confronted with technical challenges (i.e., increased key sizes or processing overhead inherent in post-quantum cryptography) and organizational hurdles associated with large-scale migration and adaptation of safety-critical equipment (Biagini, 2025).

Without timely and coordinated transition efforts, the compromise of command and control, navigational integrity, or secure communications in these environments could result in operational disruptions, data breaches, or, in extreme scenarios, catastrophic failures implicating safety and economic stability (Palleti et al., 2021). This urgency is further emphasized by leading research and regulatory advisories, which urge immediate preparedness and phased adoption of post-quantum algorithms to safeguard the aviation and aerospace digital ecosystem from emerging quantum-enabled attack vectors (Palleti et al., 2021).

BACKGROUND OF THE STUDY

The emergence of quantum computing has instigated a transformation in the theoretical and practical foundations of cryptographic security. Classical cryptography, whose resilience is grounded in the assumed intractability of problems such as integer factorization and discrete logarithms, has endured as the gold standard for securing digital transactions, identity verification, and information confidentiality for decades (Sjöberg, 2022). However, a growing body of mathematical research, epitomized by Shor's algorithm, has unequivocally demonstrated that quantum computers, once operational at scale, will be capable of compromising pivotal schemes like RSA and elliptic curve cryptography (ECC) in polynomial time, thereby undermining the security propositions that have defined modern digital infrastructure (Kong et al., 2021). According to Chen (2025), Shor's algorithm is a quantum computing process created by Peter Shor that can factor large integers proficiently by using quantum-mechanical principles (i.e., quantum parallelism, the Quantum Fourier Transform). Unlike classical algorithms, which require super-polynomial time for integer factorization, Shor's method can discern prime factors of a composite number in polynomial time, making it exponentially faster for this specific task (Shor, 1994). The algorithm has two primary stages: a classical reduction that transforms the factoring problem into an order-finding problem, and a quantum stage that uses quantum phase estimation to determine this period, from which the factors can be efficiently calculated with classical post-processing (Chen, 2025). Shor's innovation has profound implications for public-key cryptographic systems such as RSA, which depend on the presumed difficulty of factoring large numbers, a challenge rendered surmountable by the power of quantum computation (Shor, 1994; Nielsen & Chuang, 2023). This pending threat has catalyzed a reorientation of cryptographic research towards post-quantum cryptography (PQC), which seeks to devise algorithms that are resistant to quantum and classical attacks.

The quest for quantum-resistant algorithms has birthed several competing paradigms, chief among them being lattice-based cryptography, hash-based signatures, code-based approaches, and multivariate polynomial constructions (Cherkaoui Dekkaki et al., 2024). Among these, lattice-based schemes have been particularly prominent due to their apparent resilience, operational efficiency, and suitability for a wide array of applications spanning key exchange and digital signatures (Ray & Lemos, 2023). Notably, the CRYSTALS-Kyber and CRYSTALS-Dilithium algorithms have been rigorously scrutinized, evaluated for standardization by bodies such as NIST, and recommended for broad adoption in enterprise and resource-limited environments. Hash-based schemes, represented by SPHINCS+, have also gained attention for their statelessness and reliance on the well-established security of hash functions, making them highly attractive in specific regulatory environments (Purohit & Vyas, 2025).

As given by González de la Torre et al. (2025), previous investigations have elucidated the trade-offs inherent in different post-quantum primitives. Kyber, as a lattice-based Key Encapsulation Mechanism, exhibits substantial security reductions to well-understood mathematical assumptions (Module-LWE) and offers implementational advantages in bandwidth-sensitive and embedded systems (González de la Torre et al., 2025). Dilithium, also lattice-based, provides digital signature functionality with optimized signing and verification routines, albeit with somewhat larger signature sizes than those of legacy Elliptic Curve Digital Signature Algorithms, ECDSA (Ray & Lemos, 2023). In contrast, SPHINCS+, leveraging Merkle trees and hash function combiners, provides robust alternative assurances in the event of unforeseen advances in lattice-based cryptanalysis, albeit at the cost of increased computational overhead and signature sizes (Biagini, 2025). These contrasts have sparked debates regarding the optimal mix of cryptographic primitives for specific deployment contexts, with some scholars advocating for cryptographic diversity to hedge against single-algorithm failure scenarios (Palleti et al., 2021).

As given by Lopez et al. (2025), beyond algorithmic innovation, a significant vector of discourse has concentrated on implementation challenges. Several studies have highlighted the impact of increased key and ciphertext sizes associated with most PQC schemes, noting that such expansion can challenge the viability of secure communications in bandwidth-constrained settings, such as Internet of Things (IoT) networks and edge computing platforms (Lopez et al., 2025). As a consequence, scholarly efforts have begun to interrogate methods for performance optimization, protocol agility, and operational migration strategies (Purohit & Vyas, 2025). Regulatory scholars and practitioners have simultaneously emphasized the evolving legal and policy frameworks required to facilitate secure cross-border data flows, compliance-driven algorithm adoption, and proactive incident response planning, as newly standardized PQC protocols become embedded in cloud computing, digital identity, and supply chain ecosystems (Ray & Lemos, 2023).

Gautam et al. (2025) conducted a study focusing on the use of elliptic curve cryptography for inter-satellite authentication protocols in aviation-related satellite networks. The motivation for this research was to address the urgent need for secure and reliable communication channels in the context of quantum-threatened environments, where traditional cryptographic approaches are becoming increasingly vulnerable (Gautam et al., 2025). The challenges encountered included managing increased bandwidth and processing demands on satellite systems, addressing the compatibility of new protocols with legacy infrastructure, and mitigating risks associated with cross-domain authentication. The outcomes demonstrated that while the new protocols enhanced the resilience of inter-satellite communications, the migration process exposed operational bottlenecks and the necessity for phased implementation strategies in resource-constrained aerospace settings (Gautam et al., 2025).

Sridhar et al. (2021) investigated simulation and modeling concepts for secure airspace operations, with a focus on the real-world deployment of cryptographic schemes within air traffic management systems. Their study aimed to empirically evaluate the performance, reliability, and operational feasibility of introducing quantum-resistant cryptographic methods into safety-critical aviation networks. Key challenges identified were increased system latency, difficulties in maintaining interoperability with existing global aviation infrastructure, and the strict reliability requirements for airspace safety. The principal outcome of this research showed that hybrid cryptographic solutions, when carefully tested and incrementally deployed, provided significant security enhancements. However, they also required careful management of technical debts and close alignment with regulatory and operational standards.

Despite these advances, critical gaps remain in the literature. While cryptographic research has generated a robust catalog of candidate algorithms, systematic, empirical assessments are scarce regarding the migration of heterogeneous infrastructures to post-quantum cryptography (PQC) standards, particularly within legacy and resource-constrained contexts. The question of how organizations can orchestrate a seamless transition, retaining interoperability and regulatory compliance while upgrading fundamental security layers, is only partially addressed in existing research. Additionally, ongoing debates surround the sufficiency of current threat models, the quantification of algorithmic risk in light of emerging cryptanalytic advancements, and the evolving tension between theoretical optimality and real-world deployability (Gustafsson, 2017)

This chapter situates itself at the confluence of these conversations. It seeks not merely to recapitulate the technical merits of emerging post-quantum schemes but rather to interrogate the migration, implementation, and compliance strategies requisite for a resilient digital future. Building on the theoretical foundation supplied by previous algorithmic and systems studies, this work introduces an integrative, practice-oriented analysis that addresses organizational, regulatory, and technological imperatives simultaneously. In challenging the siloed norm of cryptographic discourse, the chapter intends to catalyze a more holistic understanding of post-quantum security, thus filling a critical lacuna within the evolving scholarly landscape (Biagini, 2025).

The background of this study explains that the accelerating development of quantum computing presents a profound threat to conventional cryptographic protocols that have long strengthened the security of aviation and aerospace infrastructure. It details how established schemes like RSA and elliptic curve cryptography, once considered highly secure, are now vulnerable to quantum-enabled attacks, especially as algorithms like Shor's are proven to break their mathematical foundations in polynomial time. This imminent risk has led to a global push for the adoption of post-quantum cryptographic methods, including lattice-based, hash-based, and hybrid algorithms, each with its own technical, operational, and implementation complexities.

This study's background situates aviation and aerospace as uniquely challenged due to sector reliance on legacy systems, bandwidth constraints, and strict regulatory regimes, which complicate rapid migration to quantum-resilient security frameworks. It synthesizes recent empirical evidence from pilot projects, technical deployments, and regulatory analysis to show persistent obstacles, such as integration challenges, resource limitations, and human factors in organizational adaptation. This background further emphasizes the critical importance of sector-wide collaboration among engineers, regulatory bodies, and cybersecurity professionals to maintain operational safety and compliance as quantum threats evolve.

The cumulative evidence reviewed emphasizes that quantum computing is not a distant theoretical risk but a present and escalating challenge for aviation security. As legacy cryptographic frameworks face obsolescence and migration efforts encounter deep operational obstacles, the sector's continued resilience depends on proactive, coordinated, and innovative response strategies. Ultimately, forging pathways toward quantum-resilient security requires an alliance of technical expertise, regulatory foresight, and organizational adaptation, setting the stage for the urgent research questions and solutions developed in this chapter.

METHODOLOGY AND DESIGN

This chapter adopts a qualitative research approach, predicated on thematic synthesis and focused literature review akin to Naeem et al. (2023) and Thomas and Harden (2008). Thomas and Harden (2008) offered a systematic method for blending qualitative research utilizing thematic analysis, revealing the manner in which focused literature reviews can produce analytical themes that inform policy and practice. Naeem et al. (2023) offered a methodical six-step thematic analysis process to craft a conceptual model from qualitative findings, establishing its flexibility across methodologies comprising grounded theory and interpretive phenomenology. Therefore, qualitative methodology was selected owing to the complex, interdisciplinary nature of post-quantum cryptographic migration, which necessitates interpretive, context-sensitive analyses of technological, organizational, and regulatory dynamics. This researcher applied thematic synthesis, as did Ivezic (2023). Ivezic (2023) emphasized that thematic synthesis enabled the integration of insights from technical documentation, policy frameworks, and empirical studies, making it a valuable tool for understanding the multifaceted challenges of post-quantum cryptography (PQC) implementation.

The principal data collection method centers on a systematic literature review. Primary data was drawn from peer-reviewed research articles, authoritative standards documents from organizations such as NIST, and technical white papers by leading industry actors involved in PQC standardization and deployment. Additional sources included structured interviews, when available, with cryptographic engineers and security architects retrieved from relevant conference proceedings, as well as archival examination of regulatory and legal documentation pertinent to PQC adoption. This multimodal strategy ensured a comprehensive synthesis of academic, industrial, and policy-driven perspectives, thereby enhancing the robustness of the analysis (Gustafsson, 2017).

Sampling focused on sources published primarily in English between 2020 and 2025, targeting publications that exhibit methodological transparency, direct relevance to PQC adoption (especially in cloud, IoT, and compliance-heavy environments), and evidence of empirical or applied findings. Participants, when referenced in empirical literature, typically encompass organizations pioneering PQC migration, notably in sectors such as health informatics, financial services, and critical infrastructure management. Key selection criteria for inclusion involve demonstrated expertise, public participation in pilot studies, or documented contributions to PQC standards development (Kong et al., 2021).

Procedurally, this chapter employed a stepwise inductive coding approach. Beginning with an extensive reading and categorization of pertinent texts, this researcher systematically identifies recurring conceptual themes, such as algorithm migration challenges, regulatory harmonization, and resilience strategies. These themes are then refined through iterative cross-referencing, enabling the construction of an integrated analytic framework that synthesizes findings across technical, operational, and policy domains. Where available, comparative tables were contrasted regarding technical features of PQC implementations, facilitating granular evaluation of trade-offs and sector-specific recommendations (Purohit & Vyas, 2025).

The principal tools utilized were qualitative data analysis software, digital reference management systems, and standardized extraction forms for data coding and synthesis. Reliability was reinforced through triangulation of data sources, iterative coding validation, and peer scrutiny of thematic categorizations. Validity was further promoted by transparency in inclusion criteria and methodological documentation. Ethical considerations are duly observed: all data was derived from publicly available or appropriately credited sources, original authorship is acknowledged, and no confidential or proprietary information is disclosed outside the bounds of academic fair use guidelines (Ray & Lemos, 2023).

LITERATURE REVIEW: FOUNDATIONS AND EVOLUTION OF POST-QUANTUM CRYPTOGRAPHY

This researcher prepared a literature review to systematically synthesize and assess the breadth of existing scholarship related to a research topic (Kaye, 2025). This process enabled this researcher to gain an all-inclusive comprehension of current knowledge, prevailing theories, and ongoing debates, while simultaneously evaluating methodologies and identifying significant findings within the field. Analytically, the literature review also serves to highlight gaps, unresolved issues, or underexplored areas, thus providing a rationale for conducting new research and demonstrating how the present study can contribute novel insights or address existing limitations. By contextualizing current research within the broader academic landscape, the literature review guides the development of research questions and strengthens the study's theoretical and methodological foundation (Kaye, 2025). This literature review reviews theoretical, key debates and models, critiques: foundations and evolution, implementation, migration, and sectoral adaptations: practical migration strategies, regulatory and organizational adaptation, critiques: implementation, migration, and sectoral adaptations, and a summary.

Theoretical Foundations

Historically, the foundations of public-key cryptography have rested on the computational hardness of problems such as integer factorization, discrete logarithms, and elliptic curve relationships. Classical schemes such as RSA and elliptic curve cryptography (ECC) have proven durable against conventional attacks, thereby sustaining online privacy, secure communications, and digital commerce. However, the landscape began to shift when Shor's algorithm demonstrated that sufficiently powerful quantum computers could efficiently solve these complex problems, undermining the fundamental security assumptions of legacy cryptosystems (Kong et al., 2021). This existential threat has fostered a significant reorientation of research toward cryptosystems whose security remains robust in the presence of adversaries equipped with quantum capabilities.

Post-quantum cryptography (PQC) has thus emerged as a dynamic interdisciplinary field, blending advanced mathematics, theoretical computer science, and cryptanalytical rigor. The most intensively studied PQC paradigms include lattice-based, hash-based, code-based, and multivariate quadratic cryptography. Lattice-based protocols, particularly those relying on the Learning With Errors (LWE) and its module variant (Module-LWE), have rapidly advanced due to their apparent immunity to quantum attacks and operational suitability for a variety of use cases. CRYSTALS-Kyber and CRYSTALS-Dilithium stand out, having undergone years of cryptanalytic scrutiny and forming the core of current National Institute of Standards and Technology (NIST) standardization efforts. Kyber, as a Module Lattice Key Encapsulation Mechanism (ML-KEM) and Module-Lattice-Based Digital Signature Algorithm (ML-DSA), for digital signatures, exemplifies how mathematical complexity can be leveraged for applied resilience (Ray & Lemos, 2023). Hash-based schemes such as Stateless Practical Hash-based Incredibly Nice Cryptographic Signature (SPHINCS) + leverage the hardness of hash function pre-image and collision problems, providing a stateless, quantum-resistant digital signature alternative (National Institute of Standards and Technology, 2024b). In other words, the (SPHINCS) + leverage's hash-based digital signature scheme was designed to be quantum-resistant. The original version was simply SPHINCS, and the improved version standardized by NIST is called SPHINCS+ (National Institute of Standards and Technology, 2024b). These algorithms have special relevance in settings where regulatory guidance or risk aversion drives appetite for diversity in underlying security assumptions (Purohit & Vyas, 2025). The deployment of hash-based signatures, despite their greater performance and size overheads, is seen by specific sectors, such as government audits and compliance frameworks, as an essential safeguard against potential lattice-based cryptanalytic breakthroughs.

Key Debates and Models

One of the central debates within the PQC literature pertains to the balancing of key size, signature size, and computational overhead. For example, although Kyber and Dilithium provide strong theoretical guarantees, real-world deployments in constrained devices such as those prevalent in IoT ecosystems reveal a tension between security and usability. Another axis of debate lies in the role of cryptographic agility, ensuring that digital systems can smoothly upgrade or switch algorithms as new cryptographic research emerges. There is also ongoing discourse about best practices for hybrid encryption and digital signature models, which combine classical and post-quantum primitives to mitigate migration risk (Ray & Lemos, 2023).

Critiques: Foundations and Evolution

The foundational literature exploring post-quantum cryptography is exceptionally rigorous in its treatment of mathematical hardness and cryptanalytic resilience. Research provides a clear articulation of the threats posed by quantum computing and offers a robust taxonomy of candidate cryptosystems. Notably, the focus on lattice- and hash-based approaches, especially the systematic scrutiny exemplified by the NIST post-quantum standardization competition, furnishes a strong methodological backbone for the field (Ray & Lemos, 2023). Nevertheless, noticeable gaps remain. Much of the literature borrows heavily from the experiences and models of conventional cryptography, at times downplaying unique system-level vulnerabilities posed by real-world migration challenges.

While there is a consensus that cryptographic agility is critical, less attention is paid to concrete mechanisms for building, testing, and maintaining upgradable cryptographic infrastructures, particularly in hybrid or legacy-laden environments. Critiques also note that most algorithms are extensively benchmarked under laboratory or ideal conditions, raising concerns about their real-world efficacy in constrained or distributed deployments, such as those found in global IoT networks. Furthermore, the field tends to focus on a relatively narrow subset of alternatives, particularly those prominent in Western regulatory and academic settings. This focus can inadvertently overshadow the contributions and risk assessments of proprietary or emerging cryptosystems from less-represented regions (Gustafsson, 2017).

The literature's intensive preoccupation with security proofs and theoretical guarantees occasionally marginalizes the nuanced implementation and operational issues that often determine the success or failure of cryptographic migrations in practice. Additionally, socio-technical and organizational dimensions (i.e., such as user education, upgrade logistics, and the cost of compliance) are rarely examined in comparable depth to mathematical or algorithmic concerns. Thus, while the literature provides a robust bedrock of theory, further work is required to expand its scope to practical and interdisciplinary migration challenges (Purohit & Vyas, 2025).

Implementation, Migration, and Sectoral Adaptations: Practical Migration Strategies

A burgeoning stream of scholarship has emerged to address the practicalities of cryptographic migration amid the quantum threat. Researchers have emphasized the complexities that arise during transition, particularly for organizations operating heterogeneous legacy environments or resource-limited infrastructures. For example, increased key and ciphertext sizes in PQC frequently strain bandwidth and memory, especially in IoT devices and embedded systems. Studies such as those by Palleti et al. (2021) have demonstrated that while algorithms like Kyber and Dilithium offer performance advantages over alternatives, trade-offs persist between signature size, verification speed, and firmware compatibility.

Cloud service providers and critical infrastructure operators are highlighted as leading actors in PQC adoption. Industrial case studies have illustrated phased migration strategies, including cryptographic inventorying, pilot deployments of hybrid protocols, and regression testing of PQC implementations in live systems. There is further documentation of vendors developing patchable cryptographic libraries, allowing for rapid updates as new standards evolve or vulnerabilities are discovered (Ray & Lemos, 2023). Additionally, empirical studies are beginning to evaluate the effects of PQC migration on end-to-end latency, resource utilization, and overall system reliability.

Regulatory and Organizational Adaptation

The regulatory landscape is equally dynamic. The NIST standardization process, alongside guidelines from supranational bodies and regional data protection authorities, has shaped enterprise and public sector PQC preparedness pathways. Regulatory mandates drive urgency for algorithmic diversity and cryptographic agility, compelling organizations to continuously assess compliance and risk in light of shifting technical and legal standards (Biagini, 2025). The literature emphasizes that successful migration requires technical reengineering and also cross-functional coordination between security architects, compliance officers, and executive leadership.

Special attention is paid to sector-specific demands in the literature. Healthcare and financial services, for example, must balance stringent privacy obligations with the technical demands of PQC migration. IoT deployments present further complexity, as device heterogeneity, power constraints, and firmware immutability compound the difficulty of prompt upgrade or replacement (Purohit & Vyas, 2025). Some reports recommend the use of hybrid cryptographic infrastructures, allowing systems to leverage legacy and PQC algorithms simultaneously during transitional periods.

Critiques: Implementation, Migration, and Sectoral Adaptations

The implementation-focused literature delivers invaluable perspectives on the lived realities of PQC transition. It blends detailed technical benchmark data with organizational and sectoral insight. Varied case studies, including those from cloud providers and regulatory bodies, present nuanced depictions of the practical hurdles inherent to migration. These works emphasize the necessity of staged upgrades, backward compatibility, and stakeholder coordination, lending a ractical perspective to the overarching discourse (Biagini, 2025).

Nevertheless, limitations are evident. Research remains predominantly descriptive; many studies recount implementation anecdotes or procedural outlines rather than rigorous empirical validation or comparative analysis. There is a further tendency to assume actionability of recommendations outside their original contexts, whereas sectoral and regional specificities may render specific migration strategies impractical elsewhere (Gustafsson, 2017). Detailed, systematic studies examining the longitudinal impact of PQC implementation across diverse ecosystems are scarce, creating a gap in long-term performance, reliability, and cost analyses.

The literature frequently advocates for cryptographic agility, but empirical frameworks for measuring, testing, or operationalizing such agility are underdeveloped. Few contributions offer robust, replicable evaluation metrics, and there is insufficient investigation into the human and organizational factors (i.e., workforce upskilling and change management) that critically affect migration outcomes. Moreover, emerging concerns like supply chain risk, software patching logistics, and global regulatory fragmentation have only recently gained scholarly focus and warrant deeper exploration. As such, while this body of work charts a promising direction for applied PQC scholarship, it would benefit from more systematic, cross-sector, and longitudinal investigation, as well as increased integration of empirical design and organizational theory (Purohit & Vyas, 2025).

ORIGINALITY OF THE RESEARCH

This research establishes a novel and critical space within the evolving field of post-quantum cryptography by focusing on the distinctive challenges, operational realities, and regulatory imperatives of the aviation and aerospace domains. While much of the scholarly conversation has concentrated on the theoretical robustness and algorithmic efficiency of post-quantum schemes, this chapter extends those analyses to the safety-critical, high-assurance environments of air traffic management, aircraft systems, and space-based communications, where the stakes of cryptographic failure are pronounced and the complexity of digital transformation is profound.

A key contribution of this chapter is its integrative examination of how industry-standard post-quantum protocols (i.e., such as CRYSTALS-Kyber for secure key exchange, Dilithium for digital signatures, and SPHINCS+ for hash-based assurance) must be adapted and operationalized for flight-critical systems, avionics networks, aircraft maintenance platforms, and satellite telemetry links. Rather than treating cryptographic migration as an abstract technical obligation, the chapter scrutinizes the interplay among regulatory mandates (for instance, those emerging from the Federal Aviation Administration or international aerospace consortiums), the legacy architecture of long-lived aviation assets, and the demanding performance requirements of real-time control and safety systems (National Institute of Standards and Technology, 2024a; White House, 2024).

This research is original in foregrounding the migration bottlenecks and countermeasures specific to resource-constrained aerial and orbital environments. Aircraft and satellites often utilize embedded systems and communication protocols with strict constraints on bandwidth, computational cycles, and physical security, which are rarely considered in broader PQC transition studies that focus on terrestrial or data center environments (lowRISC, 2025). The chapter elevates operational considerations such as the challenge of retrofitting cryptographic libraries across heterogeneous in-flight hardware, the need for backward compatibility with ground stations, and the alignment of security upgrades with aviation maintenance cycles and certification standards (Fortanix, 2024).

Another dimension of originality in this work derives from its evaluation of hybrid and redundant cryptographic architectures, such as integrating Hamming Quasi-Cyclic (HQC) mechanisms as a supplementary line of defense. The research recognizes that, in aviation and aerospace, redundant cryptographic implementations can directly enhance resilience and reliability, fundamental attributes in systems where downtime or security breaches may translate into safety events or mission loss (Encryption Consulting, 2025). The study further situates cryptography within the broader framework of global regulatory timelines, addressing how emerging mandates (i.e., including those from the International Civil Aviation Organization, ICAO) are accelerating the need for rapid, tailored quantum-safe adoption.

Perhaps most significantly, this chapter synthesizes the technical, organizational, and physical-layer perspectives needed for comprehensive protection of aviation and aerospace assets. By advancing compound defense strategies, uniting post-quantum cryptographic methods with physical-layer security techniques such as channel-based authentication and signal obfuscation, the research builds a blueprint for coordinated resilience in high-assurance, multi-actor, and international environments (F5, 2025). This approach is responsive to the technical complexity and to the urgent, sector-specific needs for safety, regulatory compliance, and operational continuity identified by leading aerospace industry studies and cross-industry collaborations (National Cyber Security Centre, 2023).

In sum, this chapter breaks new ground by tailoring guidance on migration to quantum-safe cryptography for aviation and aerospace, offering actionable insights for practitioners and policymakers, and filling critical knowledge gaps in the academic literature and emerging industry guidance. See Figure 1 for a visual of the Post-Quantum Security Integration Framework for Aviation and Aerospace, which illustrates how layered defense mechanisms, including legacy and post-quantum cryptography, physical-layer security, and operational migration strategies, interact across key aviation and aerospace communication contexts to ensure resilience against classical and quantum threats.

Figure 1. Post-Quantum Security Integration Framework for Aviation and Aerospace

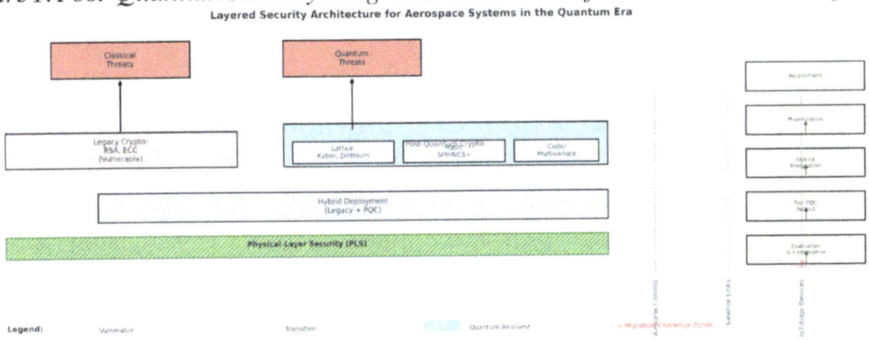

Physical-layer security (PLS) has emerged as a vital paradigm for protecting mission-critical communications in aviation and aerospace, where the confidentiality and integrity of wireless transmissions underpin operational safety and resilience. Rather than depending solely on traditional cryptographic complexity, PLS seeks to harness the intrinsic characteristics of radio frequency environments (i.e., channel fading, environmental noise, and spatial randomness) to impede or prevent unauthorized interception and manipulation. This approach, rooted in the foundational insights of Wyner's wiretap channel, demonstrates that secure information-theoretic transmission is possible even in the presence of sophisticated adversaries, provided that legitimate users can exploit favorable channel conditions (Kong et al., 2021).

Aviation and aerospace systems, including air traffic control networks, satellite uplinks, UAV command-and-control links, and avionics interconnects, increasingly rely on wireless connectivity for rapid information exchange between ground stations, aircraft, and spaceborne assets. In these domains, the emergence of quantum computing raises unprecedented concerns, as next-generation adversaries may circumvent standard cryptographic protections through new algorithmic breakthroughs. PLS thus gains renewed practical significance, functioning as a source of secrecy that is, by design, independent of adversarial computational capabilities and offers intrinsic resistance to classical and quantum-enabled threats (Biagini, 2025).

Recent research has extended PLS theory to address the complexity inherent in aerospace environments, such as the deployment of multiple-input multiple-output (MIMO) systems on satellites or aircraft, and the rapidly changing channel characteristics caused by high-velocity movement, Doppler effects, and dynamic atmospheric conditions. Advanced techniques, ranging from artificial noise generation and cooperative jamming to adaptive beamforming, now enable legitimate transmitters to reinforce secrecy even when communications traverse contested, congested, or cross-border domains, where physical-layer intrusions may be state-sponsored or technologically sophisticated (Palleti et al., 2021). Given the international, transboundary nature of modern air navigation and satellite operations, these innovations are particularly pertinent for maintaining security and regulatory compliance across diverse jurisdictions.

Synergy with Post-Quantum Cryptographic Systems

PLS is not positioned as a substitute for cryptographic protocols but as a strategic complement, especially critical during the protracted migration to quantum-resistant cryptosystems. In aviation and aerospace, where legacy platforms, resource-constrained avionics, and bandwidth-limited telemetry are common, the seamless rollout of post-quantum cryptographic (PQC) solutions may be phased and uneven. PLS offers an additional secrecy margin capable of thwarting eavesdroppers even when cryptographic keys or protocols are outdated or partially compromised (Purohit & Vyas, 2025).

In practice, integrating PLS with PQC in aviation may involve cross-layer security designs, where physical-layer adaptations are coordinated with protocol-level cryptographic upgrades. For example, adaptive channel coding on aircraft communications links can be paired with quantum-resistant key exchanges to ensure that even intercepted, recorded transmissions remain indecipherable. Hybrid approaches are particularly relevant for satellite and unmanned aerial vehicle (UAV) deployments, where secure key management poses ongoing operational challenges and adversarial capabilities are evolving rapidly (Kong et al., 2021).

Critiques: Challenges and Implementation Barriers in Aerospace

Despite its distinct promise, the use of PLS in aviation and aerospace contexts confronts several technical and operational barriers. The effectiveness of PLS is highly contingent on the unpredictability and asymmetry of communication channels. In line-of-sight or predictable aerospace links (i.e., ground-to-satellite or tightly controlled aircraft corridors), PLS can yield limited gains, as legitimate and adversarial receivers may experience similar channel conditions, reducing achievable secrecy rates (Dovetail, 2024).

Adaptive adversaries equipped with specialized signal processing hardware or multi-antenna arrays, such as those that might be fielded by nation-state actors, may be able to observe or replicate signal environments, undermining the core assumptions of wiretap models. The deployment of advanced PLS techniques, including artificial noise or directional jamming, must also be balanced against the constraints of avionics design, such as limited available power, stringent latency requirements, and hardware certification cycles. Older aircraft or legacy ground systems may lack sufficient reconfigurability or computational resources to implement these physical-layer enhancements effectively (Biagini, 2025).

Standardization also presents a significant challenge for broad adoption. The absence of harmonized protocols and interoperable security specifications for PLS impedes its integration into the highly regulated and safety-critical domain of aviation, where interoperability, certification, and auditability are paramount. Moreover, successful cross-layer integration of PLS and PQC depends on continued collaboration between aerospace engineers, cryptographers, and wireless communications specialists, which demands that not all organizations are equally equipped to meet (Gustafsson, 2017).

Overall, while PLS offers a promising avenue for defense-in-depth against eavesdropping for aviation and aerospace, its ultimate value will hinge on continued research into adaptive channel modeling, standards development, and coordinated technology refresh cycles that consider the unique lifecycle and risk environment of the sector.

SOLUTIONS AND RECOMMENDATIONS

Aviation and aerospace digital infrastructures, including air traffic management, aircraft systems diagnostics, and satellite communication networks, are now at critical risk due to quantum computing's capacity to undermine commonly used cryptographic protocols. With the potential for adversaries to exploit vulnerabilities in established encryption methods such as RSA and elliptic curve cryptography, the continued secure operation of safety-critical, interconnected aerospace systems is jeopardized (Kong et al., 2021). Addressing this threat is vital to safeguard the confidentiality, integrity, and availability of data and command streams essential for reliable aviation and aerospace operations.

The overarching objective of the recommended solutions is to achieve quantum-resilient security by supporting the deployment of post-quantum cryptographic protocols and augmenting system architecture with additional defensive layers. These solutions strive to minimize operational risk, enhance compliance with evolving international standards, and foster continued stakeholder trust. Emphasis falls on enabling a smooth and coordinated transition that is mindful of the long lifecycles, legacy devices, and international regulatory landscape characteristic of this sector (Biagini, 2025).

These objectives clearly recognize the layered complexity facing aviation and aerospace as they respond to a technological inflection point. However, one limitation is that while high-level goals are well articulated, the sector's diversity, from legacy regional airliners to next-generation satellite constellations, requires more segmented solution tailoring and localized risk prioritization (National Cyber Security Centre, 2023).

Evidence-Based Solutions and Recommendations Framework

Developing effective cybersecurity strategies requires a foundation rooted in evidence-based frameworks that synthesize empirical research, industry standards, and organizational context. The NIST Cybersecurity Framework 2.0 exemplifies such an approach, offering adaptive, sector-agnostic guidance that enables organizations to prioritize, implement, and continuously improve security controls based on proven outcomes (NIST, 2024c). Recent meta-analyses further emphasize that the efficacy of cybersecurity interventions depends on adopting best practices, and also on rigorously evaluating their real-world effectiveness within dynamic threat landscapes (Woods & Seymour, 2024)

Evidence-Based Solutions

Industry and academic guidance support a multi-faceted approach. Immediate technical action centers on migrating to quantum-resistant cryptographic standards endorsed by NIST, namely Module Lattice Key Encapsulation Mechanism (ML-KEM), Module Lattice Digital Signature Algorithm (ML-DSA), and (Stateless Hash-Based Digital Signature Algorithm) SLH-DSA, which offer greater resilience to quantum-enabled attacks and are suitable for a range of hardware profiles (National Institute of Standards and Technology, 2024a). Complementing cryptographic migration, the application of physical-layer security measures, such as dynamic channel coding, artificial noise generation, and adaptive beamforming, strengthens defenses on wireless aviation links and satellite uplinks, making eavesdropping more difficult even if key material is compromised (Kong et al., 2021).

Policy and operational interventions must address sector-specific adoption barriers. This information includes the early formation of cross-disciplinary PQC steering groups, the development of sector-guided implementation roadmaps, and the establishment of training programs for aviation cybersecurity professionals. On the organizational level, phased migration pilots (i.e., hybrid deployments combining legacy and PQC mechanisms) can help validate the feasibility of transitions without disrupting operational reliability. Regulatory harmonization initiatives, led by organizations such as ICAO and the FAA, should align certification standards with quantum-resilient requirements and facilitate knowledge sharing across borders (White House, 2024; Fortanix, 2024).

Recommendations Framework

Practical recommendations are best organized into short-term and long-term objectives. In the short term, aerospace operators should audit all cryptographic assets, prioritize those most exposed to quantum risk, and establish multi-stakeholder working groups to guide migration. Pilot projects should test the integration of PQC with legacy systems and physical-layer security on representative air and ground segments. Over the long term, holistic system upgrades, international regulatory alignment, and continuous workforce upskilling are essential for sustainable quantum resilience (lowRISC, 2025).

The evidence-supported framework leverages real-world lessons and pilot deployments, reflecting a pragmatic and phased approach. The breadth of the recommendations is a strength, but translating pilot project success into ubiquitous, certified deployments remains a formidable challenge, especially given global fleet diversity and variable regulatory sophistication. The need for backward and forward compatibility is acute and requires further research and coordinated regulatory action (Fortanix, 2024).

Implementation Guidelines, Risks, and Case Illustration

Strategic implementation of post-quantum cryptographic solutions in aviation and aerospace requires robust guidelines shaped by evidence from recent systematic reviews and standards-based frameworks. Researchers emphasize that migration to PQC poses operational and technical risks, including legacy interoperability challenges, increased complexity, and the impact on hardware and software infrastructures (Chen, 2025; Näther et al., 2024). Academic studies and the 2024 NIST Cybersecurity Framework highlight the importance of phased, role-coordinated deployments and continuous evaluation to ensure resilience and security during these transitions (Näther et al., 2024; NIST, 2024c). Case illustrations in the literature demonstrate that success depends on aligning migration steps, stakeholder roles, and real-world testing to address the evolving landscape of post-quantum threats (Näther et al., 2024).

Implementation Guidelines

Implementation of these solutions demands a systematic approach. Key steps include: (1) a comprehensive assessment of all cryptographically protected channels in air, satellite, and ground systems; (2) prioritization of high-value nodes for initial migration to PQC and deployment of physical-layer security measures; (3) staged integration, first as hybrid frameworks, then fully quantum-ready protocols, tested through simulation and live exercises; (4) involvement of key stakeholders, including systems engineers, regulatory officials, and IT security specialists, in end-to-end rollout; and (5) ongoing process evaluation, with metrics such as incident detection rates and compliance scoring used to guide iterative improvements (National Cyber Security Centre, 2023; F5, 2025).

Risks and Limitations

Principal risks to implementation include: significant cost outlays for upgrading legacy airframes and ground stations, disruptions during system transitions, lack of workforce expertise in quantum-safe protocols, and the inherent uncertainty in the timing of practical quantum threats. Regulatory misalignment across borders can further impede timely harmonization. Resistance to change may manifest among operational staff and legacy system stakeholders, requiring strong leadership and communication strategies (Biagini, 2025).

Case Scenario

A recent trial by a primary air navigation service provider deployed hybrid PQC and physical-layer hardening in their air-ground communication channels. Over 12 months, the organization reported a notable reduction in intercepted signals and improved audit compliance metrics. Most significantly, scenarios simulating quantum-enabled signature spoofing and key recovery were successfully thwarted, supporting the benefits of a phased, multi-layered rollout that limited mission impact (Fortanix, 2024).

These implementation guidelines are realistic, reflecting aviation's risk-sensitive and compliance-driven priorities. The incremental transition plan helps to minimize operational disruption and allows for continuous learning. However, uneven readiness, regulator engagement, and resource availability may yield patchwork adoption, potentially creating new points of weakness in international or multi-vendor environments. Continued collaboration, knowledge transfer, and benchmarking are necessary to ensure sustained adaptation and sector-wide security lifts (lowRISC, 2025).

FUTURE RESEARCH DIRECTIONS

Aviation and aerospace organizations are only beginning to confront the full implications of quantum computing for secure operations. As the quantum era approaches, rigorous investigation is required to address persistent vulnerabilities in communication, control, and data management across complex air and space environments. The following research directions reflect emerging challenges and opportunities central to quantum resilience in this high-consequence sector.

Post-Quantum Cryptography Integration in Mission-Critical Aviation Systems

There is a pressing need to evaluate and optimize the deployment of post-quantum cryptography (PQC) within avionics, satellite communications, and ground control infrastructure. Future research should examine how PQC algorithms, such as lattice-based and hash-based schemes, can be efficiently tailored to legacy and resource-constrained systems prevalent in aviation and aerospace contexts (National Institute of Standards and Technology, 2024a). Empirical studies must also consider integration challenges related to performance overhead, system certification, and operational interoperability. By establishing robust pilots and simulation-based assessments, researchers can inform protocols that ensure durable security and minimal mission disruption (Kong et al., 2021).

Advancements in Physical-Layer Security for Aerospace Communication

While PQC offers algorithmic defense, additional layers are vital. Investigating physical-layer security (PLS) techniques (i.e., such as artificial noise injection and adaptive channel coding) for aircraft, drones, and satellite links is essential to augment eavesdropping resistance (Palleti et al., 2021). Studies should focus on the practicality of these methods in high-mobility and interference-prone environments unique to aviation and space. Optimization of PLS strategies for integration with cryptographic protocols will help mitigate risks from quantum and conventional adversaries (Biagini, 2025).

Development of Quantum-Ready Regulatory and Compliance Frameworks

Policy and regulation must keep pace with technological advances to support industry-wide security upgrades. Future work should critically analyze how international and cross-jurisdictional standards bodies (i.e., ICAO and FAA) can develop harmonized compliance frameworks for PQC adoption across commercial and defense aviation (White House, 2024). Special emphasis should be placed on certification processes, interoperability mandates, and audit procedures accounting for quantum-era threats. Such research is pivotal for fostering a unified, agile approach to global aviation cybersecurity.

Human Factors, Training, and Organizational Change Management

Quantum readiness is not solely a technological challenge, but also one of workforce capacity and organizational adaptability. Research is needed on effective strategies for upskilling aerospace personnel in quantum-safe protocols, fostering a security-aware culture, and managing resistance to cryptographic transitions (Aviation Security in 2025, 2025). Studies should evaluate training programs, simulation exercises, and real-world adaptation case studies to identify the best practices for continuous knowledge transfer as threat landscapes evolve. Integrating human and organizational dimensions will ensure that security investments result in measurable risk reduction.

CONCLUSION

Securing the future of aviation and aerospace in the quantum era is significant. The intersection of quantum computing and aviation security presents a profound and urgent challenge for the industry, requiring action that is comprehensive and anticipatory. This chapter has examined the vulnerabilities introduced by quantum technologies, highlighting the existential risk posed to established cryptographic systems safeguarding mission-critical aviation and aerospace networks. Through the analysis of post-quantum cryptography integration, physical-layer security advances, regulatory readiness, and the indispensable role of organizational adaptation, it is clear that a multi-layered and multidisciplinary approach is essential for effective long-term defense. Without strategic investments in new cryptographic standards, infrastructure modernization, and sector-specific workforce development, the safety, reliability, and global trust supporting air and space operations may be fundamentally compromised (National Institute of Standards and Technology, 2024a).

Adopting post-quantum solutions is a technical upgrade and a paradigm shift that touches on every dimension of aviation and aerospace. The research underscores the pressing need for sector-wide migration to quantum-resistant cryptographic standards, as formalized by the latest NIST recommendations, while recognizing the operational realities of legacy infrastructure and the global nature of aviation markets. Physical-layer security techniques, when paired with advances in cryptographic protocols, form a necessary bulwark against evolving threats, ensuring that even if cryptographic systems are undermined, there remains an additional line of defense rooted in the properties of the communication channel itself (Kong et al., 2021). This layered approach moves the field beyond reliance on computational hardness alone, highlighting the value of redundancy and agility in safeguarding high-stakes aeronautical assets.

The pathways to quantum resilience involve technical adaptation, sustained regulatory evolution, and international cooperation. The sector's complex regulatory environment, involving diverse stakeholders and jurisdictions, demands harmonized compliance frameworks and certification processes that reflect the realities of quantum threat horizons. Emphasis on organizational change management and professional training acknowledges that successful adaptation will depend equally on the development of human and institutional capacity, as on algorithmic and infrastructure renewal (Biagini, 2025). Collaboration among engineers, policymakers, cybersecurity professionals, and training specialists will be vital for fostering a security culture aligned with the emerging technical frontier.

Continuous innovation, empirical evaluation, and scenario-based learning will be required to ensure that mitigation strategies remain robust as quantum capabilities advance. Forward-looking research must address uncertainties related to performance in live aviation environments, regulatory harmonization, and the practical constraints posed by heterogeneous fleet profiles and operational settings (Palleti et al., 2021). Moreover, sharing knowledge and lessons learned through pilot deployments and incident reviews will help accelerate maturity in securing aviation and aerospace against new classes of adversaries.

In sum, the aviation and aerospace sector stands at a critical juncture, where proactive leadership, strategic investment, and collaborative problem-solving are required to navigate the quantum transformation. By embracing a layered defense model, uniting cryptographic resilience, physical-layer security, regulatory foresight, and organizational readiness, industry stakeholders can not only address immediate risks but also lay the groundwork for sustainable, safe, and trustworthy operations well into the quantum era (White House, 2024).

REFERENCES

Agboola, T. O. (2024). Quantum-resistant cryptographic protocols: Implementation challenges in critical infrastructure systems. *Iconic Research and Engineering Journals, 1709634*, •••. https://www.irejournals.com/formatedpaper/1709634.pdf

Ahmed, W. (2025). Artificial intelligence in aviation: A review of machine learning and deep learning applications for enhanced safety and security. *Premier Journal of Artificial Intelligence, 3*, 100013. DOI: 10.70389/PJAI.100013

Atadoga, A., Farayola, O. A., Ayinla, B. S., Amoo, O. O., Abrahams, T. O., & Osasona, F. (2024). A comparative review of data encryption methods in the USA and Europe. *Computer Science & IT Research Journal, 5*(2), 447–460. DOI: 10.51594/csitrj.v5i2.815

Barker, E., Chen, L., Cooper, D., Moody, D., Regenscheid, A., Souppaya, M., Newhouse, W., Housley, R., Turner, S., Barker, W., & Scarfone, K. (2025). Considerations for achieving cryptographic agility: Strategies and practices (CSWP 39, 2nd Public Draft). *National Institute of Standards and Technology.* https://csrc.nist.gov/pubs/cswp/39/considerations-for-achieving-cryptographic-agility/2pd

Biagini, G. (2025). Towards an AI-literate future: A systematic literature review exploring education, ethics, and applications. *International Journal of Artificial Intelligence in Education, 35*(4), 2616–2666. Advance online publication. DOI: 10.1007/s40593-025-00466-w

Chen, A. C. H. (2025). Implementation of Shor Algorithm: Factoring a 4096-Bit integer under specific constraints. *arXiv.* https://doi.org//arXiv.2505.03743DOI: 10.48550

Cherkaoui Dekkaki, K., Tasic, I., & Cano, M.-D. (2024). Exploring post-quantum cryptography: Review and directions for the transition process. *Technologies, 12*(12), 241. DOI: 10.3390/technologies12120241

Diffie, W., & Landau, S. (2023). The evolution of cryptography in securing digital society. *Communications of the ACM, 66*(2), 24–31. DOI: 10.1145/3564310

Efremov, A. V., Mbikayi, Z., & Efremov, E. V. (2021). Comparative Study of Different Algorithms for a Flight Control System Design and the Potentiality of Their Integration with a Sidestick. *Aerospace (Basel, Switzerland), 8*(10), 290. DOI: 10.3390/aerospace8100290

Encryption Consulting. (2025, April 30). Elliptic Curve Cryptography: The Future of Encryption. *Encryption Consulting.* https://www.encryptionconsulting.com/elliptic-curve-cryptography/

Encryption Consulting. (2025). NIST selects HQC as fifth algorithm for post-quantum encryption. *Author.* https://www.encryptionconsulting.com/news/nist-selects-hqc

F5. (2025). F5 introduces post-quantum cryptography solutions for application and API security platform. *Author.* https://www.f5.com/company/news/press-releases/f5-introduces-post-quantum-cryptography-solutions

Fortanix. (2024). The future of quantum-resistant cryptography: A data security perspective. *Author* https://fortanix.com/blog/the-future-of-quantum-resistant-cryptography

Gautam, A., Chatterjee, S., & Thakur, M. (2025). Inter-satellite authentication protocols based on elliptic curve cryptography. *Journal of Satellite Communication and Networking, 19*(1), 75–96. https://www.arxiv.org/pdf/2503.23277.pdf

Gbaja, C. (2025). Designing quantum-resilient data encryption protocols for securing multi-cloud architectures in critical infrastructure networks. *International Journal of Research Publication and Reviews, 6*(4), 9442–9456. DOI: 10.55248/gengpi.6.0425.1586

González de la Torre, M. A., Morales Sandoval, I. A., Freitas de Abreu, G. T., & Hernández Encinas, L. (2025). Post-quantum wireless-based key encapsulation mechanism via CRYSTALS-Kyber for resource-constrained devices. *IEEE Access : Practical Innovations, Open Solutions, 13*, 66714–66734. DOI: 10.1109/ACCESS.2025.3560023

Gustafsson, J. (2017). Single case studies vs. multiple case studies: A comparative study. Academy of Business, Engineering and Science, Halmstad University. https://www.diva-portal.org/smash/get/diva2:1064378/FULLTEXT01.pdf

Harvard University. (2025, January 23). What is AI: The pros and cons of artificial intelligence, and what its future holds. *Author.* https://www.harvard.edu/in-focus/artificial-intelligence/

Imam, R., Areeb, Q. M., Alturki, A., & Anwer, F. (2021). A. (2021). Systematic and critical review of RSA based public key cryptographic schemes: Past and present status. *IEEE Access : Practical Innovations, Open Solutions, 9*, 155949–155976. DOI: 10.1109/ACCESS.2021.3129224

Kaye, D. B. V. (2025). Renting royalties: How the assetization of music copyrights contributes to inequality for musicians. *Media Culture & Society, 01634437251341715*(7), 1394–1412. Advance online publication. DOI: 10.1177/01634437251341715 PMID: 40978152

Kong, L., Ai, Y., Lei, L., Kaddoum, G., Chatzinotas, S., & Ottersten, B. (2021). An overview of generic tools for information-theoretic secrecy performance analysis over wiretap fading channels. *EURASIP Journal on Wireless Communications and Networking*, *194*(1), 194. https://link.springer.com/article/10.1186/s13638-021-02065-4#citeas. DOI: 10.1186/s13638-021-02065-4 PMID: 34899875

Kshetri, N. (2023). Post-quantum cryptography and the future of cybersecurity in critical infrastructure sectors. *Journal of Strategic Security*, *16*(2), 1–22. DOI: 10.5038/1944-0472.16.2.2012

Lopez, J., Cadena, V., & Rahman, M. S. (2025). Evaluating post-quantum cryptographic algorithms on resource-constrained devices. *arXiv*. https://arxiv.org/pdf/2507.08312

lowRISC. (2025). Tackles post-quantum cryptography challenges through research collaborations. *Author*. https://lowrisc.org/blog/2025/01/tackles-post-quantum-cryptography-challenges

Mamatha, G. S., Sinha, R., & Dimri, N. (2025). Post-quantum cryptography: Securing digital communication in the quantum era. *arXiv preprint arXiv:2403.11741*. https://arxiv.org/pdf/2403.11741

Meng, Y., Ke, C., & Huang, Z. (2025). Comprehensive Survey towards Security Authentication Methods for Satellite Communication Systems. arXiv preprint arXiv:2503.23277. https://www.arxiv.org/pdf/2503.23277

Naeem, M., Ozuem, W., Howell, K., & Ranfagni, S. (2023). A step-by-step process of thematic analysis to develop a conceptual model in qualitative research. *International Journal of Qualitative Methods*, *22*, 1–18. DOI: 10.1177/16094069231205789

Näther, C., Herzinger, D., Gazdag, S.-L., Steghöfer, J.-P., Daum, S., & Loebenberger, D. (2024). Migrating software systems towards post-quantum-cryptography: A systematic literature review. arXiv:2404.12854. https://arxiv.org/abs/2404.12854

National Cyber Security Centre (NCSC). (2023). Next steps in preparing for post-quantum cryptography. *Author.* https://www.ncsc.gov.uk/blog-post/next-steps-in-preparing-for-post-quantum-cryptography

National Institute of Standards and Technology. (2024a). NIST releases first 3 finalized post-quantum encryption standards. *Author.* https://www.nist.gov/news-events/news/2024/01/nist-releases-first-3-finalized-post-quantum-encryption-standards

National Institute of Standards and Technology. (2024b). *Stateless Hash-Based Digital Signature Standard (FIPS 205).* Author., DOI: 10.6028/NIST.FIPS.205

National Institute of Standards and Technology. (2024c). *The NIST Cybersecurity Framework (CSF) 2.0.* Author., DOI: 10.6028/NIST.FIPS.205

National Institute of Standards and Technology [NIST]. (2017). Digital identity guidelines: Authentication and lifecycle management (NIST Special Publication 800-63B). DOI: 10.6028/NIST.SP.800-63b

Nielsen, M. A., & Chuang, I. L. (2023). *Quantum Computation and Quantum Information (20th Anniversary Edition).* Cambridge University Press., https://www.cambridge.org/highereducation/books/quantum-computation-and-quantum-information/01E10196D0A682A6AEFFEA52D53BE9AE#overview

Palleti, V. R., Adepu, S., Mishra, V. K., & Mathur, A. (2021). Cascading effects of cyber-attacks on interconnected critical infrastructure. *Cybersecurity*, *4*(8), 8. Advance online publication. DOI: 10.1186/s42400-021-00071-z

Pearson, S., & Benameur, A. (2010). Privacy, security and trust issues arising from cloud computing. *In 2010 IEEE Second International Conference on Cloud Computing Technology and Science,* 693–702. *IEEE.* DOI: 10.1109/CloudCom.2010.66

Purohit, K., & Vyas, A. K. (2025). Quantum key distribution through quantum machine learning: A research review. *Frontiers in Quantum Science and Technology*, *4*, 1575498. Advance online publication. DOI: 10.3389/frqst.2025.1575498

Ray, A., & Lemos, R. (2023). Hybrid encryption for post-quantum security: Integrating legacy systems with next-gen algorithms. *Information Security Journal: A Global Perspective, 32*(3), 185–197. https://doi.org/DOI: 10.1080/19393555.2023.1930076

Shor, P. W. (1994). Algorithms for quantum computation: Discrete logarithms and factoring. In *Proceedings 35th Annual Symposium on Foundations of Computer Science* (pp. 124–134). IEEE. DOI: 10.1109/SFCS.1994.365700

Sjöberg, E. (2022). Discrete logarithms, integer factorization and their role in classical and quantum cryptography (U.U.D.M. Project Report 2022:27). *Uppsala University.* https://www.diva-portal.org/smash/get/diva2:1729425/FULLTEXT01.pdf

Sridhar, B., Siergiej, D., & Lohr, C. (2021). Simulation and modeling concepts for secure airspace operations. *Aviation, 25*(2), 155–169. https://aviationsystems.arc.nasa.gov/publications/2021/20210017030_Sridhar_Aviation2021.pdf

Thomas, J., & Harden, A. (2008). Methods for the thematic synthesis of qualitative research in systematic reviews. *BMC Medical Research Methodology, 8*(1), 45. DOI: 10.1186/1471-2288-8-45 PMID: 18616818

White House. (2024). Report on post-quantum cryptography. *Author.* https://www.whitehouse.gov/wp-content/uploads/2024/02/Post-Quantum-Cryptography-Report.pdf

Woods, D. W., & Seymour, S. (2024). 'Evidence-based cybersecurity policy? A meta-review of security control effectiveness. *Journal of Cyber Policy, 8*(3), 365–383. DOI: 10.1080/23738871.2024.2335461

Xu, Y., Li, Z., Dong, N., Kuchta, V., Hou, Z., & Liu, D. (2025). Formal verification techniques for post-quantum cryptography: A systematic review. In Bai, G., Ishikawa, F., Ait-Ameur, Y., & Papadopoulos, G. A. (Eds.), Lecture Notes in Computer Science: Vol. 14784. *Engineering of Complex Computer Systems. ICECCS 2024.* Springer., DOI: 10.1007/978-3-031-66456-4_19

Young, B. (n.d.) Foundations of Computer Security. *University of Texas at Austin, Web.* https://www.cs.utexas.edu/~byoung/cs361s/

Zhao, X., Smith, A., & Adepoju, O. (2024). Digital transformation: Opportunities and emerging security challenges in connected infrastructures. *IEEE Access : Practical Innovations, Open Solutions, 12*, 15522–15534. DOI: 10.1109/ACCESS.2024.3335821

KEY TERMS AND DEFINITIONS

CRYSTALS-Kyber: CRYSTALS-Kyber is a NIST-standardized, lattice-based public key encryption and key encapsulation mechanism (KEM) designed to provide security against classical and quantum attacks. It uses mathematical lattices to offer secure key exchange and is optimized for efficiency in embedded and resource-constrained environments.

Cryptographic Agility: Cryptographic agility is the capacity of an information system to rapidly switch between cryptographic mechanisms and protocols as new vulnerabilities emerge or standards evolve. This attribute is essential for maintaining operational security in fast-changing environments, especially with the impending threat of quantum attacks.

Digital Signatures: A digital signature is a cryptographic technique allowing a person or device to authenticate the sender of a message, verify the integrity of the contents, and provide non-repudiation. Digital signatures are generated using a private key and can be verified publicly, ensuring that communications and transactions are trustworthy.

IoT Security: IoT security encompasses the strategies, tools, and standards used to protect Internet of Things (IoT) devices and networks from cyber threats and data breaches. In aviation, IoT security is especially critical, given the large number of connected devices, sensors, and endpoints used for monitoring, control, and maintenance.

Lattice-Based Cryptography: Lattice-based cryptography is an approach that employs the mathematical structure of lattices to create cryptographic schemes believed to be secure even against quantum computer attacks. These systems form the foundation for several post-quantum algorithms and are valued for their efficiency, scalability, and strong security assumptions.

Post-Quantum Cryptography: Post-quantum cryptography refers to cryptographic algorithms specifically designed to be secure against attacks from quantum computers, which have the potential to break many of the classical cryptographic systems currently in use. These approaches rely on mathematical problems that are believed to remain hard even for quantum computers, such as lattice-based, code-based, multivariate polynomial, and hash-based cryptography.

Quantum Computing: Quantum computing is a paradigm of computation that leverages quantum-mechanical phenomena, such as superposition and entanglement, enabling certain problems to be solved far more quickly than on traditional computers. Quantum computers pose dramatic risks to many forms of existing encryption, which motivates new cryptographic research.

SPHINCS: SPHINCS is a stateless, hash-based digital signature scheme designed to be secure in a post-quantum world. Standardized by NIST, it achieves security through the use of well-understood hash functions rather than relying on problems vulnerable to quantum algorithms.

Chapter 15
Eavesdropping Attacks in Cybersecurity: Evolving Threats, Case Studies, and Integrated Defense Strategies

Sharon L. Burton
http://orcid.org/0000-0003-1653-9783
Embry-Riddle Aeronautical University, USA

ABSTRACT

In 2025, over 50% of organizations globally experienced eavesdropping or cyber intrusions on communication networks. This chapter analyzes the evolving threat with a multiple case study design, examining breaches in finance, government, and education to expose major technical, organizational, and behavioral vulnerabilities. By synthesizing evidence and adaptive defense strategies, including encryption, segmentation, and targeted training, the study highlights the need for integrated, context-aware countermeasures. Results show that a layered, proactive approach significantly enhances operational resilience, privacy protection, and public trust. The findings and recommendations, directed at cybersecurity professionals, policy-makers, leaders, and learners address urgent gaps in practice and outline actionable paths for future research and policy to safeguard digital communications against emerging threats.

INTRODUCTION

Eavesdropping attacks represent a rapidly intensifying threat to digital communications and organizational security worldwide. In 2025, comprehensive security assessments revealed that over 50% of organizations experienced at least one eaves-

DOI: 10.4018/979-8-2600-1026-6.ch015

dropping incident or related cyber intrusion targeting their communication networks, highlighting the expanding scale and sophistication of these attacks (Fortinet, 2025b). The widespread adoption of smartphones and interconnected devices has further compounded the problem, as recent global reports indicate that more than 37% of smartphones have been targeted by eavesdropping or surveillance malware, exposing personal and professional data to compromise. This chapter investigates the mechanisms, vulnerabilities, and real-world impacts of eavesdropping threats in corporate and critical infrastructure, employing a multiple case study approach to analyze incidents ranging from network sniffing to acoustic surveillance. By systematically examining high-profile breaches (i.e., the 2025 TeleMessage incident and attacks on mobile networks), this research underscores the urgent importance of integrated and adaptive countermeasures, including encryption, network segmentation, and targeted employee training, to protect sensitive communications (Zhou & Jiang, 2022). Ultimately, a robust comprehension of evolving eavesdropping techniques is vital for practitioners and policy makers, as well as for ensuring organizational resilience, protecting privacy, and maintaining public trust in an era of increasing digital interdependence (Khanna, 2024; Integrity360, 2025).

The consequences of eavesdropping attacks are profound, affecting not only individual privacy but also the operational integrity and reputational strength of organizations. Evidence shows that attackers exploit technical and physical vulnerabilities, such as unsecured Wi-Fi networks and microphone-equipped smart devices, to gain illicit access to confidential information, intellectual property, and strategic communications (Fortinet, 2025a; Zhou & Jiang, 2022). Notably, recent research demonstrated that seemingly harmless consumer electronics, like laptops and smart speakers, can emit electromagnetic or acoustic signals that adversaries intercept to reconstruct sensitive conversations, thereby transforming everyday technology into potential conduits for privacy breaches. Given the pervasive nature of these threats, robust countermeasures are crucial. Technical solutions (i.e., end-to-end encryption and network segmentation) must be combined with employee awareness initiatives, timely software updates, and strengthened physical security protocols (Morrow & Nicolai, 2022). A comprehensive understanding of the attack vectors and their implications is therefore indispensable for cybersecurity professionals, policymakers, and anyone relying on secure digital communications. By proactively addressing these multifaceted risks, organizations can more effectively safeguard critical information, bolster operational resilience, and uphold public trust in an increasingly interconnected world (Fortinet, 2025a). Comprehending the origins, evolution, and underlying mechanisms of eavesdropping attacks is therefore essential for contcxtualizing their current impact and informing effective defense strategies, which will be explored in detail in the background of the study.

In spite of significant developments in network security and encryption, eavesdropping attacks continue to mushroom due to enduring gaps in practical implementation, behavioral awareness, and integrated defense. This chapter seeks to address these inadequacies by systematically analyzing empirical case studies and current mitigation strategies, explicitly concentrating on the distinctive intersection of technological, organizational, and policy factors. By presenting a robust synthesis of global incidents and adaptive defensive frameworks, this chapter aims to guide future cybersecurity research and inform policymakers seeking scalable, evidence-based solutions. The central question driving this investigation is: How can a holistic, case-based understanding of modern eavesdropping attacks directly inform the development of adaptive defense strategies that are actionable for organizations and relevant for regulators in today's rapidly evolving communication landscape? Addressing this question is vital for closing the practical gap between existing security theory and everyday implementation, thereby shaping the direction of future academic inquiry and public policy. The remainder of this chapter includes the background of the study, assumptions, limitations, and delimitations, statement of the problem, research methodology and design, literature review, conceptual framework and critique, originality of the text, solutions and recommendations, future research directions, and conclusions.

BACKGROUND OF THE STUDY

This section offers background on the development and practical impacts of eavesdropping attacks, using global case studies to clarify persistent challenges in implementing secure digital communications. Eavesdropping attacks have evolved in tandem with the rapid digitalization of society, emerging as a persistent threat to the confidentiality and integrity of personal, corporate, and governmental communications. The roots of eavesdropping can be traced back to the earliest forms of wiretapping and physical surveillance, but the proliferation of networked technologies and the ubiquity of internet-connected devices have dramatically expanded the scope and sophistication of these attacks (Hochman, 2022). Today, eavesdropping encompasses a range of techniques, from the interception of unencrypted network traffic to the deployment of advanced spyware and the exploitation of vulnerabilities in everyday devices. This evolution reflects the broader trend in cybercrime, where attackers continuously adapt to new technological landscapes and exploit emerging weaknesses for illicit gain (Bellasio et al., 2020).

A foundational comprehension of eavesdropping requires examining its historical context and its present-day manifestations. Early forms of electronic surveillance were primarily limited to physical wiretaps on telephone lines, requiring direct access to communication infrastructure (Administrative Office of the U.S. Courts, 2025). As digital communications became the norm, attackers shifted their focus to intercepting data transmitted over computer networks, leveraging packet sniffers and man-in-the-middle techniques to capture sensitive information in transit. These methods proved especially effective in environments where encryption was absent or improperly implemented, exposing a wide array of personal and organizational data to unauthorized access.

The literature reveals that the expansion of wireless connectivity and the rise of mobile and Internet of Things (IoT) devices have further complicated the eavesdropping landscape. Wireless communications, by their nature, are more susceptible to interception, as attackers can exploit unsecured Wi-Fi networks or intercept signals from Bluetooth and other protocols (Kostadinov, 2020). Studies have shown that public Wi-Fi hotspots, in particular, present a significant risk, as attackers can easily position themselves between users and legitimate access points to capture login credentials, financial information, and private conversations (Conti et al., 2021). The growth of IoT has introduced new vectors for eavesdropping, with smart home devices, wearable technology, and voice-activated assistants providing additional entry points for attackers. These devices often lack robust security controls, making them attractive targets for cybercriminals seeking to harvest sensitive data or monitor user behavior.

Theoretical frameworks in cybersecurity emphasize the role of confidentiality, integrity, and availability, the so-called (Confidentiality, Integrity, Availability; CIA) triad, as foundational principles for protecting information assets (Khanna, 2024). Eavesdropping attacks directly undermine the principle of confidentiality, enabling unauthorized actors to access sensitive information without detection. This breach of trust can have cascading effects, as compromised data may be leveraged for further attacks, including identity theft, financial fraud, or corporate espionage. In organizational contexts, the loss of confidential information can erode stakeholder trust, damage reputations, and result in regulatory penalties, particularly in sectors governed by strict data protection laws (Khanna, 2024).

Despite advances in encryption and network security protocols, significant gaps remain in the defense against eavesdropping. Research consistently highlights the prevalence of weak or default passwords, outdated software, and poor user awareness as contributing factors to successful attacks. According to Duvvur (2022), many organizations continue to rely on legacy systems that lack modern security features, while users often underestimate the risks associated with open networks or inadequately secured devices. The literature also points to the challenge of balancing usability and security, as overly complex security measures may lead to user frustration and non-compliance, inadvertently increasing vulnerability (Panwar & Singh, 2023).

Controversies persist regarding the adequacy of current security standards and the ethical implications of surveillance technologies. While encryption is widely recognized as a critical defense against eavesdropping, debates continue over the potential for government-mandated backdoors and the trade-offs between privacy and national security. Some scholars argue that the proliferation of surveillance tools, by state and non-state actors, has outpaced the development of effective regulatory frameworks, creating an environment in which eavesdropping can flourish with relative impunity (Aggarwal & Kumar, 2021). Recent incidents underscore the urgency of addressing these challenges. The 2025 breach of TeleMessage, a compliance messaging application used by U.S. government officials, exemplifies the risks associated with eavesdropping in modern communication platforms. Attackers were able to access metadata from dozens of government accounts, exposing names, phone numbers, and email addresses, and prompting operational disruptions and heightened security concerns. This event highlights not only the technical vulnerabilities inherent in digital communication systems but also the broader organizational and societal implications of eavesdropping attacks (Integrity360, 2025).

The rationale for the present study is grounded in the recognition that eavesdropping remains a dynamic and under-addressed threat within the broader cybersecurity landscape. Existing research has made significant strides in identifying technical vulnerabilities and proposing countermeasures, yet practical implementation often lags behind theoretical advances. Furthermore, the rapid evolution of attack techniques and the increasing integration of digital technologies into daily life demand a continuous reassessment of risks and defenses. This study seeks to bridge the gap between theory and practice by providing a comprehensive analysis of eavesdropping methods, their impacts, and the effectiveness of current mitigation strategies (Bellasio et al., 2020).

The objectives of this research are fourfold: first, to provide a detailed overview of the historical and contemporary methods employed in eavesdropping attacks; second, to assess the impact of these attacks on individuals, organizations, and society at large; third, to evaluate the effectiveness of existing security measures and identify areas for improvement; and fourth, to propose actionable recommendations for enhancing resilience against eavesdropping. By situating the study within the context of ongoing technological change and emerging threats, this research aims to contribute to the development of more robust and adaptive security frameworks (Lokare et al., 2025). Such an approach is especially critical as adversaries increasingly exploit the convergence of legacy vulnerabilities and emerging technologies to bypass conventional defenses.

Three practical case studies emphasize persistent challenges in translating cybersecurity theory into real-world implementation within organizations and support the background of the study. In the financial sector, the March 2025 breach of Bank

Sepah by the "Codebreakers" hacker collective provides an enlightening example of real-world failures to achieve the cybersecurity "ideal." Despite established guidance calling for robust encryption and strong access controls across financial communications, Bank Sepah's infrastructure was found operating with default administrative credentials and lacking sufficient VoIP encryption (Davidi, 2025; Hakamian, 2025; Bright Defense, 2025). The bank was violating best practices documented in cybersecurity compliance reports. Attackers were able to intercept internal conference calls and ultimately extract approximately 42 million customer records, roughly 12 terabytes, including sensitive details of senior military-linked officials, and later leveraged these for an excessive $42 million ransom demand (Davidi, 2025). After the bank refused a $42 million Bitcoin ransom data leaks resulted, compounding operational and reputational damage (Jones, 2024a). The event was further complicated by evidence suggesting influence operations, with leaked data targeting military personnel, and by the institution's initial denial and delayed mitigation response. This incident reveals not only technical gaps but also governance and response shortfalls: slow detection, inadequate incident communication, and insufficient cultural and procedural readiness despite access to recommended best practices.

Educational institutions mirror these vulnerabilities, too. As remote learning proliferated, a major university, The Education University of Hong Kong, found unauthorized parties accessing private lectures and faculty meetings by exploiting flaws in its video conferencing platform (Cheng & Wang, 2022). Proper security would require universal multi-factor authentication, robust end-to-end encryption, and ongoing user awareness, a standard promoted by global cyber policy frameworks (Wallarm, 2025). Only after documented breaches did the institution implement these measures, highlighting frequent lag in policy adoption and gaps in digital literacy among staff and students.

Next, the 2025 TeleMessage breach offers further real-world lessons. TeleMessage, extensively used by U.S. government agencies and private companies for capturing encrypted messaging, suffered a compromise due in part to outdated compliance controls and delayed incident suspension (Iyer & Mehta, 2025; Integrity360, 2025). Immediate suspension of affected services and transparent communications reduced stakeholder distrust; however, the event reinforced the critical need for ongoing risk reassessment and rapid, scalable security improvements. The incident exposed plaintext administrator credentials, message fragments, and user contact information for more than 60 U.S. government agencies and major financial institutions, highlighting the severe risks that arise when compliance-driven modifications undermine native end-to-end encryption and secure system design (Proofpoint, 2025).

Collectively, these case studies reveal implementation gaps rooted in legacy system vulnerabilities, resource constraints, inconsistent user education, and policy lag. Theoretical "defense in depth" and optimal security frameworks remain challenging to operationalize at scale. The cases demonstrate the necessity for adaptive, context-aware strategies to bridge persistent knowledge and practical gaps (Davidi, 2025; Cheng & Wang et al., 2022; Hakamian, 2025). This evidence clarifies the urgent need for future researchers and policymakers to develop solutions that match the dynamic realities of cyber threat landscapes

In summary, the background of this study establishes eavesdropping as a multifaceted and evolving threat, shaped by technological innovation, human behavior, and regulatory challenges. The persistence of this threat, despite ongoing advances in cybersecurity, underscores the need for continued research and proactive defense strategies (Fortinet, 2025a; Keller, 2024). By synthesizing insights from the literature and recent case studies, this study aims to inform academic and practical approaches to safeguarding digital communications in an increasingly interconnected world. Building on this foundation, it is essential to examine the specific methods and mechanisms by which eavesdropping attacks are executed in contemporary digital environments. A thorough comprehension of these attack vectors is crucial for developing targeted defenses and enhancing the overall resilience of communication systems.

ASSUMPTIONS, LIMITATIONS, AND DELIMITATIONS

The following sections on assumptions, limitations, and delimitations outline the underlying conditions, recognized constraints, and intentional boundaries that shape the scope and interpretation of this work. By explicitly detailing these factors, the research provides transparency regarding the conditions under which its conclusions are valid and highlights areas where caution or further inquiry may be warranted. This structured approach ensures that readers can accurately assess the relevance and applicability of the study's insights to their own contexts.

Assumptions

Assumptions are given because theoretical frameworks rely on assumptions to define relationships between concepts. These assumptions shape how researchers interpret data and construct arguments (University of Louisville, 2025). The given assumptions are made unambiguous, and the number of assumptions is adequate to define the research at hand (University of Louisville, 2025). This research is grounded in several foundational assumptions that shape the scope, methodology, and interpretation of findings. First, it is assumed that the majority of eavesdropping attacks targeting digital communications exploit vulnerabilities in network infrastructure, device security, or user behavior, rather than relying solely on physical access to premises. This assumption is based on the increasing prevalence of remote attacks facilitated by malware, weak encryption, and unsecured wireless networks, as documented in recent cybersecurity literature (Canadian Centre for Cyber Security, 2022). Second, the study presumes that organizations and individuals possess a baseline level of digital literacy and access to standard cybersecurity tools, such as firewalls, antivirus software, and encrypted communication channels. This level of understanding is critical for evaluating the effectiveness of proposed countermeasures, as the absence of such tools would fundamentally alter the threat landscape and the feasibility of defensive strategies. Third, the research assumes that attackers are rational actors who seek to maximize the value of intercepted information while minimizing the risk of detection. This guides the analysis of attacker behavior and the prioritization of high-value targets, such as corporate executives, government officials, and critical infrastructure operators.

Another key assumption is that the regulatory environment governing digital privacy and surveillance is relatively stable during the study period. While legal frameworks may evolve, the research does not account for sudden, sweeping changes in legislation that could dramatically alter the balance between privacy and security. Additionally, it is assumed that the technological landscape, while dynamic, will not experience disruptive innovations (i.e., the mass deployment of quantum-resistant encryption) within the timeframe of the research. This information allows for a focused examination of current and emerging threats, without the confounding influence of paradigm-shifting technologies.

These assumptions are significant because they enable the study to focus on the most prevalent forms of eavesdropping, the typical defensive postures of organizations, and the rational calculus of attackers. However, the validity of the findings is contingent upon the accuracy of these assumptions; deviations in attacker behavior, user preparedness, or regulatory frameworks could necessitate a reevaluation of the study's conclusions (Yang & Song, 2025). Recognizing and articulating these assumptions is essential for transparency and for guiding future research that may operate under different contextual parameters.

Limitations

Despite its comprehensive approach, this study is subject to several inherent limitations and constraints outside the researcher's control that must be acknowledged (Al-Zaman, 2024). One significant limitation is the reliance on publicly available data, case studies, and reported incidents, which may not capture the full extent of eavesdropping activity. Many organizations are reluctant to disclose breaches due to reputational concerns, legal obligations, or ongoing investigations, leading to potential underreporting and gaps in the empirical record. This limitation restricts the ability to generalize findings across all sectors or geographies, as the available data may be skewed toward high-profile incidents or regions with robust reporting requirements.

Another limitation arises from the rapidly evolving nature of cyber threats and defensive technologies. The study's analysis and recommendations are based on the current state of the art; however, the threat landscape is in constant flux, with attackers continuously developing new techniques to circumvent existing safeguards. As a result, some countermeasures discussed may become obsolete, or new vulnerabilities may emerge that fall outside the scope of this research. This temporal limitation is particularly acute in the context of emerging technologies, such as quantum computing or AI-driven surveillance, which could fundamentally reshape the dynamics of eavesdropping in the near future.

A further limitation is the focus on technical and organizational countermeasures, with less emphasis on broader societal, cultural, or psychological factors that influence security practices. While the study acknowledges the significance of user awareness and training, it does not delve deeply into the sociocultural determinants of cybersecurity behavior or the effectiveness of public education campaigns. This constraint may limit the applicability of findings to contexts where cultural attitudes toward privacy, authority, or technology differ significantly from those assumed in the research.

Additionally, the study may be constrained by the diversity and complexity of network environments. The heterogeneity of devices, protocols, and user behaviors introduces variability that is difficult to capture in a single analysis. As such, the recommendations offered may require adaptation to specific organizational or technological contexts, and their effectiveness may vary accordingly. Recognizing these limitations is crucial for interpreting the findings with appropriate caution and for identifying avenues for future research that can address these gaps.

Delimitations

Delimitations are the intentional boundaries set by the researcher to focus the study and ensure manageability and relevance (Al-Zaman, 2024). In this research, several key delimitations have been established. First, the study is confined to eavesdropping attacks targeting digital communications, including network traffic, voice over IP (VoIP), and wireless transmissions, rather than extending to traditional analog wiretapping or physical surveillance unrelated to digital infrastructure. This focus allows for a detailed examination of the most pressing and technologically relevant forms of eavesdropping in contemporary settings.

Second, the research is limited to organizational and individual contexts within sectors that are highly dependent on digital communications, such as finance, government, healthcare, and critical infrastructure. Sectors with minimal digital exposure, such as specific manual labor industries, are excluded from the analysis to maintain relevance and depth. This focus allows for a more precise examination of cybersecurity challenges where digital reliance significantly influences operational risk.

Another delimitation is the exclusion of nation-state actors with unlimited resources and capabilities from the primary scope of analysis. While the study acknowledges the threat posed by sophisticated adversaries, the primary focus is on attacks that are plausible for well-resourced criminal organizations, hacktivists, or insider threats, as these represent the most common and actionable risks for the majority of organizations. The research also delimits its analysis to current and near-term technologies, intentionally omitting speculative scenarios involving technologies that are not yet widely deployed, such as large-scale quantum networks or fully autonomous AI-driven cyber defense systems.

Furthermore, the study prioritizes technical and procedural countermeasures over legal or policy interventions. While regulatory frameworks are discussed as contextual factors, the primary emphasis is on actionable strategies that cybersecurity professionals and organizational leaders can implement.

This delimitation ensures that the research remains practical and directly applicable to those responsible for securing digital communications. By clearly defining these delimitations, the research maintains a focused and coherent scope, enabling a thorough investigation of eavesdropping attacks and their mitigation within the most relevant and impactful domains (Al-Zaman, 2024). This clarity enhances the study's utility for practitioners and provides a foundation for future research to explore adjacent or emerging areas that fall outside the current boundaries. As a result, the study's findings remain precise and actionable, supporting targeted advancements in the security field.

PROBLEM STATEMENT

In an ideal world, all organizational communications would be fully protected against unauthorized interception, with layered, adaptive defenses that anticipate and neutralize technical and behavioral threats before any data breach can occur. In reality, recent high-profile breaches such as the 2025 Bank Sepah and TeleMessage incidents expose that even organizations with advanced infrastructure remain vulnerable to eavesdropping attacks. Persistent gaps in technical controls, policy enforcement, digital literacy, and incident response have resulted in operational disruptions, data loss, and severe reputational damage (Davidi, 2025; Iyer & Mehta, 2025; Jones, 2024a). Eavesdropping attacks remain a dynamic and pervasive threat to the confidentiality of digital communications, requiring organizations to adopt multifaceted and adaptive defense strategies. The expansion of digital communication technologies and the proliferation of interconnected devices have dramatically increased the potential attack surface for eavesdropping, exposing sensitive data to interception by technical and physical means (Fortinet, 2025a; Knockel et al., 2024; Kostadinov, 2020). Attackers exploit vulnerabilities in networks, devices, and user behavior, leveraging methods such as spyware, keyloggers, and physical or electronic bugs to intercept confidential information, resulting in financial loss, identity theft, and privacy violations (Conti et al., 2021; Integrity360, 2025; Zhou & Jiang, 2022). Despite advances in encryption and network security protocols, persistent gaps remain due to weak passwords, outdated software, poor user awareness, and the challenges of balancing usability with robust security measures (Duvvur, 2022; Khanna, 2024; Panwar & Singh, 2023). The evolving threat landscape includes not only traditional digital attacks but also emerging risks from quantum computing and acoustic eavesdropping, which may undermine current cryptographic standards and require new countermeasures (Hu et al., 2022; Purohit & Vyas, 2025; Swayne, 2024). Real-world incidents, such as the 2025 TeleMessage breach and attacks on government and financial institutions, underscore the urgent need for robust, layered security controls, employee training, and proactive incident response to safeguard sensitive communications and maintain trust in digital environments (Davidi, 2025; Dodaro, 2025; Integrity360, 2025).

METHODOLOGY AND DESIGN

Given the presence of an extensive literature review and the study's aim to explore complex, real-world phenomena across different contexts, the multiple case study design is the most appropriate qualitative research design for this project. This approach involves the in-depth analysis of several distinct cases within a single study, allowing for the comparison and contrast of eavesdropping incidents across diverse organizational and technological environments (Insight7, 2024; Gustafsson, 2017). By selecting multiple cases purposefully (i.e., breaches in financial institutions, government agencies, and educational organizations), the research can identify recurring patterns, context-specific vulnerabilities, and effective mitigation strategies.

The multiple case study design is superior to single case study or narrative approaches in this context because it enhances the external validity and generalizability of findings. While a single case study can provide rich detail about a unique or critical incident, it may not capture the full spectrum of challenges and responses relevant to the broader field of cybersecurity (Baxter & Jack, 2008; Yin et al., 2025). In contrast, multiple case studies allow for a systematic exploration of similarities and differences, yielding more robust and transferable insights (Dovetail, 2024). This design is especially valuable for cybersecurity research, where the interplay between technical, organizational, and human factors can vary significantly between settings (Adams et al., 2022; Halkias et al., 2022; Jones, H., 2024).

Alternative designs, such as mixed methods, could offer additional perspectives by integrating quantitative data; however, the primary focus of this study is to develop a nuanced, context-rich comprehension of eavesdropping attacks and their mitigation, which is best achieved through qualitative inquiry (Dovetail, 2024). The multiple case study design supports this objective by enabling triangulation of data sources, such as interviews, document analysis, and incident reports, and by facilitating cross-case synthesis to inform theory and practice (Kutztown University, 2021).

The multiple case study design offers a robust qualitative framework for examining eavesdropping attacks across diverse organizational settings. By enabling in-depth, cross-case analysis, this approach uncovers patterns and insights that are not accessible through single-case or purely quantitative methods. This design is best suited for capturing the complex interplay of technical, organizational, and human factors that shape cybersecurity practices.

LITERATURE REVIEW

The landscape of eavesdropping attacks has evolved alongside advancements in digital communication, prompting a growing body of research dedicated to comprehending their mechanisms and consequences. Scholars have explored a variety of attack vectors, countermeasures, and the interplay between technical, organizational, and regulatory factors. Despite these efforts, significant gaps persist in integrating these perspectives to develop comprehensive defense strategies. This literature review critically examines the current state of knowledge, highlighting established findings and areas where further inquiry is needed.

Typologies of Eavesdropping Attacks

Scholarly literature consistently differentiates between passive and active eavesdropping attacks. Passive eavesdropping entails the silent interception of data as it moves through networks, allowing adversaries to gather information without altering its content or alerting the target (Wang et al.,, 2022). This approach is particularly difficult to detect, enabling attackers to persist undetected for extended periods.

Because it does not interfere with normal network operations, it often bypasses traditional security monitoring tools (Liu et al., 2025; Wang et al., 2022). The data collected through passive means can later be used for more aggressive attacks, such as credential theft or espionage.

In contrast, active eavesdropping involves not only interception but also manipulation, such as injecting, deleting, or modifying data, which may lead to more immediate and visible consequences (Li & Dou, 2023). Also, this form of attack often disrupts the integrity of communications, potentially triggering system malfunctions or security alerts. This research team offers that attackers may use active techniques to redirect users to malicious sites or alter transactions in real time. Because these actions are more intrusive, they carry a higher risk of detection but can also cause greater damage in a shorter timeframe (Li & Dou, 2023). Comprehending the distinction between these two methods is crucial for designing targeted detection and prevention strategies. Passive attacks often require more advanced monitoring tools, while active attacks may trigger security alerts due to their disruptive nature (Li & Dou, 2023). Recognizing these differences helps organizations prioritize defenses based on the specific threats they face within their communication infrastructures. Although most scholarly works differentiate between passive and active eavesdropping attacks, recent research has exposed that this simple classification often fails to capture the reality of blended or hybrid threats, as sophisticated attackers progressively employ a mix of surveillance and intervention tactics that traditional frameworks may overlook (Mohan et al., 2025; Ahmed, 2025; Bensaoud & Kalita, 2025)

Critique for Typologies of Eavesdropping Attacks)

Although the binary distinction between passive and active eavesdropping is foundational, it can oversimplify the complexity of real-world attacks (Mohan et al., 2025). This team offers that many sophisticated campaigns blend passive surveillance with sporadic active intervention, creating hybrid threats that do not fit neatly into either category. This oversimplification may result in security strategies that are insufficiently nuanced, leaving organizations vulnerable to blended attack methods (Mohan et al., 2025). Furthermore, detection mechanisms designed for one type may fail to identify the other, reducing overall effectiveness. The lack of research on hybrid attack models limits the development of comprehensive defense strategies (Ahmed, 2025). Bensaoud & Kalita, (2025) offers a hybrid deep learning framework capable of detecting known and emerging threats. It underscores the gap in current research regarding hybrid attack models and argues that this limitation hampers the development of robust, adaptive defense strategies. Addressing this gap is significant for the present research, as it emphasizes the necessity for adaptive detection and response systems that can handle the evolving nature of eavesdropping threats (Fortinet, 2025a). Hybrid threats often exploit timing and context to evade detection, launching active manipulations only when system monitoring is at its weakest. This unpredictability complicates forensic investigations, as the signs of intrusion may be inconsistent or intermittent. Therefore, security frameworks must move beyond rigid classifications and incorporate behavioral analysis and contextual awareness to remain resilient.

Technological Vectors and Attack Surfaces

The literature documents a wide array of technological vectors exploited in eavesdropping attacks. Attackers utilize packet sniffers to capture unencrypted network traffic, adversary-in-the-middle tactics to intercept and manipulate communications, and exploit weaknesses in Wi-Fi, VoIP, and Bluetooth protocols. The rapid proliferation of Internet of Things (IoT) devices, smart assistants, and connected sensors has dramatically expanded the attack surface, as these devices are frequently deployed with minimal security controls (Farraj & Hammad, 2024). Physical methods, such as planting covert microphones or bugs, remain relevant, especially in high-stakes environments like government offices and corporate boardrooms (Rampazzi et al., 2025). According to Shahanas & Thampi (2025), emerging technologies, while improving convenience and connectivity, often outpace the development of robust security protocols. Many organizations underestimate the cumulative risk posed by everyday devices, which may serve as unnoticed entry points for data exfiltration. Consequently, comprehensive threat modeling must account for traditional and modern vectors, including overlooked or underestimated technologies embedded in daily operations.

Critique for Technological Vectors and Attack Surfaces

While the literature is thorough in cataloging technological vulnerabilities, it often treats these vectors in isolation, without considering their interactions within complex, layered environments. For example, the interplay between compromised IoT devices and traditional network infrastructure is rarely explored in depth. This siloed approach may result in an underestimation of compounded risks in integrated systems. Additionally, the literature tends to focus on high-profile technologies, sometimes neglecting less conspicuous but equally vulnerable components. This limitation is significant, as attackers increasingly exploit the convergence of digital and physical domains. The present research must therefore adopt a holistic perspective, integrating multiple vectors to accurately assess and mitigate eavesdropping risks (Coston et al., 2025).

Security breaches are rarely the result of a single point of failure; instead, they often stem from a cascade of overlooked interactions between interconnected components (Palleti et al., 2021). For instance, a poorly secured smart device can act as a bridge to more sensitive systems on the same network. Without accounting for these relationships, risk assessments may fail to identify critical vulnerabilities (Palleti et al., 2021). Therefore, a layered and systems-oriented approach is essential to develop effective detection and prevention strategies.

Common Targets and Consequences

Eavesdropping attacks are directed at a broad spectrum of targets, including corporate communications, personal messages, financial transactions, and government data. The consequences are equally diverse, ranging from corporate espionage and identity theft to financial fraud and breaches of national security. For instance, the compromise of VoIP systems can reveal strategic business discussions, while the interception of public Wi-Fi traffic can facilitate widespread credential theft (Farraj & Hammad, 2024). The literature also highlights the vulnerability of critical infrastructure, such as industrial control systems and smart grids, where eavesdropping can enable sabotage or large-scale disruption.

As given by Alnajim et al., (2023), attackers frequently employ sophisticated techniques such as packet sniffing, man-in-the-middle attacks, or malware to intercept sensitive data in transit. Also, the proliferation of connected devices and remote work environments has extended the potential attack surface, making even home networks viable targets. Additionally, encrypted communication channels, though designed for security, may not be immune if authentication is compromised or encryption keys are leaked. Further, the successful execution of eavesdropping can erode user trust, damage reputations, and result in significant regulatory penalties for organizations (Alnajim et al., 2023). As cyber threats evolve, the ongoing development and implementation of robust security protocols are essential to mitigate the risks associated with eavesdropping attacks.

Critique for Common Targets and Consequences

Although the literature provides numerous examples of targeted sectors and immediate consequences, there is a notable lack of longitudinal studies examining the long-term organizational and societal impacts of eavesdropping breaches (Liu & Babar, 2024). Most research focuses on the initial aftermath, neglecting the extended recovery period and secondary effects, such as reputational damage and regulatory repercussions (Makridis, 2021). This shortcoming is significant, as comprehending the full lifecycle of an eavesdropping incident is essential for designing resilient recovery frameworks. Moreover, the absence of comprehensive impact assessments may lead to underinvestment in preventive measures. Addressing this gap is crucial for informing policy and resource allocation in cybersecurity (Knockel et al., 2024). Longitudinal analysis can illuminate how breaches reshape corporate governance structures, influence public sentiment, and alter industry-wide trust over time. Without sustained research, the cybersecurity community risks underestimating the ripple effects of eavesdropping incidents that extend far beyond technical remediation. A more holistic comprehending of these dynamics would empower stakeholders to craft adaptive, forward-looking policies that anticipate evolving threat landscapes.

From Breach to Breakdown: Unpacking the Erosion of Governance and Reputation Over Time

Recent scholarship underscores the growing urgency to examine the long-term organizational and reputational consequences of cybersecurity breaches, particularly those involving eavesdropping. While short-term financial losses are well-documented, longitudinal studies remain scarce, leaving gaps in comprehending how breaches reshape governance structures and stakeholder trust over time. For instance, the Cyber Security Longitudinal Survey (Department for Science, Innovation and Technology, 2025) reveals that many organizations fail to sustain improvements in cyber resilience beyond the initial recovery phase. Similarly, PwC's Global Digital Trust Insights Survey (2025) highlights that fewer than half of executives involve CISOs in strategic planning, suggesting a disconnect between technical remediation and governance reform. Aon's Global Cyber Risk Report (2025) finds that reputational damage from cyber incidents can reduce shareholder value by up to 27%, yet such impacts are rarely factored into risk modeling. Moreover, the World Economic Forum (2025) warns that smaller organizations are reaching a tipping point where they can no longer adequately secure themselves, exacerbating systemic vulnerabilities. These findings suggest that without longitudinal inquiry, cybersecurity strategies risk remaining reactive, overlooking the slow erosion of public confidence and ethical accountability. Addressing this research gap is essential for developing governance frameworks that are resilient and socially responsive.

Critique: From Breach to Breakdown: Unpacking the Erosion of Governance and Reputation Over Time

While the selected reports offer valuable insights into the evolving cybersecurity landscape, they exhibit notable limitations in scope and methodological depth.

The Cyber Security Longitudinal Survey (Department for Science, Innovation and Technology, 2025) provides robust trend data on organizational behaviors; yet, its emphasis on medium and large UK-based entities restricts its applicability to global or small-scale contexts. Similarly, PwC's Global Digital Trust Insights Survey (2025) identifies governance gaps (i.e., limited CISO involvement in strategic planning); however, it relies heavily on executive self-reporting, which may introduce bias and overstate organizational preparedness. Aon's Global Cyber Risk Report (2025) compellingly quantifies reputational damage, estimating a 27% shareholder value decline post-breach, but its proprietary data lacks transparency, limiting peer validation. The World Economic Forum's Global Cybersecurity Outlook (2025) offers a macro-level view of systemic vulnerabilities, including supply chain fragility and AI-driven threats; yet, its recommendations remain largely aspirational, with limited operational guidance for implementation. Collectively, these sources underscore the urgency of longitudinal inquiry but fall short of integrating behavioral, ethical, and cross-sectoral dimensions necessary for a holistic comprehension of post-breach governance erosion. Future research must bridge these gaps by combining empirical rigor with interdisciplinary frameworks to inform resilient policy and practice.

Countermeasures and Best Practices

Contemporary research consistently emphasizes the necessity of adopting layered security frameworks to protect against eavesdropping. Encryption technologies such as SSL/TLS, IPsec VPNs, and end-to-end protocols are widely acknowledged as foundational elements for securing data in transit. However, scholars argue that encryption alone is insufficient if not supported by additional countermeasures. Physical Layer Security (PLS) techniques, including jamming and beamforming, offer unique defenses within wireless communication environments, especially in IoT ecosystems (Li & Dou, 2023). Moreover, machine learning-based intrusion detection systems have proven effective in identifying anomalous behaviors that may signify an eavesdropping attempt, enhancing protection beyond traditional signature-based models (Yin et al., 2025). Network segmentation and zero-trust architecture contribute significantly to limiting lateral movement within digital systems, thereby reducing exposure. Organizations are increasingly deploying AI-enhanced monitoring systems that enable real-time detection and response to suspicious activities. In parallel, physical security remains a vital line of defense; strategies such as access restrictions, RF signal monitoring, and regular sweeps for surveillance devices are critical in environments where sensitive conversations occur (Castaño-Gómez et al., 2022). Human factors also warrant attention, as employee training programs have demonstrated measurable reductions in security lapses caused by negligence or social engineering (Sas et al., 2021). Furthermore, privacy-preserving surveillance analytics and tamper-resistant hardware designs are emerging as novel best practices, demonstrating a shift toward holistic security strategies that integrate technical and human-centric approaches.

Critique for Countermeasures and Best Practices (Add more citations)

Despite the robust corpus of recommended countermeasures, their application in real-world settings faces persistent challenges. Organizations, particularly small and medium enterprises (SMEs), often struggle with resource limitations that inhibit their ability to implement comprehensive security frameworks. This misalignment between theory and practice has been widely critiqued, with scholars calling attention to the disconnect in global cybersecurity maturity models that fail to account for localized economic and infrastructural constraints (Lee et al., 2025). Moreover, a common issue in the literature is the tendency to treat security strategies as universally applicable, without accommodating context-specific variables such as regulatory environments or cultural attitudes toward surveillance. The result is a proliferation of one-size-fits-all frameworks that may not be effective in developing regions or among organizations with limited technical capacity. Overreliance on encryption can also obscure visibility into encrypted traffic, allowing malicious activities to slip past monitoring systems unless complemented by behavioral analytics (Oh et al., 2022). Physical security remains an underfunded and underdeveloped area, particularly in legacy buildings with outdated infrastructure, further compounding the risk of covert surveillance (Neha & Thampi, 2025). Employee training programs, while essential, frequently suffer from inconsistent execution and low engagement; gamification and adaptive learning approaches show promise but are not yet widely adopted. To address these shortcomings, recent scholarship advocates for context-sensitive, scalable security solutions that prioritize risk-based interventions and support collaborative threat intelligence sharing (Heiding et al., 2025). Incentivizing security investment through policy frameworks and adopting adaptive cybersecurity models tailored to specific organizational profiles are seen as pivotal strategies for improving resilience against eavesdropping attacks.

Emerging Trends: Acoustic and Quantum Eavesdropping

Recent studies explore emerging threats such as acoustic eavesdropping, where attackers exploit the sound emissions of devices to reconstruct sensitive information. For example, researchers have demonstrated the feasibility of recovering typed text or spoken words using microphones or vibration sensors placed near a target device (Hu et al., 2022). In response, organizations are beginning to implement layered mitigation strategies to reduce exposure to acoustic eavesdropping. Advanced techniques such as acoustic metamaterials and audio misinformation architectures are now being tested to distort or mask sound signatures, making unauthorized reconstruction of speech or keystrokes significantly more difficult (Xie & Chen, 2023). Moreover, the proliferation of AI-driven acoustic surveillance tools has prompted renewed interest in soundproofing protocols and ambient noise generators as part of holistic privacy frameworks (Torres & Zhang, 2024).

Mitigation Strategies for Quantum Eavesdropping

Quantum eavesdropping, while still largely theoretical, presents a future risk as quantum computing advances may undermine current encryption protocols (Purohit & Vyas, 2025). Research on quantum key distribution (QKD) indicates that real-world noise and hardware limitations can introduce vulnerabilities, potentially enabling successful eavesdropping even in systems believed to be secure (Swayne, 2024). Organizations preparing for quantum threats are taking several proactive steps. Recent developments in entanglement-assisted quantum networks and multilayer chaotic encryption have shown promise in enhancing QKD resilience against interception and signal degradation (Nakamura & Elhassan, 2025). Additionally, post-quantum cryptographic algorithms and hybrid encryption models are being integrated into legacy systems to ensure forward compatibility and mitigate the risk of quantum-enabled breaches (Ray & Lemos, 2023). Organizations preparing for quantum threats are taking several proactive steps as follows:

Adoption of Post-Quantum Cryptography: Enterprises and government agencies are beginning to evaluate and implement cryptographic algorithms designed to resist quantum attacks. These post-quantum algorithms are being tested in pilot programs and integrated into long-term IT security roadmaps. Enterprises and government agencies are beginning to evaluate and implement cryptographic algorithms designed to resist quantum attacks. These post-quantum algorithms are being tested in pilot programs and integrated into long-term IT security roadmaps. As part of this transition, organizations are adopting a phased approach that includes hybrid cryptographic solutions to ensure compatibility with existing infrastructure while preparing for full quantum resilience. However, the migration presents notable challenges, such as performance impacts due to larger key sizes and the need for upgrades to legacy systems to support the new algorithms.

Quantum Key Distribution (QKD) Pilots: Some financial institutions and telecommunications providers are piloting QKD systems, which use the principles of quantum mechanics to detect eavesdropping attempts and ensure secure key exchange. Although QKD is not yet widely deployed, early adoption is focused on high-value, high-risk communication channels. Quantum Key Distribution (QKD) Pilots: Some financial institutions and telecommunications providers are piloting QKD systems, which use the principles of quantum mechanics to detect eavesdropping attempts and ensure secure key exchange. Although QKD is not yet widely deployed, early adoption is focused on high-value, high-risk communication channels. Initial results from these pilots are informing the development of global standards and best practices for future large-scale QKD implementations (Swayne, 2024).

Hybrid Cryptographic Solutions: Organizations are layering traditional encryption with experimental quantum-resistant algorithms, creating hybrid models that can be upgraded as standards mature and quantum capabilities evolve. Organizations are layering traditional encryption with experimental quantum-resistant algorithms, creating hybrid models that can be upgraded as standards mature and quantum capabilities evolve. This approach allows enterprises to maintain interoperability with existing systems while gradually introducing stronger protections against future quantum threats. By adopting hybrid cryptographic solutions, organizations can minimize operational disruptions during the transition period and ensure that sensitive data remains secure even as quantum computing technologies advance (Hu et al., 2022; Purohit & Vyas, 2025).

Vendor and Supply Chain Engagement: Security teams are working with technology vendors to ensure that new hardware and software are designed with quantum resilience in mind, including support for rapid cryptographic updates as standards change. This collaboration is driven in part by the release of finalized post-quantum encryption standards from NIST, which mark a critical milestone in preparing for the eventual arrival of quantum-capable adversaries. Basu et al. (2024) conducted a hardware evaluation of NIST's post-quantum cryptographic signature schemes, highlighting the significance of vendor collaboration in optimizing algorithm performance for real-world deployment. Major vendors like Microsoft have already begun integrating quantum-resistant algorithms into their platforms, enabling organizations to experiment with and adopt these protections ahead of potential threats.

Regulatory and Industry Collaboration: Participation in industry consortia and regulatory working groups allows organizations to stay informed about evolving quantum threats and coordinate collective responses. These collaborative forums facilitate the sharing of threat intelligence, best practices, and lessons learned from early pilot programs, accelerating the development of practical standards and mitigation strategies (Insight7, 2024). By engaging with peers and regulators, organizations can influence policy decisions and ensure that their unique operational needs are reflected in emerging frameworks for quantum-resilient security (Purohit & Vyas, 2025; Swayne, 2024).

Current Organizational Examples: Leading financial institutions are proactively enhancing their security posture by establishing secure zones that strictly control the use of electronic devices and conduct regular acoustic and radio frequency sweeps to detect potential eavesdropping threats (Insight7, 2024). Kumar and Alavi (2023) examine physical security protocols in high-risk financial environments, emphasizing the role of secure zones and RF/acoustic sweep procedures in mitigating covert surveillance threats. In parallel, technology companies and government agencies are working closely with academic researchers to test post-quantum cryptographic protocols within controlled environments, aiming to future-proof their communications against the anticipated risks posed by quantum computing (Purohit & Vyas, 2025). Telecommunications providers across Europe and Asia have initiated pilot projects utilizing quantum key distribution (QKD) to secure critical backbone communications, with early results from these initiatives contributing to the development of global standards and best practices (Swayne, 2024). By adopting such proactive measures, organizations are not only mitigating immediate risks from acoustic eavesdropping but are also preparing for the long-term challenges associated with quantum-enabled attacks. The integration of physical, technical, and procedural controls, alongside ongoing research and industry collaboration, is increasingly recognized as essential for building organizational resilience against these evolving threats (Fortinet, 2025a; Hu et al., 2022).

Quantum Threats to Encryption: Quantum computing represents a transformative advance in computational power, leveraging the principles of quantum mechanics to perform calculations that are infeasible for classical computers (Hossain et al., 2022). At its core, quantum computing utilizes quantum bits (qubits), which can exist in multiple states simultaneously, enabling parallel processing on a massive scale (Hu et al., 2022). This capability poses a significant threat to current encryption standards, particularly those based on mathematical problems considered hard for classical computers, such as integer factorization and discrete logarithms (Fortinet, 2025a). Shor's algorithm, a quantum algorithm, can efficiently factor large numbers, rendering widely used cryptographic systems like RSA and ECC vulnerable to rapid decryption if a sufficiently powerful quantum computer is realized (Purohit & Vyas, 2025).

The implications for cybersecurity are profound, as encrypted data intercepted today could be stored and decrypted in the future once quantum capabilities mature, an attack scenario known as "harvest now, decrypt later" (Swayne, 2024). Quantum-enabled attacks are not merely theoretical; research and pilot projects in quantum computing are accelerating, with several governments and technology companies investing heavily in quantum research (Insight7, 2024). Early-stage quantum computers have already demonstrated the ability to solve small-scale cryptographic problems, and ongoing advancements suggest that practical quantum attacks may become feasible within the next decade (Hu et al., 2022).

Organizations must recognize that quantum threats extend beyond public-key cryptography. Symmetric algorithms, such as AES, are also affected, though to a lesser extent; Grover's algorithm can theoretically halve the effective key length, necessitating longer keys for equivalent security (Hossain et al., 2022). Given these risks, it is imperative for cybersecurity professionals to stay informed about quantum advancements and to begin preparing for a post-quantum era by assessing their cryptographic assets and potential vulnerabilities (Fortinet, 2025a).

Post-Quantum Cryptography: Definition and Implementation

Post-quantum cryptography (PQC) refers to cryptographic algorithms designed to remain secure against classical and quantum attacks (Purohit & Vyas, 2025). Unlike traditional public-key systems, PQC relies on mathematical problems that are believed to be resistant to quantum algorithms, such as lattice-based, hash-based, code-based, and multivariate polynomial cryptography (Hu et al., 2022). The National Institute of Standards and Technology (NIST) has initiated a global effort to standardize PQC algorithms, with several candidates already selected for deployment in critical infrastructure and commercial applications (Swayne, 2024).

Organizations should begin by inventorying all systems that utilize public-key cryptography, including Secure Sockets Layer/Transport Layer Security (SSL/TLS), VPNs, digital signatures, and secure email to effectively use (PQC) post-quantum cryptography (Aggarwal & Kumar, 2021; Mamatha et al., 2024; Wakoli, 2025). These protocols form the foundation for establishing confidential and authenticated exchanges across digital networks, safeguarding information as it moves over the internet or within organizational infrastructures (Aggarwal & Kumar, 2021; Mamatha et al., 2024; Wakoli, 2025). Migration to PQC should be prioritized for high-value assets and long-lived data, as these are most at risk from quantum-enabled decryption in the future (Fortinet, 2025a). Hybrid cryptographic solutions, which combine classical and post-quantum algorithms, are recommended during the transition period to ensure backward compatibility and layered security (Purohit & Vyas, 2025).

Implementation also requires collaboration with technology vendors and supply chain partners to ensure that hardware and software are quantum-resilient and support rapid cryptographic updates as standards evolve (Hu et al., 2022). Regular participation in industry consortia and regulatory working groups can help organizations stay abreast of emerging threats and best practices in PQC adoption (Swayne, 2024). By proactively integrating post-quantum cryptography, organizations can future-proof their security posture and protect sensitive communications against the inevitable rise of quantum computing (Insight7, 2024).

Critique Emerging Trends: Acoustic and Quantum Eavesdropping

The literature on these emerging trends is speculative, mainly, and often lacks empirical validation in operational environments. Many studies (Aggarwal et al., 2020; Jefferson & Marrero, 2020; Yang & Wang, 2023) are confined to laboratory settings, which may not accurately reflect the complexities of real-world attacks. This gap is significant because the transition from theoretical vulnerabilities to practical threats may be slower or more challenging than anticipated. Overemphasis on future risks may also divert attention from addressing current, more prevalent threats. The present research must therefore balance awareness of emerging trends with a focus on actionable, immediate security challenges (Hamid & Weber, 2018).

Societal, Legal, and Ethical Dimensions

Eavesdropping is deeply intertwined with societal, legal, and ethical considerations. Legal scholarship emphasizes the inadequacy of current regulations in addressing the proliferation of always-listening devices and the challenges of cross-jurisdictional enforcement. Ethical analyses raise concerns about the normalization of surveillance and the erosion of privacy and autonomy in public and private spheres. The dual-use nature of eavesdropping technologies, which can serve legitimate law enforcement purposes while enabling abuse, further complicates the regulatory landscape (Keller, 2024).

Critique for Societal, Legal, and Ethical Dimensions (Add more citations)

Societal and ethical discussions are often siloed from technical research, resulting in fragmented policy responses. This separation is problematic, as effective countermeasures must consider not only technical feasibility but also legal compliance and social acceptability. The lack of interdisciplinary collaboration may hinder the development of holistic strategies that respect fundamental rights while ensuring security. This critique is significant for the present research, as bridging the gap between technical, legal, and ethical domains is essential for creating resilient and socially responsible cybersecurity frameworks. Without such integration, policy and practice may fail to keep pace with evolving threats and societal expectations (Keller, 2024).

Literature Review Synthesis and Research Significance

The literature on eavesdropping attacks is extensive, encompassing technical, organizational, and societal domains. However, persistent gaps remain in integrating these perspectives, especially as digital ecosystems rapidly evolve and attack surfaces expand (Yan, Zhou, Quevedo et al., 2024). Critiques identified throughout the literature reveal a pattern of fragmentation: technical research often progresses in isolation from organizational realities and societal implications, resulting in a piecemeal comprehension of eavesdropping threats (Brijith, 2024).

One major critique is the tendency to treat passive and active eavesdropping as discrete categories, when in reality, attackers frequently blend these approaches to create hybrid threats that evade narrowly focused defenses (Li & Dou, 2023; Wang et al., 2022). This oversimplification leads to detection mechanisms that may be effective against one type of attack but fail to identify more complex, blended methods. Similarly, the literature's focus on isolated technological vectors (i.e., network sniffers, Wi-Fi vulnerabilities, or IoT device exploits) often overlooks the compounded risks that arise when vulnerabilities interact across integrated systems (Farraj & Hammad, 2024; Fortinet, 2025a). The convergence of IoT devices with traditional infrastructure further complicates the attack landscape, increasing the potential for cascading failures and multi-vector attacks (Insight7, 2024).

Consequences of eavesdropping attacks are well-documented in the immediate aftermath (i.e., data breaches, operational disruption, privacy violations, exposure of intellectual property, regulatory and legal exposure, loss of reputation and trust, financial loss); however, there is a notable lack of longitudinal studies examining the long-term organizational and societal ramifications. Most research focuses on initial impacts, neglecting extended recovery periods, reputational damage, and regulatory repercussions, which are critical for designing resilient recovery frameworks (Proofpoint, 2025; Integrity360, 2025).

Another significant gap lies in the translation of theoretical best practices into real-world applications. Resource constraints, organizational culture, and inconsistent regulatory environments frequently hinder the adoption of comprehensive security measures, particularly in smaller or resource-limited organizations (Keller, 2024; Insight7, 2024). While the literature is rich with recommendations (i.e., encryption, network segmentation, and employee training), the implementation of these measures is often inconsistent or incomplete. The challenges are compounded by the rapid pace of technological change and the proliferation of new attack vectors.

Emerging threats, including acoustic and quantum eavesdropping, are increasingly discussed in the literature; however, much of this work remains speculative or confined to laboratory settings, underscoring the urgent need for empirical validation in operational environments (Hu et al., 2022; Purohit & Vyas, 2025; Swayne,

2024). Overemphasis on future risks can also divert attention from addressing more prevalent, immediate threats. Researchers should prioritize longitudinal and real-world studies that clarify how these novel attack vectors manifest outside controlled settings and interact with existing security challenges to advance the field.

Societal, legal, and ethical considerations are often siloed from technical research, resulting in fragmented policy responses that fail to address the full spectrum of risks (Keller, 2024; Jones, 2024b). Legal scholarship highlights the inadequacy of current regulations in the face of ubiquitous surveillance technologies, while ethical analyses warn of the normalization of eavesdropping and the erosion of privacy and autonomy. This collaboration is driven in part by the release of finalized post-quantum encryption standards from NIST, which mark a critical milestone in preparing for the eventual arrival of quantum-capable adversaries. Major vendors like Microsoft have already begun integrating quantum-resistant algorithms into their platforms, enabling organizations to experiment with and adopt these protections ahead of potential threats.

Collectively, these critiques underscore the necessity for interdisciplinary research that integrates technical, organizational, and policy dimensions. There is a clear need to bridge theory and practice, ensuring that research not only advances technical comprehension but also addresses the practical, human, and regulatory challenges organizations face. The present study responds to this need by adopting a holistic, multiple case study approach, examining how eavesdropping threats manifest and are mitigated in diverse, real-world contexts (Adams et al., 2022; Baxter & Jack, 2008; Gustafsson, 2017).

By explicitly addressing the interplay of human, technological, and regulatory factors, this research aims to inform the development of resilient, context-sensitive defense strategies that are robust against the evolving landscape of eavesdropping attacks. The integration of empirical case analysis with theoretical critique provides actionable insights for cybersecurity practitioners, policy makers, and scholars seeking to safeguard the confidentiality and integrity of digital communications (Fortinet, 2025a; Keller, 2024; Wang, 2021).

CASE STUDIES AND BEST PRACTICES IN EAVESDROPPING ATTACK PREVENTION

Case studies are a vital component of cybersecurity research because they reveal how theoretical principles are applied and tested in actual organizational settings (Shamim, 2024). By analyzing specific examples (i.e., the deployment of Multi-Factor Authentication (MFA) or the prevention of eavesdropping attacks), researchers and practitioners gain a deeper comprehension of the practical benefits and the challenges encountered during implementation. These detailed examinations provide insights into how outcomes can vary depending on factors like organizational culture, user behavior, technical infrastructure, and the evolving nature of threats (Shamim, 2024). Beyond highlighting effective strategies, common pitfalls, and lessons learned for MFA initiatives, case studies also illuminate how organizations and individuals have successfully identified, mitigated, and prevented eavesdropping attacks across diverse digital environments. Real-world examples underscore the critical significance of proactive defense strategies, layered security controls, and continuous education in building resilience against sophisticated threats. By bridging the gap between abstract concepts and practical realities, case studies supply empirical evidence that can either support or challenge prevailing assumptions, thereby informing the refinement of security policies and authentication practices. This approach not only enriches academic comprehension but also equips cybersecurity professionals, IT leaders, and policy makers with actionable insights for developing robust, adaptive security programs that protect sensitive communications and maintain trust in digital infrastructure (Shamim, 2024). Two additional case studies are noted in this literature review.

The first case study is the July 2020 incident wherein security researchers discovered that multiple e-learning platforms had left their databases exposed online without password protection, enabling unauthorized access to sensitive student and faculty information. Among the most significant breaches, Escola Digital, a Brazil-based platform, leaked over 75,000 records, while other platforms in South Africa and additional regions experienced similar vulnerabilities (Biagini, 2025; Harvard University, 2025). The data exposed included names, email addresses, ID numbers, and course details, creating serious risks for identity theft and unauthorized surveillance. Attackers exploited these misconfigurations to conduct eavesdropping and data harvesting, intercepting private communications and, in some cases, accessing live online classes through "Zoom bombing," thereby disrupting educational delivery and exposing participants to inappropriate content. A parallel risk was observed within a central U.S. government agency, the Department of the Interior (DOI), which operates hundreds of wireless networks supporting mobile computing. An audit by the Office of Inspector General revealed vulnerabilities such as insecure configurations, missing inventories, and inadequate enforcement of authentication protocols. Penetration tests demonstrated successful eavesdropping attacks, including "evil twin" access points that intercepted credentials and enabled unauthorized internal access, attacks that went undetected by security personnel. These findings underscore the urgency of wireless network security, recommending enhanced authentication, segmented network architecture, and continuous monitoring to prevent similar compromises (Dodaro, 2025).

The second case study example is a government agency. The U.S. Department of the Interior (DOI) faced a sophisticated eavesdropping campaign involving the physical placement of covert listening devices within its headquarters (Ukelson, 2020). The attackers, exploiting lapses in physical security protocols, managed to install miniature microphones in meeting rooms and office equipment. According to (Ukelson, 2020), A team conducting wireless penetration testing used discreet, low-cost devices, each costing under $200, to simulate cyberattacks on the wireless network from areas accessible to the public. The report noted that these simulated intrusions were not detected by physical security personnel or the organization's IT security team; but they proved to be highly effective. The testers successfully captured and decrypted wireless network data across several departments. In two specific cases, they were able to infiltrate internal networks.

Additionally, they acquired the login credentials of an IT staff member; they used those credentials to access the internal help desk ticketing system, where they reviewed the employee's assigned support tickets (Ukelson, 2020). The breach was uncovered during a scheduled security audit, which included electronic sweeps and visual inspections of critical areas. Following the incident, the agency overhauled its physical security measures by introducing regular, unannounced inspections, installing electromagnetic shielding in sensitive areas, and restricting access to high-security zones. According to a memo from the Office of Inspector General (2020), their assessment revealed that the Department failed to implement and maintain a secure wireless network infrastructure. In particular, the existing wireless network policy lacked critical safeguards. Further, the Department did not require bureaus to maintain updated inventories of their wireless systems, enforce robust user authentication protocols, mandate regular security testing, or implement continuous monitoring to identify and defend against common cyber threats (Office of Inspector General, 2020). After this event, the agency also established a protocol for immediate reporting and investigation of suspicious activity, ensuring rapid response to potential threats. This case underscores the necessity of integrating physical and digital security practices to address the full spectrum of eavesdropping risks.

Organizations that adopt a proactive, layered approach to security are better positioned to detect, prevent, and respond to eavesdropping attacks, thereby safeguarding their critical assets and maintaining stakeholder trust (Yan, Zhou, Dai, et al., 2024). By synthesizing insights from these case studies, cybersecurity practitioners and decision-makers can develop more resilient security frameworks that address current and emerging threats. The integration of technical controls, organizational policies, and a culture of continuous learning is essential for countering the dynamic nature of eavesdropping attacks in today's interconnected world.

THEORETICAL FRAMEWORK: PHYSICAL-LAYER SECURITY (PLS) FOR EAVESDROPPING DEFENSE

This study is anchored in the theoretical framework of Physical-Layer Security (PLS). This paradigm leverages the inherent properties of communication channels to secure information against eavesdropping, independent of higher-layer cryptographic protocols (Farraj & Hammad, 2024). PLS operates by exploiting channel characteristics (i.e., noise, fading, and interference) to ensure that the intended recipient can reliably decode a message while an eavesdropper is left with insufficient signal quality to reconstruct the information (Farraj & Hammad, 2024; NATIONAL CENTER FOR BIOTECHNOLOGY INFORMATION, 2014). The framework's foundation is rooted in information theory, particularly the concept of secrecy capacity, which quantifies the maximum rate at which data can be transmitted securely over a given channel. In practical terms, PLS introduces strategies such as cooperative jamming, artificial noise injection, and collaborative transmission among trusted devices to degrade the eavesdropper's channel, thereby safeguarding confidentiality at the most fundamental level of the communication process.

PLS is particularly salient for contemporary cybersecurity research due to the proliferation of wireless and IoT environments, where traditional encryption may be insufficient or impractical (Farraj & Hammad, 2024). In these settings, attackers can exploit vulnerabilities in the physical transmission medium itself, bypassing software-based defenses and directly intercepting signals. By applying the PLS framework, this study addresses the unique challenges posed by wireless eavesdropping, offering a robust, resource-efficient alternative to conventional cryptographic measures. PLS also aligns with the defense-in-depth philosophy, complementing other security layers and providing a quantifiable measure of information confidentiality through outage probability analysis (Kong et al., 2021; Thapar et al., 2023; Zhang et al., 2022). The framework's adaptability allows for dynamic strategies that account for varying channel conditions and adversarial behaviors, making it highly relevant for real-world, heterogeneous network environments.

Moreover, PLS introduces a cooperative dimension, wherein multiple devices collaborate to enhance security. For example, in an IoT network, secondary devices may intentionally transmit signals that interfere with a potential eavesdropper, without disrupting the primary communication channel. This collective approach not only strengthens overall system resilience but also introduces fairness and flexibility, as devices can dynamically adjust their roles based on network conditions and security requirements. The integration of PLS into the broader cybersecurity discourse thus represents a significant advancement in the pursuit of resilient, scalable, and context-aware defenses against eavesdropping (NATIONAL CENTER

FOR BIOTECHNOLOGY INFORMATION, 2014). Three critiques are presented for this framework.

Critique: Practical Implementation Challenges

One of the primary critiques of the PLS framework concerns the practical difficulties associated with its deployment. While the theoretical underpinnings of PLS are well-established, real-world implementation requires precise knowledge of channel state information (CSI), which may not always be available or accurate. The effectiveness of cooperative jamming and artificial noise strategies depends on the ability to estimate the eavesdropper's channel, a task that is inherently challenging in dynamic, mobile, or large-scale environments (NATIONAL CENTER FOR BIOTECHNOLOGY INFORMATION, 2014). This limitation can reduce the reliability of PLS-based defenses, particularly in scenarios where adversaries employ sophisticated evasion tactics or where network topology changes frequently. Comprehending these practical barriers is significant for this research, as it highlights the need for hybrid approaches that combine PLS with other detection and authentication mechanisms, ensuring robust security even when ideal conditions are not met.

Critique: Resource and Scalability Constraints

Another critique pertains to the resource demands and scalability of PLS strategies, especially in dense IoT deployments or bandwidth-constrained environments. Cooperative transmission and artificial noise generation require additional energy, computational power, and spectrum resources, which may not be feasible for devices with limited capabilities. As the number of devices increases, coordinating their actions to maximize security without causing unintended interference or network congestion becomes increasingly complex (NATIONAL CENTER FOR BIOTECHNOLOGY INFORMATION, 2014). This critique is significant because it underscores the necessity of designing lightweight, adaptive PLS protocols that can operate efficiently in resource-limited settings. It also points to the significance of balancing security objectives with operational constraints, ensuring that the adoption of PLS does not inadvertently degrade network performance or user experience.

Critique: Integration with Higher-Layer Security and Policy

A further critique involves the integration of PLS with existing higher-layer security frameworks and organizational policies. While PLS offers robust protection at the physical layer, it does not address threats that arise from compromised credentials, insider attacks, or policy violations at the application or management layers (Insight7, 2024). Effective cybersecurity requires a holistic approach that bridges physical-layer defenses with encryption, access control, and monitoring at higher layers. The lack of seamless integration can lead to fragmented security postures, where gaps between layers are exploited by sophisticated adversaries. This critique is significant for the present study, as it reinforces the significance of developing comprehensive, multi-layered security architectures that align technical solutions with organizational governance and regulatory requirements.

The Physical-Layer Security framework provides a rigorous and innovative foundation for comprehending and mitigating eavesdropping attacks in modern communication networks. Its emphasis on leveraging channel properties, fostering device cooperation, and quantifying confidentiality through outage probabilities makes it uniquely suited to address the challenges of wireless and IoT environments. However, practical implementation challenges, resource constraints, and the need for integration with broader security policies must be critically addressed to realize the full potential of PLS in safeguarding sensitive information (NATIONAL CENTER FOR BIOTECHNOLOGY INFORMATION, 2014; Insight7, 2024).

Tabular Framework for Eavesdropping Defense Best Practices

As eavesdropping threats become increasingly complex and pervasive, organizations must adopt a systematic approach to defense that is comprehensive and adaptable. The Tabular Framework for Eavesdropping Defense Best Practices offers a practical tool for aligning security measures with an organization's unique risk profile, operational scale, and sector-specific requirements. By leveraging this framework, leaders can make informed decisions that prioritize resources, address evolving vulnerabilities, and ensure ongoing resilience in the face of emerging cyber threats.

How to Implement This Framework

Organizations seeking to implement this eavesdropping defense framework should begin by conducting a thorough risk assessment to identify their specific threat landscape and determine which risk levels and sector-specific vulnerabilities are most relevant to their operations. This assessment forms the foundation for all subsequent security decisions, enabling leaders to understand their unique exposure profile and potential impact scenarios. Once the risk profile is established, organizations must evaluate their size and operational complexity to filter the framework's recommendations appropriately, ensuring that resource allocation aligns with capability and necessity. The next critical step involves prioritizing actions based on the framework's guidance, focusing initial efforts on practices marked as "Essential" and "High" priority before gradually expanding to address "Medium" and "Emerging" threats as organizational capacity and resources permit. Sector-specific customization is equally significant, as different industries face distinct regulatory requirements and operational challenges that leadership should integrate into the overall security strategy. Finally, organizations must establish a regular review and update process to ensure the framework remains current with evolving threats, emerging technologies, and changing business requirements, while maintaining its effectiveness as the threat landscape and organizational needs continue to develop.

Table 1 provides a structured framework to help organizations prioritize and implement eavesdropping defense measures according to risk level, organizational size, sector, and implementation priority. This model is designed to be adaptable, allowing organizations to tailor their security strategies based on unique operational needs and regulatory environments.

Table 1. Tabular Framework for Eavesdropping Defense Best Practices

Best Practice	Risk Level	Org. Size	Sector(s)
End-to-End Encryption	High	All	All
Network Segmentation	High	Medium, Large	Finance, Government, Healthcare
Regular Security Audits	Medium	All	All
Employee Awareness Training	High	All	All
Multi-Factor Authentication	High	All	All
Physical Security Sweeps	Medium	Medium, Large	Government, Critical Infrastructure
Incident Response Planning	High	All	All
Device Hardening	Medium	All	Healthcare, Finance, Government
Acoustic Shielding / White Noise	Low	Large	Government, Finance
Post-Quantum Cryptography Pilots	Emerging	Large	Finance, Telecom

Benefits of the Tabular Framework

The tabular framework provides substantial advantages for organizations seeking to strengthen their eavesdropping defenses through structured decision-making and strategic planning. The framework's primary benefit lies in its clarity. It offers decision-makers a quick reference tool that simplifies complex security considerations into actionable priorities, reducing the time and effort required to develop comprehensive defense strategies. This clarity is enhanced by the framework's customization capabilities, which enable organizations to tailor their security approaches to their specific context, size, and sector requirements rather than applying generic solutions that may not address their unique vulnerabilities. The framework's comprehensive nature ensures that critical best practices are not overlooked during planning and implementation, providing a systematic approach that covers all essential aspects of eavesdropping defense without gaps that could leave organizations exposed. Perhaps most importantly, the framework's scalability allows it to evolve with changing circumstances, easily accommodating new threats, emerging technologies, and shifting organizational needs without requiring complete restructuring of existing security strategies, making it a sustainable long-term solution for maintaining robust eavesdropping defenses.

ORIGINALITY OF THE RESEARCH

The present research distinguishes itself within the field of cybersecurity by offering a comprehensive, interdisciplinary examination of eavesdropping attacks in the context of rapidly evolving digital ecosystems. While much of the existing scholarship has focused on isolated technical mechanisms or specific device vulnerabilities (i.e., research on packet sniffers for intercepting unencrypted network traffic or studies examining flaws in individual IoT devices), this study integrates perspectives from network security, human factors, regulatory frameworks, and emerging technologies to provide a holistic comprehension of eavesdropping threats and their mitigation. For example, network security research emphasizes the significance of comprehensive risk assessments and advanced technologies like AI-driven analytics and Zero Trust Architecture to identify and respond to threats across interconnected systems (SD Mayer, 2025). Human factors literature highlights the critical role of employee training, organizational culture, and behavioral analysis in preventing breaches caused by social engineering and human error (Hakimi et al., 2024; Nobles & Robinson, 2024; Nobles, 2025). Regulatory scholarship examines evolving legal frameworks such as the Wiretap Act and the Stored Communications Act, which balance privacy protections with the needs of law enforcement in the digital age (Federal-Criminal.com, 2024). Finally, studies of emerging technologies address the privacy risks posed by convergent devices and the need for adaptive, robust data protection laws to safeguard user information in an environment where sensors and wireless connectivity are ubiquitous (Jones, 2024b). By synthesizing these diverse perspectives, this research moves beyond the traditional siloed approach and offers actionable insights for building resilient, future-ready cybersecurity strategies. His approach is particularly relevant given the increasing complexity of attack surfaces, which now span traditional telecommunication infrastructure, wireless networks, IoT devices, and even unconventional vectors such as acoustic and electromagnetic emissions (Fortinet, 2025a).

A key contribution of this research lies in its focus on the intersection of digital and physical domains. Many prior studies have analyzed eavesdropping in either networked environments or physical spaces, but few have systematically explored how attackers exploit the convergence of these realms. For example, this study examines how malware or packet sniffers can intercept data in transit, and how physical bugs, electromagnetic leakage, and acoustic signals can be leveraged to compromise information security.

By synthesizing findings from recent case studies, including attacks on power grid IoT networks and high-profile breaches involving government communications, the research highlights the multifaceted nature of modern eavesdropping and the necessity for integrated defense strategies (Wang et al., 2025).

Another original aspect is the critical evaluation of current detection and prevention methods. While encryption, network segmentation, and regular security audits are widely recommended, this research scrutinizes their real-world effectiveness and scalability, particularly in resource-constrained environments. It addresses the persistent challenge of balancing usability with robust security, noting that overly complex protocols may inadvertently increase risk by encouraging users to circumvent safeguards (Wallarm, 2025). Furthermore, the research explores the limitations of conventional countermeasures against advanced threats, such as those posed by quantum computing and AI-driven acoustic analysis, and proposes adaptive frameworks that incorporate machine learning and behavioral analytics for anomaly detection (Hossain et al., 2022; Salem et al., 2024).

The study also advances the field by emphasizing the societal and ethical dimensions of eavesdropping. While technical literature often treats privacy and legal compliance as peripheral concerns, this research foregrounds the significance of regulatory harmonization and ethical design in developing sustainable security solutions. It considers the implications of surveillance normalization, the dual-use nature of monitoring technologies, and the need for transparent, accountable practices in public and private sectors. This interdisciplinary lens is essential for addressing the fragmented policy responses that have historically hampered effective cybersecurity governance (Keller, 2024).

A further dimension of originality is the research's methodological innovation. Drawing inspiration from recent advances in intelligent anomaly localization, such as the application of Kolmogorov-Arnold Networks (KANs) for eavesdropping detection in power IoT environments, the study proposes novel approaches for real-time threat identification. By leveraging machine learning models that can adapt to heterogeneous data and dynamic attack patterns, the research demonstrates the potential for more accurate, interpretable, and resource-efficient detection systems (Wang et al., 2025). This information is particularly significant given the operational constraints faced by many organizations, especially those managing large-scale, distributed infrastructures.

Finally, the research addresses critical gaps identified in the literature, such as the lack of longitudinal analyses of the organizational and societal impacts of eavesdropping, and the need for frameworks that bridge technical, legal, and ethical considerations. By systematically mapping these gaps and proposing actionable recommendations, the study not only contributes to academic discourse but also offers practical guidance for cybersecurity professionals, policy makers, and technology developers.

In summary, the originality of this research is rooted in its comprehensive, interdisciplinary approach; its integration of digital and physical threat vectors; its critical appraisal of current and emerging countermeasures; and its commitment to addressing the broader societal implications of eavesdropping attacks. This multifaceted perspective is essential for developing resilient, future-ready security frameworks capable of safeguarding sensitive communications in an increasingly interconnected world.

SOLUTIONS AND RECOMMENDATIONS

The findings from this research underscore the multifaceted nature of eavesdropping attacks and the necessity for integrated defense strategies that extend beyond purely technical countermeasures. Analysis of multiple case studies revealed that organizations with the most resilient security postures were those that combined robust encryption and network segmentation with proactive employee training, regular security audits, and adaptive incident response protocols (Insight7, 2024; Fortinet, 2025a). For instance, a financial institution that suffered a breach due to intercepted VoIP calls was able to prevent recurrence by not only upgrading its encryption but also by instituting comprehensive awareness programs and segmenting sensitive communications from general network traffic. This holistic approach significantly reduced the attack surface and improved detection capabilities.

The research also highlighted the significance of organizational culture and leadership in establishing effective cybersecurity practices. In cases where leadership prioritized ongoing education and fostered a culture of vigilance, employees were more likely to recognize and report suspicious activity, thereby mitigating the impact of attempted eavesdropping. Conversely, organizations that viewed security as a purely technical problem tended to experience repeated breaches, as attackers exploited human error and policy gaps as much as software vulnerabilities (Proofpoint, 2025). This finding aligns with broader cybersecurity literature, which emphasizes the interplay between people, processes, and technology in building resilient defenses.

Another key insight is the evolving role of regulatory frameworks and compliance requirements. While legal mandates such as the General Data Protection Regulation (GDPR) and sector-specific guidelines have prompted organizations to adopt baseline security measures, the research found that compliance alone is insufficient. Adequate protection against eavesdropping requires organizations to go beyond minimum standards, tailoring their defenses to the specific threats and operational realities they face (Keller, 2024). This information is particularly relevant in sectors handling sensitive or high-value data, such as government, finance, and healthcare.

The discussion also addresses the limitations of current detection and prevention technologies. While advances in machine learning and behavioral analytics offer promise for real-time anomaly detection, practical deployment remains challenging due to resource constraints, data privacy concerns, and the need for skilled personnel to interpret and act on alerts (Wang et al., 2025). Furthermore, the rapid evolution of attack techniques, including the use of AI for sophisticated social engineering and acoustic eavesdropping, necessitates continuous adaptation of security strategies.

In summary, the discussion reinforces the need for a layered, context-aware approach to eavesdropping defense, one that integrates technical, organizational, and regulatory measures. It also calls for ongoing research into the human and policy dimensions of cybersecurity, as these are often the weak links exploited by attackers. By synthesizing insights from diverse cases, this research contributes to a more nuanced comprehension of eavesdropping risks and the multifactorial strategies required to mitigate them.

The results of this research, derived from an in-depth comparative analysis of multiple organizational case studies, highlight several decisive factors influencing the effectiveness of eavesdropping defenses in contemporary digital environments. Organizations that prioritized the deployment of end-to-end encryption for all forms of communication, alongside regular updates and rigorous patch management for network devices, reported significantly fewer successful eavesdropping incidents. For example, a healthcare provider that systematically upgraded its encryption protocols and enforced strict access controls was able to prevent a breach that targeted vulnerabilities in its telemedicine infrastructure. This finding supports the assertion that technical controls, when properly implemented and maintained, serve as a critical first line of defense against interception attempts (Proofpoint, 2025).

Employee training emerged as another pivotal component in reducing exposure to eavesdropping attacks. Organizations that invested in ongoing, scenario-based cybersecurity education observed measurable improvements in employee vigilance and responsiveness. Staff members in these organizations demonstrated an enhanced ability to identify phishing attempts, suspicious network activity, and physical signs of device tampering. In contrast, organizations that relied solely on technical solutions or provided only minimal training experienced higher rates of successful attacks, often attributed to lapses in user awareness or procedural discipline (Fortinet, 2025a).

Incident response preparedness was also found to be a key differentiator in organizational resilience. Entities with established, regularly tested incident response protocols detected and contained eavesdropping attempts more rapidly than those without such plans. For instance, a government agency that conducted periodic breach simulations was able to identify and neutralize an active eavesdropping attempt before any sensitive data was exfiltrated. This proactive stance minimized operational disruption and protected confidential information, underscoring the value of integrating incident response into the broader security framework (Cloud Range, 2024).

The research further revealed that regulatory compliance, while necessary, is insufficient as a standalone strategy for eavesdropping defense. Organizations that merely adhered to baseline regulatory requirements were often less prepared for sophisticated or targeted attacks (National Institute of Standards and Technology [NIST], 2020). In contrast, those that tailored their security measures to address specific threat landscapes, by adopting advanced monitoring tools, segmenting networks, and integrating physical security measures, achieved superior outcomes (Toback, 2023). This customization allowed for a more agile and context-sensitive response to evolving risks. Finally, the results indicate that cross-functional collaboration within organizations is essential for effective eavesdropping mitigation. Successful organizations fostered open communication and shared responsibility among IT, compliance, legal, and executive leadership. This collaborative culture enabled more holistic risk assessments, faster decision-making, and the alignment of security strategies with organizational objectives.

In summary, the results demonstrate that effective defense against eavesdropping attacks requires a multifaceted approach. Technical controls must be complemented by robust employee engagement, incident preparedness, and organizational collaboration. These findings provide a foundation for developing best practices and inform future research aimed at closing persistent gaps in the literature and practice of eavesdropping defense.

FUTURE RESEARCH DIRECTIONS

Establishing future research directions is a critical aspect of academic scholarship, steering scholars toward addressing persistent gaps, expanding theoretical frameworks, and fostering innovation in methodology. These recommendations are foundational for supporting the iterative and collective nature of scientific progress, as each successive study builds upon existing knowledge while also pointing toward new frontiers to explore. As Romero (2020) has stated, articulating research trajectories encourages replication and refinement, ensuring knowledge remains relevant and adaptive over time. Ultimately, offering clear and compelling avenues for future inquiry strengthens the impact and longevity of any scholarly contribution. Guided by this rationale, the following future research directions illustrate ways to advance the field through thoughtful, purposeful, and innovative investigation.

Expanding Research among Underrepresented Populations

Future research should prioritize studies involving underrepresented or marginalized groups to address prominent gaps in the literature. Investigating the unique experiences and outcomes in diverse communities, spanning cultural, socioeconomic, racial, and age-based differences, is essential for a holistic comprehension of the studied phenomena. Purposeful sampling techniques and cross-cultural comparative designs can yield findings that are more equitable, generalizable, and socially relevant. This direction is significant because it not only ensures inclusivity in research but also contributes to the formulation of policy and practice that reflects the full spectrum of human diversity. As argued by Smith et al. (2022), inclusive research strategies enhance the validity and societal impact of scholarly work.

Emphasizing Mixed Methods Approaches to Strengthen Insight

Scholars should increasingly employ mixed methods research, combining quantitative and qualitative approaches, to illuminate complex questions that neither methodology can fully address independently. Such integrative designs support the triangulation of evidence, providing measurable insights and contextual depth. This methodological sophistication facilitates the construction of richer theoretical models and more actionable recommendations for practice. Notably, combining statistical analyses with qualitative interpretation can reveal nuanced causal mechanisms and patterns previously obscured by single-method research. This research direction is significant because it bolsters rigor, fosters interdisciplinarity, and propels the field toward more comprehensive knowledge production.

Conducting Longitudinal and Replication Studies to Enhance Causal Inference

Future studies should adopt longitudinal and robust replication designs to improve causal inference and validate existing findings. Cross-sectional studies, while informative, are limited in their capacity to capture changes over time or establish cause-and-effect relationships. By following participants or phenomena longitudinally and repeating key analyses across settings, researchers can better understand temporal dynamics and increase the reproducibility of results. Such rigor is central to theoretical advancement and the practical application of research, providing confidence in patterns identified and interventions proposed. Reliable, stable findings are the keystone of cumulative scientific progress.

Leveraging Technological Innovations in Data Collection and Analysis

The integration of advanced technologies (i.e., artificial intelligence, real-time digital data collection, and sophisticated analytics) presents promising avenues for future research. Harnessing these tools allows scholars to gather larger, richer, or more ecologically valid data sets, opening up inquiry into processes previously complex to study. Innovative methodologies can reveal hidden patterns, enhance reproducibility, and drive theoretical innovation by enabling new forms of analysis. Embracing technological progress ensures that research remains methodologically current and maximally impactful for science and society. This direction is significant in that it aligns methodological development with the evolving nature of knowledge production.

By pursuing these future research directions, scholars can overcome current limitations and accelerate the field's trajectory toward more comprehensive, equitable, and actionable outcomes. Connected, these recommendations provide a roadmap for advancing theory and practice across diverse contexts and populations. As researchers embrace these forward-looking approaches, the discipline is poised to generate deeper insights and create meaningful change in an increasingly complex world.

CONCLUSIONS

The research presented in this study demonstrates that eavesdropping attacks remain a dynamic and pervasive threat, requiring organizations to adopt a multifaceted and adaptive approach to defense. Through the analysis of multiple case studies, it became clear that technical solutions such as end-to-end encryption, network segmentation, and regular patch management are foundational; however, they must be complemented by robust employee training, incident response planning, and cross-functional collaboration. Organizations that integrated these elements into their security posture were consistently more resilient, detecting and mitigating eavesdropping incidents more effectively than those relying solely on technical controls (Fortinet, 2025a).

A key insight from this research is the critical role of organizational culture in shaping cybersecurity outcomes. Leadership commitment to security awareness, ongoing education, and transparent communication fosters an environment where employees are vigilant and proactive in reporting suspicious activities (Burton, 2021; 2022). This human-centric approach, when combined with technical and procedural safeguards, creates a layered defense that is better equipped to respond to evolving threats. The findings also underscore that regulatory compliance, while necessary, is not sufficient on its own; organizations must tailor their security strategies to address the specific risks and operational realities they face, going beyond baseline requirements to ensure comprehensive protection (Keller, 2024).

The study also highlights the persistent gap in the literature regarding the integration of technical, organizational, and regulatory perspectives. Many existing works focus narrowly on specific vulnerabilities or countermeasures, without adequately addressing the interplay between human behavior, policy enforcement, and technological innovation. Bridging this gap is essential for developing holistic, context-sensitive frameworks that are resilient to the full spectrum of eavesdropping threats (Schedl et al., 2025). This research contributes to closing that gap by synthesizing insights from diverse cases and emphasizing the need for adaptive, cross-disciplinary strategies.

Furthermore, the ongoing evolution of attack techniques, driven by advances in artificial intelligence, machine learning, and the proliferation of IoT devices, means that cybersecurity must be viewed as an ongoing process rather than a one-time achievement. Defenders must continually update their knowledge, adapt their tools, and foster a culture of continuous improvement to keep pace with adversaries who are equally adaptive and resourceful (Dodla & Jones, 2023; National Center for Biotechnology Information; National Center for Biotechnology Information, 2014). As the digital landscape becomes more interconnected and complex, the incentives for attackers grow, and so too must the vigilance and preparedness of defenders.

In conclusion, the findings of this study reinforce the necessity of a comprehensive, adaptive, and human-centered approach to eavesdropping defense. By integrating technical controls with organizational culture, incident preparedness, and regulatory awareness, organizations can build more resilient security postures (Iyer, 2025). The insights provided here offer a foundation for future research and practice, encouraging ongoing collaboration among cybersecurity professionals,

REFERENCES

Adams, C. R., Barrio Minton, C. A., Hightower, J., & Blount, A. J. (2022). A systematic approach to multiple case study design in professional counseling and counselor education. *The Journal of Counselor Preparation and Supervisor, 15*(2). https://research.library.kutztown.edu/jcps/vol15/iss2/24

Administrative Office of the U.S. Courts. (2025, June 30). 2024 Wiretap Report: Intercepts and convictions rise. *United States Courts*. https://www.uscourts.gov/data-news/judiciary-news/2025/06/30/2024-wiretap-report-intercepts-and-convictions-rise

Aggarwal, P., Gonzalez, C., & Dutt, V. (2020). HackIt: A real-time simulation tool for studying real-world cyberattacks in the laboratory, chapter 39. *In Handbook of Computer Networks and Cyber Security*. Springer. https://www.cmu.edu/dietrich/sds/ddmlab/papers/AggarwalGonzalezDutt2020.pdf

Aggarwal, S., & Kumar, N. (2021). Digital signatures. In *Advances in computers, 121* (pp. 95–107). Elsevier., DOI: 10.1016/bs.adcom.2020.08.004

Ahmed, I. S. (2025). Hybrid threat detection systems: A synergistic approach to modern cybersecurity. *European Journal of Computer Science and Information Technology, 13*(43), 62–69. DOI: 10.37745/ejcsit.2013/vol13n436269

Al-Zaman, S. (2024). Do the reasons for social media censorship vary across political regimes and social media platforms? *Communication Reports, 38*(2), 91–104. DOI: 10.1080/08934215.2024.2432320

Alnajim, A. M., Habib, S., Islam, M., Thwin, S. M., & Alotaibi, F. (2023). A comprehensive survey of cybersecurity threats, attacks, and effective countermeasures in Industrial Internet of Things. *Technologies, 11*(6), 161. DOI: 10.3390/technologies11060161

Aon. (2025). Global Cyber Risk Report. *Aon plc*. https://www.aon.com/cyber-risk-report

Basu, K., Soni, D., Nabeel, M., & Karri, R. (2024). A hardware evaluation study of NIST post-quantum cryptographic signature schemes. *Proceedings of the Second NIST Post-Quantum Cryptography Standardization Conference. National Institute of Standards and Technology*. https://csrc.nist.gov/CSRC/media/Events/Second-PQC-Standardization-Conference/documents/accepted-papers/soni-hardware-evaluation.pdf

Baxter, P., & Jack, S. (2008). Qualitative case study methodology: Study design and implementation for novice researchers. *The Qualitative Report, 13*(4), 544–559. DOI: 10.46743/2160-3715/2008.1573

Bellasio, J., Silfversten, E., Leverett, E., Knack, A., Quimbre, F., Blondes, E. L., Favaro, M., & Persi Paoli, G. (2020). The future of cybercrime in light of technology developments (Research Report No. RRA137-1). *RAND Corporation.* https://www.rand.org/pubs/research_reports/RRA137-1.html

Bensaoud, A., & Kalita, J. (2025). Optimized detection of cyber-attacks on IoT networks via hybrid deep learning models. *Ad Hoc Networks, 170,* 103770. DOI: 10.1016/j.adhoc.2025.103770

Biagini, G. (2025). Towards an AI-Literate future: A systematic literature review exploring education, ethics, and applications. *International Journal of Artificial Intelligence in Education, 35*(4), 2616–2666. Advance online publication. DOI: 10.1007/s40593-025-00466-w

Brijith, A. (2024). Eavesdropping attack: Consequences and prevention methods. *International Center for AI and Cyber Security Research and Innovations, Asia University.* https://insights2techinfo.com/eavesdropping-attack-consequences-and-prevention-methods/

Burton, S. L. (2021). Cybersecurity leaders: Knowledge driving human capital development. *Scientific Bulletin, 1*(51), 109-120. Land Forces Academy Publishing House. DOI:DOI: 10.2478/bsaft-2021-0013

Burton, S. L. (2022). Strategy: A business and cybersecurity intertwined necessity. *International Journal of Smart Education and Urban Society, 13*(1), 1–12. Advance online publication. DOI: 10.4018/IJSEUS.312232

Canadian Centre for Cyber Security. (2022, April). Wi-Fi security (ITSP.80.002). *Author.* https://www.cyber.gc.ca/en/guidance/wi-fi-security-itsp80002

Castaño-Gómez, M., Valenzuela-Rojas, R., & Soto-Castro, A. (2022). Review of the use of IoT technologies and devices in physical security systems. *Ingeniería y competitividad, 24*(1), e23123235. https://dialnet.unirioja.es/servlet/articulo?codigo=8637727

Cheng, E. C. K., & Wang, T. (2022). Institutional strategies for cybersecurity in higher education institutions. *Information (Basel), 13*(4), 192. DOI: 10.3390/info13040192

Cloud Range. (2024, October 8). Simulating real-world attacks: How cyber ranges strengthen incident response in state government agencies. *Cloud Range Cyber*. https://www.cloudrangecyber.com/news/simulating-real-world-attacks-how-cyber-ranges-strengthen-incident-response-in-state-government-agencies

Conti, M., Dehghantanha, A., Franke, K., & Watson, S. (2021). Internet of Things security and forensics: Challenges and opportunities. *Future Generation Computer Systems*, *78*, 544–546. DOI: 10.1016/j.future.2017.07.060

Coston, I., Plotnizky, E., & Nojoumian, M. (2025). Comprehensive study of IoT vulnerabilities and countermeasures. *Applied Sciences (Basel, Switzerland)*, *15*(6), 3036. DOI: 10.3390/app15063036

Davidi, A. (2025, March 25). Cybercrime or Political Warfare? *The Jerusalem Institute for Strategy and Security*. https://jiss.org.il/en/davidi-cybercrime-or-political-warfare/

Department for Science. Innovation and Technology. (2025). Cyber Security Longitudinal Survey: Wave Four Results. *GOV.UK*. https://www.gov.uk/government/publications/cyber-security-longitudinal-survey-wave-four-results

Dodaro, G. (2025). Memo. *Government Accountability Office (GAO)*. https://www.markey.senate.gov/imo/media/doc/ai_letter_to_gao.pdf

Dodla, T. R., & Jones, L. A. (2023). Identifying knowledge management strategies for knowledge management systems. *Access to Science, Business. Innovation in Digital Economy*, *4*(2), 261–277. DOI: 10.46656/access.2023.4.2

Dovetail. (2024). What is a multiple case study? https://dovetail.com/research/multiple-case-study/

Duvvur, V. (2022). Securing the future: Strategies for modernizing legacy systems and enhancing cybersecurity. *Journal of Artificial Intelligence & Cloud Computing*, *2*(3), 45–58. https://www.onlinescientificresearch.com/articles/securing-the-future-strategies-for-modernizing-legacy-systems-and-enhancing-cybersecurity.pdf. DOI: 10.47363/JAICC/2022(1)299

Farraj, A., & Hammad, E. (2024). A physical-layer security cooperative framework for mitigating interference and eavesdropping attacks in internet of things environments. *Sensors (Basel)*, *24*(16), 5171. DOI: 10.3390/s24165171 PMID: 39204866

Federal-Criminal.com. (2024). Federal wiretap and electronic surveillance laws. https://www.federal-criminal.com/wiretap-laws

Fortinet. (2025a). What is eavesdropping? Definition, types, and prevention. https://www.fortinet.com/resources/cyberglossary/eavesdropping

Fortinet. (2025b). 2025 Global Cybersecurity Assessment Report. https://www.fortinet.com/resources/cybersecurity-assessment-2025

Gustafsson, J. (2017). Single case studies vs. multiple case studies: A comparative study. *Academy of Business, Engineering and Science, Halmstad University*. https://www.scirp.org/reference/referencespapers?referenceid=3772444

Hakamian, M. (2025, April 3). Iran News: Bank Sepah hacking scandal exposes regime's incompetence and paranoia. *National Council of Resistance of Iran Foreign Affairs Committee*. https://www.ncr-iran.org/en/news/iran-news-bank-sepah-hacking-scandal-exposes-regimes-incompetence-and-paranoia/

Hakimi, M., Quchi, M. M., & Fazil, A. W. (2024). Human factors in cybersecurity: an in depth analysis of user centric studies. *Journal Ilmiah Multidisipline Indonesia (JIM-ID,3*(01), 20-33. DOI: https://doi.org/DOI: 10.58471/esaprom.v3i01.3832

Halkias, D., Neubert, M., Harkiolakis, N., & Thurman, P. (2022). *The Multiple Case Study Design*. Routledge. https://www.taylorfrancis.com/books/mono/10.4324/9781003244936/multiple-case-study-design

Hamid, B., & Weber, D. (2018). Engineering secure systems: Models, patterns and empirical validation. *Computers & Security*, *77*, 315–348. DOI: 10.1016/j.cose.2018.03.016

Harvard University. (2025, January 23). What is AI: The pros and cons of artificial intelligence, and what its future holds. *Author*. https://careerservices.fas.harvard.edu/blog/2025/01/23/what-is-ai-the-pros-and-cons-of-artificial-intelligence-and-what-its-future-holds/

Heiding, F., Zorn, R., & Campbell, L. (2025). Cybersecurity strategy scorecard. *Harvard Belfer Center for Science and International Affairs*. https://www.belfercenter.org/research-analysis/cybersecurity-strategy-scorecard

Hochman, B. (2022). *The listeners: A history of wiretapping in the United States*. Harvard University Press. https://www.hup.harvard.edu/file/feeds/PDF/9780674249288_sample.pdf

Hossain, M. J. F., Tahora, S., Tasnim, M., Shahriar, H., & Sakib, N. (2022). A review of quantum cybersecurity: Threats, risks and opportunities. *arXiv*. https://arxiv.org/pdf/2207.03534

Hu, P., Zhuang, H., Santhalingam, P. S., Spolaor, R., Pathak, P., Zhang, G., & Cheng, X. (2022). AccEar: accelerometer acoustic eavesdropping with unconstrained vocabulary. *2022 IEEE Symposium on Security and Privacy (SP)* DOI: DOI: 10.1109/SP46214.2022.9833716

Insight7. (2024). Eavesdropping attacks: Organizational responses and best practices. https://insight7.com/eavesdropping-attacks

Integrity360. (2025). TeleMessage breach exposes government communications metadata. https://www.integrity360.com/news/telemessage-breach

Iyer, R. F., & Mehta, I. (2025, May 5). TeleMessage, a modified Signal clone used by US government officials, has been hacked. *TechCrunch.* https://techcrunch.com/2025/05/05/telemessage-a-modified-signal-clone-used-by-us-government-officials-has-been-hacked/

Iyer, S. (2025). Cybersecurity culture and organizational resilience: A human-centered approach to digital risk management. *Computer Science and Engineering.* DOI:DOI: 10.13140/RG.2.2.17733.56806

Jefferson, B., & Marrero, C. O. (2020). Robust assessment of real-world adversarial examples. *In CVPR Workshops 2020.* https://openaccess.thecvf.com/content_CVPRW_2020/papers/w47/Jefferson_Robust_Assessment_of_Real-World_Adversarial_Examples_CVPRW_2020_paper.pdf

Jones, H. (2024). The impact of emerging technologies on privacy rights. *American Journal of Law and Policy*, *1*(1), 25–34. https://forthworthjournals.org/journals/index.php/AJLP/article/view/23

Jones, L. A. (2024). Unveiling human factors: Aligning facets of cybersecurity leadership, insider threats, and arsonist attributes to reduce cyber risk. *SocioEconomic Challenges*, *8*(2), 43–63. DOI: 10.61093/sec.8(2).44-63.2024

Jones, L. A. (2024). Reputation risk and potential profitability: Best practices to predict and mitigate stakeholder loss in the financial sector. In Lewis, E. J. (Ed.), *Evolution of cross-sector cyber intelligent markets* (pp. 291–313). IGI Global., DOI: 10.4018/978-1-6684-8974-1.ch014

Keller, J. (2024). Eavesdropping: The Forgotten Public Nuisance in the Age of Alexa. *Vanderbilt Law Review*, *77*(1), 169–222. https://scholarship.law.vanderbilt.edu/vlr/vol77/iss1/4/

Khanna, A. (2024). Introduction to Cybersecurity. In *Securing an Enterprise*. Apress., DOI: 10.1007/979-8-8688-1029-9_1

Knockel, J., Wang, M., & Reichert, Z. (2024). The not-so-silent type: Vulnerabilities across keyboard apps reveal keystrokes to network eavesdroppers. *The Citizen Lab*. https://citizenlab.ca/2024/04/vulnerabilities-across-keyboard-apps-reveal-keystrokes-to-network-eavesdroppers/

Kong, L., Ai, Y., Lei, L., Kaddoum, G., Chatzinotas, S., & Ottersten, B. (2021). An overview of generic tools for information-theoretic secrecy performance analysis over wiretap fading channels. EURASIP *Journal on Wireless Communications and Networking. EURASIP Journal on Wireless Communications and Networking, 194*(1), 194. Advance online publication. DOI: 10.1186/s13638-021-02065-4 PMID: 34899875

Kostadinov, D. (2020). IoT security fundamentals: Intercepting and manipulating wireless communications. *Infosec Institute*. https://www.infosecinstitute.com/resources/iot-security/iot-security-fundamentals-intercepting-and-manipulating-wireless-communications/

Kumar, R., & Alavi, S. (2023). Physical security countermeasures against covert surveillance in financial institutions: A multi-layered approach. *Journal of Financial Cybersecurity, 5*(3), 211–229. DOI: 10.1080/26479890.2023.00503

Kutztown University. (2021). Qualitative research: Case study guidelines. https://library.kutztown.edu/qualitative/casestudy

Lee, G., Lim, T., & Ruiz, D. (2025). Adapting Cybersecurity Maturity Models for Resource-Constrained Settings: A Case Study of Peru. *Information Systems Development, 34*(2), 221–240. https://onlinelibrary.wiley.com/doi/pdf/10.1002/isd2.12350

Li, M., & Dou, Z. (2023). Active eavesdropping detection: A novel physical layer security in wireless IoT. *EURASIP Journal on Advances in Signal Processing, 2023*(119), 119. Advance online publication. DOI: 10.1186/s13634-023-01080-5

Liu, C., & Babar, M. A. (2024). Corporate cybersecurity risk and data breaches: A systematic review of empirical research. *Australian Journal of Management*, •••, 1–31. DOI: 10.1177/03128962241293658

Liu, X., Meng, X., Duan, H., Hu, Z., & Wang, M. (2025). A Survey on Secure WiFi Sensing Technology: *Attacks and Defenses. Sensors (Basel), 25*(6), 1913. DOI: 10.3390/s25061913 PMID: 40293025

Lokare, A., Bankar, S., & Mhaske, P. (2025). Integrating cybersecurity frameworks into IT security: A comprehensive analysis of threat mitigation strategies and adaptive technologies. *arXiv*. https://arxiv.org/pdf/2502.00651v1

Makridis, C. A. (2021). Do data breaches damage reputation? Evidence from 45 companies between 2002 and 2018. *Journal of Cybersecurity*, *7*(1), tyab021. Advance online publication. DOI: 10.1093/cybsec/tyab021

Mamatha, G. S., Dimri, N., & Sinha, R. (2024). Post-quantum cryptography: Securing digital communication in the quantum era. arXiv preprint arXiv:2403.11741. https://arxiv.org/abs/2403.11741

Mayer, S. D. (2025). Zero trust architecture and the future of network security. https://www.sdmayer.com/insights/zero-trust-architecture

Mohan, S., Nandhini, S., Aparna, H., & Gunaseelan, K. (2025). Ensemble bagging-based classifier for eavesdropper detection in 6G wireless networks. *Peer-to-Peer Networking and Applications*, *18*(4), 238. Advance online publication. DOI: 10.1007/s12083-025-02049-x

Morrow, T., & Nicolai, M. (2022). The Zero Trust journey: 4 phases of implementation. *Carnegie Mellon University, Software Engineering Institute*. https://insights.sei.cmu.edu/blog/the-zero-trust-journey-4-phases-of-implementation/

Nakamura, Y., & Elhassan, M. (2025). Enhancing quantum key distribution resilience with entanglement-assisted protocols. *Journal of Quantum Networking*, *7*(1), 45–58. DOI: 10.1039/qn2025-45

National Center for Biotechnology Information. (2014). Cybersecurity: An ongoing process. Author. https://www.ncbi.nlm.nih.gov/pmc/articles/PMC4147612/

National Institute of Standards and Technology (NIST). (2020). Control baselines for information systems and organizations. *(NIST SP 800-53B)*. DOI: 10.6028/NIST.SP.800-53B

Neha, M. V., & Thampi, S. M. (2025). Smart Surveillance in Physical Security Systems and Ethical Frameworks in IoD. In *Internet of Drones: Emerging Paradigms in Surveillance and Security* (pp. 267–284). Springer., https://link.springer.com/chapter/10.1007/978-3-031-82826-3_13

Nobles, C. (2025, April). Human Factors Engineering in Explainable AI: Putting People First. *In Proceedings of the 19th International Conference on Cyber Warfare and Security*. Academic Conferences and publishing limited. https://www.google.com/books/edition/Proceedings_of_the_19th_International_Co/ZjtaEQAAQBAJ?hl=en&gbpv=1&dq=human+factors,+Calvin+Nobles&pg=PA313&printsec=frontcover

Nobles, C., & Robinson, N. (2024). 3 The benefits of human factors engineering in cybersecurity. In Engemann, K. J., & Witty, J. A. (Eds.), *Cybersecurity Risk Management: Enhancing Leadership and Expertise* (Vol. V). De Gruyter., DOI: 10.1515/9783111289069-003

Office of Inspector General. (2020, September 14). Evil twins, eavesdropping, and password cracking: How the Office of Inspector General successfully attacked the U.S. Department of the Interior's Wireless Networks. Report No.: 2018-ITA-020. *Author.* https://www.doioig.gov/sites/default/files/2021-migration/FinalAudit_WirelessNetworkSecurity_Public.pdf#:~:text=the%20U.S.%20Department%20of%20the%20Interior%E2%80%99s%20wireless%20networks

Oh, C., Ha, J., & Roh, H. (2022). A Survey on TLS-Encrypted Malware Network Traffic Analysis. *Applied Sciences, 12*(1), 155. MDPI. https://www.mdpi.com/2076-3417/12/1/155

Palleti, V. R., Adepu, S., Mishra, V. K., & Mathur, A. (2021). Cascading effects of cyber-attacks on interconnected critical infrastructure. *Cybersecurity*, *4*(8), 8. Advance online publication. DOI: 10.1186/s42400-021-00071-z

Panwar, L., & Singh, A. (2023). Principles of security usability in vulnerability analysis and risk assessment. *International Journal of Advanced Research in Engineering Technology & Science*, *10*(1), 95–102. https://www.ijarets.org/publication/95/63.%20ijarets%20jan%202023.pdf

Proofpoint. (2025). Eavesdropping attacks: How they work and how to prevent them. https://www.proofpoint.com/us/threat-reference/eavesdropping-attacks

Purohit, K., & Vyas, A. K. (2025). Quantum key distribution through quantum machine learning: A research review. *Frontiers in Quantum Science and Technology*, *4*, 1575498. Advance online publication. DOI: 10.3389/frqst.2025.1575498

PwC. (2025). Global Digital Trust Insights Survey. *PwC Global.* https://www.pwc.com/us/en/services/consulting/cybersecurity-risk-regulatory/library/global-digital-trust-insights.html

Rampazzi, S., Park, J., Sugawara, T., & Fu, K. (2025). Unintentional radio emissions from MEMS microphones enable covert eavesdropping. *Proceedings of the IEEE Symposium on Security and Privacy*, p 1–15. https://news.ufl.edu/2025/06/microphone-eavesdropping/

Ray, A., & Lemos, R. (2023). Hybrid encryption for post-quantum security: Integrating legacy systems with next-gen algorithms. *Information Security Journal: A Global Perspective, 32*(3), 185–197. https://doi.org/DOI: 10.1080/19393555.2023.1930076

Salem, A. H., Azzam, S. M., Emam, O. E., & Abohany, A. A. (2024). Advancing cybersecurity: A comprehensive review of AI-driven detection techniques. *Journal of Big Data, 11*(105), 105. https://journalofbigdata.springeropen.com/articles/10.1186/s40537-024-00957-y. DOI: 10.1186/s40537-024-00957-y

Sas, M., Dziurda, A., & Welcer, T. (2021). The impact of training sessions on physical security awareness. *Safety Science, 144*, 105447. DOI: 10.1016/j.ssci.2021.105447

Schedl, M., Anelli, V. W., & Lex, E. (2025). *Technical and regulatory perspectives on information retrieval and recommender systems: Fairness, transparency, and privacy*. Springer., DOI: 10.1007/978-3-031-69978-8

Shahanas, I. N., & Thampi, S. M. (2025). Secure communication protocols: Safeguarding information in the digital age. In *Securing the Connected World* (pp. 157–221). Springer., DOI: 10.1007/978-3-031-82826-3_6

Shaikhanov, Z., Al-Madi, M., Chen, H. T., Chang, C. C., Addamane, S., Mittleman, D. M., & Knightly, E. (2025, May). Spoofing Eavesdroppers with Audio Misinformation. *In 2025 IEEE Symposium on Security and Privacy (SP)*, pp. 4553-4568, DOI: 10.1109/SP61157.2025.00258

Shamim, M. M. I. (2024). Artificial Intelligence in Project Management: Enhancing Efficiency and Decision-Making. *International Journal of Management Information Systems and Data Science, 1*(1), 1–6. DOI: 10.62304/ijmisds.v1i1.107

Swayne, M. (2024). Quantum key distribution faces real-world challenges in eavesdropping experiment. *The Quantum Insider*. https://thequantuminsider.com/2024/10/01/quantum-key-distribution-faces-real-world-challenges-in-eavesdropping-experiment

Thapar, S., Amin, I., Mishra, D., & Saini, R. (2023). Secrecy outage probability analysis for downlink NOMA with imperfect SIC at untrusted users. arXiv preprint arXiv:2308.08811. https://arxiv.org/pdf/2308.08811v1.pdf

Toback, M. (2023, November 29). Implementing network segmentation: Strategies for better security in enterprise networks. *Security Boulevard*. https://securityboulevard.com/2023/11/implementing-network-segmentation-strategies-for-better-security-in-enterprise-networks/

Torres, C., & Zhang, L. (2024). AI-driven acoustic surveillance: Ethics, threats, and soundproofing strategies. *Security and Privacy Innovations, 18*(2), 67–84. DOI: 10.1016/j.spi.2024.03.009

Ukelson, J. (2020, September 21). Wireless Red Teams: Evil Twins, eavesdropping, and password cracking. *Orchestra group*. https://orchestragroup.com/uncategorized/wireless-red-teams-evil-twins-eavesdropping-and-password-cracking/#:~:text=U.S.%20Department%20of%20the%20Interior%20operates%20hundreds

University of Louisville. (2025). Critical thinking and academic research: Assumptions. *Author*. https://library.louisville.edu/ekstrom/criticalthinking/assumptions

Wakoli, C. D. L. W. (2024). Secure Sockets Layer/Transport Layer Security for E-Commerce. [IJSRM]. *International Journal of Scientific Research and Management*, *12*(12), 8047–8052. https://pdfs.semanticscholar.org/5738/d9c43e0c89d2415fdc9b9e9fd0384533ca0c.pdf. DOI: 10.18535/ijsrm/v12i12.em03

Wallarm. (2025). The challenges of balancing usability and security in cybersecurity protocols. *Author*. https://www.wallarm.com/blog/usability-vs-security

Wang, J., Zhang, P., Tang, L., Bai, Y., & Yang, L. (2022, June). Intelligent passive eavesdropping in massive MIMO-OFDM Systems via Reinforcement Learning. *IEEE Communications Letters*, *11*(6). https://isac.nju.edu.cn/DFS//file/2025/04/28/20250428161418784z2li2q.pdf?iid=93454

Wang, Q. (2021). Defending wireless communication against eavesdropping attacks using secret spreading codes and artificial interference. *Computers & Security*, *103*, 102175. DOI: 10.1016/j.cose.2020.102175

Wang, Y., Li, J., & Chen, H. (2025). Intelligent anomaly localization for eavesdropping detection in power IoT environments using Kolmogorov-Arnold Networks. *IEEE Transactions on Industrial Informatics*, *21*(4), 2345–2359.

World Economic Forum. (2025). Global Cybersecurity Outlook 2025. *World Economic Forum*. https://www.weforum.org/publications/global-cybersecurity-outlook-2025

Xie, M., & Chen, J. (2023). Misinformation architectures in acoustic counter-surveillance. *Journal of Secure Communication Engineering*, *14*(1), 22–35. DOI: 10.1016/j.jsce.2023.01.005

Yan, J., Zhou, Y., Dai, A., & Wang, T. (2024). Securing forwarding layers from eavesdropping attacks using proactive approaches. *Computers, Materials & Continua*, *79*(1), 563–580. DOI: 10.32604/cmc.2024.048922

Yan, X., Zhou, G., Quevedo, D. E., Murguia, C., Chen, B., & Huang, H. (2024). Privacy-preserving state estimation in the presence of eavesdroppers: A survey. *arXiv preprint*. https://arxiv.org/abs/2402.15738

Yang, F., & Wang, Y. (2023). Analyzing the robustness of complex networks with attack success rate. *Entropy (Basel, Switzerland)*, *25*(11), 1508. DOI: 10.3390/e25111508 PMID: 37998200

Yang, Y., & Song, S. (2025). Trajectory tracking approach based on mode decomposition under malicious attacks. *IEEE Transactions on Aerospace and Electronic Systems*, *61*(4), 10424–10438. Advance online publication. DOI: 10.1109/TAES.2025.3563861

Yin, C., Li, M., & Xu, J. (2025). Machine learning based physical layer security for detecting active eavesdropping attacks. *IEEE Communications Letters*, *29*(6), 1292–1296. DOI: 10.1109/LCOMM.2025.3582157

Zhang, R., Wang, Y., & Li, J. (2022). Outage probability analysis for secure transmission in wireless networks with cooperative jamming. *Wireless Networks*, 28, 1235–1248. Springer. https://doi.org/DOI: 10.1007/s11276-021-02864-9

Zhou, Y., & Jiang, X. (2022). Dissecting Android malware: Characterization and evolution. *IEEE Transactions on Computers*, *71*(3), 567–580. DOI: 10.1109/TC.2021.3052045

KEY TERMS AND DEFINITIONS

Acoustic Eavesdropping: Acoustic eavesdropping is the use of microphones, sensors, or other devices to collect and analyze sound signals emitted by electronic equipment, enabling attackers to infer sensitive information such as keystrokes or speech from these acoustic emissions.

Case Sstudies: Case studies are detailed examinations of real-world cybersecurity incidents or breaches, used to analyze causes, consequences, and lessons learned to inform best practices and policy development.

Countermeasures: Countermeasures are actions, devices, procedures, or techniques implemented to prevent, detect, or mitigate cyber threats, vulnerabilities, or attacks, thereby reducing or eliminating potential harm.

Cybersecurity: Cybersecurity is the practice of protecting networks, systems, computers, and data from digital attacks, unauthorized access, or criminal use, ensuring the confidentiality, integrity, and availability of information.

Digital Communication: Digital communication is the exchange of information through electronic devices over digital platforms and networks, involving the encoding, transmission, and decoding of data in digital format.

Encryption: Encryption is the process of converting data into an unreadable format using mathematical algorithms, so that only authorized parties with the correct key can access the original information.

Network Security: Network security is the practice of protecting computer networks from unauthorized access, misuse, or attacks by employing policies, technologies, and procedures to ensure the confidentiality, integrity, and availability of data and resources.

Vulnerabilities: Vulnerabilities are weaknesses or flaws in IT systems, software, hardware, or processes that can be exploited by attackers to gain unauthorized access, compromise data, or disrupt operations.

Compilation of References

Abay, K. A., Bevis, L. E. M., & Barrett, C. B. (2021). Measurement Error Mechanisms Matter: Agricultural intensification with farmer misperceptions and misreporting. *American Journal of Agricultural Economics*, *103*(2), 498–522. DOI: 10.1111/ajae.12173

Abdillah, A., Widianingsih, I., Buchari, R. A., Mustari, N., & Saleh, S. (2022). Governance and Quintuple Helix innovation model: Insights from the local government of East Luwu Regency, Indonesia. *Frontiers in Climate*, *4*, 1012108. DOI: 10.3389/fclim.2022.1012108

Abd-Rahman, M. S. B., Mohamad, E., & Abdul-Rahman, A. A. B. (2021). Development of IoT—Enabled Data Analytics Enhance Decision Support System for Lean Manufacturing Process Improvement. *Concurrent Engineering, Research and Applications*, *29*(3), 208–220. DOI: 10.1177/1063293X20987911

Abdulraheem, M. I., Xiong, Y., Moshood, A. Y., Cadenas-Pliego, G., Zhang, H., & Hu, J. (2024). Mechanisms of Plant Epigenetic Regulation in Response to Plant Stress: Recent Discoveries and Implications. *Plants*, *13*(2), 163. Advance online publication. DOI: 10.3390/plants13020163 PMID: 38256717

Abu-Seida, A. M., Abdulkarim, A., & Hassan, M. H. (2024). Veterinary telemedicine: A new era for animal welfare. *Open Veterinary Journal*, *14*(4), 952–961. DOI: 10.5455/OVJ.2024.v14.i4.2 PMID: 38808291

Adams, C. R., Barrio Minton, C. A., Hightower, J., & Blount, A. J. (2022). A systematic approach to multiple case study design in professional counseling and counselor education. *The Journal of Counselor Preparation and Supervisor*, *15*(2). https://research.library.kutztown.edu/jcps/vol15/iss2/24

Adams, J., Hofman, K., Moubarac, J. C., & Thow, A. M. (2020, June 26). Public health response to ultra-processed food and drinks. *BMJ (Clinical Research Ed.)*, *369*, m2391. DOI: 10.1136/bmj.m2391 PMID: 32591348

Addy, C., Chorengel, M., Collins, M., & Etzel, M. (2019, Jan-Feb). Calculating the Value of Impact Investing. *Harvard Business Review*. Retrieved 5 March 2020: https://hbr.org/2019/01/calculating-the-value-of-impact-investing

Administrative Office of the U.S. Courts. (2025, June 30). 2024 Wiretap Report: Intercepts and convictions rise. *United States Courts*. https://www.uscourts.gov/data-news/judiciary-news/2025/06/30/2024-wiretap-report-intercepts-and-convictions-rise

Adula, M., Birbirsa, Z. A., & Kant, S. (2023). The effect of interpersonal, problem solving and technical training skills on performance of Ethiopia textile industry: Continuance, normative and affective commitment as mediators. *Cogent Business & Management*, *10*(3), 2286672. DOI: 10.1080/23311975.2023.2286672

Agboola, T. O. (2024). Quantum-resistant cryptographic protocols: Implementation challenges in critical infrastructure systems. *Iconic Research and Engineering Journals*, *1709634*, •••. https://www.irejournals.com/formatedpaper/1709634.pdf

Aggarwal, P., Gonzalez, C., & Dutt, V. (2020). HackIt: A real-time simulation tool for studying real-world cyberattacks in the laboratory, chapter 39. *In Handbook of Computer Networks and Cyber Security*. Springer. https://www.cmu.edu/dietrich/sds/ddmlab/papers/AggarwalGonzalezDutt2020.pdf

Aggarwal, S., & Kumar, N. (2021). Digital signatures. In *Advances in computers, 121* (pp. 95–107). Elsevier., DOI: 10.1016/bs.adcom.2020.08.004

Ahmed, I. S. (2025). Hybrid threat detection systems: A synergistic approach to modern cybersecurity. *European Journal of Computer Science and Information Technology*, *13*(43), 62–69. DOI: 10.37745/ejcsit.2013/vol13n436269

Ahmed, W. (2025). Artificial intelligence in aviation: A review of machine learning and deep learning applications for enhanced safety and security. *Premier Journal of Artificial Intelligence*, *3*, 100013. DOI: 10.70389/PJAI.100013

Ahtisham, M., & Obaid, Z. (2025). *Harnessing Epigenetic Mechanisms for Crop Resilience: A Comprehensive Review of Plant Responses to Biotic and Abiotic Stresses*. https://doi.org/DOI: 10.70389/PJPB.100007

Ajunwa, I., Crawford, K., & Schultz, J. (2017). Limitless Worker Surveillance. *California Law Review*, *105*(3), 735–776.

Al-Abdwani, A., Al-Jahwari, F., Al-Mawali, A., & Al-Zakwani, I. (2023). Exploring patient satisfaction with telemedicine services and its influencing factors. *International Journal of Environmental Research and Public Health*, *20*(4), 3165. DOI: 10.3390/ijerph20043165

Alam, M. Z., Hu, W., & Uddin, M. A. (2020). Digital Transformation In Healthcare Services Sector of Bangladesh: Current Status, Chalenges and Future Dırection. *Journal on Innovation and Sustainability*, *11*(1), 30–38. DOI: 10.23925/2179-3565.2020v11i1p30-38

Alanazi, F., Ushaw, G., & Morgan, G. (2024). Improving Detection of DeepFakes through Facial Region Analysis in Images. *Electronics (Basel)*, *13*(1), 126. Advance online publication. DOI: 10.3390/electronics13010126

Al-Ansari, M., Nabeel, H., Abdella, G. M., Mekkawy, T. E., & Kutty, A. A. (2024). A Non- Parametric Approach-Based Trade-Off between Food System Efficiency and Robustness. *Sustainability (Basel)*, *16*(15), 6480. DOI: 10.3390/su16156480

Alghamdi, A. (2022). *Cybersecurity Threats to Healthcare Sectors during Covid-19. 2022 2nd International Conference on Computing and Information Technology*. ICCIT., https://ieeexplore.ieee.org/abstract/document/9711659?casa_token=e9-Ic-EucqUAAAAA:0UXqAIBHns9A_1pZz9A-vtSgjnWDCTnxNTCpqGkpEotKTDVXs6SDa9-QnT6jKyeUvJ0F4jstnSo, DOI: 10.1109/ICCIT52419.2022.9711659

Almuhaideb, M., Bandaru, D., Abou Chawareb, E., Yafi, F. A., & Hammad, M. A. M. (2025). Ethical perspective on telemedicine usage in sexual medicine: A friend or a foe? *UroPrecision*, *3*(1), 47–53. DOI: 10.1002/uro2.111

Almulihi, A. H., Alassery, F., Khan, A. I., Shukla, S., Gupta, B. K., & Kumar, R. (2022). Analyzing the implications of healthcare data breaches through computational technique. *Intelligent Automation & Soft Computing*, *32*(3). Advance online publication. https://d1wqtxts1xzle7.cloudfront.net/84205476/pdf.pdf?1738492817=&response-content-disposition=inline%3B+filename%3DAnalyzing_the_Implications_of_Healthcare.pdf&Expires=1770151649&Signature=PeIfKoCrbVIFzdbmQ4IYLJy7OKJ9oZ1JtAc6uJiVUit3C189poqRF9Wvi8ZG7PKAK~7zOs6RwkItWbjJzJ0WgV7OdkIPHvp71PVt2ecZ85Tbc-cahxMVx3kiqAXVPZ1ku~UDF0dP5iJmtznTOAC9qr4Pu7Mdc9gS8HJWhga5yzEXc5YCZmCF18kCaw91dyJEbeMDx7LOu83Rrl-dpQrOFrMYfQ7Plk8iHWxweA~cdYI~SKWlPLyxS3IvqD~v43gETpLPH1R6U4quC7vhbKJBoz0LPYcn-Ol08qiPtzzpHHIH~eyDGO~vdU7IHS~8O~puGbfpMTM9OGwH7MSwZBPVjA__&Key-Pair-Id=APKAJLOHF5GGSLRBV4ZA. DOI: 10.32604/iasc.2022.023460

Alnajim, A. M., Habib, S., Islam, M., Thwin, S. M., & Alotaibi, F. (2023). A comprehensive survey of cybersecurity threats, attacks, and effective countermeasures in Industrial Internet of Things. *Technologies*, *11*(6), 161. DOI: 10.3390/technologies11060161

Al-Zaman, S. (2024). Do the reasons for social media censorship vary across political regimes and social media platforms? *Communication Reports, 38*(2), 91–104. DOI: 10.1080/08934215.2024.2432320

Amaan, A., Prekshi, G., & Prachi, S. (2024). Unlocking the Transformative Power of Synthetic Biology. *Archives of Biotechnology and Biomedicine*. https://doi.org/ DOI: 10.29328/journal.abb.1001039

American College of Obstetricians and Gynecologists. (2020, June 3). *ACOG seeks to expand access, increase quality, and improve outcomes for maternal health in rural communities*. https://www.acog.org/news/news-articles/2020/06/acog-seeks-to-expand-access-increase-quality-and-improve-outcomes-for-maternal-health-in-rural-communities

American Hospital Association. (2025, February 6). *Fact sheet: Telehealth*. https://www.aha.org/fact-sheets/2025-02-07-fact-sheet-telehealth

American Society for Reproductive Medicine Ethics Committee. (2024). *Ethical considerations for telemedical delivery of fertility care: An Ethics Committee opinion*. American Society for Reproductive Medicine. https://www.asrm.org/practice-guidance/ethics-opinions/ethical-considerations-for-telemedical/

American Veterinary Medical Association & American Animal Hospital Association. (2018). *The real-life rewards of virtual care: How to turn your hospital into a digitally connected practice with telehealth*. Retrieved March 2, 2025, from https://www.aaha.org/globalassets/05-pet-health-resources/virtual_care.pdf

American Veterinary Medical Association. (2021). *AVMA guidelines for the use of telehealth in veterinary practice: Implementing connected care*. American Veterinary Medical Association. Retrieved from https://www.avma.org/sites/default/files/2021-01/AVMA-Veterinary-Telehealth-Guidelines.pdf

American Veterinary Medical Association. (2023, November 2). *VCPR requirements fuel state legislative activity*. Retrieved from https://www.avma.org/news/vcpr-requirements-fuel-state-legislative-activity

Amodu, O. (2024). Security auditors' perspective in tackling cyber-threats. *Computer, 11*(5). https://www.researchgate.net/profile/Oluwatosin-Amodu/publication/381280763_Security_Auditors'_Perspective_in_Tackling_Cyber-Threats/links/6664c731a54c5f0b94580cd5/Security-Auditors-Perspective-in-Tackling-Cyber-Threats.pdf

Aon. (2025). Global Cyber Risk Report. *Aon plc*. https://www.aon.com/cyber-risk-report

Apio, A. T., Djiby, R. T., & Dinar, A. (2023). Farming under drought: An analysis of the factors influencing farmers' multiple adoption of water conservation practices to mitigate farm-level water scarcity. *Journal of Agricultural and Applied Economics, 55*(3), 432–470. DOI: 10.1017/aae.2023.23

Appelbaum, S. H., Habashy, S., Malo, J., & Shafiq, H. (2012). Back to the future: Revisiting Kotter's 1996 change model. *Journal of Management Development, 31*(8), 764–782. DOI: 10.1108/02621711211253231

Argyris, C., & Schön, D. A. (1978). *Organizational learning: A theory of action perspective*. Addison-Wesley.

Association of American Medical Colleges. (2021, June 11). *The complexities of physician supply and demand: Projections from 2019 to 2034* [Press release]. https://www.aamc.org/news/press-releases/aamc-report-reinforces-mounting-physician-shortage, https://digirepo.nlm.nih.gov/master/borndig/9918417887306676/9918417887306676.pdf

Atadoga, A., Farayola, O. A., Ayinla, B. S., Amoo, O. O., Abrahams, T. O., & Osasona, F. (2024). A comparative review of data encryption methods in the USA and Europe. *Computer Science & IT Research Journal, 5*(2), 447–460. DOI: 10.51594/csitrj.v5i2.815

Atashov, B., Abbasov, A., & Gurbanzade, A. (2025). Government policy on ensuring food security: Bibliometric analysis. *Problems and Perspectives in Management, 23*(1), 496–514. DOI: 10.21511/ppm.23(1).2025.38

Attia, I., Dong, Y., & Ignea, C. (2024). Synthetic biology advances towards a bio-based society in the era of artificial intelligence. *Current Opinion in Biotechnology, 87*, 103143. DOI: 10.1016/j.copbio.2024.103143 PMID: 38781699

Awan, H. M., & Zaurez, S. (2025). Strategic evaluation using PESTLE and SWOT frameworks: Public sector perspective. *International Journal of Research in Business. Economics and Management, 14*(2), 109–120. DOI: 10.5281/zenodo.14854362

Axios. (2024). *Pets in California now have wider access to telehealth options.* Retrieved May 29, 2025, from https://www.axios.com/local/san-francisco/2024/01/03/telehealth-pets-california-law

Baccarne, B., Logghe, S., Schuurman, D., & De Marez, L. (2016, March). Governing Quintuple Helix Innovation: Urban Living Labs and Socio-Ecological Entrepreneurship. Technology *Innovation Management Review*. Retrieved 5 March 2020, from: https://timreview.ca/article/972

Bai, Y., Herforth, A., & Masters, W. A. (2022). Global variation in the cost of a nutrient-adequate diet by population group: An observational study. *The Lancet. Planetary Health*, *6*(1), e19–e28. DOI: 10.1016/S2542-5196(21)00285-0 PMID: 34998455

Balanced Scorecard Institute. (2025, August 7). *Balanced scorecard basics.* https://balancedscorecard.org/bsc-basics-overview/

Ball, K. (2010). Workplace surveillance: An overview. *Labor History*, *51*(1), 87–106. DOI: 10.1080/00236561003654776

Bandura, A. (1997). *Self-efficacy: The exercise of control.* W.H. Freeman.

Bang, D., & Frith, C. D. (2017). Making better decisions in groups. *Royal Society Open Science*, *4*(8), 170193. DOI: 10.1098/rsos.170193

Barker, E., Chen, L., Cooper, D., Moody, D., Regenscheid, A., Souppaya, M., Newhouse, W., Housley, R., Turner, S., Barker, W., & Scarfone, K. (2025). Considerations for achieving cryptographic agility: Strategies and practices (CSWP 39, 2nd Public Draft). *National Institute of Standards and Technology.* https://csrc.nist.gov/pubs/cswp/39/considerations-for-achieving-cryptographic-agility/2pd

Barretta, R., Taherdoost, H., & Madanchian, M. (2023). Multi-Criteria Decision Making (MCDM) Methods and Concepts. *Encyclopedia 2023, Vol. 3, Pages 77-87, 3*(1), 77–87. DOI: 10.3390/encyclopedia3010006

Barth, T. (2011). The idea of a green new deal in a Quintuple Helix Model of knowledge, know-how and innovation. *International Journal of Social Ecology and Sustainable Development*, *1*(2), 1–14. DOI: 10.4018/jsesd.2011010101

Basu, K., Soni, D., Nabeel, M., & Karri, R. (2024). A hardware evaluation study of NIST post-quantum cryptographic signature schemes. *Proceedings of the Second NIST Post-Quantum Cryptography Standardization Conference. National Institute of Standards and Technology.* https://csrc.nist.gov/CSRC/media/Events/Second-PQC-Standardization-Conference/documents/accepted-papers/soni-hardware-evaluation.pdf

Baxter, P., & Jack, S. (2008). Qualitative case study methodology: Study design and implementation for novice researchers. *The Qualitative Report*, *13*(4), 544–559. DOI: 10.46743/2160-3715/2008.1573

Bejarano, G., Csiernik, B., Young, J. J., Stuber, K., & Zadro, J. R. (2022). Healthcare students' attitudes towards patient centred care: A systematic review with meta-analysis. *BMC Medical Education*, *22*(1), 1–22. DOI: 10.1186/s12909-022-03371-1 PMID: 35477455

Bellasio, J., Silfversten, E., Leverett, E., Knack, A., Quimbre, F., Blondes, E. L., Favaro, M., & Persi Paoli, G. (2020). The future of cybercrime in light of technology developments (Research Report No. RRA137-1). *RAND Corporation.* https://www.rand.org/pubs/research_reports/RRA137-1.html

Bensaoud, A., & Kalita, J. (2025). Optimized detection of cyber-attacks on IoT networks via hybrid deep learning models. *Ad Hoc Networks, 170*, 103770. DOI: 10.1016/j.adhoc.2025.103770

Bergstrom, J. C. R., & Schall, A. (2014). *Eye tracking in user experience design.* https://api.semanticscholar.org/CorpusID:40789776

Beyene, S. D., & Demissew, A. (2023). The impact of food insecurity on health outcomes: Empirical evidence from Sub-Saharan African countries. *BMC Public Health, 23*(1), 338. DOI: 10.1186/s12889-023-15244-3 PMID: 36793014

Bhardwaj, R. L., Parashar, A., Parewa, H. P., & Vyas, L. (2024). An alarming decline in the nutritional quality of foods: The biggest challenge for future generations' health. *Foods, 13*(6), 877. DOI: 10.3390/foods13060877 PMID: 38540869

Bhosale, K., Nenova, M., & Iliev, G. (2021). *A study of cyber attacks: In the health-care sector, 2021 Sixth Junior Conference on Lighting*. Lighting., https://ieeexplore.ieee.org/abstract/document/9598947?casa_token=LA0C0P-X-xsAAAAA:vBmONsY-b7MaKYdYJzRsVWUpRrvOA5KjpFphZGFKg3aAQ3antcnYS5angKmYQQ2GBlAd6JDblx8, DOI: 10.1109/Lighting49406.2021.9598947

Bhuvaneswari, R., Saravanan, K., Vennila, S., & Suganthi, S. (2020). The Role of Epigenetics in Plant Breeding Understanding Heritable Changes beyond DNA Sequence. *Plant Science Archives, 5*(1), 22–25. DOI: 10.51470/PSA.2020.5.1.22

Biagini, G. (2025). Towards an AI-literate future: A systematic literature review exploring education, ethics, and applications. *International Journal of Artificial Intelligence in Education, 35*(4), 2616–2666. Advance online publication. DOI: 10.1007/s40593-025-00466-w

Białowąs, S., & Szyszka, A. (2019). *Eye-tracking in Marketing Research.*, DOI: 10.12657/9788379862771-6

Bishop, G. T., Evans, B. A., Kyle, K. L., & Kogan, L. R. (2018). Owner satisfaction with use of videoconferencing for recheck examinations following routine surgical sterilization in dogs. *Journal of the American Veterinary Medical Association, 253*(9), 1151–1157. DOI: 10.2460/javma.253.9.1151 PMID: 30311532

Bivins, A., North, D., Ahmad, A., Ahmed, W., Alm, E., Been, F., Bhattacharya, P., Bijlsma, L., Boehm, A. B., Brown, J., Buttiglieri, G., Calabro, V., Carducci, A., Castiglioni, S., Cetecioglu Gurol, Z., Chakraborty, S., Costa, F., Curcio, S., de los Reyes, F. L.III, & Bibby, K. (2020). Wastewater-Based Epidemiology: Global Collaborative to Maximize Contributions in the Fight Against COVID-19. *Environmental Science & Technology*, *54*(13), 7754–7757. DOI: 10.1021/acs.est.0c02388 PMID: 32530639

Björkdahl, J., & Kronblad, C. (2021). Getting on track for digital work: Digital transformation in an administrative court before and during COVID-19. *Journal of Professions and Organization*, *8*(3), 374–393. DOI: 10.1093/jpo/joab015

Blakely, T., Cleghorn, C., Mizdrak, A., Waterlander, W., Nghiem, N., Swinburn, B., Wilson, N., & Ni Mhurchu, C. (2020). The effect of food taxes and subsidies on population health and health costs: A modelling study. *The Lancet. Public Health*, *5*(7), e404–e413. DOI: 10.1016/S2468-2667(20)30116-X PMID: 32619542

Blanco, E. E., & Sheffi, Y. (2017). Green logistics. *Sustainable supply chains: A research-based textbook on operations and strategy*, 147-187.

Blanco, E. E., & Sheffi, Y. (2024). Green logistics. In *Sustainable supply chains: A research-based textbook on operations and strategy* (pp. 101–141). Springer International Publishing. DOI: 10.1007/978-3-031-45565-0_5

Blum, J., & Parnell, J. (2021). Importance of nutrient availability and metabolism for skeletal muscle regeneration. *Frontiers in Physiology*, *12*, 696018. DOI: 10.3389/fphys.2021.696018 PMID: 34335302

Bolsunovskaya, M. V., Kudryavtseva, T. Y., Rudskaya, I. A., Gintciak, A. M., Zhidkov, D. O., Fedyaevskaya, D. E., & Burlutskaya, Z. V. (2023). Digital Platform for Modeling the Development of Regional Innovation Systems of Russian Federation. *International Journal of Technology*, *14*(8), 1779. DOI: 10.14716/ijtech.v14i8.6843

Boone, G., Bain, M., Cutler, J., & Moody, C. M. (2023). Evaluating video telemedicine for providing virtual health care for cats via mock spay recheck examinations. https://doi.org/DOI: 10.1016/j.applanim.2023.106061

Boshnjaku, A., & Plasari, E. (2023). The role of e-agriculture in promoting sustainable agricultural practices and rural development in Albania. *Management, Enterprise and Benchmarking in the 21st Century*, 64–73. https://login.captechu.idm.oclc.org/login?url=https://www.proquest.com/scholarly-journals/role-e-agriculture-promoting-sustainable/docview/2906894933/sc-2

Bouis, H., & Saltzman, A. (2017). Improving nutrition through biofortification: A review of evidence from HarvestPlus, 2003 through 2016. *Global Food Security, 12*, 49–58. DOI: 10.1016/j.gfs.2017.01.009 PMID: 28580239

Bouktif, S. & N. K. (2024). Real-Time Detection of Student Attention in Online Learning: A Scalable Solution for Crisis-Driven Remote Education. *2024 IEEE International Conference on Progress in Informatics and Computing (PIC)*, 75-79. DOI: 10.1109/PIC62406.2024.10892772

Bowles, T. M., Mooshammer, M., Socolar, Y., Schmer, M. R., Strock, J. A., & Grandy, A. S.. (2020). Long-term evidence shows that crop-rotation diversification increases agricultural resilience to adverse growing conditions in North America. *One Earth, 3*(4), 464–474. DOI: 10.1016/j.oneear.2020.02.007

Brijith, A. (2024). Eavesdropping attack: Consequences and prevention methods. *International Center for AI and Cyber Security Research and Innovations, Asia University*. https://insights2techinfo.com/eavesdropping-attack-consequences-and-prevention-methods/

Brown, A. W., Aslibekyan, S., Bier, D., Ferreira da Silva, R., Hoover, A., Klurfeld, D., & Pavela, G. (2023). Toward more rigorous and informative nutritional epidemiology: The rational space between dismissal and defense of the status quo. *Critical Reviews in Food Science and Nutrition, 63*(18), 3150–3167. DOI: 10.1080/10408398.2021.1985427 PMID: 34678079

Bughin, J., LaBerge, L., & Mellbye, A. (2019, October 29). *Unlocking success in digital transformations*. McKinsey & Company. https://www.mckinsey.com/capabilities/people-and-organizational-performance/our-insights/unlocking-success-in-digital-transformations

Bunnell, B., Barrera, J., Paige, S., Turner, D., & Welch, B. (2020). Acceptability of telemedicine features to promote its uptake in practice: A survey of community telemental health providers. *International Journal of Environmental Research and Public Health, 17*(22), 8525. DOI: 10.3390/ijerph17228525

Buolamwini, J., & Gebru, T. (2018). Gender Shades: Intersectional Accuracy Disparities in Commercial Gender Classification. In S. A. Friedler & C. Wilson (Eds.), *Proceedings of the 1st Conference on Fairness, Accountability and Transparency* (Vol. 81, pp. 77-91). PMLR. https://proceedings.mlr.press/v81/buolamwini18a.html

Burrell, D. N. (2022). Creating inclusive, diverse, and psychologically safe workplaces for African-American women working in public health response supply chain and logistics. *Land Forces Academy Review, 27*(2), 134-140. https://www.armyacademy.ro/reviste/rev2_2022/Art_Burrell_Raft_2_2022.pdf

Burrell, D. N. (2024). Understanding healthcare cybersecurity risk management complexity. *Land Forces Academy Review, 29*(1), 38-49. https://reference-global.com/download/article/10.2478/raft-2024-0004.pdf

Burrell, D. N., Burton, S. L., Nobles, C., Springs, D., Huff, A. J., Brown-Jackson, K. L., . . . Jones, A. J. (2023). The managerial ethical and operational challenges of hospital cybersecurity. In *Transformational Interventions for Business, Technology, and Healthcare* (pp. 444-458). IGI Global. https://www.igi-global.com/chapter/the-managerial-ethical-and-operational-challenges-of-hospital-cybersecurity/332759

Burrell, D. N. (2021). Creating Diverse and Religiously Inclusive Workplace Cultures in Hyper-Connected, Technical, and Cyber-Driven Organizations. [IJSKD]. *International Journal of Sociotechnology and Knowledge Development, 13*(3), 17–32. https://www.igi-global.com/article/creating-diverse-and-religiously-inclusive-workplace-cultures-in-hyper-connected-technical-and-cyber-driven-organizations/281051. DOI: 10.4018/IJSKD.2021070102

Burrell, D. N. (2022). Supply chain and logistics management in public health surveillance and protection during COVID-19. *Science Bulletin, 27*, 12–20.

Burrell, D. N., Aridi, A. S., McLester, Q., Shufutinsky, A., Nobles, C., Dawson, M., & Muller, S. R. (2021). Exploring system thinking leadership approaches to the healthcare cybersecurity environment. [IJEACH]. *International Journal of Extreme Automation and Connectivity in Healthcare, 3*(2), 20–32. https://www.igi-global.com/article/exploring-system-thinking-leadership-approaches-to-the-healthcare-cybersecurity-environment/284462. DOI: 10.4018/IJEACH.2021070103

Burrell, D. N., Burton, S., Jones, L. A., Nobles, C., & Dawson, M. (2026). Healthcare administration in the age of complex dynamics in hospital and health system cybersecurity risk mitigation. *Brazilian Journal of Business, 8*(1), e85317. DOI: 10.34140/bjbv8n1-015

Burrell, D. N., Lu, S., Lindsay, P. V. L., Burton, S. L., French, R., Finch, A., McLester, Q., & Springs, D. (2021). Sexual Harassment Training Focused Solely on Protecting Organizations From Lawsuits Won't Change a Toxic Culture for Women to an Inclusive One. [IJPSS]. *International Journal of Public Sociology and Sociotherapy, 1*(1), 12–25. https://www.igi-global.com/article/sexual-harassment-training-focused-solely-on-protecting-organizations-from-lawsuits-wont-change-a-toxic-culture-for-women-to-an-inclusive-one/269514. DOI: 10.4018/IJPSS.2021010102

Burton, S. L. (2021). Cybersecurity leaders: Knowledge driving human capital development. *Scientific Bulletin, 1*(51), 109-120. Land Forces Academy Publishing House. DOI:DOI: 10.2478/bsaft-2021-0013

Burton, S. L. (2023). Change management and cybersecurity in healthcare: Mitigating human factors and risks. In *Transformational Interventions for Business, Technology, and Healthcare* (pp. 426-443). IGI Global. https://www.igi-global.com/chapter/change-management-and-cybersecurity-in-healthcare/332758

Burton, S. L. (2022). Strategy: A business and cybersecurity intertwined necessity. *International Journal of Smart Education and Urban Society*, *13*(1), 1–12. Advance online publication. DOI: 10.4018/IJSEUS.312232

Cai, Y., & Lattu, A. (2022). Triple Helix or Quadruple Helix: Which Model of Innovation to Choose for Empirical Studies? *Minerva*, *60*(2), 257–280. DOI: 10.1007/s11024-021-09453-6

Çakmakçı, S.. (2024). Foods of the Future: Challenges, Opportunities, Trends in Sustainable Foods, Alternative Protein Sources and Novel Efficient Food Systems. *Foods*, *13*(17), 2663. DOI: 10.3390/foods13172663 PMID: 39272427

Calvo, R. A., & D'Mello, S. (2010). Affect Detection: An Interdisciplinary Review of Models, Methods, and Their Applications. *IEEE Transactions on Affective Computing*, *1*(1), 18–37. DOI: 10.1109/T-AFFC.2010.1

Campra, M., Esposito, P., & Brescia, V. (2023). corporate social responsibility and technological perspectives in healthcare: An exploratory analysis of the evolution of the anti-corruption system through multiple case studies. *Corporate Social Responsibility and Environmental Management*, *30*(6), 2816–2829. DOI: 10.1002/csr.2517

Canadian Centre for Cyber Security. (2022, April). Wi-Fi security (ITSP.80.002). *Author*. https://www.cyber.gc.ca/en/guidance/wi-fi-security-itsp80002

Cantú, C., Franco, C., & Frost, J. (2024). The economic implications of services in the metaverse. In *Global perspectives in the metaverse: Law, economics, and finance* (pp. 83–118). Springer Nature Switzerland., https://link.springer.com/chapter/10.1007/978-3-031-54802-4_6 DOI: 10.1007/978-3-031-54802-4_6

Capatina, A., Bleoju, G., Matos, F., & Vairinhos, V. (2017). Leveraging intellectual capital through Lewin's Force Field Analysis: The case of software development companies. *Journal of Innovation & Knowledge*, *2*(3), 125–133. DOI: 10.1016/j.jik.2016.07.001

Carayannis, E. G., Barth, T. D., & Campbell, D. F. (2012). The Quintuple Helix innovation model: Global warming as a challenge and driver for innovation. *Journal of Innovation and Entrepreneurship*, *1*, 1–12. DOI: 10.1186/2192-5372-1-1

Carayannis, E. G., & Campbell, D. F. (2021). Democracy of climate and climate for democracy: The evolution of quadruple and quintuple helix innovation systems. *Journal of the Knowledge Economy, 12*(4), 2050–2082. DOI: 10.1007/s13132-021-00778-x

Carayannis, E. G., & Campbell, D. F. (2022). Towards an Emerging Unified Theory of Helix Architectures (EUTOHA): Focus on the Quintuple innovation helix framework as the integrative device. *Triple Helix (Heidelberg), 9*(1), 65–75. DOI: 10.1163/21971927-bja10028

Carayannis, E. G., & Campbell, D. F. J. (2009). "Mode 3" and "Quadruple Helix": Toward a 21st century fractal innovation ecosystem. *International Journal of Technology Management, 46*(3/4), 201–234. DOI: 10.1504/IJTM.2009.023374

Carayannis, E. G., Grigoroudis, E., Stamati, D., & Valvi, T. (2019). Social business model innovation: A quadruple/quintuple helix-based social innovation ecosystem. *IEEE Transactions on Engineering Management, 68*(1), 235–248. DOI: 10.1109/TEM.2019.2914408

Carayannis, E. G., & Morawska, J. (2023). University and education 5.0 for emerging trends, policies and practices in the concept of industry 5.0 and society 5.0. In *Industry 5.0: Creative and Innovative Organizations* (pp. 1–25). Springer International Publishing. DOI: 10.1007/978-3-031-26232-6_1

Carayannis, E. G., & Rakhmatullin, R. (2014). The Quadruple Innovation Helix (QIH): Application for the study of university-industry-government-community interfaces. *International Journal of Innovative Research and Development, 5*(3-4), 191–212.

Carelli, L., Solca, F., Tagini, S., Torre, S., Verde, F., Ticozzi, N., Ferrucci, R., Pravettoni, G., Aiello, E. N., Silani, V., & Poletti, B. (2022). Gaze-Contingent Eye-Tracking Training in Brain Disorders: A Systematic Review. *Brain Sciences, 12*(7), 931. Advance online publication. DOI: 10.3390/brainsci12070931 PMID: 35884737

Carrera, A., Zoccarato, F., Mazzeo, M., Lettieri, E., Toletti, G., Bertoli, S., Castelnuovo, G., & Fresa, E. (2023). What drives patients' acceptance of digital therapeutics? establishing a new framework to measure the interplay between rational and institutional factors. *BMC Health Services Research, 23*(1), 145. Advance online publication. DOI: 10.1186/s12913-023-09090-7

Casillas, A., Valdovinos, C., Wang, E., Abhat, A., Mendez, C., Gutierrez, G., Portz, J., Brown, A., & Lyles, C. R. (2022). Perspectives from leadership and frontline staff on telehealth transitions in the Los Angeles safety net during the COVID-19 pandemic and beyond. *Frontiers in Digital Health, 4*, 944860. Advance online publication. DOI: 10.3389/fdgth.2022.944860

Castaño-Gómez, M., Valenzuela-Rojas, R., & Soto-Castro, A. (2022). Review of the use of IoT technologies and devices in physical security systems. *Ingeniería y competitividad, 24*(1), e23123235. https://dialnet.unirioja.es/servlet/articulo?codigo=8637727

Cavallo, M., Morgana, G., Dozzani, I., Gatti, A., Vandoni, M., Pippi, R., Pucci, G., Vaudo, G., & Fanelli, C. G. (2023). Unraveling barriers to a healthy lifestyle: Understanding barriers to diet and physical activity in patients with chronic non-communicable diseases. *Nutrients, 15*(15), 3473. DOI: 10.3390/nu15153473 PMID: 37571410

Cengil, A. B., Eksioglu, S., Eksioglu, B., Eswaran, H., Hayes, C. J., & Bogulski, C. A. (2024). Statistical analysis of telehealth use and pre- and postpandemic insurance coverage in selected health care specialties in a large health care system in Arkansas: Comparative cross-sectional study. *Journal of Medical Internet Research, 26*, e49190. DOI: 10.2196/49190

Chakravarthy, B. K., Krishna Alaparthi, G., Sawant Jitendra, R., Prasad Rao, P., Shetty, V., & Chandrasekaran, B. (2017). *"We are so close; yet too far": perceived barriers to smartphone-based telerehabilitation among healthcare providers and patients with Chronic Obstructive Pulmonary Disease in India.* https://doi.org/DOI: 10.1016/j.heliyon.2021.e07857

Cheng, E. C. K., & Wang, T. (2022). Institutional strategies for cybersecurity in higher education institutions. *Information (Basel), 13*(4), 192. DOI: 10.3390/info13040192

Cheng, S., Li, J., Luo, Y., & Li, X. (2024). Family Quality of Life for Rehabilitated Individuals: Evidence Based on Demographic, Psychological, and Rehabilitation Factors. *SAGE Open, 14*(2), 21582440241258300. Advance online publication. DOI: 10.1177/21582440241258300

Chen, W., Men, Y., Fuster, N., Osorio, C., & Juan, A. A. (2024). Artificial intelligence in logistics optimization with sustainable criteria: A review. *Sustainability (Basel), 16*(21), 9145. DOI: 10.3390/su16219145

Cherkaoui Dekkaki, K., Tasic, I., & Cano, M.-D. (2024). Exploring post-quantum cryptography: Review and directions for the transition process. *Technologies, 12*(12), 241. DOI: 10.3390/technologies12120241

Chesbrough, H. W. (2003). *Open innovation: The new imperative for creating and profiting from technology*. Harvard Business Press.

Clark, T. (2020). *The 4 Stages of Psychological Safety: Defining the Path to Inclusion and Innovation*. Berrett-Koehler Publishers.

Cloud Range. (2024, October 8). Simulating real-world attacks: How cyber ranges strengthen incident response in state government agencies. *Cloud Range Cyber*. https://www.cloudrangecyber.com/news/simulating-real-world-attacks-how-cyber-ranges-strengthen-incident-response-in-state-government-agencies

Code World. (2024). *Learn Business Management*. Retrieved May 30, 2024 from: https://play.google.com/store/apps/details?id=com.codeworld.learnbusiness management&hl=pt

Constantinides, S. V., Turner, C., Frongillo, E. A., Bhandari, S., Reyes, L. I., & Blake, C. E. (2021). Using a global food environment framework to understand relationships with food choice in diverse low-and middle-income countries. Global Food Security, 29, 100511.

Conti, M., Dehghantanha, A., Franke, K., & Watson, S. (2021). Internet of Things security and forensics: Challenges and opportunities. *Future Generation Computer Systems*, *78*, 544–546. DOI: 10.1016/j.future.2017.07.060

Contreras-Manzano, A., Jáuregui, A., Nieto, C., Pineda-López, L. A., Romero-Martínez, M., Medina, C., & Barquera, S. (2024). Self-reported decreases in the purchases of selected unhealthy food categories due to front-of-package warning labels in Mexico. *The International Journal of Behavioral Nutrition and Physical Activity*, *21*(1), 74. DOI: 10.1186/s12966-024-01609-3 PMID: 38987796

Correia, M. I. T. D., & Waitzberg, D. L. (2003). The impact of malnutrition on morbidity, mortality, length of hospital stay and costs evaluated through a multivariate model analysis. *Clinical Nutrition (Edinburgh, Lothian)*, *22*(3), 235–239. DOI: 10.1016/S0261-5614(02)00215-7 PMID: 12765661

Coston, I., Plotnizky, E., & Nojoumian, M. (2025). Comprehensive study of IoT vulnerabilities and countermeasures. *Applied Sciences (Basel, Switzerland)*, *15*(6), 3036. DOI: 10.3390/app15063036

Coventry, L., & Branley, D. (2018). Cybersecurity in healthcare: A narrative review of trends, threats and ways forward. *Maturitas*, *113*, 48–52. https://www.sciencedirect.com/science/article/abs/pii/S0378512218301658?casa_token=DtWzhCy47E8AAAAA:9P3rU8CS-fZb9fBgnFoqfanjIjmuq_u6L15NNIXowFkA2zfi0VlN_NS6K9MintplUTqKccQZRNc. DOI: 10.1016/j.maturitas.2018.04.008

Dargin, M. (2020, October 29). 'Credible threat': How to protect networks from ransomware: With recent warnings of ransomware attacks targeting US healthcare organizations, this 7-step plan including backup and recovery, network monitoring, and antivirus may help protect your company. *Network World (Online).* Retrieved from https://www.networkworld.com/article/964167/how-to-protect-your-network-from-ransomware-attacks.html

Darmon, N., & Drewnowski, A. (2015). Contribution of food prices and diet cost to socioeconomic inequalities in diet quality. *Nutrition Reviews, 73*(10), 643–660. DOI: 10.1093/nutrit/nuv027 PMID: 26307238

Davidi, A. (2025, March 25). Cybercrime or Political Warfare? *The Jerusalem Institute for Strategy and Security.* https://jiss.org.il/en/davidi-cybercrime-or-political-warfare/

Davis, F. D. (1989). Perceived Usefulness, Perceived Ease of Use, and User Acceptance of Information Technology. *MIS Quarterly, 13*(3), 319-340. JSTOR. DOI: 10.2307/249008

Davis, Karen,R.D.H., B.S.D.H. (2011). Sugar obsession in America! Rdh,, 46-49. journals/sugar-obsession-america/docview/927664206/se-2

Davis, R. (2024, March 25). *A healthcare transformation success story: A case study in applying Kotter's 8-step framework.* LinkedIn. https://www.linkedin.com/pulse/healthcare-transformation-success-story-case-study-russ-722dc

de Oliveira Comini, L., Lopes, S. O., Rocha, D. M. U. P., & Silva, M. M. (2025). The effects of subsidies for healthy foods on food purchasing behaviors, consumption patterns, and obesity/overweight: A systematic review. *Nutrition Reviews, 83*(7), e1722–e1739. DOI: 10.1093/nutrit/nuae153 PMID: 39468774

DeChristopher, L. R., Auerbach, B. J., & Tucker, K. L. (2020). High fructose corn syrup, excess-free-fructose, and risk of coronary heart disease among African Americans – the Jackson Heart Study. *BMC Nutrition, 6*(1), 70. DOI: 10.1186/s40795-020-00396-x PMID: 33292663

Deci, E. L., & Ryan, R. M. (2000). The "what" and "why" of goal pursuits: Human needs and the self-determination of behavior. *Psychological Inquiry, 11*(4), 227–268. DOI: 10.1207/S15327965PLI1104_01

DeFries, R. S., Fanzo, J., Remans, R., Palm, C. A., Wood, S., & Anderman, T. L. (2015). Metrics for land-scarce agriculture. *Science, 349*(6245), 238–240. DOI: 10.1126/science.aaa5766 PMID: 26185232

Deguchi, A., Hirai, C., Matsuoka, H., Nakano, T., Oshima, K., Tai, M., & Tani, S. (2020). What is society 5.0. *Society*, *5*, 1–23.

Deloitte. (2024). *Maternal health inequities persist. Can digital tools be part of the solution?* Deloitte Insights. https://www.deloitte.com/us/en/insights/industry/health-care/how-digital-tools-can-help-the-maternal-health-crisis.html

Dempewolf, S., Mouser, B., Rupe, M., Owen, E. C., Reider, L., & Willey, M. C. (2023). What Are the Barriers to Incorporating Nutrition Interventions Into Care of Older Adults With Femoral Fragility Fractures? *The Iowa Orthopaedic Journal*, *43*(2), 172–182. PMID: 38213858

Department for Science. Innovation and Technology. (2025). Cyber Security Longitudinal Survey: Wave Four Results. *GOV.UK*. https://www.gov.uk/government/publications/cyber-security-longitudinal-survey-wave-four-results

Dereso, C. W., Kant, S., Muthuraman, M., & Tufa, G. (2023). Effect of Point of Service on Health Department Student's Creativity in Comprehensive Universities of Ethiopia: Moderating Role of Public-Private Partnership and Mediating Role of Work Place Learning. In Jain, S., Groppe, S., & Mihindukulasooriya, N. (Eds.), *Proceedings of the International Health Informatics Conference* (Vol. 990, pp. 145–157). Springer. DOI: 10.1007/978-981-19-9090-8_13

Devriendt, T., Borry, P., & Shabani, M. (2021). F*actors that influence data sharing through data sharing platforms*: A qualitative study on the views and experiences of cohort holders and platform developers. *PLoS One*, *16*(7), 1–14. DOI: 10.1371/journal.pone.0254202 PMID: 34214146

Diffie, W., & Landau, S. (2023). The evolution of cryptography in securing digital society. *Communications of the ACM*, *66*(2), 24–31. DOI: 10.1145/3564310

Dixit, P. (2023). Nutrition education and behaviour change strategies: Effectiveness and implications for promoting healthy eating habits. *Journal of Sports Science and Nutrition*, *4*(2), 127–132. https://www.allsportsjournal.com/article/197/4-2-27-693.pdf. DOI: 10.33545/27077012.2023.v4.i2b.197

Dixon, T. A., Curach, N., & Pretorius, I. S. (2020). Bio-informational futures: The convergence of artificial intelligence and synthetic biology. *EMBO Reports*, *21*(3), 50036. DOI: 10.15252/embr.202050036 PMID: 32043291

Dodaro, G. (2025). Memo. *Government Accountability Office (GAO)*. https://www.markey.senate.gov/imo/media/doc/ai_letter_to_gao.pdf

Dodla, T. R., & Jones, L. A. (2023). Identifying knowledge management strategies for knowledge management systems. *Access to Science, Business. Innovation in Digital Economy*, *4*(2), 261–277. DOI: 10.46656/access.2023.4.2

Donham, B., & Wickett, M. L. (2018). Novel use of FaceTime video calling in a deployed setting to assist with the care of a military working dog. *Canadian Journal of Surgery*, •••, S232–S234. DOI: 10.1503/CJS.015118 PMID: 30418010

Dovetail. (2024). What is a multiple case study? https://dovetail.com/research/multiple-case-study/

Do, W. L., Coughlin, J. W., & Grant, F. D.. (2020). Consumption of foods derived from subsidized crops and cardiometabolic risk among US adults. *JAMA Internal Medicine*, *180*(6), 787–795.

Doz, Y., & Kosonen, M. (2008). The dynamics of strategic agility: Nokia's rollercoaster experience. *California Management Review*, *50*(3), 95–118. DOI: 10.2307/41166447

Drame, P. S. (2019). *Ransomware attacks in the healthcare industry: Attack methods and preventive steps* (Utica College). Retrieved from https://www.proquest.com/openview/6fc2060ab53329f97deec56acd244c37/1?pq-origsite=gscholar&cbl=18750&diss=y

Duan, C. (2023). A survey of facial expression recognition in the wild. *Applied and Computational Engineering*, *6*(1), 173–181. DOI: 10.54254/2755-2721/6/20230760

Duchowski, A. (2007). *Eye tracking methodology: Theory and practice*. Springer.

Dudley, J., McLaughlin, S., & Lee, T. H. (2022). *Why So Many Women Physicians Are Quitting*. Harvard Business Review. https://hbr.org/2022/01/why-so-many-women-physicians-are-quitting

Dugstad, J., Eide, T., Nilsen, E. R., & Eide, H. (2019). Towards successful digital transformation through co-creation: A longitudinal study of a four-year implementation of digital monitoring technology in residential care for persons with dementia. *BMC Health Services Research*, *19*(1), 366. DOI: 10.1186/s12913-019-4191-1

Durga, P., Godavarthi, D., Kant, S., et al. (2024). Aspect-based drug review classification through a hybrid model with ant colony optimization using deep learning. Discover Computing, 27(19).

Duvvur, V. (2022). Securing the future: Strategies for modernizing legacy systems and enhancing cybersecurity. *Journal of Artificial Intelligence & Cloud Computing, 2*(3), 45–58. https://www.onlinescientificresearch.com/articles/securing-the-future-strategies-for-modernizing-legacy-systems-and-enhancing-cybersecurity.pdf. DOI: 10.47363/JAICC/2022(1)299

Dwiputra, B., Ambari, A. M., Desandri, D. R., Fatrin, S., Zuhdi, N., & Radi, B. (2024). The Future of Digital Healthcare in Cardiac Prevention Programme in Indonesia: From Bedside to Boardroom. *Journal of Asian Pacific Society of Cardiology, 3*, e24. Advance online publication. DOI: 10.15420/japsc.2023.46

Dwyer, S. J., III, Weaver, A. C., & Hughes, K. K. (2004). Health insurance portability and accountability act. *Security Issues in the Digital Medical Enterprise, 72*(2), 9-18. https://web.archive.org/web/20060915001735id_/http://www.cs.virginia.edu/~acw/security/doc/Publications/HIPAA.pdf

Eastburn, J., Fowkes, J., & Kellner, K. (with Swanson, B.) (2022). *Digital transformation: Health systems' investment priorities.* McKinsey & Company. https://www.mckinsey.com/industries/healthcare/our-insights/digital-transformation-health-systems-investment-priorities

Edmondson, A. (2018). *The Fearless Organization: Creating Psychological Safety in the Workplace for Learning, Innovation, and Growth.* Wiley.

Edmondson, A. C. (2018). *The fearless organization: Creating psychological safety in the workplace for learning, innovation, and growth.* John Wiley & Sons.

Efe, M. N. (2024). Dijital dönüşüm çağında sağlık sektöründe metaverse: Swot analizi. *International Refereed Journal*, (24), 50–60.

Efremov, A. V., Mbikayi, Z., & Efremov, E. V. (2021). Comparative Study of Different Algorithms for a Flight Control System Design and the Potentiality of Their Integration with a Sidestick. *Aerospace (Basel, Switzerland), 8*(10), 290. DOI: 10.3390/aerospace8100290

Ekman, P., & Friesen, W. V. (1978). *Facial action coding system: A technique for the measurement of facial movement.* Consulting Psychologists Press.

Encryption Consulting. (2025). NIST selects HQC as fifth algorithm for post-quantum encryption. *Author.* https://www.encryptionconsulting.com/news/nist-selects-hqc

Encryption Consulting. (2025, April 30). Elliptic Curve Cryptography: The Future of Encryption. *Encryption Consulting.* https://www.encryptionconsulting.com/elliptic-curve-cryptography/

Endler, M., Lavelanet, A., Cleeve, A., Ganatra, B., Gomperts, R., & Gemzell-Danielsson, K. (2020). Telemedicine for medical abortion: A systematic review. *BJOG, 127*(9), 1094–1102. DOI: 10.1111/1471-0528.15684

Eslami, R., & Khoveyni, M. (2024). Optimizing Sustainable Green Logistics with AI-Enhanced RAM-Hybrid Models: Addressing Economic and Crisis Management Challenges. *Journal of Industrial Engineering International, 20*, 34.

Etzkowitz, H., & Leydesdorff, L. (2000). The dynamics of innovation: From National Systems and "Mode 2" to a Triple Helix of university–industry–government relations. *Research Policy, 29*(2), 109–1231. DOI: 10.1016/S0048-7333(99)00055-4

F5. (2025). F5 introduces post-quantum cryptography solutions for application and API security platform. *Author.* https://www.f5.com/company/news/press-releases/f5-introduces-post-quantum-cryptography-solutions

Fagerberg, J., & Mowery, D. (2015). *Recent research on innovation and economic change. Innovation, Technology and Economic Change.* Edward Elgar. DOI: 10.4337/9781784718633

Faleide, A. G. H., & Inderhaug, E. (2023). It is time to target psychological readiness (or lack of readiness) in return to sports after anterior cruciate ligament tears. *Journal of Experimental Orthopaedics, 10*(94), 94. Advance online publication. DOI: 10.1186/s40634-023-00657-1 PMID: 37728618

Fal, K., Tomkova, D., Vachon, G., Chaboute, M.-E., Berr, A., & Carles, C. C. (2021). Chromatin Manipulation and Editing: Challenges, New Technologies and Their Use in Plants. *International Journal of Molecular Sciences, 22*(2), 512. DOI: 10.3390/ijms22020512 PMID: 33419220

Fan, S., Yosef, S., & Pandya-Lorch, R. (2020). Linking agriculture to nutrition: The evolution of policy. *China Agricultural Economic Review, 12*(4), 595–604. DOI: 10.1108/CAER-03-2020-0040

Fanzo, J., Hawkes, C., & Udomkesmalee, E. (2015). Nutrition-sensitive agriculture and food systems in practice: Options for intervention. Food and Agriculture Organization of the United Nations. https://openknowledge.fao.org/bitstream/7055154c-1ac0-494a-a566- a0b1fbae10f9/download

Fanzo, J., Davis, C., McLaren, R., & Choufani, J. (2021). Sustainable food systems and nutrition in the 21st century. *The Lancet. Planetary Health, 5*(9), e501–e512. DOI: 10.1016/S2542-5196(21)00184-0

Fanzo, J., Davis, C., McLaren, R., & Choufani, J. (2022). Global food systems and sustainable nutrition in the 21st century: Insights from the Harvard Nutrition Obesity Symposium. *The American Journal of Clinical Nutrition, 115*(1), 18–33. DOI: 10.1093/ajcn/nqab315 PMID: 34523669

Fanzo, J., Davis, C., McLaren, R., & Choufani, J. (2023). Challenges and opportunities for increasing the use and impacts of food reformulation to improve diets and nutrition. *Critical Reviews in Food Science and Nutrition, 63*(5), 1165–1177. DOI: 10.1016/j.foodpol.2023.102515

Fanzo, J., Haddad, L., Schneider, K. R., Béné, C., Covic, N. M., Guarin, A., Herforth, A. W., Herrero, M., Sumaila, U. R., Aburto, N. J., Amuyunzu-Nyamongo, M., Barquera, S., Battersby, J., Beal, T., Bizzotto Molina, P., Brusset, E., Cafiero, C., Campeau, C., Caron, P., & Rosero Moncayo, J. (2021). Viewpoint: Rigorous monitoring is necessary to guide food system transformation in the countdown to the 2030 global goals. *Food Policy, 104*, 102163. Advance online publication. DOI: 10.1016/j.foodpol.2021.102163

Farkas, K., Hillary, L., Malham, S., McDonald, J., & Jones, D. (2020). Wastewater and public health: the potential of wastewater surveillance for monitoring COVID-19, Current Opinion in Environmental Science & Health, Volume 17, Pages 14-20, ISSN 2468-5844.

Farraj, A., & Hammad, E. (2024). A physical-layer security cooperative framework for mitigating interference and eavesdropping attacks in internet of things environments. *Sensors (Basel), 24*(16), 5171. DOI: 10.3390/s24165171 PMID: 39204866

Fasel, B., & Luettin, J. (2003). Automatic facial expression analysis: A survey. *Pattern Recognition, 36*(1), 259–275. DOI: 10.1016/S0031-3203(02)00052-3

Faujdar, D. S., Singh, T., Kaur, M., Sahay, S., & Kumar, R. (2021). Stakeholders' Perceptions of the Implementation of a Patient-Centric Digital Health Application for Primary Healthcare in India. *Healthcare Informatics Research, 27*(4), 315–324. DOI: 10.4258/hir.2021.27.4.315

Federal-Criminal.com. (2024). Federal wiretap and electronic surveillance laws. https://www.federal-criminal.com/wiretap-laws

Fields, S. (2004, October). The fat of the land: Do agricultural subsidies foster poor health? *Environmental Health Perspectives, 112*(14), A820–A823. DOI: 10.1289/ehp.112-a820 PMID: 15471721

Finger, R., Kaphengst, T., El Benni, N., & Walter, A. (2023). Digital innovations for sustainable and resilient agricultural systems. *European Review of Agriculture Economics*, *50*(4), 1277–1309. DOI: 10.1093/erae/jbad021

Floridi, L., Cowls, J., Beltrametti, M., Chatila, R., Chazerand, P., Dignum, V., Luetge, C., Madelin, R., Pagallo, U., Rossi, F., Schafer, B., Valcke, P., & Vayena, E. (2018). AI4People—An Ethical Framework for a Good AI Society: Opportunities, Risks, Principles, and Recommendations. *Minds and Machines*, *28*(4), 689–707. DOI: 10.1007/s11023-018-9482-5 PMID: 30930541

Food and Agriculture Organization (FAO). (2024). The state of agricultural commodity markets 2024. Rome: FAO. https://www.fao.org/publications

Food and Agriculture Organization of the United Nations (FAO). (2025). Transforming agrifood systems with digital public infrastructure: One of the paths to sustainable development. Retrieved from https://www.fao.org/e-agriculture/news/transforming-agrifood-systems-digital-public-infrastructure-one-paths-sustainable-development

Food and Agriculture Organization of the United Nations. (2023a). Price inflation of nutrient-dense foods relative to other foods. In *The state of food security and nutrition in the world 2023*. FAO., DOI: 10.4060/cc0639en

Food and Agriculture Organization of the United Nations. (2023b). *World food and agriculture: Statistical yearbook 2023*. FAO Statistics Division. DOI: 10.4060/cc8166en

Fortanix. (2024). The future of quantum-resistant cryptography: A data security perspective. *Author* https://fortanix.com/blog/the-future-of-quantum-resistant-cryptography

Fortinet. (2025a). What is eavesdropping? Definition, types, and prevention. https://www.fortinet.com/resources/cyberglossary/eavesdropping

Fortinet. (2025b). 2025 Global Cybersecurity Assessment Report. https://www.fortinet.com/resources/cybersecurity-assessment-2025

Franck, C., Grandi, S. M., & Eisenberg, M. J. (2013). Agricultural subsidies and the American obesity epidemic. *American Journal of Preventive Medicine*, *45*(3), 327–333. DOI: 10.1016/j.amepre.2013.04.010 PMID: 23953360

Friedman, N., Pancer, Z., Savic, R., Tseng, F., Lee, M. S., Mclean, L., Bagli, D. J., & Tessaro, M. O. (2024). Evaluation of remote assistance for point-of-care ultrasonography in a large animal hospital: A controlled randomized trial. *Journal of the American Veterinary Medical Association*, *262*(5), 589–595. DOI: 10.2460/javma.23.09.0511 PMID: 38452487

Friis, R. H. (2018). *Epidemiology 101* (2nd ed.). Jones & Bartlett Learning.

Fu, W., Wang, Q., & Zhao, X. (2018). Platform-based service innovation and system design: A literature review. *Industrial Management & Data Systems, 118*(5), 946–974. DOI: 10.1108/IMDS-03-2017-0129

Gallegos-Rejas, V. M., De Guzman, K. R., Kelly, J. T., Smith, A. C., & Thomas, E. E. (2025). Strategies to improve telehealth access for culturally and linguistically diverse communities: A systematic review. *Journal of Public Health (Oxford, England), 47*(3), e374–e390. DOI: 10.1093/pubmed/fdaf030

Gamage, A., Gangahagedara, R., Subasinghe, S., Gamage, J., Guruge, C., Senaratne, S., Randika, T., Rathnayake, C., Hameed, Z., Madhujith, T., & Merah, O. (2024). Advancing sustainability: The impact of emerging technologies in agriculture. *Current Plant Biology, 40*, 100420. DOI: 10.1016/j.cpb.2024.100420

Gans, K. M., Risica, P. M., Wylie-Rosett, J., Yuan, Q., Ross, E. S., & Strolla, L. O. (2023). Local food system approaches to address food and nutrition security among low-income populations: A systematic review. *Advances in Nutrition, 15*(4), 100156. DOI: 10.1016/j.advnut.2023.100156 PMID: 38616069

Garg, D., Majumdar, R., Chauhan, S., Preenja, R., Parihar, J., Saluja, A., & Dhamija, R. K. (2021). Teleneurorehabilitation among person with parkinson's disease in India: The initial experience and barriers to implementation. *Annals of Indian Academy of Neurology, 24*(4), 536–541. DOI: 10.4103/aian.AIAN_127_21

Garg, M., Sharma, N., Sharma, S., Kapoor, P., Kumar, A., Chunduri, V., & Arora, P. (2018). Biofortified crops generated by breeding, agronomy, and transgenic approaches are improving lives of millions of people around the world. *Frontiers in Nutrition, 5*, 12. DOI: 10.3389/fnut.2018.00012 PMID: 29492405

Gautam, A., Chatterjee, S., & Thakur, M. (2025). Inter-satellite authentication protocols based on elliptic curve cryptography. *Journal of Satellite Communication and Networking, 19*(1), 75–96. https://www.arxiv.org/pdf/2503.23277.pdf

Gbaja, C. (2025). Designing quantum-resilient data encryption protocols for securing multi-cloud architectures in critical infrastructure networks. *International Journal of Research Publication and Reviews, 6*(4), 9442–9456. DOI: 10.55248/gengpi.6.0425.1586

Gedikli, E. (2024). Future digital health trends. In Demir Uslu, Y., & Zülfikar, H. (Eds.), *Digitalization in Health (ss* (pp. 181–191). İstanbul University Press., DOI: 10.26650/B/LSB40.2024.035.14

Gehrer, N. A., Svaldi, J., & Duchowski, A. (2024). Assessment of body-related attention processes via mobile eye tracking: A pilot study to validate an automated analysis pipeline. *Proceedings of the 2024 Symposium on Eye Tracking Research and Applications*. DOI: 10.1145/3649902.3653354

Geissdoerfer, M., Savaget, P., Bocken, N. M., & Hultink, E. J. (2017). The Circular Economy–A new sustainability paradigm? *Journal of Cleaner Production, 143*, 757–768. DOI: 10.1016/j.jclepro.2016.12.048

Geng, X., & Xia, Y. (2014). Head Pose Estimation Based on Multivariate Label Distribution. *2014 IEEE Conference on Computer Vision and Pattern Recognition*, 1837-1842. DOI: 10.1109/CVPR.2014.237

Gibbons, M., Limoges, C., Scott, P., Schwartzman, S., & Nowotny, H. (1994). *The new production of knowledge: The dynamics of science and research in contemporary societies.*

GII. (2018). *Global Innovation Index 2018.* Energizing the world with innovation. WIPO. Available at: https://www.wipo.int/edocs/pubdocs/en/wipo_pub_gii_2018.pdf

GII. (2023). *Global Innovation Index 2023.* Innovation in the face of uncertainty. WIPO. Available at: https://www.wipo.int/edocs/pubdocs/en/wipo-pub-2000-2023-en-main-report-global-innovation-index-2023-16th-edition.pdf

Goleman, D., Boyatzis, R. E., & McKee, A. (2002). *Primal leadership: Realizing the power of emotional intelligence.* Harvard Business School Press.

González de la Torre, M. A., Morales Sandoval, I. A., Freitas de Abreu, G. T., & Hernández Encinas, L. (2025). Post-quantum wireless-based key encapsulation mechanism via CRYSTALS-Kyber for resource-constrained devices. *IEEE Access : Practical Innovations, Open Solutions, 13*, 66714–66734. DOI: 10.1109/ACCESS.2025.3560023

Goodfellow, I. J., Erhan, D., Carrier, P. L., Courville, A., Mirza, M., Hamner, B., Cukierski, W., Tang, Y., Thaler, D., Lee, D.-H., Zhou, Y., Ramaiah, C., Feng, F., Li, R., Wang, X., Athanasakis, D., Shawe-Taylor, J., Milakov, M., Park, J., & Bengio, Y. (2013). Challenges in Representation Learning: A Report on Three Machine Learning Contests. In Lee, M., Hirose, A., Hou, Z.-G., & Kil, R. M. (Eds.), *Neural Information Processing* (pp. 117–124). Springer Berlin Heidelberg. DOI: 10.1007/978-3-642-42051-1_16

Gottlieb, S., & Rand, J. (2018). Managing feline diabetes: Current perspectives. *Veterinary Medicine (Auckland), 9*, 33–42. DOI: 10.2147/VMRR.S125619 PMID: 30050865

Granja, C., Janssen, W., & Johansen, M. A. (2018). Factors Determining the Success and Failure of eHealth Interventions: Systematic Review of the Literature. *Journal of Medical Internet Research, 20*(5), e10235. DOI: 10.2196/10235

Green, R., Cornelsen, L., Dangour, A. D., Turner, R. J., Shankar, B., Mazzocchi, M., & Smith, R. D. (2013). The effect of rising food prices on food consumption: Systematic review with meta-regression. *BMJ (Clinical Research Ed.), 346*(jun17 1), f3703. https://www.bmj.com/content/346/bmj.f3703. DOI: 10.1136/bmj.f3703 PMID: 23775799

Groseclose, S. L., & Buckridge, D. L. (2016, December 15). *Public Health Surveillance systems: Recent advances in their use and evaluation*. Annual Reviews. Retrieved September 27, 2021, from https://www.annualreviews.org/doi/10.1146/annurev-publhealth-031816-044348

Guintard, A., Assié, S., Lefèvre, L., Delsart, M., Dile, B., & Millar, N. (2024). Barriers and facilitators to implementing veterinary telemedicine in animal production. *Frontiers in Veterinary Science, 11*, 1452653. DOI: 10.3389/fvets.2024.1452653 PMID: 39634757

Gujral, K., Illarmo, S., Jacobs, J., & Wagner, T. (2025). The economics of telehealth: An overview. *Telemedicine Journal and e-Health, 31*(9), 1074–1095. Advance online publication. DOI: 10.1089/tmj.2025.0073

Gün, R., & Karaoğlu, M. M. (2024). Detection of honey adulteration by characterization of the physico-chemical properties of honey adulterated with the addition of glucose–fructose and maltose corn syrups. European Food Research and Technology = Zeitschrift Für Lebensmittel-Untersuchung Und-Forschung.A, 250(8), 2255-2272.

Gustafsson, J. (2017). Single case studies vs. multiple case studies: A comparative study. Academy of Business, Engineering and Science, Halmstad University. https://www.diva-portal.org/smash/get/diva2:1064378/FULLTEXT01.pdf

Gustafsson, J. (2017). Single case studies vs. multiple case studies: A comparative study. *Academy of Business, Engineering and Science, Halmstad University*. https://www.scirp.org/reference/referencespapers?referenceid=3772444

Guyo, K., & Kero, C. A. (2023). Mediation of Marketing Intermediaries and AI Adoption Between Livestock Products Marketing and Economic Status of Pastoralist in Ethiopia. In 2023 International Conference on Communication, Security and Artificial Intelligence (ICCSAI) (pp. 311–316). IEEE. DOI: 10.1109/ICCSAI59793.2023.10421261

Haghi, M., Danyali, S., Ayasseh, S., Wang, J., Aazami, R., & Deserno, T. M. (2021). Wearable devices in health monitoring from the environmental towards multiple domains: A survey. *Sensors (Basel)*, *21*(6), 2130. DOI: 10.3390/s21062130 PMID: 33803745

Hakamian, M. (2025, April 3). Iran News: Bank Sepah hacking scandal exposes regime's incompetence and paranoia. *National Council of Resistance of Iran Foreign Affairs Committee*. https://www.ncr-iran.org/en/news/iran-news-bank-sepah-hacking-scandal-exposes-regimes-incompetence-and-paranoia/

Hakimi, M., Quchi, M. M., & Fazil, A. W. (2024). Human factors in cybersecurity: an in depth analysis of user centric studies. *Journal Ilmiah Multidisipline Indonesia (JIM-ID,3*(01), 20-33. DOI: https://doi.org/DOI: 10.58471/esaprom.v3i01.3832

Halkias, D., Neubert, M., Harkiolakis, N., & Thurman, P. (2022). *The Multiple Case Study Design*. Routledge. https://www.taylorfrancis.com/books/mono/10.4324/9781003244936/multiple-case-study-design

Hall, B. H., & Rosenberg, N. (2010). *Handbook of the Economics of Innovation*. Elsevier.

Hamdy, R., Omar, Y. M. K., & Maghraby, F. A. (2023). DeepEpi: Deep Learning Model for Predicting Gene Expression Regulation Based on Epigenetic Histone Modifications. *Current Bioinformatics*, *18*. Advance online publication. DOI: 10.2174/1574893618666230818121046

Hamid, B., & Weber, D. (2018). Engineering secure systems: Models, patterns and empirical validation. *Computers & Security*, *77*, 315–348. DOI: 10.1016/j.cose.2018.03.016

Hammond, R. A., & Dubé, L. (2012). A systems science perspective and transdisciplinary models for food and nutrition security. *Proceedings of the National Academy of Sciences of the United States of America*, *109*(31), 12356–12363. DOI: 10.1073/pnas.0913003109 PMID: 22826247

Hao, H., & Padman, R. (2016). An empirical study of opinion leader effects on mobile technology implementation by physicians in an american community health system. *Health Informatics Journal*, *24*(3), 323–333. DOI: 10.1177/1460458216675499

Harris, J., Tan, W., Mitchell, B., & Zayed, D. (2021). Equity in agriculture-nutrition-health research: A scoping review. *Nutrition Reviews*, *80*(1), 78–90. Advance online publication. DOI: 10.1093/nutrit/nuab001 PMID: 33907802

Harst, L., Lantzsch, H., & Scheibe, M. (2019). Theories predicting end-user acceptance of telemedicine use: Systematic review. *Journal of Medical Internet Research*, *21*(5), e13117. DOI: 10.2196/13117

Harvard University. (2025, January 23). What is AI: The pros and cons of artificial intelligence, and what its future holds. *Author*. https://careerservices.fas.harvard.edu/blog/2025/01/23/what-is-ai-the-pros-and-cons-of-artificial-intelligence-and-what-its-future-holds/

Harvard University. (2025, January 23). What is AI: The pros and cons of artificial intelligence, and what its future holds. *Author*. https://www.harvard.edu/in-focus/artificial-intelligence/

Harvie, A., & Wise, T. A. (2009). Sweetening the pot: Implicit subsidies to corn sweeteners and the U.S. obesity epidemic. Global Development and Environment Institute Working Paper.

Hawkes, C. (2012, May 15). Food policies for healthy populations and healthy economies. *BMJ (Clinical Research Ed.)*, *344*(may15 2), e2801. DOI: 10.1136/bmj.e2801 PMID: 22589521

Hayes, T. O'Neill, & Kerska, K. (2021, November 3). Primer: Agriculture Subsidies and Their Influence on the Composition of U.S. Food Supply and Consumption. American Action Forum

Headey, D. D., & Alderman, H. (2019). The relative caloric prices of healthy and unhealthy foods differ systematically across income levels and continents. *The Journal of Nutrition*, *149*(11), 2020–2033. DOI: 10.1093/jn/nxz158 PMID: 31332436

Headey, D., & Ecker, O. (2013). Rethinking the measurement of food security: From first principles to best practice. *Food Security*, *5*(3), 327–343. DOI: 10.1007/s12571-013-0253-0

Hedley, C. B. (2015). The role of precision agriculture for improved nutrient management on farms. *Journal of the Science of Food and Agriculture*, *95*(1), 12–19. DOI: 10.1002/jsfa.6734 PMID: 24816925

Heiding, F., Zorn, R., & Campbell, L. (2025). Cybersecurity strategy scorecard. *Harvard Belfer Center for Science and International Affairs*. https://www.belfercenter.org/research-analysis/cybersecurity-strategy-scorecard

Heilweil, R. (2021, December 21). How Omicron broke Covid-19 testing. Vox.

Hennessey, C., Noureddin, B., & Lawrence, P. (2006). A single camera eye-gaze tracking system with free head motion. *Proceedings of the 2006 Symposium on Eye Tracking Research & Applications*, 87-94. DOI: 10.1145/1117309.1117349

Herforth, A., Wiesmann, D., & Coates, J. (2022). Connecting the food and agriculture sector to nutrition interventions: A framework for integration. *Global Food Security, 30*, 100593. DOI: 10.1016/j.gfs.2021.100593

Hibbard, D., Hibbard, M., & Stockman, J. (2006). The Canoe Theory: A Business Success Strategy for Leaders and Associates . iUniverse.

Hibbard, D. R., Hibbard, M. S., & Stockman, J. W. (2001). *The canoe theory: A secret to building people and profit*. Thomson Learning Custom Publishing.

Hill, C., Martin, J. L., Thomson, S., Scott-Ram, N., Penfold, H., & Creswell, C. (2017). Navigating the challenges of digital health innovation: Considerations and solutions in developing online and smartphone-application-based interventions for mental health disorders. *The British Journal of Psychiatry, 211*(2), 65–69. DOI: 10.1192/bjp.bp.115.180372

Hochman, B. (2022). *The listeners: A history of wiretapping in the United States*. Harvard University Press. https://www.hup.harvard.edu/file/feeds/PDF/9780674249288_sample.pdf

Hossain, M. J. F., Tahora, S., Tasnim, M., Shahriar, H., & Sakib, N. (2022). A review of quantum cybersecurity: Threats, risks and opportunities. *arXiv*. https://arxiv.org/pdf/2207.03534

Houser, S. H., Flite, C. A., & Foster, S. L. (2023). Privacy and security risk factors related to telehealth services—A systematic review. *Perspectives in Health Information Management, 20*(1), Article 1f. https://pmc.ncbi.nlm.nih.gov/articles/PMC9860467/

Huangfu, P., Adams, J., Monsivais, P., & Black, C. (2024). Impact of price reductions, subsidies, or financial incentives on fruit and vegetable intake: A systematic review and meta-analysis. The Lancet Planetary Health, 8(4), e156–e168. 5196(24)00004-4

Hudgins, T., Brown, K. D., Layne, D., & Maggard Stephens, T. (2022). The Effect of Academic Nurse Leaders' Toxic Behaviors. *The Journal of Nursing Education, 61*(2), 88–92. https://journals.healio.com/doi/abs/10.3928/01484834-20211213-02. DOI: 10.3928/01484834-20211213-02 PMID: 35112950

Huey, S. L., Krisher, J. T., Bhargava, A., Friesen, V. M., Konieczynski, E. M., Mbuya, M. N. N., Mehta, N. H., Monterrosa, E., Nyangaresi, A. M., & Mehta, S. (2022). Review of the impact pathways of biofortified foods and food products. *Nutrients*, *14*(6), 1200. DOI: 10.3390/nu14061200 PMID: 35334857

Hughes, H. (2021). *The use of emergency department electronic health data for syndromic surveillance to enhance public health surveillance programmes in england* (Order No. 28827644). Available from ProQuest Dissertations & Theses Global.

Hügle, T., & Grek, J. (2023). Digital transformation of an academic hospital department: A case study on strategic planning using the balanced scorecard. *PLOS Digital Health*, *2*(11), e0000385. Advance online publication. DOI: 10.1371/journal.pdig.0000385

Hu, P., Zhuang, H., Santhalingam, P. S., Spolaor, R., Pathak, P., Zhang, G., & Cheng, X. (2022). AccEar: accelerometer acoustic eavesdropping with unconstrained vocabulary. *2022 IEEE Symposium on Security and Privacy (SP)* DOI: DOI: 10.1109/SP46214.2022.9833716

Hussain, K. (2025). Revolutionizing Route Optimization Systems with Artificial Intelligence for a Smarter, Sustainable Logistics Ecosystem. *International Journal of Computer Science and Mobile Computing*, *14*(2), 66–68. DOI: 10.47760/ijcsmc.2025.v14i02.008

Huynh, M. (2019). Smartphone-based device in exotic pet medicine. *The Veterinary Clinics of North America. Exotic Animal Practice*, *22*(3), 349–366. DOI: 10.1016/j.cvex.2019.05.001 PMID: 31395319

Ibrahim, A. M., Abdel-Aziz, H. R., Mohamed, H. A. H., Zaghamir, D. E. F., Wahba, N. M. I., Hassan, G. A., Shaban, M., El-Nablaway, M., Aldughmi, O. N., & Aboelola, T. H. (2024). Balancing confidentiality and care coordination: Challenges in patient privacy. *BMC Nursing*, *23*(1), 564. DOI: 10.1186/s12912-024-02231-1

Iizumi, T., Shin, Y., Kim, W., Kim, M., & Choi, J. (2020). The global dataset of historical yields for major crops 1981–2016. *Scientific Data*, *7*(1), 97. Advance online publication. DOI: 10.1038/s41597-020-0433-7 PMID: 32198349

Imam, R., Areeb, Q. M., Alturki, A., & Anwer, F. (2021). A. (2021). Systematic and critical review of RSA based public key cryptographic schemes: Past and present status. *IEEE Access : Practical Innovations, Open Solutions*, *9*, 155949–155976. DOI: 10.1109/ACCESS.2021.3129224

Inampudi, S., Rajkumar, E., Gopi, A., Vany Mol, K. S., & Sruthi, K. S. (2024). Barriers to implementation of digital transformation in the Indian health sector: a systematic review. In *Humanities and Social Sciences Communications* (Vol. 11, Issue 1). Springer Nature. DOI: 10.1057/s41599-024-03081-7

Innovation Markets. (2024). *Income Markets*. Retrieved May 30, 2024 from: https://play.google.com/store/apps/details?id=com.innovationmarkets.android&hl=pt

Insight7. (2024). Eavesdropping attacks: Organizational responses and best practices. https://insight7.com/eavesdropping-attacks

Integrity360. (2025). TeleMessage breach exposes government communications metadata. https://www.integrity360.com/news/telemessage-breach

Intergovernmental Panel on Climate Change. (2022). *Climate Change 2022: Mitigation of Climate Change. Contribution of Working Group III to the Sixth Assessment Report of the IPCC*. Cambridge University Press., DOI: 10.1017/9781009157926

International Institute for Sustainable Development (IISD). (2009). Biofuels—at what cost? Government support for ethanol and biodiesel in the United States: 2007 update.

Isakov, T.-M., Härkönen, H., Atkova, I., Wang, F., Vesty, G., Hyvämäki, P., & Jansson, M. (2024). From challenges to opportunities: Digital transformation in hospital-at-home care. *International Journal of Medical Informatics*, *192*, 192. DOI: 10.1016/j.ijmedinf.2024.105644

Ivanova, J., Shah, B., Foote, C., & Cummins, M. R. (2025). Assessing the costs and savings of telemedicine: Insights from the consolidated framework for implementation research. *Digital Health*, *11*, 20552076251314552. Advance online publication. DOI: 10.1177/20552076251314552

Iyer, R. F., & Mehta, I. (2025, May 5). TeleMessage, a modified Signal clone used by US government officials, has been hacked. *TechCrunch*. https://techcrunch.com/2025/05/05/telemessage-a-modified-signal-clone-used-by-us-government-officials-has-been-hacked/

Iyer, M. S., Way, D. P., MacDowell, D. J., Overholser, B. M., Spector, N. D., & Jagsi, R. (2023). Bullying in Academic Medicine: Experiences of Women Physician Leaders. *Academic Medicine*, *98*(2), 255–263. https://journals.lww.com/academicmedicine/fulltext/2023/02000/bullying_in_academic_medicine__experiences_of.27.aspx. DOI: 10.1097/ACM.0000000000005003 PMID: 36484542

Iyer, S. (2025). Cybersecurity culture and organizational resilience: A human-centered approach to digital risk management. *Computer Science and Engineering*. DOI:DOI: 10.13140/RG.2.2.17733.56806

Jackson, R. J., Minjares, R., Naumoff, K. S., Shrimali, B. P., & Martin, L. K. (2009). Agriculture policy is health policy. *Journal of Hunger & Environmental Nutrition*, *4*(3-4), 393–408. DOI: 10.1080/19320240903321367 PMID: 23144677

Janis, I. L. (1982). *Groupthink: Psychological studies of policy decisions and fiascoes* (2nd ed.). Houghton Mifflin.

Jefferson, B., & Marrero, C. O. (2020). Robust assessment of real-world adversarial examples. *In CVPR Workshops 2020*. https://openaccess.thecvf.com/content_CVPRW_2020/papers/w47/Jefferson_Robust_Assessment_of_Real-World_Adversarial_Examples_CVPRW_2020_paper.pdf

Jergler, D. (2018, May 7). *High-tech remote health monitoring*. Veterinary Practice News. https://www.veterinarypracticenews.com/high-tech-remote-health-monitoring/

Johansson, E., & Kuktaite, R.. (2025). Plant proteins for human consumption – from local to global supply chains. *One Health*, •••, 100836.

Johnson, E. E., Kruis, R., Verdin, R., Wells, E., Ford, D. W., & Sterba, K. R. (2023). Development of an implementation science telehealth toolkit to promote research capacity in evaluation of telehealth programs. *Telemedicine Reports*, *4*(1), 286–291. DOI: 10.1089/tmr.2023.0039 PMID: 37817872

Johnson, L., & Martinez, R. (2025). Incorporating video telehealth for improving at-home management of cats with chronic mobility issues. *Frontiers in Veterinary Science*, *12*, 1510006. DOI: 10.3389/fvets.2025.1510006 PMID: 40235573

Jones, H. (2024). The impact of emerging technologies on privacy rights. *American Journal of Law and Policy*, *1*(1), 25–34. https://forthworthjournals.org/journals/index.php/AJLP/article/view/23

Jones, L. A. (2024). Reputation risk and potential profitability: Best practices to predict and mitigate stakeholder loss in the financial sector. In Lewis, E. J. (Ed.), *Evolution of cross-sector cyber intelligent markets* (pp. 291–313). IGI Global., DOI: 10.4018/978-1-6684-8974-1.ch014

Jones, L. A. (2024). Unveiling human factors: Aligning facets of cybersecurity leadership, insider threats, and arsonist attributes to reduce cyber risk. *SocioEconomic Challenges*, *8*(2), 43–63. DOI: 10.61093/sec.8(2).44-63.2024

Jones, L. A., & Burrell, D. N. (2025). Illegal Cybersecurity Threats Created by Organizational Arsonists in Healthcare Organizations. *Law. Economics and Society, 1*(1), 93–p93. https://j.ideasspread.org/index.php/les/article/view/1427

Kaboré, S. S., Ngangue, P., Soubeiga, D., Barro, A., Pilabré, A. H., Bationo, N., Pafadnam, Y., Drabo, K. M., Hien, H., & Savadogo, G. B. L. (2022). Barriers and facilitators for the sustainability of digital health interventions in low and middle-income countries: A systematic review. *Frontiers in Digital Health, 4*, 1014375. DOI: 10.3389/fdgth.2022.1014375

Kaiser Permanente. (n.d.). *24/7 virtual care*. https://healthy.kaiserpermanente.org/learn/how-to-use-telehealth

Kaiser, J. L., Morhe, E., & Abern, L. R. (2022). Quantitative and qualitative impact of one key question on primary care providers' contraceptive counseling at routine preventive health visits. *Contraception, 109*, 73–79. DOI: 10.1016/j.contraception.2022.01.004

Kale, A. A. (2025). Modern AI in Green Logistics: Smart Systems for Sustainability and Carbon Reduction. *Green Logistics: Smart Systems for Sustainability and Carbon Reduction (March 01, 2025). Global Journal of Emerging AI and Computing, 1*(2), 147–169. DOI: 10.70445/gjeac.1.2.2025.147-169

Kanade, P., David, F., & Kanade, S. (2021). Convolutional Neural Networks(CNN) based Eye-Gaze Tracking System using Machine Learning Algorithm. *European Journal of Electrical Engineering and Computer Science*, *5*(2), 36–40. DOI: 10.24018/ejece.2021.5.2.314

Kane, C. K., & Gillis, K. (2018). The use of telemedicine by physicians: Still the exception rather than the rule. *Health Affairs, 37*(12), 1923–1930. DOI: 10.1377/hlthaff.2018.05077 PMID: 30633670

Kanter, R., Walls, H. L., Tak, M., Roberts, F., & Waage, J. (2015). A conceptual framework for understanding the impacts of agriculture and food system policies on nutrition and health. *Food Security, 7*(4), 767–777. DOI: 10.1007/s12571-015-0473-6

Katre, A., & Raddatz, B. (2023). Low-income families' direct participation in food-systems innovation to promote healthy food behaviors. *Nutrients, 15*(5), 1271. DOI: 10.3390/nu15051271 PMID: 36904271

Kaufman, A., Boren, J., Koukel, S., Ronquillo, F., Davies, C., & Naranjo, J. (2017). Agriculture and health sectors collaborate in addressing population health. *Annals of Family Medicine, 15*(5), 475–480. DOI: 10.1370/afm.2087 PMID: 28893819

Kaye, D. B. V. (2025). Renting royalties: How the assetization of music copyrights contributes to inequality for musicians. *Media Culture & Society, 01634437251341715*(7), 1394–1412. Advance online publication. DOI: 10.1177/01634437251341715 PMID: 40978152

Keller, T. (2023). *It is time to change the way we change* [Master's thesis, University of Central Florida]. STARS. https://purls.library.ucf.edu/go/DP0027553

Keller, J. (2024). Eavesdropping: The Forgotten Public Nuisance in the Age of Alexa. *Vanderbilt Law Review*, *77*(1), 169–222. https://scholarship.law.vanderbilt.edu/vlr/vol77/iss1/4/

Kellnhofer, P., Recasens, A., Stent, S., Matusik, W., & Torralba, A. (2019). Gaze360: Physically Unconstrained Gaze Estimation in the Wild. *2019 IEEE/CVF International Conference on Computer Vision (ICCV)*, 6911-6920. DOI: 10.1109/ICCV.2019.00701

Kesarwani, A., & Gochhayat, S. (2023). Ransomware attacks in the healthcare industry. *Journal of Student Research*, *12*(4). Advance online publication. https://www.jsr.org/hs/index.php/path/article/view/5799. DOI: 10.47611/jsrhs.v12i4.5799

Keshaviah, A., Hu, X., & Henry, M. (2021). Developing a Flexible National Wastewater Surveillance System for COVID-19 and Beyond. *Environmental Health Perspectives*, *129*(4), 045002. Advance online publication. DOI: 10.1289/EHP8572 PMID: 33877858

Kester, D. T. (2021). *Proactive text messaging for public health surveillance during the COVID-19 pandemic* (Order No. 28652914). Available from ProQuest Dissertations & Theses Global.

Khalil, L., Serhier, Z., & Othmani, M. B. (2024). Exploring patients' acceptance of mental health e-services in Morocco: A quantitative approach. *Cureus*. Advance online publication. DOI: 10.7759/cureus.76143

Khanna, A. (2024). Introduction to Cybersecurity. In *Securing an Enterprise*. Apress., DOI: 10.1007/979-8-8688-1029-9_1

Kho, J., Gillespie, N., & Martin-Khan, M. (2020). A systematic scoping review of change management practices used for telemedicine service implementations. *BMC Health Services Research*, *20*(1), 815. Advance online publication. DOI: 10.1186/s12913-020-05657-w

Kholiavko, N., Grosu, V., Safonov, Y., Zhavoronok, A., & Cosmulese, C. G. (2021). Quintuple helix model: Investment aspects of higher education impact on sustainability. *Management Theory and Studies for Rural Business and Infrastructure Development*, *43*(1), 111–128. DOI: 10.15544/mts.2021.10

Kim, J., Cai, Z. R., Chen, M. L., Onyeka, S., Ko, J. M., & Linos, E. (2024). Telehealth utilization and associations in the United States during the third year of the COVID-19 pandemic: Cross-sectional survey study. *JMIR Public Health and Surveillance, 10*, e51279. https://publichealth.jmir.org/2024/1/e51279/. DOI: 10.2196/51279

Klerkx, L., Jakku, E., & Labarthe, P. (2019). Social science on digital agriculture, smart farming, and actor-networks: A review. *NJAS Wageningen Journal of Life Sciences, 90–91*, 100315. DOI: 10.1016/j.njas.2019.100315

Klotz, S. A., Chan, C. B., Bianchi, S., Egurrola, C., & York, L. D. (2023). The genie is out of the bottle: Telemedicine is more effective than brick-and-mortar clinics in the care of HIV-infected outpatients. *The American Journal of Medicine, 136*(4), 360–364. https://pubmed.ncbi.nlm.nih.gov/36495936/. DOI: 10.1016/j.amjmed.2022.11.012

Knai, C., Petticrew, M., Mays, N., Capewell, S., Cassidy, R., Cummins, S., Eastmure, E., Fafard, P., Hawkins, B., Jensen, J. D., Katikireddi, S. V., Mwatsama, M., Orford, J., & Weishaar, H. (2018, September). Systems Thinking as a Framework for Analyzing Commercial Determinants of Health. *The Milbank Quarterly, 96*(3), 472–498. DOI: 10.1111/1468-0009.12339 PMID: 30277610

Knockel, J., Wang, M., & Reichert, Z. (2024). The not-so-silent type: Vulnerabilities across keyboard apps reveal keystrokes to network eavesdroppers. *The Citizen Lab.* https://citizenlab.ca/2024/04/vulnerabilities-across-keyboard-apps-reveal-keystrokes-to-network-eavesdroppers/

Kobylińska, M., Antosik, K., Decyk, A., & Kurowska, K. (2022). Malnutrition in obesity: Is it possible? *Obesity Facts, 15*(1), 19–25. DOI: 10.1159/000519503 PMID: 34749356

Kogan, L. R., Hellyer, P. W., Ruch-Gallie, R., Rishniw, M., & Schoenfeld-Tacher, R. (2016). Veterinarians' use and perceptions of information and communication technologies. *Medical Research Archives, 4*(2). https://journals.ke-i.org/mra/article/view/502

Kognou, A. L. M., Shrestha, S., Jiang, Z., Xu, C., Sun, F., & Qin, W. (2022). High-fructose corn syrup production and its new applications for 5-hydroxymethylfurfural and value-added furan derivatives: Promises and challenges. Journal of Bioresources and Bioproducts. DOI: 10.1016/j.jobab.2022.03.004

Kong, L., Ai, Y., Lei, L., Kaddoum, G., Chatzinotas, S., & Ottersten, B. (2021). An overview of generic tools for information-theoretic secrecy performance analysis over wiretap fading channels. *EURASIP Journal on Wireless Communications and Networking*, *194*(1), 194. https://link.springer.com/article/10.1186/s13638-021-02065-4#citeas. DOI: 10.1186/s13638-021-02065-4 PMID: 34899875

Kostadinov, D. (2020). IoT security fundamentals: Intercepting and manipulating wireless communications. *Infosec Institute*. https://www.infosecinstitute.com/resources/iot-security/iot-security-fundamentals-intercepting-and-manipulating-wireless-communications/

Kotter, J. P. (1995). Leading change: Why transformation efforts fail? *Harvard Business Review*, *73*(2), 59–67. https://hbr.org/1995/05/leading-change-why-transformation-efforts-fail-2

Kotter, J. P. (1996). *Leading change*. Harvard Business School Press.

Koyame-Marsh, R. O., & Marsh, J. L. (2014). Data breaches and identity theft: Costs and responses. *IOSR Journal of Economics and Finance (IOSR-JEF), 5*(3), 36-45. https://www.researchgate.net/profile/Rita-Koyame-Marsh/publication/274881413_Data_Breaches_and_Identity_Theft_Costs_and_Responses_Rita_O_Koyame-Marsh_and_John_L_Marsh/links/552b71d70cf2779ab7930f83/Data-Breaches-and-Identity-Theft-Costs-and-Responses-Rita-O-Koyame-Marsh-and-John-L-Marsh.pdf

Kraus, S., Schiavone, F., Pluzhnikova, A., & Invernizzi, A. C. (2021). Digital transformation in healthcare: Analyzing the current state-of-research. *Journal of Business Research*, *123*, 557–567. DOI: 10.1016/j.jbusres.2020.10.030

Kreier, F. (2021). The Myriad Ways Sewage Surveillance Is Helping Fight COVID Around the World. *Nature*. Advance online publication. DOI: 10.1038/d41586-021-01234-1 PMID: 33972790

Kruse, C. S., Frederick, B., Jacobson, T., & Monticone, D. K. (2017). Cybersecurity in healthcare: A systematic review of modern threats and trends. *Technology and Health Care*, *25*(1), 1–10. https://journals.sagepub.com/doi/abs/10.3233/THC-161263. DOI: 10.3233/THC-161263

Kshetri, N. (2018). The Economics of Cyber-Insurance. vol. 20, no. 6, pp. 9-14, 1 IEEE. Nov.-Dec. doi.1109/MITP.2018.287421010._https://ieeexplore.ieee.org/abstract/document/8617758?casa_token=CPydYW0yQ0cAAAAA:IL6cOMxY2LMDQ9uIY8wP-Ed23cZ3shVMwmQD6XFn-4CGawlYHDmivjTjOvTTzVT5iOs0lPPRFfs

Kshetri, N. (2023). Post-quantum cryptography and the future of cybersecurity in critical infrastructure sectors. *Journal of Strategic Security, 16*(2), 1–22. DOI: 10.5038/1944-0472.16.2.2012

Kumar, A. (2015). Green Logistics for sustainable development: An analytical review. *IOSRD International Journal of Business, 1*(1), 7–13.

Kumar, R. (2024). A comprehensive review of mcdm methods, applications, and emerging trends. *Decision Making Advances, 3*(1), 185–199. DOI: 10.31181/dma31202569

Kumar, R., & Alavi, S. (2023). Physical security countermeasures against covert surveillance in financial institutions: A multi-layered approach. *Journal of Financial Cybersecurity, 5*(3), 211–229. DOI: 10.1080/26479890.2023.00503

Kumar, S. (2025). Innovations in agricultural inputs and services markets in India: A pathway to sustainable agricultural growth. *Indian Journal of Agricultural Economics, 80*(1), 91–112. DOI: 10.63040/25827510.2025.01.004

Kumar, S., & Joshi, D. (2018). Role of agricultural infrastructure and climate change on agricultural efficiency in Uttar Pradesh: A panel data analysis. *Economic Affairs, 63*(4), 871–882. DOI: 10.30954/0424-2513.4.2018.10

Kumera, D., Amentie, C., & Bali, N. (2024). Effect of technological innovation on firm's performance: mediating effect of competitive advantage: a study on manufacturing firms operating in Ethiopian industrial parks. *Brazilian Journal of Operations & Production Management, 21*(3), e20242146. DOI: 10.14488/BJOPM.2146.2024

Kutnjak, A. (2021). Covid-19 Accelerates Digital Transformation in Industries: Challenges, Issues, Barriers and Problems in Transformation. *IEEE Access : Practical Innovations, Open Solutions, 9*, 79373–79388. DOI: 10.1109/ACCESS.2021.3084801

Kutztown University. (2021). Qualitative research: Case study guidelines. https://library.kutztown.edu/qualitative/casestudy

Lakkakula, P., Schmitz, A., & Ripplinger, D. (2016). U.S. Sweetener Demand Analysis: A QUAIDS Model Application. *Journal of Agricultural and Resource Economics, 41*(3), 533–548. https://login.captechu.idm.oclc.org/login?url=https://www.proquest.com/scholarly-journals/u-s-sweetener-demand-analysis-quaids-model/docview/1826924823/se-2

Lamé, G., Jouini, O., & Stal-Le Cardinal, J. (2017). Combining the viable system model and Kotter's 8 steps for multidepartment integration in hospitals. *IDEAS Working Paper Series from RePEc*. Retrieved from https://hal.science/hal-01519593

Lara-Castor, L., Micha, R., Cudhea, F., Miller, V., & Ali, J. (2023). Sugar-sweetened beverage intakes among adults between 1990 and 2018 in 185 countries. *Nature Communications, 14*(1), 5957. Advance online publication. DOI: 10.1038/s41467-023-41269-8 PMID: 37788998

Lawn, J. E., Rohde, J., Rifkin, S., Were, M., Paul, V. K., & Chopra, M. (2018). Alma-Ata 40 years on: Primary health care for the 21st century. *Lancet, 372*(9642), 917–927. DOI: 10.1016/S0140-6736(08)61402-6 PMID: 18790315

Lawrence, M. A., Baker, P. I., Pulker, C. E., & Pollard, C. M. (2019). Sustainable, resilient food systems for healthy diets: The transformation agenda. *Public Health Nutrition, 22*(1), 87–96. https://www.cambridge.org/core/journals/public-health-nutrition/article/abs/sustainable-resilient-food-systems-for-healthy-diets-the-transformation-agenda/3D12E7C3679EE2BDC6EA97D1CE82C1E8. DOI: 10.1017/S1368980019003112 PMID: 31407652

Lee, G., Lim, T., & Ruiz, D. (2025). Adapting Cybersecurity Maturity Models for Resource-Constrained Settings: A Case Study of Peru. *Information Systems Development, 34*(2), 221–240. https://onlinelibrary.wiley.com/doi/pdf/10.1002/isd2.12350

Legler, S., Diehl, M., Hilliard, B., Olson, A., Markowitz, R., Tignanelli, C., Melton, G. B., Broccard, A., Kirsch, J., & Usher, M. (2021). Evaluation of an Intrahospital Telemedicine Program for Patients Admitted With COVID-19: Mixed Methods Study. *Journal of Medical Internet Research, 23*(4), e25987. DOI: 10.2196/25987

Lemmettylä, A., & Kinnunen, U.-M. (2025). Driving and restraining forces in the implementation of information systems in public sector organizations: Scoping review using Lewin's change theory. *JMIR Human Factors, 12*, e71575. https://humanfactors.jmir.org/2025/1/e71575. DOI: 10.2196/71575

Lewin, K. (1951). *Field theory in social science: Selected theoretical papers.* Harper & Row.

Li, C. H., Yang, W. G., & Shih, I. T. (2021). Exploration on the gap of single- and double-loop learning of balanced scorecard and organizational performance in a health organization. *Heliyon, 7*(12), e08553. DOI: 10.1016/j.heliyon.2021.e08553

Li, J., Zhao, B., & Zhang, H. (2009). Face Recognition Based on PCA and LDA Combination Feature Extraction. *2009 First International Conference on Information Science and Engineering*, 1240-1243. DOI: 10.1109/ICISE.2009.581

Li, M., & Dou, Z. (2023). Active eavesdropping detection: A novel physical layer security in wireless IoT. *EURASIP Journal on Advances in Signal Processing, 2023*(119), 119. Advance online publication. DOI: 10.1186/s13634-023-01080-5

Liu, F., Shahzad, M. A., Feng, Z., Wang, L., & He, J. (2024). An analysis of the effect of agriculture subsidies on technical efficiency: Evidence from rapeseed production in China. Heliyon, 10(13).

Liu, C., & Babar, M. A. (2024). Corporate cybersecurity risk and data breaches: A systematic review of empirical research. *Australian Journal of Management*, •••, 1–31. DOI: 10.1177/03128962241293658

Liu, J., Qin, Y., Liu, H., Liu, Y., Yang, Y., Ning, Y., & Ye, H. (2025). Global, regional, and national burden of female infertility and trends from 1990 to 2021 with projections to 2050 based on the GBD 2021 analysis. *Scientific Reports*, *15*(1), 17559. Advance online publication. DOI: 10.1038/s41598-025-01498-x

Liu, W., & Park, E. (2012). E-healthcare security solution framework. *International Conference on Computer Communications and Networks (ICCCN)*. https://doi.org/ DOI: 10.1109/ICCCN.2012.6289239

Liu, X., Meng, X., Duan, H., Hu, Z., & Wang, M. (2025). A Survey on Secure WiFi Sensing Technology: *Attacks and Defenses. Sensors (Basel)*, *25*(6), 1913. DOI: 10.3390/s25061913 PMID: 40293025

Liyanage, S. I. H., & Netswera, F. G. (2021). Greening universities with mode 3 and quintuple helix model of innovation–production of knowledge and innovation in knowledge-based economy, Botswana. *Journal of the Knowledge Economy*, •••, 1–31.

Lo Sasso, A. T., Armstrong, D., Forte, G., & Gerber, S. E. (2020). Differences In Starting Pay For Male And Female Physicians Persist; Explanations For The Gender Gap Remain Elusive. *Health Affairs*, *39*(2), 256–263. https://pubmed.ncbi.nlm.nih.gov/31967925/. DOI: 10.1377/hlthaff.2019.00664 PMID: 31967925

Lokare, A., Bankar, S., & Mhaske, P. (2025). Integrating cybersecurity frameworks into IT security: A comprehensive analysis of threat mitigation strategies and adaptive technologies. *arXiv*. https://arxiv.org/pdf/2502.00651v1

Lopez, J., Cadena, V., & Rahman, M. S. (2025). Evaluating post-quantum cryptographic algorithms on resource-constrained devices. *arXiv*. https://arxiv.org/pdf/2507.08312

Lowell, J. (2004). The food industry and its impact upon increasing global obesity: A case study. *British Food Journal*, *106*(2), 238–248. DOI: 10.1108/00070700410528817

lowRISC. (2025). Tackles post-quantum cryptography challenges through research collaborations. *Author*. https://lowrisc.org/blog/2025/01/tackles-post-quantum-cryptography-challenges

Lugaresi, C., Tang, J., Nash, H., McClanahan, C., Uboweja, E., Hays, M., Zhang, F., Chang, C.-L., Yong, M., Lee, J., Chang, W.-T., Hua, W., Georg, M., & Grundmann, M. (2019). *MediaPipe: A Framework for Building Perception Pipelines*. https://doi.org//arXiv.1906.08172DOI: 10.48550

Lundy, D., Burrell, D. N., McCargo, T., Lima, F. M., Harris, N., Eke, H., & Prunella, K. (2025, November). Understanding the Psychology of Pregnancy Discrimination as a Challenging Woman's Mental Health Workplace Dynamic. In *RAIS Conference Proceedings 2022-2025* (No. 0608). Research Association for Interdisciplinary Studies. https://rais.education/wp-content/uploads/0608.pdf

Lynn, J. (2024, May 16). Reactions to the Ascension Healthcare Ransomware Attack and Suggestions for Healthcare Organizations. *Healthcare Technology Today*. Retrieved from: https://www.healthcareittoday.com/2024/05/16/reactions-to-the-ascension-healthcare-ransomware-attack-and-suggestions-for-healthcare-organizations/

Lyons, N. B., Bernardi, K., Olavarria, O. A., Shah, P., Dhanani, N., Loor, M., Holihan, J. L., & Liang, M. K. (2021). Gender Disparity Among American Medicine and Surgery Physicians: A Systematic Review. *The American Journal of the Medical Sciences*, *361*(2), 151–168. https://www.sciencedirect.com/science/article/abs/pii/S0002962920304705. DOI: 10.1016/j.amjms.2020.10.017 PMID: 33526213

MacKenzie, I. S. (2010). An eye on input: Research challenges in using the eye for computer input control. *Proceedings of the 2010 Symposium on Eye-Tracking Research & Applications*, 11-12. DOI: 10.1145/1743666.1743668

Magalhães-Sant'Ana, M., Peleteiro, M. C., & Stilwell, G. (2020). Opinions of Portuguese veterinarians on telemedicine: A policy delphi study. *Frontiers in Veterinary Science*, *7*, 549. Advance online publication. DOI: 10.3389/fvets.2020.00549 PMID: 32974405

Maganinho, M., Almeida, C., & Padrão, P. (2024). Industrially Produced Plant-Based Food Products: Nutritional Value and Degree of Processing. *Foods*, *13*(11), 1752. DOI: 10.3390/foods13111752 PMID: 38890980

Mahardhani, A. J. (2023). The Role of Public Policy in Fostering Technological Innovation and Sustainability. [ADMAN]. *Journal of Contemporary Administration and Management*, *1*(2), 47–53. DOI: 10.61100/adman.v1i2.22

Makridis, C. A. (2021). Do data breaches damage reputation? Evidence from 45 companies between 2002 and 2018. *Journal of Cybersecurity*, *7*(1), tyab021. Advance online publication. DOI: 10.1093/cybsec/tyab021

Malila, Y.. (2024). Current challenges of alternative proteins as future foods. *Nature Food*, •••, 00291. PMID: 39147771

Mamatha, G. S., Dimri, N., & Sinha, R. (2024). Post-quantum cryptography: Securing digital communication in the quantum era. arXiv preprint arXiv:2403.11741. https://arxiv.org/abs/2403.11741

Mamatha, G. S., Sinha, R., & Dimri, N. (2025). Post-quantum cryptography: Securing digital communication in the quantum era. *arXiv preprint arXiv:2403.11741.* https://arxiv.org/pdf/2403.11741

Marklund, M., Martinez-Steele, E., Barquera, S., & L'Abbé, M. (2020). Health impact and cost-effectiveness of financing fruit and vegetable subsidies for a vulnerable population in the United States. *PLoS One*, *15*(2), e0229269. PMID: 32084196

Marshall, K. (2019, February 7). *Ethical issues for today's veterinarian in the digital age*. Veterinary Practice News. https://www.veterinarypracticenews.com/ethical-issues-for-todays-veterinarian-in-the-digital-age/

Martin, G., Martin, P., Hankin, C., Darzi, A., & Kinross, J. (2017). Cybersecurity and healthcare: How safe are we? *BMJ (Clinical Research Ed.)*, *358*, j3179. https://www.bmj.com/content/358/bmj.j3179.full. DOI: 10.1136/bmj.j3179

Ma, Y., Zhao, C., Zhao, Y., Lu, J., Jiang, H., Cao, Y., & Xu, Y. (2022). Telemedicine application in patients with chronic disease: A systematic review and meta-analysis. *BMC Medical Informatics and Decision Making*, *22*(1), 105. Advance online publication. DOI: 10.1186/s12911-022-01845-2

Mayer, S. D. (2025). Zero trust architecture and the future of network security. https://www.sdmayer.com/insights/zero-trust-architecture

Mbunge, E., Muchemwa, B., Jiyane, S., & Batani, J. (2021). Sensors and healthcare 5.0: Transformative shift in virtual care through emerging digital health technologies. *Global Health Journal (Amsterdam, Netherlands)*, *5*(4), 169–177. DOI: 10.1016/j.glohj.2021.11.008

McClintock, H. F., & Imel, B. E. (2024). Food insecurity and medication restricting behavior among persons with diabetes in the United States. *Nutrition and Health (Berkhamsted, Hertfordshire)*, *30*(2), 341–347. DOI: 10.1177/02601060221115588 PMID: 35876349

McKeon, J. (2023, January 11). Global Cyberattacks Increased By 38% Last Year, Healthcare Hit Hard. Health IT Security. https://www.techtarget.com/healthtechsecurity/news/366594501/Global-Cyberattacks-Increased-By-38-Last-Year-Healthcare-Hit-Hard

McKeown, P. C., & Spillane, C. (2014). *Landscaping plant epigenetics* (Vol. 1112). Humana Press., DOI: 10.1007/978-1-62703-773-0_1

McKinnon, A., Browne, M., Whiteing, A., & Piecyk, M. (Eds.). (2015). *Green logistics: Improving the environmental sustainability of logistics*. Kogan Page Publishers.

McNabb, S. J. N., Conde, J. M., Ferland, L., Memish, Z. A., Okutani, S., Park, M. M., Ryland, P., Shaikh, A. T., & Singh, V. (Eds.). (2016). *Transforming Public Health Surveillance: Proactive measures for prevention, detection, and response*. Elsevier.

Meister, S., Burmann, A., & Deiters, W. (2019). Digital Health Innovation Engineering: Enabling Digital Transformation in Healthcare: Introduction of an Overall Tracking and Tracing at the Super Hospital Aarhus Denmark. *Management for Professionals*, (Part F518), 329–341. DOI: 10.1007/978-3-319-95273-4_17

Mekyassi, H., & Kızıldeniz, T. (2024). The role of nutrition-sensitive climate-smart agriculture in ensuring global food security. BIO Web of Conferences, 85, 01055. https://doi.org/DOI: 10.1051/bioconf/20248501055

Mendieta, J. P., Sangra, A., Yan, H., Minow, M. A. A., & Schmitz, R. J. (2023). Exploring Plant Cis-Regulatory Elements at Single-Cell Resolution: Overcoming Biological and Computational Challenges to Advance Plant Research. *The Plant Journal, 115*(6), 1486–1499. Advance online publication. DOI: 10.1111/tpj.16351 PMID: 37309871

Mendoza-Velázquez, A., Guzmán-Rodríguez, M., Lara-Arévalo, J., & Drewnowski, A. (2023). The nutrient rich food price index: A nutrition-relevant adaptation of the Laspeyres price index to track the cost of affordable nutrient density. *Frontiers in Nutrition, 10*, 1107573. DOI: 10.3389/fnut.2023.1107573 PMID: 37275634

Meng, Y., Ke, C., & Huang, Z. (2025). Comprehensive Survey towards Security Authentication Methods for Satellite Communication Systems. arXiv preprint arXiv:2503.23277. https://www.arxiv.org/pdf/2503.23277

Mennella, C., Maniscalco, U., De Pietro, G., & Esposito, M. (2024). Ethical and regulatory challenges of AI technologies in healthcare: A narrative review. *Heliyon, 10*(4), e26297. DOI: 10.1016/j.heliyon.2024.e26297

Menon, A. K., Adhya, S., & Kanitkar, M. (2021). Health technology assessment of telemedicine applications in Northern borders of India. *Medical Journal, Armed Forces India, 77*(4), 452–458. DOI: 10.1016/j.mjafi.2021.03.007

Mertler, C. A. (2017). *Action research communities: Professional learning, empowerment, and improvement through collaborative action research* (1st ed.). Taylor & Francis Group., DOI: 10.4324/9781315164564

Miletic, B., Sutter, Y., Starčević-Klasan, G., Courteney, U., Jelovica, L., Lekić, A., & Šegulja, S. (2024). Malnutrition among the older adult: An additional challenge for the rehabilitation team-insights of a small population of Switzerland. *Frontiers in Public Health, 12*, 1436566. DOI: 10.3389/fpubh.2024.1436566 PMID: 39045162

Miller, V., Webb, P., Cudhea, F., Shi, P., Zhang, J., Reedy, J., Erndt-Marino, J., Coates, J., Mozaffarian, D., Bas, M., Ali, J. H., Abumweis, S., Krishnan, A., Misra, P., Hwalla, N. C., Janakiram, C., Liputo, N. I., Musaiger, A., Pourfarzi, F., & Hakeem, R. (2022). Global dietary quality in 185 countries from 1990 to 2018 show wide differences by nation, age, education, and urbanicity. *Nature Food, 3*(9), 694–702. DOI: 10.1038/s43016-022-00594-9 PMID: 37118151

Mohamad, N. I., Abd Rahman, I., Sanusi, S., Abdullah, S., & Ibrahim, N. (2023). An empirical study of the relationship between leadership practice in training programs and skill development: Motivational climate as mediating. *TEM Journal, 12*(4), 2112–2122. https://doi.org/DOI: 10.18421/TEM124-22

Mohan, S., Nandhini, S., Aparna, H., & Gunaseelan, K. (2025). Ensemble bagging-based classifier for eavesdropper detection in 6G wireless networks. *Peer-to-Peer Networking and Applications, 18*(4), 238. Advance online publication. DOI: 10.1007/s12083-025-02049-x

Moisey, L. L., Merriweather, J. L., & Drover, J. W. (2022). The role of nutrition rehabilitation in the recovery of survivors of critical illness: Underrecognized and underappreciated. *Critical Care, 26*(1), 1–17. DOI: 10.1186/s13054-022-04143-5 PMID: 36076215

Morimoto, C. H., & Mimica, M. R. M. (2005). Eye gaze tracking techniques for interactive applications. *Computer Vision and Image Understanding, 98*(1), 4–24. DOI: 10.1016/j.cviu.2004.07.010

Morrow, T., & Nicolai, M. (2022). The Zero Trust journey: 4 phases of implementation. *Carnegie Mellon University, Software Engineering Institute.* https://insights.sei.cmu.edu/blog/the-zero-trust-journey-4-phases-of-implementation/

Mouloudj, K., Oanh, V. L., Bouarar, A., Bouarar, A. C., Asanza, D. M., & Srivastava, M. (2024). *Adopting Artificial Intelligence in Healthcare: A Narrative Review. S. Teixeira, & J. Remondes içinde, The Use of Artificial Intelligence in Digital Marketing: Competitive Strategies and Tactics.* IGI Global Scientific Publishing.

Muhammad, S. K., Ansari, T. A., Shabbir, B., Almagharbeh, W. T., Rehman, A., Arjmand, R., & Ghulam, A. (2025). THE ROLE OF ARTIFICIAL INTELLIGENCE IN PUBLIC HEALTH SURVEILLANCE: A POST-PANDEMIC PERSPECTIVE. *Insights-Journal of Life and Social Sciences, 3*(3 (Social)), 74–80. DOI: 10.71000/pa7ab080

Muheriwa-Matemba, S. R., Alcena-Stiner, D. C., Glazier, A., & LeBlanc, N. M. (2024). Telehealth use for sexual and reproductive health promotion and care during the early phase of the COVID-19 pandemic: A descriptive-interpretive qualitative study of healthcare providers' perspectives and experiences in Western—Central New York State. *PLOS Global Public Health, 4*(12), e0003259. Advance online publication. DOI: 10.1371/journal.pgph.0003259

Mumtaz, H., Riaz, M. H., Wajid, H., Saqib, M., Zeeshan, M. H., Khan, S. E., Chauhan, Y. R., Sohail, H., & Vohra, L. I. (2023). Current challenges and potential solutions to the use of digital health technologies in evidence generation: A narrative review. *Frontiers in Digital Health, 5*, 1203945. DOI: 10.3389/fdgth.2023.1203945

Murray-Kolb, L. E., Wenger, M. J., Scott, S. P., Rhoten, S. E., Lung'aho, M. G., & Haas, J. D. (2017). Consumption of iron-biofortified beans positively affects cognitive performance in 18- to 27-year-old Rwandan female college students in an 18-week randomized controlled efficacy trial. *The Journal of Nutrition, 147*(11), 2109–2117. DOI: 10.3945/jn.117.255356 PMID: 28954841

Muruccia, Y., Solazzo, G., Del Vecchio, P., & Passiante, G. (2020). Evidence from network analysis application to innovation systems and quintuple helix. *Technological Forecasting and Social Change, 161*, 120306. DOI: 10.1016/j.techfore.2020.120306

Mwogosi, A. (2023). *Digital Transformation in Tanzania's Healthcare Sector: A Systematic Review of Robust Electronic Health RecordsSystems' Critical Success Factors*. DOI: 10.21203/rs.3.rs-3034281/v1

Naamati-Schneider, L. (2023). The effect of digitalization on service orientation and service perception among Israeli healthcare professionals: A qualitative study. *Digital Health, 30*(9), 20552076231191892. DOI: 10.1177/20552076231191892

Nadhamuni, S., John, O., Kulkarni, M., Nanda, E., Venkatraman, S., Varma, D., Balsari, S., Gudi, N., Samantaray, S., Reddy, H., & Sheel, V. (2021). Driving digital transformation of comprehensive primary health services at scale in India: An enterprise architecture framework. *BMJ Global Health, 6*(Suppl 5), e005242. DOI: 10.1136/bmjgh-2021-005242

Naeem, M., Ozuem, W., Howell, K., & Ranfagni, S. (2023). A step-by-step process of thematic analysis to develop a conceptual model in qualitative research. *International Journal of Qualitative Methods*, *22*, 1–18. DOI: 10.1177/16094069231205789

Nakamura, Y., & Elhassan, M. (2025). Enhancing quantum key distribution resilience with entanglement-assisted protocols. *Journal of Quantum Networking*, *7*(1), 45–58. DOI: 10.1039/qn2025-45

Namin, F. S., Ghadi, A., & Saki, F. (2022). A literature review of Multi Criteria Decision-Making (MCDM) towards mining method selection (MMS). *Resources Policy*, *77*, 102676. DOI: 10.1016/j.resourpol.2022.102676

NASDAQ OMX's News Release Distribution Channel. (2022, January 24). The global precision farming market is expected to grow by $3.52 bn during 2022–2026, progressing at a CAGR of 8.41% during the forecast period [Press release]. *Nasdaq*. https://www.proquest.com/wire-feeds/global-precision-farming-market-is-expected-grow/docview/2622232350/se-2

Näther, C., Herzinger, D., Gazdag, S.-L., Steghöfer, J.-P., Daum, S., & Loebenberger, D. (2024). Migrating software systems towards post-quantum-cryptography: A systematic literature review. arXiv:2404.12854. https://arxiv.org/abs/2404.12854

National Academies of Sciences, Engineering, and Medicine. (2021). *Implementing high-quality primary care: Rebuilding the foundation of health care.* The National Academies Press. https://doi.org/DOI: 10.17226/25983

National Center for Biotechnology Information. (2014). Cybersecurity: An ongoing process. Author. https://www.ncbi.nlm.nih.gov/pmc/articles/PMC4147612/

National Cyber Security Centre (NCSC). (2023). Next steps in preparing for post-quantum cryptography. *Author.* https://www.ncsc.gov.uk/blog-post/next-steps-in-preparing-for-post-quantum-cryptography

National Institute of Standards and Technology (NIST). (2020). Control baselines for information systems and organizations. *(NIST SP 800-53B).* DOI: 10.6028/NIST.SP.800-53B

National Institute of Standards and Technology [NIST]. (2017). Digital identity guidelines: Authentication and lifecycle management (NIST Special Publication 800-63B). DOI: 10.6028/NIST.SP.800-63b

National Institute of Standards and Technology. (2024a). NIST releases first 3 finalized post-quantum encryption standards. *Author.* https://www.nist.gov/news-events/news/2024/01/nist-releases-first-3-finalized-post-quantum-encryption-standards

National Institute of Standards and Technology. (2024b). *Stateless Hash-Based Digital Signature Standard (FIPS 205)*. Author., DOI: 10.6028/NIST.FIPS.205

National Rural Health Association. (2024, January 31). *Maternal health in rural America* [White paper]. https://www.ruralhealth.us/nationalruralhealth/media/documents/advocacy/white%20paper/maternal-health-in-rural-america-white-paper-final.pdf

Navas de Solis, C., Bevevino, K., Doering, A., O'Gan, D., Teller, L., & Underwood, C. (2019). Real-time telehealth using ultrasonography is feasible in equine practice. *Equine Veterinary Education, 32*(4), 218–222. DOI: 10.1111/eve.13177

Naveen, R., Sundaram, T. G., Agarwal, V., & Gupta, L. (2021). Teleconsultation experience with the idiopathic inflammatory myopathies: A prospective observational cohort study during the COVID-19 pandemic. *Rheumatology International, 41*(1), 67–76. DOI: 10.1007/s00296-020-04737-8

Neal, S. M., & Greenberg, M. J. (2024). Identifying policy alternatives to enable the virtual establishment of a veterinarian–client–patient relationship. *Journal of Feline Medicine and Surgery, 26*(3), 1098612X231224167. Advance online publication. DOI: 10.1177/1098612X231224167 PMID: 38478919

Neha, M. V., & Thampi, S. M. (2025). Smart Surveillance in Physical Security Systems and Ethical Frameworks in IoD. In *Internet of Drones: Emerging Paradigms in Surveillance and Security* (pp. 267–284). Springer., https://link.springer.com/chapter/10.1007/978-3-031-82826-3_13

Neill, C. L., & Morgan, K. L. (2025). Few differences in rural and urban adoption of veterinary telehealth services. *Journal of the American Veterinary Medical Association, 263*(2), 1–7. DOI: 10.2460/javma.24.08.0508 PMID: 39642457

Nelson-Pratt, A. (2018). Telemedicine debate is taking far too long. *The Veterinary Record, 183*(21), 663–663. DOI: 10.1136/vr.k5073 PMID: 30504170

Nicoletti, B., & Appolloni, A. (2024). Green Logistics 5.0: A review of sustainability-oriented innovation with foundation models in logistics. *European Journal of Innovation Management, 27*(9), 542–561. DOI: 10.1108/EJIM-07-2024-0787

Nielsen, M. A., & Chuang, I. L. (2023). *Quantum Computation and Quantum Information (20th Anniversary Edition)*. Cambridge University Press., https://www.cambridge.org/highereducation/books/quantum-computation-and-quantum-information/01E10196D0A682A6AEFFEA52D53BE9AE#overview

Niessen, S., & Forcada, Y. (2019, March 11). *Telemedicine advances: Putting specialist help in all practices.* Vet Times. https://www.vettimes.co.uk/article/telemedicine-advances-putting-specialist-help-in-all-practices/

Nobles, C. (2018). Botching human factors in cybersecurity in business organizations. *HOLISTICA Journal of Business and Public Administration, 9*(3), 71-88. https://holisticajournal.ro/docs/1589fb80c59c854cabe168f3b7b8c1d7.pdf

Nobles, C. (2025, April). Human Factors Engineering in Explainable AI: Putting People First. *In Proceedings of the 19th International Conference on Cyber Warfare and Security*. Academic Conferences and publishing limited. https://www.google.com/books/edition/Proceedings_of_the_19th_International_Co/ZjtaEQAAQBAJ?hl=en&gbpv=1&dq=human+factors,+Calvin+Nobles&pg=PA313&printsec=frontcover

Nobles, C., Rangarajan, A., & Burrell, D. N. (2025). Human Factors Engineering-as-a-Service in Cybersecurity. In *New Horizons in Leadership: Inclusive Explorations in Health, Technology, and Education* (pp. 315-342). IGI Global Scientific Publishing. https://www.igi-global.com/chapter/human-factors-engineering-as-a-service-in-cybersecurity/371041

Nobles, C., & Robinson, N. (2024). 3 The benefits of human factors engineering in cybersecurity. In Engemann, K. J., & Witty, J. A. (Eds.), *Cybersecurity Risk Management: Enhancing Leadership and Expertise* (Vol. V). De Gruyter., DOI: 10.1515/9783111289069-003

Nozari, H. (2024). Green supply chain management based on artificial intelligence of everything. *Journal of Economics and Management, 46*, 171–188. DOI: 10.22367/jem.2024.46.07

Nuako, A., Liu, J., Pham, G., Smock, N., James, A., Baker, T., Bierut, L., Colditz, G., & Chen, L.-S. (2022). Quantifying rural disparity in healthcare utilization in the United States. *PLoS One, 17*(2), e0263718. Advance online publication. DOI: 10.1371/journal.pone.0263718

Nunes, S., Lopes, R. & Fuller-Love, N (2017, June). Networking, Innovation, and Firms' Performance: Portugal as Illustration. *Journal of the Knowledge Economy* 10, 899–920. DOI: 10.1007/s13132-017-0508-7 -

Odoms-Young, A., Brown, A. G. M., Agurs-Collins, T., & Glanz, K. (2024). Food insecurity, neighborhood food environment, and health disparities: State of the science, research gaps and opportunities. *The American Journal of Clinical Nutrition, 119*(3), 850–861. DOI: 10.1016/j.ajcnut.2023.12.019 PMID: 38160801

Ofe, M. A. (2023). A qualitative study exploring security practices healthcare providers need to reduce the risk of successful ransomware attacks on electronic health record systems (Order No. 30523518). Colorado Technical University. Retrieved from https://www.proquest.com/openview/675d545f322aecdce179edd0438b9879/1?pq-origsite=gscholar&cbl=18750&diss=y

Office for Civil Rights. (2025, June 27). *Regulatory initiatives*. U.S. Department of Health and Human Services. https://www.hhs.gov/hipaa/for-professionals/regulatory-initiatives/index.html

Office of Inspector General. (2020, September 14). Evil twins, eavesdropping, and password cracking: How the Office of Inspector General successfully attacked the U.S. Department of the Interior's Wireless Networks. Report No.: 2018-ITA-020. *Author.* https://www.doioig.gov/sites/default/files/2021-migration/FinalAudit_WirelessNetworkSecurity_Public.pdf#:~:text=the%20U.S.%20Department%20of%20the%20Interior%E2%80%99s%20wireless%20networks

Ofori, K. F., Antoniello, S., English, M. M., & Aryee, A. N. A. (2022). Improving nutrition through biofortification: A systematic review. *Frontiers in Nutrition, 9,* 1043655. DOI: 10.3389/fnut.2022.1043655 PMID: 36570169

Ogawa, T., Nakazawa, A., & Nishida, T. (2017). Point of gaze estimation using corneal surface reflection and omnidirectional camera image. *2017 Fifteenth IAPR International Conference on Machine Vision Applications (MVA)*, 227-230. DOI: 10.23919/MVA.2017.7986842

Oh, C., Ha, J., & Roh, H. (2022). A Survey on TLS-Encrypted Malware Network Traffic Analysis. *Applied Sciences, 12*(1), 155. MDPI. https://www.mdpi.com/2076-3417/12/1/155

Olyaeemanesh, A., Takian, A., Mostafavi, H., Mobinizadeh, M., Bakhtiari, A., Yaftian, F., Vosoogh-Moghaddam, A., & Mohamadi, E. (2023, November 18). Health equity impact assessment (HEIA) reporting tool: Developing a checklist for policymakers. *International Journal for Equity in Health, 22*(1), 241. https://pmc.ncbi.nlm.nih.gov/articles/PMC10657117/. DOI: 10.1186/s12939-023-02031-0

Orero-Blat, M., Leal-Rodríguez, A. L., Albort-Morant, G., & Ariza-Montes, A. (2025). The strategic advantage of innovative organizational culture: An exploratory analysis in digital transformation and big data analytics capabilities. *Journal of Management & Organization, 31*(2), 443–462. DOI: 10.1017/jmo.2024.74

Otter, V., & Robinson, D. (2024). Transparency and changing stakeholder roles in the digital age of sustainable agri-food supply chain networks. *Frontiers in Sustainable Food Systems, 8,* 1449684. DOI: 10.3389/fsufs.2024.1449684

Otto, L. (2019). *Implementation Barriers for Telemedicine Initiatives Investigating Barriers for the Implementation of Telemedicine Initiatives: A Systematic Review of Reviews Completed Research.*

Pacifico Silva, H., Lehoux, P., Miller, F. A., Wong, G., & Denis, J.-L. (2018). Introducing responsible innovation in health: A policy-oriented framework. *Health Research Policy and Systems, 16*(1), 90. Advance online publication. DOI: 10.1186/s12961-018-0362-5 PMID: 30200985

Pakravan-Charvadeh, M. R.. (2024). Understanding the interplay between subsidy reforms, food prices, and nutrition outcomes: Empirical evidence from Iran. *Frontiers in Public Health, 12*, 2024.

Palar, K., Sheira, L. A., Frongillo, E. A., Kushel, M., Wilson, T. E., Conroy, A. A., Adedimeji, A., Merenstein, D., Cohen, M. H., Wentz, E. L., Adimora, A. A., Ofotokun, I., Metsch, L. R., Turan, J. M., Tien, P. C., & Weiser, S. D. (2023). Longitudinal relationship between food insecurity, engagement in care, and ART adherence among US women living with HIV. *AIDS and Behavior, 27*(10), 3345–3355. DOI: 10.1007/s10461-023-04053-9 PMID: 37067613

Palleti, V. R., Adepu, S., Mishra, V. K., & Mathur, A. (2021). Cascading effects of cyber-attacks on interconnected critical infrastructure. *Cybersecurity, 4*(8), 8. Advance online publication. DOI: 10.1186/s42400-021-00071-z

Panwar, L., & Singh, A. (2023). Principles of security usability in vulnerability analysis and risk assessment. *International Journal of Advanced Research in Engineering Technology & Science, 10*(1), 95–102. https://www.ijarets.org/publication/95/63.%20ijarets%20jan%202023.pdf

Pattison-Gordon, J. (2023, November 27). Ransomware impacts healthcare systems in six states. *Government Technology*. Retrieved from https://www.govtech.com/security/ransomware-impacts-health-care-systems-in-six-states

Paul, P. K., & Jena, S. K. (2023). Economical, financial and allied concerns in healthy agricultural information systems practice: The context of developing countries. *Economic Affairs, 68*(4), 2161–2170. DOI: 10.46852/0424-2513.4.2023.25

Pauszek, J. R. (2023). An introduction to eye tracking in human factors healthcare research and medical device testing. *Human Factors in Healthcare, 3*, 100031. DOI: 10.1016/j.hfh.2022.100031

Pearson, S., & Benameur, A. (2010). Privacy, security and trust issues arising from cloud computing. *In 2010 IEEE Second International Conference on Cloud Computing Technology and Science*, 693–702. *IEEE*. DOI: 10.1109/CloudCom.2010.66

Peltier, T. R. (2016). *Information security policies, procedures, and standards: Guidelines for effective information security management.* CRC Press. https://books.google.com/books/about/Information_Security_Policies_Procedures.html?id=mM_LsS-W4f4C

Picard, R. W. (1997). *Affective computing.* MIT Press. DOI: 10.7551/mit-press/1140.001.0001

Picha, K. J., DiSanti, J. S., & Uhl, T. (2025, August 25). A sport medicine clinician's guide to improving patient adherence through tenets of Social Cognitive Theory. *Medical Research Archives, 13*(8). Advance online publication. DOI: 10.18103/mra.v13i8.6846

Pinto, H., & Nogueira, C. (2018) Mapping na Entrepreneurial, Innovative and Sustainable Ecosystem Using Social Network Analysis: An Exploratory Approach of Publicly Funded Innovative Project Data. In: Leitão J., Alves H., Krueger N., Park J. (eds) *Entrepreneurial, Innovative and Sustainable Ecosystems. Applying Quality of Life Research* (Best Practices). Springer, Cham. DOI: 10.1007/978-3-319-71014-3_12

Plank-Bazinet, J. L., Heggeness, M. L., Lund, P. K., & Clayton, J. A. (2017). Women's Careers in Biomedical Sciences: Implications for the Economy, Scientific Discovery, and Women's Health. *Journal of women's health (2002), 26*(5), 525–529. https://www.liebertpub.com/doi/abs/10.1089/jwh.2016.6012

Plant Epigenomics (pp. 263–286). (2023). Elsevier eBooks. https://doi.org/DOI: 10.1016/B978-0-323-91909-8.00007-4

Poeran, J., Cho, L., Wilson, L., Zhong, H., Mazumdar, M., Liu, J., & Memtsoudis, S. (2021). Pre-existing disparities and potential implications for the rapid expansion of telemedicine in response to the coronavirus disease 2019 pandemic. *Medical Care, 59*(8), 694–698. DOI: 10.1097/MLR.0000000000001585

Popkin, B. M., & Ng, S. W. (2022). The nutrition transition to a stage of high obesity and noncommunicable disease prevalence dominated by ultra-processed foods. *Obesity Reviews, 23*(1), e13366. DOI: 10.1111/obr.13366 PMID: 34632692

Porter, M. E. (1985). *Competitive advantage: Creating and sustaining superior performance.* Free Press.

Pratt, H., Moroney, T., & Middleton, R. (2021). The influence of engaging authentically on nurse–patient relationships: A scoping review. *Nursing Inquiry, 28*(2), e12388. Advance online publication. DOI: 10.1111/nin.12388

Preko, M., Odame, F., & Anim-Yeboah, S. (2022). Doctorpreneurship in the digital age. *International Journal of E-Entrepreneurship and Innovation*, *12*(2), 1–19. DOI: 10.4018/IJEEI.314464

Pricopoaia, O., Busila, A. V., Cristache, N., Susanu, I., & Matis, C. (2023). Challenges for entrepreneurial innovation: Startups as tools for a better knowledge-based economy. *The International Entrepreneurship and Management Journal*, •••, 1–42.

Proofpoint. (2025). Eavesdropping attacks: How they work and how to prevent them. https://www.proofpoint.com/us/threat-reference/eavesdropping-attacks

Punuri, S. B., Kuanar, S. K., & Mishra, T. K. (2023). Facial Emotion Recognition in Unconstrained Environments through Rank-Based Ensemble of Deep Learning Models using 1-Cycle Policy. *2023 International Conference on the Confluence of Advancements in Robotics, Vision and Interdisciplinary Technology Management (IC-RVITM)*, 1-8. DOI: 10.1109/IC-RVITM60032.2023.10435159

Purohit, K., & Vyas, A. K. (2025). Quantum key distribution through quantum machine learning: A research review. *Frontiers in Quantum Science and Technology*, *4*, 1575498. Advance online publication. DOI: 10.3389/frqst.2025.1575498

Putra, Z. D. W., & van der Knaap, W. G. (2018). Urban innovation system and the role of an open web-based platform: The case of Amsterdam smart city. *Journal of Regional and City Planning*, *29*(3), 234–249. DOI: 10.5614/jrcp.2018.29.3.4

PwC. (2025). Global Digital Trust Insights Survey. *PwC Global*. https://www.pwc.com/us/en/services/consulting/cybersecurity-risk-regulatory/library/global-digital-trust-insights.html

Qmarkets (2024). *End-to-End Innovation Management Platform.* Retrieved 30 may, 2024 from https://www.qmarkets.net/lp/innovation-management-software/?hsa_qmt=y&utm_term=innovation%20platform&utm_campaign=Tier+1+Countries+%7C+Gold+KWs+%26+Silver+KWs+%7C+Exact+%26+Phrase&utm_source=adwords&utm_medium=ppc&hsa_acc=3759777188&hsa_cam=8690653365&hsa_grp=139195627185&hsa_ad=696764641653&hsa_src=g&hsa_tgt=kwd-6668275533&hsa_kw=innovation%20platform&hsa_mt=p&hsa_net=adwords&hsa_ver=3&gad_source=1&gclid=CjwKCAjwx-CyBhAqEiwAeOcTde0vSdyeHGevzm0u29NihVgwUZcEE_6P3UWSO69UWyysLMQuWORLExoCPoUQAvD_BwE

Quain, A., Ward, M. P., & Mullan, S. (2021). Ethical challenges posed by advanced veterinary care in companion animal veterinary practice. *Animals (Basel)*, *11*(11), 3010. DOI: 10.3390/ani11113010 PMID: 34827742

Radu, G., Solomon, M., Gheorghe, C. M., Hostiuc, M., Bulescu, I. A., & Purcarea, V. L. (2017). The adaptation of health care marketing to the digital era. *Journal of Medicine and Life, 10*(1), 44–50. https://pmc.ncbi.nlm.nih.gov/articles/PMC5304370/

Rahman, M. S., Wu, O. Y., Battaglia, K., Blackstone, N. T., Economos, C. D., & Mozaffarian, D. (2024). Integrating food is medicine and regenerative agriculture for planetary health. *Frontiers in Nutrition, 11*, 1508530. DOI: 10.3389/fnut.2024.1508530 PMID: 39726870

Ramos, F. D., Fontanilla, J. A., & Lat, R. E. (2021). Association between degrees of malnutrition and clinical outcomes among non-critically ill hospitalized adult patients with type 2 diabetes mellitus. *Journal of the ASEAN Federation of Endocrine Societies, 36*(2), 172–179. DOI: 10.15605/jafes.036.02.12 PMID: 34966202

Rampazzi, S., Park, J., Sugawara, T., & Fu, K. (2025). Unintentional radio emissions from MEMS microphones enable covert eavesdropping. *Proceedings of the IEEE Symposium on Security and Privacy*, p 1–15. https://news.ufl.edu/2025/06/microphone-eavesdropping/

Ray, A., & Lemos, R. (2023). Hybrid encryption for post-quantum security: Integrating legacy systems with next-gen algorithms. *Information Security Journal: A Global Perspective, 32*(3), 185–197. https://doi.org/DOI: 10.1080/19393555.2023.1930076

Reis, J., Rosado, D. P., Ribeiro, D. F., & Melão, N. (2022). Quintuple Helix Innovation Model for the European Union Defense Industry—An Empirical Research. *Sustainability (Basel), 14*(24), 16499. DOI: 10.3390/su142416499

Roberts, E. T., & Mehrotra, A. (2020). Assessment of disparities in digital access among Medicare beneficiaries. *JAMA Internal Medicine, 180*(10), 1386–1389. DOI: 10.1001/jamainternmed.2020.2666

Rodrigue, J. P., Slack, B., & Comtois, C. (2017). Green logistics. In *Handbook of logistics and supply-chain management* (Vol. 2, pp. 339–350). Emerald Group Publishing Limited.

Rogers, K. (2022). *The effects of remote work on organizational culture: Examining the effects of remote work on communication, social connectedness, and culture* (Master's thesis, University of Tennessee). TRACE: Tennessee Research and Creative Exchange. https://trace.tennessee.edu/utk_gradthes/6395

Rogers, E. M. (2003). *Diffusion of innovations* (5th ed.). Free Press., Retrieved from https://teddykw2.wordpress.com/wp-content/uploads/2012/07/everett-m-rogers-diffusion-of-innovations.pdf

Romanosky, S., Hoffman, D., & Acquisti, A. (2014). Empirical analysis of data breach litigation. *Journal of Empirical Legal Studies*, *11*(1), 74–104. https://onlinelibrary.wiley.com/doi/abs/10.1111/jels.12035?casa_token=7Ns-u_519yQAAAAA%3AOd9MgbWjMTDergbVLTX4Ue1PsvrMoakb7tuIlsdMoJJ1KMs2BG1rGfE13Kj9U6L84mFPdvemYZ6P_d6K. DOI: 10.1111/jels.12035

Rosenstock, I. M., Strecher, V. J., & Becker, M. H. (1988). Social learning theory and the Health Belief Model. *Health Education Quarterly*, *15*(2), 175–183. DOI: 10.1177/109019818801500203 PMID: 3378902

Rudko, I., Bashirpour Bonab, A., Fedele, M., & Formisano, A. V. (2025). New institutional theory and AI: Toward rethinking of artificial intelligence in organizations. *Journal of Management History*, *31*(2), 261–284. DOI: 10.1108/JMH-09-2023-0097

Ruel, M. T., & Alderman, H.Maternal and Child Nutrition Study Group. (2013). Nutrition-sensitive interventions and programmes: How can they help to accelerate progress in improving maternal and child nutrition? *Lancet*, *382*(9891), 536–551. DOI: 10.1016/S0140-6736(13)60843-0 PMID: 23746780

Ruel, M. T., Quisumbing, A. R., & Balagamwala, M. (2018). Nutrition-sensitive agriculture: What have we learned so far? *Global Food Security*, *17*, 128–153. DOI: 10.1016/j.gfs.2018.01.002

Ruiz Cano, A. (2024). *Cybersecurity strategies information technology leaders use to protect healthcare information systems from ransomware* (Order No. 30993688). Walden University. https://www.proquest.com/openview/d536d1866854baf6598de9ab8c74eaf8/1?pq-origsite=gscholar&cbl=18750&diss=y

Saaty, T. L. (2001). *Fundamentals of the Analytic Hierarchy Process*. 15–35. https://doi.org/DOI: 10.1007/978-94-015-9799-9_2

Saeed, S., Altamimi, S. A., Alkayyal, N. A., Alshehri, E., & Alabbad, D. A. (2023). Digital Transformation and Cybersecurity Challenges for Businesses Resilience: Issues and Recommendations. *Sensors (Basel)*, *23*(15), 6666. DOI: 10.3390/s23156666

Said, Y., Saidani, T., Atri, M., Alsheikhy, A. A., & Shawly, T. (2026). Computational intelligence for emotion recognition in autism spectrum disorder: A systematic review of signal-based modeling, simulation, and clinical potential. *Biomedical Signal Processing and Control*, *111*, 108367. DOI: 10.1016/j.bspc.2025.108367

Salem, A. H., Azzam, S. M., Emam, O. E., & Abohany, A. A. (2024). Advancing cybersecurity: A comprehensive review of AI-driven detection techniques. *Journal of Big Data*, *11*(105), 105. https://journalofbigdata.springeropen.com/articles/10.1186/s40537-024-00957-y. DOI: 10.1186/s40537-024-00957-y

Salisi, J. C., & Faller, E. M. (2024). Application of Force Field Analysis in healthcare: A systematic review of change management strategies and performance improvement initiatives. *International Journal of Research Publication and Reviews*, *5*(7), 3256–3263. https://ijrpr.com/uploads/V5ISSUE7/IJRPR31644.pdf

Sánchez Hernández, M. I.. (2024). Exploring Responsible Research and Innovation in agri-food cooperatives. *Journal of Responsible Innovation*. Advance online publication. DOI: 10.1080/23299460.2024.2322756

Santamaria-Granados, L., Munoz-Organero, M., Ramirez-González, G., Abdulhay, E., & Arunkumar, N. (2019). Using Deep Convolutional Neural Network for Emotion Detection on a Physiological Signals Dataset (AMIGOS). *IEEE Access : Practical Innovations, Open Solutions*, *7*, 57–67. DOI: 10.1109/ACCESS.2018.2883213

Santarsiero, F., Schiuma, G., Carlucci, D., & Helander, N. (2023). Digital transformation in healthcare organisations: The role of innovation labs. *Technovation*, *122*, 102640. DOI: 10.1016/j.technovation.2022.102640

Sas, M., Dziurda, A., & Welcer, T. (2021). The impact of training sessions on physical security awareness. *Safety Science*, *144*, 105447. DOI: 10.1016/j.ssci.2021.105447

Schedl, M., Anelli, V. W., & Lex, E. (2025). *Technical and regulatory perspectives on information retrieval and recommender systems: Fairness, transparency, and privacy*. Springer., DOI: 10.1007/978-3-031-69978-8

Schein, E. H. (2010). *Organizational culture and leadership* (4th ed.). Jossey-Bass.

Scholkmann, A. B. (2021). *Resistance to (Digital)*. Change Individual, Systemic and Learning-Related Perspectives., DOI: 10.1007/978-3-030-55878-9_13

Scott, W. R. (2001). *Institutions and organizations* (2nd ed.). Sage Publications.

Shahanas, I. N., & Thampi, S. M. (2025). Secure communication protocols: Safeguarding information in the digital age. In *Securing the Connected World* (pp. 157–221). Springer., DOI: 10.1007/978-3-031-82826-3_6

Shah, S. M., Sun, Z., Zaman, K., Hussain, A., Shoaib, M., & Pei, L. (2022). A Driver Gaze Estimation Method Based on Deep Learning. *Sensors (Basel)*, *22*(10), 3959. Advance online publication. DOI: 10.3390/s22103959 PMID: 35632365

Shaikhanov, Z., Al-Madi, M., Chen, H. T., Chang, C. C., Addamane, S., Mittleman, D. M., & Knightly, E. (2025, May). Spoofing Eavesdroppers with Audio Misinformation. *In 2025 IEEE Symposium on Security and Privacy (SP)*, pp. 4553-4568, DOI: 10.1109/SP61157.2025.00258

Shamim, M. M. I. (2024). Artificial Intelligence in Project Management: Enhancing Efficiency and Decision-Making. *International Journal of Management Information Systems and Data Science, 1*(1), 1–6. DOI: 10.62304/ijmisds.v1i1.107

Sharma, M., Mittal, A., Rewliya, A., & Mittal, H. (2024). Comparative Analysis of Deep Learning Models for Facial Emotion Recognition. *2024 IEEE International Conference on Intelligent Signal Processing and Effective Communication Technologies (INSPECT)*, 1-6. DOI: 10.1109/INSPECT63485.2024.10896027

Shaver, J. (2022). The state of telehealth before and after the COVID-19 pandemic. *Primary Care, 49*(4), 517–530. DOI: 10.1016/j.pop.2022.04.002

Shirey, M. R. (2011). Establishing a sense of urgency for leading transformational change. *The Journal of Nursing Administration, 41*(4), 145–148. DOI: 10.1097/NNA.0b013e3182118550

Shi, Y., Liang, Z., Zhang, Y., Zhou, Y., Dong, S., Li, J., & Gao, G. (2025). Latent profiles of psychological resilience in patients with chronic disease and their association with social support and activities of daily living. *BMC Psychology, 13*(1), 1–13. DOI: 10.1186/s40359-025-03115-y PMID: 40665388

Shoolin, J. S. (2010). Change management - recommendations for successful electronic medical records implementation. *Applied Clinical Informatics, 1*(3), 286–292. DOI: 10.4338/ACI-2010-01-R-0001

Shor, P. W. (1994). Algorithms for quantum computation: Discrete logarithms and factoring. In *Proceedings 35th Annual Symposium on Foundations of Computer Science* (pp. 124–134). IEEE. DOI: 10.1109/SFCS.1994.365700

Siegel, K. R., Bullard, K. M., Imperatore, G., Kahn, H. S., Stein, A. D., Ali, M. K., & Narayan, K. M. V. (2016). Association of higher consumption of foods derived from subsidized commodities with adverse cardiometabolic risk among US adults. *JAMA Internal Medicine, 176*(8), 1124–1132. DOI: 10.1001/jamainternmed.2016.2410 PMID: 27379488

Siegel, K. R., McKeever Bullard, K., Ali, M. K., Stein, A. D., Kahn, H. S., Mehta, N. K., Webb Girard, A., Narayan, K. M., & Imperatore, G. (2016, June). The contribution of subsidized food commodities to total energy intake among US adults. *Public Health Nutrition, 19*(8), 1348–1357. DOI: 10.1017/S1368980015002414 PMID: 26322920

Sigut, J., & Sidha, S.-A. (2011). Iris Center Corneal Reflection Method for Gaze Tracking Using Visible Light. *IEEE Transactions on Biomedical Engineering, 58*(2), 411–419. DOI: 10.1109/TBME.2010.2087330 PMID: 20952326

Silvestri, F., Odisho, N., Kumar, A., & Grigoriadis, A. (2024). Examining gaze behavior in undergraduate students and educators during the evaluation of tooth preparation: An eye-tracking study. *BMC Medical Education*, *24*(1), 1030. DOI: 10.1186/s12909-024-06019-4 PMID: 39300488

Singh, V. K., Ahmed, S., Saini, D. K., Gahlaut, V., Chauhan, S., Khandare, K. P., Kumar, A., Sharma, P. K., & Kumar, J. (2023). *Manipulating epigenetic diversity in crop plants: Techniques, challenges and opportunities.* 130544. https://doi.org/ DOI: 10.1016/j.bbagen.2023.130544

Sipahi, S., & Timor, M. (2010). The analytic hierarchy process and analytic network process: An overview of applications. *Management Decision*, *48*(5), 775–808. DOI: 10.1108/00251741011043920

Sjöberg, E. (2022). Discrete logarithms, integer factorization and their role in classical and quantum cryptography (U.U.D.M. Project Report 2022:27). *Uppsala University*. https://www.diva-portal.org/smash/get/diva2:1729425/FULLTEXT01.pdf

Slåtten, T., Mutonyi, B. R., & Lien, G. (2021). Does organizational vision really matter? An empirical examination of factors related to organizational vision integration among hospital employees. *BMC Health Services Research*, *21*(1), 483. Advance online publication. DOI: 10.1186/s12913-021-06503-3

Snoswell, C. L., Stringer, H., Taylor, M. L., Caffery, L. J., Smith, A. C., & Haydon, H. M. (2021). The effect of telehealth on hospital services use: Systematic review and meta-analysis. *Journal of Medical Internet Research*, *23*(7), e25987. Advance online publication. DOI: 10.1177/1357633X211023700

Solar, O., & Irwin, A. (2010). A conceptual framework for action on the social determinants of health. World Health Organization. https://apps.who.int/iris/handle/10665/44489

Solow, R. M. (1956). A contribution to the theory of economic growth. *The Quarterly Journal of Economics*, *70*(1), 65–94. DOI: 10.2307/1884513

Sood, S., Garg, N., & Singh, S. (2023). Biofortification: An approach to eradicate micronutrient deficiency. *Frontiers in Nutrition*, *10*, 1233070. DOI: 10.3389/fnut.2023.1233070 PMID: 37789898

Southwick, R. (2023, February 23). Nearly 50 million Americans impacted by health data breaches in 2022. Chief Healthcare Executive. https://www.chiefhealthcareexecutive.com/view/nearly-50-million-americans-impacted-by-health-data-breaches-in-2022

Sparling, T. M., White, H., & Kadiyala, S. (2021). Understanding pathways between agriculture, food systems, and nutrition: An evidence and gap map of research tools, metrics, and methods. *Advances in Nutrition, 12*(4), 1122–1136. DOI: 10.1093/advances/nmaa158 PMID: 33395472

Spence, N., Paul, D. P.III, & Coustasse, A. (2017). Ransomware in healthcare facilities: The future is now. *Perspectives in Health Information Management, 14*(1), 1–10. https://marshall.elsevierpure.com/ws/portalfiles/portal/40008403/Ransomware%20in%20Healthcare%20Facilities_%20The%20Future%20is%20Now.pdf

Springer, S., Lund, T. B., Corr, S. A., & Sandøe, P. (2023). Seeing the benefits, but not taking advantage of them: Dog and cat owners' beliefs about veterinary telemedicine. *The Veterinary Record, 194*(5), e3312. Advance online publication. DOI: 10.1002/vetr.3312 PMID: 37733831

Springmann, M., & Freund, F. (2022). Options for reforming agricultural subsidies from health, climate, and economic perspectives: Integrated modelling shows often-conflicting objectives can be aligned. *Nature Communications, 13*(1), 82. DOI: 10.1038/s41467-021-27645-2 PMID: 35013212

Sridhar, B., Siergiej, D., & Lohr, C. (2021). Simulation and modeling concepts for secure airspace operations. *Aviation, 25*(2), 155–169. https://aviationsystems.arc.nasa.gov/publications/2021/20210017030_Sridhar_Aviation2021.pdf

Srivastava, S., Vikram Singh, S., Bhadur Singh, R., & Kumar Shukla, H. (2021). Digital Transformation of Healthcare: A Blockchain study. In *IJISET-International Journal of Innovative Science, Engineering & Technology* (Vol. 8). www.ijiset.com

Stads, G.-J., Nin-Pratt, A., Omot, N., & Pham, N. T. (2020). *Agricultural research in Southeast Asia: A cross-country analysis of resource allocation, performance, and impact on productivity (ASTI Regional Report).* Asia Pacific Association of Agricultural Research Institutions & International Food Policy Research Institute. https://www.asti.cgiar.org/pdf/SoEA-Ag-Regional-Report-2020.pdf

Stone, P. W., Harrison, M. I., Feldman, P., Lin, S. X., Purrington, A., Ringel, M., Arnold, S. B., Tietjen, J., Cohen, L., Fahey, L., Rosen, B., Wald, H. L., Curry, L., & Schnipper, J. L. (2018). How guiding coalitions promote positive culture change in hospitals: A longitudinal mixed methods interventional study. *BMJ Quality & Safety, 27*(3), 218–225. DOI: 10.1136/bmjqs-2017-006574

Sun, M. (2025). Advances in Plant Epigenetic Regulation of Abiotic Stress Response. *Theoretical and Natural Science, 90*(1), 81–87. DOI: 10.54254/2753-8818/2025.GU20455

Suwanda, S., & Nugroho, B. Y. (2022). Literature reviews: McKinsey 7S model to support organizational performance. *Technium Social Sciences Journal, 38*, 1–9. DOI: 10.47577/tssj.v38i1.7744

Swayne, M. (2024). Quantum key distribution faces real-world challenges in eavesdropping experiment. *The Quantum Insider*. https://thequantuminsider.com/2024/10/01/quantum-key-distribution-faces-real-world-challenges-in-eavesdropping-experiment

Swinburn, B., Kraak, V. I., Allender, S., Atkins, V. J., Baker, P. I., Bogard, J. R., & Dietz, W. H. (2019). The Global Syndemic of obesity, undernutrition, and climate change: The Lancet Commission report. The Lancet, 393(10173), 791–846. 6736(18)32822-8

Tahir, M., Norouzi, M., Khan, S. S., Davie, J., Yamanaka, S., & Ashraf, A. (2024). Artificial intelligence and deep learning algorithms for epigenetic sequence analysis: A review for epigeneticists and AI experts. *Computers in Biology and Medicine, 183*, 109302. DOI: 10.1016/j.compbiomed.2024.109302 PMID: 39500240

Tan, S. S. X., & Kuebbing, S. E. (2023). A synthesis of the effect of regenerative agriculture on soil carbon sequestration in Southeast Asian croplands. *Agriculture, Ecosystems & Environment, 349*, 108450. DOI: 10.1016/j.agee.2023.108450

Teller, L. M. (2022). Exploring the challenges and opportunities presented by veterinary telemedicine. *The Veterinary Record, 191*(5), 210–212. DOI: 10.1002/vetr.2209 PMID: 36083053

Teran, E., Chetty, M., & Touirsi, S. (2025). Agriculture and health: Synergizing for a sustainable future. *Frontiers in Public Health, 13*, 1558198. DOI: 10.3389/fpubh.2025.1558198 PMID: 40241943

Thapar, S., Amin, I., Mishra, D., & Saini, R. (2023). Secrecy outage probability analysis for downlink NOMA with imperfect SIC at untrusted users. arXiv preprint arXiv:2308.08811. https://arxiv.org/pdf/2308.08811v1.pdf

Thiébaut, R., & Cossin, S. (2019). Artificial intelligence for surveillance in public health. *Yearbook of Medical Informatics, 28*(01), 232–234. DOI: 10.1055/s-0039-1677939 PMID: 31419837

Thomas, J., & Harden, A. (2008). Methods for the thematic synthesis of qualitative research in systematic reviews. *BMC Medical Research Methodology, 8*(1), 45. DOI: 10.1186/1471-2288-8-45 PMID: 18616818

Tiffin, R., & Arnoult, M. (2010). The demand for a healthy diet: Estimating the almost ideal demand system with infrequency of purchase. *European Review of Agriculture Economics*, *37*(4), 501–521. DOI: 10.1093/erae/jbq038

Tkachenko, I. N., & Chesnyukova, L. K. (2023). Digital technologies in the sphere of health care as a way to ensure the quality of human capital. Izvestiya of Saratov University. Economics. DOI: 10.18500/1994-2540-2023- 23-1-121-128

Toback, M. (2023, November 29). Implementing network segmentation: Strategies for better security in enterprise networks. *Security Boulevard*. https://securityboulevard.com/2023/11/implementing-network-segmentation-strategies-for-better-security-in-enterprise-networks/

Tornatzky, L. G., & Fleischer, M. (1990). The technology–organization–environment framework. In L. G. Tornatzky & M. Fleischer, The processes of technological innovation (pp. 153–178). Lexington, MA: Lexington Books. Retrieved September 20, 2025, from https://www.researchgate.net/publication/226145805_The_Technology-Organization-Environment_Framework

Torres, C., & Zhang, L. (2024). AI-driven acoustic surveillance: Ethics, threats, and soundproofing strategies. *Security and Privacy Innovations*, *18*(2), 67–84. DOI: 10.1016/j.spi.2024.03.009

Tosh, P. K., Feldman, H., Christian, M. D., Devereaux, A. V., Kissoon, N., & Dichter, J. R. (2014). Business and continuity of operations: care of the critically ill and injured during pandemics and disasters: CHEST consensus statement. *Chest*, *146*(4), e103S–e117S. https://www.sciencedirect.com/science/article/abs/pii/S0012369215519934. DOI: 10.1378/chest.14-0739

Turner, C., Aggarwal, A., Walls, H., Herforth, A., Drewnowski, A., Coates, J., Kalamatianou, S., & Kadiyala, S. (2018). Concepts and critical perspectives for food environment research: A global framework with implications for action in low- and middle-income countries. *Global Food Security*, *18*, 93–101. DOI: 10.1016/j.gfs.2018.08.003

U. S. Department of Health and Human Services. (2024). HIPAA and telehealth: Guidance for providers. https://www.hhs.gov/hipaa/for-professionals/special-topics/telehealth/index.html

U.S. Department of Agriculture, Economic Research Service. (2025). *International agricultural productivity.* https://www.ers.usda.gov/data-products/international-agricultural-productivity

U.S. Department of Defense. (2022) 2022 Department of Defense Sustainability Plan. Retrieved from: https://www.sustainability.gov/pdfs/dod-2022-sustainability-plan.pdf

U.S. Department of Defense. (2024). *Green Procurement Compilation*. Retrieved from https://sftool.gov/greenprocurement/dod

U.S. Department of Health & Human Services. (2023). HIPAA for professionals: Breach notification. Retrieved July 20, 2023, from https://www.hhs.gov/hipaa/for-professionals/breach-notification/index.html

U.S. Department of Health & Human Services. (n.d.). Telehealth. https://www.hhs.gov/hipaa/for-professionals/faq/telehealth/index.html

Uddin, J., & Fariha, T. (2024). *Bridging the gap: Expanding telehealth services to address rural health disparities* [Unpublished manuscript]. Social and Behavioral Health Department, University of Nevada.

Ukelson, J. (2020, September 21). Wireless Red Teams: Evil Twins, eavesdropping, and password cracking. *Orchestra group*. https://orchestragroup.com/uncategorized/wireless-red-teams-evil-twins-eavesdropping-and-password-cracking/#:~:text=U.S.%20Department%20of%20the%20Interior%20operates%20hundreds

United Nations. (2024). The Sustainable Development Goals. Retrieved from: https://www.un.org/sustainabledevelopment/

University of Louisville. (2025). Critical thinking and academic research: Assumptions. *Author*. https://library.louisville.edu/ekstrom/criticalthinking/assumptions

Usigbe, M. J., Asem-Hiablie, S., Uyeh, D. D., Iyiola, O., Park, T., & Mallipeddi, R. (2024). Enhancing resilience in agricultural production systems with AI-based technologies. *Environment, Development and Sustainability*, *26*(9), 21955–21983. DOI: 10.1007/s10668-023-03588-0

Uzule, K., & Verina, N. (2023). Digital Barriers in Digital Transition and Digital Transformation: Literature Review. *Economics and Culture*, *20*(1), 125–143. DOI: 10.2478/jec-2023-0011

Valle, R., Buenaposada, J. M., Valdés, A., & Baumela, L. (2019). Face alignment using a 3D deeply-initialized ensemble of regression trees. *Computer Vision and Image Understanding*, *189*, 102846. DOI: 10.1016/j.cviu.2019.102846

van Buul, V. J., Tappy, L., & Brouns, F. (2014). Misconceptions about fructose-containing sugars and their role in the obesity epidemic. *Nutrition Research Reviews*, *27*(2), 119–130. DOI: 10.1017/S0954422414000067 PMID: 24666553

Van Dijk, J. (2020). The digital divide. *Polity.*

Venkatesh, V., Morris, M. G., Davis, G. B., & Davis, F. D. (2003). User Acceptance of Information Technology: Toward A Unified View1. *Management Information Systems Quarterly, 27*(3), 425–478. DOI: 10.2307/30036540

Venkatesh, V., Thong, J. Y. L., & Xu, X. (2012). Consumer Acceptance and Use of Information Technology: Extending the Unified Theory of Acceptance and Use of Technology1. *Management Information Systems Quarterly, 36*(1), 157–178. DOI: 10.2307/41410412

Vignieri, V. (2021). Crowdsourcing as a mode of open innovation: Exploring drivers of success of a multisided platform through system dynamics modelling. *Systems Research and Behavioral Science, 38*(1), 108–124. DOI: 10.1002/sres.2636

Waage, J., Hawkes, C., Turner, R., Ferguson, E., Johnston, D., Shankar, B., Homans, H., & Harris, J. (2018). Integrating agriculture and health research for development: LCIRAH as an interdisciplinary programme to address a global challenge. *Global Challenges (Hoboken, NJ), 3*(4), 1700104. DOI: 10.1002/gch2.201700104 PMID: 31565369

Wakabayashi, H., & Sashika, H. (2021). Malnutrition is associated with poor rehabilitation outcome in elderly inpatients with hospital-associated deconditioning. *Journal of Rehabilitation Medicine, 53*(7), jrm00150. Advance online publication. DOI: 10.2340/16501977-1258 PMID: 24213734

Wakeman, M., Buckman, D. W., & El-Toukhy, S. (2025). Disparities in digital health care use in 2022. *JAMA Network Open, 8*(4), e255359. Advance online publication. DOI: 10.1001/jamanetworkopen.2025.5359

Wakoli, C. D. L. W. (2024). Secure Sockets Layer/Transport Layer Security for E-Commerce. [IJSRM]. *International Journal of Scientific Research and Management, 12*(12), 8047–8052. https://pdfs.semanticscholar.org/5738/d9c43e0c89d2415fdc9b9e9fd0384533ca0c.pdf. DOI: 10.18535/ijsrm/v12i12.em03

Walker, A. E., Wattick, R. A., & Olfert, M. D. (2021). The application of systems science in nutrition-related behaviors and outcomes implementation research: Scoping review. *Current Developments in Nutrition, 5*(9), nzab105. Advance online publication. DOI: 10.1093/cdn/nzab105 PMID: 34522835

Wallarm. (2025). The challenges of balancing usability and security in cybersecurity protocols. *Author*. https://www.wallarm.com/blog/usability-vs-security

Wang, J., Zhang, P., Tang, L., Bai, Y., & Yang, L. (2022, June). Intelligent passive eavesdropping in massive MIMO-OFDM Systems via Reinforcement Learning. *IEEE Communications Letters*, *11*(6). https://isac.nju.edu.cn/DFS//file/2025/04/28/20250428161418784z2li2q.pdf?iid=93454

Wang, Q. (2021). Defending wireless communication against eavesdropping attacks using secret spreading codes and artificial interference. *Computers & Security*, *103*, 102175. DOI: 10.1016/j.cose.2020.102175

Wang, S., von Huben, A., Sivaprakash, P. P., Saurman, E., Norris, S., & Wilson, A. (2025). Addressing health service equity through telehealth: A systematic review of reviews. *Digital Health*, *11*, 20552076251326233. Advance online publication. DOI: 10.1177/20552076251326233

Wang, Y., Li, J., & Chen, H. (2025). Intelligent anomaly localization for eavesdropping detection in power IoT environments using Kolmogorov-Arnold Networks. *IEEE Transactions on Industrial Informatics*, *21*(4), 2345–2359.

Waterman, R. H.Jr, Peters, T. J., & Phillips, J. R. (1980). Structure is not organization. *Business Horizons*, *23*(3), 14–26. DOI: 10.1016/0007-6813(80)90027-0

Watson, K., Mardhekar, N. M., Pandey, U., Wells, J., Gruszynski, K., Wilkerson, A. H., Dascanio, J. J., Quadri, S. S., Leo, J. T., Samuel, K. D., Nahar, V. K., & Davis, R. E. (2024). Veterinary students' perceptions toward incorporating and expanding telehealth service delivery knowledge in curricula and future practice: A cross-sectional study. *American Journal of Veterinary Research*, *85*(11). Advance online publication. DOI: 10.2460/ajvr.24.02.0032 PMID: 39137815

Webb, P., & Kennedy, E. (2014). Impacts of agriculture on nutrition: Nature of the evidence and research gaps. *Food and Nutrition Bulletin*, *35*(1), 126–132. DOI: 10.1177/156482651403500113 PMID: 24791585

Weffort, V. R. S., & Lamounier, J. A. (2023). Hidden hunger—A narrative review. *Journal of Pediatrics,* Article e202300010. https://doi.org/DOI: 10.1590/0031-400x-2023-00010

Weitoish, T., & Burrell, D. N. (2024). Nuclear Power Organizations as Learning Organizations Around Cybersecurity, Public Health, Public Safety, and Critical Infrastructure Protection. In *Evolution of Cross-Sector Cyber Intelligent Markets* (pp. 333–355). IGI Global Scientific Publishing., https://www.igi-global.com/chapter/nuclear-power-organizations-as-learning-organizations-around-cybersecurity-public-health-public-safety-and-critical-infrastructure-protection/338619

Welch, T. D., & Smith, T. B. (2023). Strategic planning: Moving from plan to implementation. *Nursing Administration Quarterly*, *47*(4), 283–288. https://pubmed.ncbi.nlm.nih.gov/37643226/. DOI: 10.1097/NAQ.0000000000000597

White House. (2024). Report on post-quantum cryptography. *Author.* https://www.whitehouse.gov/wp-content/uploads/2024/02/Post-Quantum-Cryptography-Report.pdf

White, J. S. (2008). Straight talk about high-fructose corn syrup: What it is and what it ain't. *The American Journal of Clinical Nutrition*, *88*(6), 1716S–1721S. DOI: 10.3945/ajcn.2008.25825B PMID: 19064536

Whitman, M. E., & Mattord, H. J. (2018). *Principles of information security* (6th ed.). Cengage Learning.

Wieslander, L., Bäckström, I., & Häggström, M. (2024). Participation in the digital transformation of healthcare: a review of qualitative studies. *International journal of health care quality assurance, ahead-of-print*(ahead-of-print), 68–84. https://doi.org/DOI: 10.1108/IJHCQA-03-2024-0021

Wilber, K., Yan, M., & Zhu, Y. (2022, December 30). *Unanticipated outcomes of a telehealth pilot project in skilled nursing facilities care: Don't put the cart before the course.* California Emerging Technology Fund & University of Southern California Leonard Davis School of Gerontology. https://s42263.pcdn.co/wp-content/uploads/2023/03/230307_CETF_SNF_Telehealth_Pilot_Project_Final-Evaluation-Report_2023.pdf

William Bridges Associates. (n.d.). *What is transition? Bridges Transition Model.* https://wmbridges.com/about/what-is-transition/

Wood, E., Baltrušaitis, T., Morency, L.-P., Robinson, P., & Bulling, A. (2016). Learning an appearance-based gaze estimator from one million synthesised images. *Proceedings of the Ninth Biennial ACM Symposium on Eye Tracking Research & Applications*, 131-138. DOI: 10.1145/2857491.2857492

Woods, D. W., & Seymour, S. (2024). 'Evidence-based cybersecurity policy? A meta-review of security control effectiveness. *Journal of Cyber Policy*, *8*(3), 365–383. DOI: 10.1080/23738871.2024.2335461

World Bank. (2025). Cereal yield (kg per hectare) [Data set]. *World Development Indicators.* World Bank DataBank. https://databank.worldbank.org/source/world-development-indicators/series/AG.YLD.CREL.KG

World Economic Forum. (2025). Global Cybersecurity Outlook 2025. *World Economic Forum.* https://www.weforum.org/publications/global-cybersecurity-outlook-2025

World Health Organization. (2022). *WHO manual on sugar-sweetened beverage taxation policies to promote healthy diets*. WHO.

World Health Organization. (2025). *Obesity and overweight: Key facts*. WHO.

Wright, J. B., & Burrell, D. N. (2023). Cybersecurity leadership ethics in healthcare. In *Handbook of Research on Cybersecurity Risk in Contemporary Business Systems* (pp. 137–148). IGI Global., https://www.igi-global.com/chapter/cybersecurity-leadership-ethics-in-healthcare/321016 DOI: 10.4018/978-1-6684-7207-1.ch007

Wunderlich, A. C., & Kohler, A. (2022). The impact of agricultural subsidies on obesity. *Journal of Agriculture and Food Research*, *10*, 100439. DOI: 10.1016/j.jafr.2022.100439

Xiao, F., Zheng, D., Huang, K., Qiu, Y., & Shen, H. (2018). A Single-Camera Gaze Tracking System Under Natural Light. *Journal of Eye Movement Research*, *11*(4), 1–14. DOI: 10.16910/jemr.11.4.5 PMID: 33828707

Xie, M., & Chen, J. (2023). Misinformation architectures in acoustic counter-surveillance. *Journal of Secure Communication Engineering*, *14*(1), 22–35. DOI: 10.1016/j.jsce.2023.01.005

Xiong, H. (2023). *Epigenetic regulation in plant salt stress response*. https://doi.org/DOI: 10.54254/2753-8818/6/20230296

Xu, H., Zhang, J., Sun, H., Qi, M., & Kong, J. (2023). Analyzing students' attention by gaze tracking and object detection in classroom teaching. *Data Technologies and Applications*, *57*(5), 643–667. DOI: 10.1108/DTA-09-2021-0236

Xu, Y., & Cheng, L. (2025). Face emotion recognition based on Gabor wavelet and particle swarm optimization algorithm. *Multimedia Systems*, *31*(6), 435. Advance online publication. DOI: 10.1007/s00530-025-01976-2

Xu, Y., Li, Z., Dong, N., Kuchta, V., Hou, Z., & Liu, D. (2025). Formal verification techniques for post-quantum cryptography: A systematic review. In Bai, G., Ishikawa, F., Ait-Ameur, Y., & Papadopoulos, G. A. (Eds.), Lecture Notes in Computer Science: Vol. 14784. *Engineering of Complex Computer Systems. ICECCS 2024*. Springer., DOI: 10.1007/978-3-031-66456-4_19

Yan, X., Zhou, G., Quevedo, D. E., Murguia, C., Chen, B., & Huang, H. (2024). Privacy-preserving state estimation in the presence of eavesdroppers: A survey. *arXiv preprint*. https://arxiv.org/abs/2402.15738

Yang, F., & Wang, Y. (2023). Analyzing the robustness of complex networks with attack success rate. *Entropy (Basel, Switzerland)*, *25*(11), 1508. DOI: 10.3390/e25111508 PMID: 37998200

Yang, M., Zhang, X., & Wang, Y. (2023). Does Internet use connect smallholder farmers to a healthy diet? Evidence from rural China. *Frontiers in Nutrition*, *10*, 1122677. DOI: 10.3389/fnut.2023.1122677 PMID: 37153910

Yang, Y., & Song, S. (2025). Trajectory tracking approach based on mode decomposition under malicious attacks. *IEEE Transactions on Aerospace and Electronic Systems*, *61*(4), 10424–10438. Advance online publication. DOI: 10.1109/TAES.2025.3563861

Yan, J., Zhou, Y., Dai, A., & Wang, T. (2024). Securing forwarding layers from eavesdropping attacks using proactive approaches. *Computers, Materials & Continua*, *79*(1), 563–580. DOI: 10.32604/cmc.2024.048922

Yarborough, P. D. (2013). Change management – recommendations for successful electronic medical records implementation. *The Permanente Journal*, *17*(2), 83–85. DOI: 10.7812/TPP/12-111

Yarbus, A. L. (1967). *Eye movements and vision*. Plenum Press. DOI: 10.1007/978-1-4899-5379-7

Yin, C., Li, M., & Xu, J. (2025). Machine learning based physical layer security for detecting active eavesdropping attacks. *IEEE Communications Letters*, *29*(6), 1292–1296. DOI: 10.1109/LCOMM.2025.3582157

Yingfei, Y., Mengze, Z., Zeyu, L., Ki-Hyung, B., Avotra, A. A. R. N., & Nawaz, A. (2022). Green logistics performance and infrastructure on service trade and environment-measuring firm's performance and service quality. *Journal of King Saud University. Science*, *34*(1), 101683. DOI: 10.1016/j.jksus.2021.101683

Yinusa, G., Scammell, J., Murphy, J., Ford, G., & Baron, S. (2021). Multidisciplinary provision of food and nutritional care to hospitalized adult in-patients: A scoping review. *Journal of Multidisciplinary Healthcare*, *14*, 459–491. DOI: 10.2147/JMDH.S255256 PMID: 33654405

Young, B. (n.d.) Foundations of Computer Security. *University of Texas at Austin, Web*. https://www.cs.utexas.edu/~byoung/cs361s/

Zanaboni, P., Ngangue, P., Mbemba, G. I. C., Schopf, T. R., Bergmo, T. S., & Gagnon, M. P. (2018). Methods to evaluate the effects of internet-based digital health interventions for citizens: Systematic review of reviews. *Journal of Medical Internet Research*, *20*(6), e10202. Advance online publication. DOI: 10.2196/10202

Zanganeh, K. S., Burrell, D. N., & Richardson, K. (2023). Improving Hospital Diversity Through Management Consulting Interventions. In *Transformational Leadership Styles for Global Leaders: Management and Communication Strategies* (pp. 377-390). IGI Global. https://www.igi-global.com/chapter/improving-hospital-diversity-through-management-consulting-interventions/331376

Zarif, A. (2021). The ethical challenges facing the widespread adoption of digital healthcare technology. *Health and Technology*, *12*(1), 175–179. DOI: 10.1007/s12553-021-00596-w

Zeng, D., Cao, Z., & Neill, D. B. (2021). Artificial intelligence–enabled public health surveillance—from local detection to global epidemic monitoring and control. In *Artificial intelligence in medicine* (pp. 437–453). Academic Press. DOI: 10.1016/B978-0-12-821259-2.00022-3

Zen, I. S., & Shibakawa, H. (2022). Quintuple Helix Lens for Transformation: An Okayama Model of Education for Sustainable Development. *Frontiers in Sustainability*, *3*, 798330. DOI: 10.3389/frsus.2022.798330

Zerga, A. A., Tadesse, S. E., Ayele, F. Y., & Ayele, S. Z. (2022). Impact of malnutrition on the academic performance of school children in Ethiopia: A systematic review and meta-analysis. *SAGE Open Medicine*, *10*, 20503121221122398. Advance online publication. DOI: 10.1177/20503121221122398 PMID: 36161209

Zhai, S. (2008). On the ease and efficiency of human-computer interfaces. *Proceedings of the 2008 Symposium on Eye Tracking Research & Applications*, 9-10. DOI: 10.1145/1344471.1344473

Zhang, R., Wang, Y., & Li, J. (2022). Outage probability analysis for secure transmission in wireless networks with cooperative jamming. *Wireless Networks*, 28, 1235–1248. Springer. https://doi.org/DOI: 10.1007/s11276-021-02864-9

Zhang, D., Xu, F., Wang, F., Le, L., & Pu, L. (2024). Synthetic biology and artificial intelligence in crop improvement. *Plant Communications*, *101220*. Advance online publication. DOI: 10.1016/j.xplc.2024.101220 PMID: 39668563

Zhang, G., Qiu, Y., Boireau, P., Zhang, Y., Ma, X., Jiang, H., Xin, T., Zhang, M., Tadesse, Z., Wani, N. A., Song, J., & Ding, J. (2024). Modern agriculture and One Health. *Infectious Diseases of Poverty*, *13*(1), 74. DOI: 10.1186/s40249-024-01240-1 PMID: 39385259

Zhang, H., Yang, Z., Wang, Y., Ankrah Twumasi, M., & Chandio, A. A. (2023). Impact of agricultural mechanization level on farmers' health status in western China: Analysis based on CHARLS data. *International Journal of Environmental Research and Public Health*, *20*(5), 4654. DOI: 10.3390/ijerph20054654 PMID: 36901664

Zhang, X., Sugano, Y., Fritz, M., & Bulling, A. (2015). Appearance-based gaze estimation in the wild. *2015 IEEE Conference on Computer Vision and Pattern Recognition (CVPR)*, 4511-4520. DOI: 10.1109/CVPR.2015.7299081

Zhao, G., Cheng, Y., & Gatbonton, R. R. G. (2024). Biotechnological innovation in athlete health management: From nutrition to rehabilitation. *Journal of Commercial Biotechnology*, *29*(3), 427–438. https://commercialbiotechnology.com/article-detail/?id=2243

Zhao, X., Smith, A., & Adepoju, O. (2024). Digital transformation: Opportunities and emerging security challenges in connected infrastructures. *IEEE Access : Practical Innovations, Open Solutions*, *12*, 15522–15534. DOI: 10.1109/ACCESS.2024.3335821

Zhao, X., Tanaka, R., Mandour, A. S., Shimada, K., & Hamabe, L. (2025). Remote vital sensing in clinical veterinary medicine: A comprehensive review of recent advances, accomplishments, challenges, and future perspectives. *Animals (Basel)*, *15*(7), 1033. DOI: 10.3390/ani15071033 PMID: 40218426

Zhao, Y., Cui, S., & Xing, Y. (2025). The impact of agricultural technological innovation on residents' health and its spatial effects. *Frontiers in Public Health*, *13*, 1633413. DOI: 10.3389/fpubh.2025.1633413 PMID: 40636876

Zheng, F. (2022). Facial Expression Recognition Based on LDA Feature Space Optimization. *Computational Intelligence and Neuroscience*, *2022*(1), 9521329. DOI: 10.1155/2022/9521329 PMID: 36072737

Zhou, X., Cheng, G., & Zhang, B. (2025). Does farmland transfer promote digital agriculture? Based on panel data from 30 provinces in China. *Land (Basel)*, *14*(3), 481. DOI: 10.3390/land14030481

Zhou, Y., & Jiang, X. (2022). Dissecting Android malware: Characterization and evolution. *IEEE Transactions on Computers*, *71*(3), 567–580. DOI: 10.1109/TC.2021.3052045

Zoccarato, F., Manzoni, M., Minotti, D., Lettieri, E., & Boaretto, A. (2024). Unveiling the interplay between rational, psychological and functional factors in continuous glucose monitoring early adoption: Novel evidence from the dexcom one case in italy. *BMC Health Services Research*, *24*(1), 747. Advance online publication. DOI: 10.1186/s12913-024-11195-6

About the Contributors

Darrell Norman Burrell is a visiting scholar at the Samuel DeWitt Proctor Institute for Leadership, Equity, and Justice at Rutgers University, where he is currently focusing his research on sustainability, health equity, and inclusive leadership. His work as a visiting researcher at the Pellegrino Center for Clinical Bioethics at the Georgetown University Medical Center in Washington, D.C., and as a post-doctoral public health researcher at the University of Maryland School of Pharmacy in Baltimore, MD, has provided him with a unique perspective on issues related to health disparities, health technology access, telehealth, adolescent health, and health literacy. For more than 12 years, he has also served in various roles as a doctoral faculty member, dissertation chair, and research panel participant at Marymount University, the Florida Institute of Technology, Capitol Technology University, the University of Liverpool in the U.K., and The Chicago School of Professional Psychology in Washington D.C. Dr. Burrell is a Certified Diversity Professional and a Certified Executive Coach. Dr. Burrell is an alumnus of the Presidential Management Fellows Program with the U.S. Federal Government and U.S. Nuclear Regulatory Commission, where he managed and developed leadership and diversity programs for professionals in Public Health, Environmental Health, Emergency Response, Technology, and Engineering. Academically, Dr. Burrell has three doctorate degrees and six graduate degrees, including a Master of Research in Artificial Intelligence Biotechnology Applications and Healthcare, along with additional graduate degrees in (H.R. Management, Management, Health Leadership, Sales and Marketing Management, Higher Education Administration, and Interfaith Action and Conflict Management). Dr. Burrell received his first doctoral degree in health education (DHEd), focusing on environmental public health and executive leadership coaching, at A.T. Still University in 2010. Dr. Burrell completed his 2nd doctorate, a Doctor of Philosophy (Ph.D.) in Cybersecurity Leadership at Capitol Technology University. In 2022, Dr. Burrell completed his 3rd Doctor of Business Administration (DBA) in Supply Chain Management at Capitol Technology

University. He has over 20 years of management, teaching, and training experience in academia, government, and private industry. Dr. Burrell has over 200 peer-reviewed publications and more than 1200 Google Scholar citations.

***

Senthil Kumar A.V. is working as a Professor and Principal, Nehru Institute of Information Technology and Management, Coimbatore, India. He has worked as Professor and Director, PG and Research Department of Computer Applications, Hindusthan College of Arts & Science, Coimbatore, Tamilnadu for 15 years and as Senior Grade Lecturer in CMS College of Science and Commerce for 14 years. To his credit he has industrial experience for five years and teaching experience of 29 years. He has also received his Doctor of Science (D.Sc in Computer Science). He has to his credit 86 Book Chapters, 234 papers in International and National Journals, 85 papers in International Conferences in International and National Conferences, and edited 17 books and 3 Text books. He is as Associate Editor of IEEE Access. He is an Editor-in-Chief for many journals and Key Member for India, Machine Intelligence Research Lab (MIR Labs). He is an Editorial Board Member and Reviewer for various International Journals. He is also a Committee member for various International Conferences. He has guided 16 Ph.D scholars and guiding 3 Ph. D scholars now.

S Nagakishore Bhavanam having 16 years of Experience in Teaching and Research. In which 12 Years of Administrative experience as Registrar, Mangalayatan University Jabalpur (MUJ), Jabalpur, Madhya Pradesh, Assistant Director, IQAC, Acharya Nagarjuna University (ANU), Coordinator, Rankings, ANU, Head of the Department, ECE, University College of Engineering & Technology, ANU and many more positions at Acharya Nagarjuna University, Guntur, Andhra Pradesh. So far Published 190 research & review articles, 30 Patents are Published and 2 Innovation and 11 Design Patent grants were secured from National & International (UK & Canada) Level. Contributed 11 Text Books, 23 Book chapters in Electronics and Computer Science Engineering field and received 12 awards. Seven (07) doctorate degrees are awarded under my supervision. During my administrative experience more than 60 State level, National, International level Seminars/ Conferences/Workshops are organized. More than 150 ranks received by ANU and 57 ranks received by Mangalayatan University, Jabalpur under my supervision. Delivered more than 70 lectures as resource person/speaker at reputed organizations, Universities and institutions. I am the Potential reviewer for various Elsevier Journals and IEEE Transactions on VLSI Journal.

Arthur D. Bradford is a finance and operations executive with over 30 years of experience in higher education, Fortune 100 companies, and global retail organizations. Most recently, he served as the Chief Financial Officer and Executive Vice President for Administration at the University of North Texas at Dallas. His career encompasses significant leadership roles at Walmart, JCPenney, Conagra Brands, and Frito-Lay, where he excelled in financial planning and analysis, process improvement, and strategy. As a Certified Public Accountant, Arthur's financial acumen has driven measurable improvements in return on investment, operational efficiency, and organizational growth. He is also a dedicated mentor, committed to fostering talent and guiding teams toward achieving strategic goals. Arthur holds a Doctor of Business Administration from Marymount University, a Master of Business Administration from the McCombs School of Business at the University of Texas at Austin, a Master of Biblical and Theological Studies from Dallas Theological Seminary, and a Bachelor of Science in Accounting from Grambling State University. Arthur's leadership philosophy is rooted in his Christian faith, integrity, lifelong learning, and a passion for driving meaningful change. His diverse experience across sectors uniquely positions him as a thought leader in finance, strategy, and operations.

Sharon L. Burton stands out as an eminent figure in cybersecurity leadership, information security, business, and education with a distinguished career spanning over 25 years at the intersection of academia, industry, and government. Her expertise is grounded in a robust foundation of technical acumen and strategic oversight, enabling her to fortify information security and enhance organizational resilience effectively. Dr. Burton's academic credentials include a Ph.D. in Cybersecurity Leadership, a DBA in Quality Systems Management, two MBA degrees (Human Resources Management, and Management, alongside multiple certifications in in-formation security management, project management, change management, strate-gic workforce planning, and Lean Six Sigma Black Belt, among others. A prolific thought leader, Dr. Burton has contributed over 110 publications to the field, including peer-reviewed journal articles, book chapters, and conference pro-ceedings, showcasing her dedication to advancing knowledge and practice in cyber-security and leadership. Her work is recognized and indexed in prestigious data-bases such as SCOPUS (56943042000), ORCID (0000-0003-1653-9783), and Web of Science (B-2916-2019). Dr. Burton's commitment to education and mentorship is evident in her roles as a doctoral dissertation chair, adjunct faculty, and senior executive consultant, where she has significantly impacted STEM education and supported the next generation of cybersecurity professionals.

Gabrielle Cugini is pursuing her Doctor of Business Administration at Marymount University, with a focus on financial literacy and retirement preparedness. She holds a Bachelor of Arts in Politics from The Catholic University of America, a Master of Science in Management from the University of Maryland University College, and a Master of Business Administration from Aspen University. Additionally, she serves as the director of strategic initiatives at a financial services organization, where she provides strategic leadership, project planning, implementation, analysis, and reporting. With her academic background and professional experience, she has extensive expertise in building and streamlining processes, emphasizing increased value and reducing administrative burdens. Ms. Cugini can be reached at CuginiG@gmail.com.

Hilda Eke is a nonprofit executive, certified Human Resources professional (SHRM-CP), and doctoral candidate at Marymount University. Her scholarly work examines leadership sustainability, psychological safety, and executive attrition in nonprofit organizations. She brings extensive experience in organizational operations, human resources administration, and systems development within mission-driven and community-based contexts. Hilda holds a Master's degree in Healthcare Administration from California State University, Long Beach, and a Bachelor's degree in Linguistics from the University of Benin. Her work integrates theory and practice to advance organizational effectiveness, workforce sustainability, and strategic leadership in nonprofit settings. Her contributions have been recognized through membership in Phi Beta Kappa and the Golden Key International Honour Society, as well as an award for service to nonprofit organizations in Northern Nigeria.

Mary Magdalene Jane F is Dean of Computer Science at Dr. N.G.P. Arts and Science College, Coimbatore, with over two decades of academic experience. She is a recipient of the prestigious Tata Research Fellowship and served as a full-time Research Fellow at PSG College of Technology for four years. She also has industry experience, complementing her academic career Her research interests lie in data science, intelligent data systems, mobile and distributed computing, machine learning and customer analytics. She has secured funding from DRDO, ACM, TNSCST and TCS for research and academic initiatives. Her publications with Springer, IEEE and ACM reflect her contributions to mobile computing, intelligent data systems and emerging data-driven applications. She has presented her work at international conferences in China, Greece, France, Jordan and Malaysia . She has served as an editor and reviewer for journals and conferences. She is an active member of academic councils, editorial board and professional societies including ACM, ASI and ISTE.

Vanishree G has rich experience, 28 years in Academics & administration. Having worked as HOD, Principal, and vice principal in Vignan Institutions, Institute of Aeronautical Engineering, Badruka, St. Patricks, St.Joseph College Kristu Jayanti College, etc, she also worked as Director for Kempe Gowda Institute of Management Studies and Research Bangalore. Presently, she is working as a professor at IBS-B Bangalore. Prof has been a credit participator in several National seminars & conferences and published journal articles. She has won the Best Teacher, parent, and Adarsh Yuvathi awards. She makes herself suited to the position placed. She is responsible and inspiring, has a good attitude, and believes in ethical norms. Having a good record of good results, she believes in the statement, "Learning is a continuous process."

Erman Gedikli received the B.S. degree in economics from Karadeniz Technical University, in 2013, the master's degree in health management from University of Eurasia, in 2016 and the Ph.D. degree in health management from Istanbul Medipol University, in 2021. The whole time his academic life, he has authored various international proceeding, articles and book chapters. Mr. Gedikli has been working as Assoc. Prof. at Department of Health Management, Istanbul Medipol University. His research interests include health management, facilities management, heath tourism, organizational behaviour, strategic management and mixed research methods.

Natasha Harris is a doctoral researcher in Business Intelligence at Marymount University. Her interdisciplinary research examines human behavior and systems design across finance, organizational contexts, sustainability, and emerging technologies, with a focus on human-centered, capability-enhancing approaches to institutional decision-making and resilience across complex, evolving systems.

Garrett Hart is a Doctor of Business Administration (Business Intelligence) student at Marymount University and the Director of Training and Sales at ConEquip Parts and Equipment, where he leads organization-wide sales enablement and leadership development initiatives. His work sits at the intersection of performance coaching, sales leadership, and organizational development, with eleven research articles published or accepted and an ongoing research focus on Generation Z, motivation, and autonomy in customer-facing sales roles. He has designed and delivered training programs for large sales teams and built coaching frameworks and toolkits for leaders, suggesting a practical, systems-minded approach to improving capability and retention. He is also a U.S. Army veteran with experience training peers in high-stakes operational environments.

Allison Huff focuses on technology to address complex disorders, particularly focusing on how technology impacts SDOH. Previously, she was a behavioral health clinician working with individuals with serious mental illnesses and substance use disorders. Dr. Huff is PI and Co-I on several NSF, HRSA, and industry sponsored grants and leads clinical trials and research programs. She also is a mentor for several underserved pre-professional health and pre-med students each summer and through the academic year. Dr. Huff's Doctor of Health Education is from A.T. Still University, her Master of Education in Instructional Technology is from University of Oklahoma, and her Bachelor's in Psychology is from University of West Florida.

Laura Jones is an award-winning risk management thought leader, strategist, and practitioner. She specializes in cybersecurity risk, quality assurance, and artificial intelligence. She currently serves as an adjunct faculty member at Carnegie Mellon University, Wake Forest University, and Capitol Technology University. Dr. Jones earned a PhD in Cyber Leadership, a PhD in Product Management, an MS in Quality Systems Management, and a BS in Management Studies. She has completed executive education programs at Carnegie Mellon University (Chief Risk Officer Program), Harvard Kennedy School (Leadership Decision-Making), and UC Berkeley Haas (Business Strategies and Artificial Intelligence). Dr. Jones has numerous publications on cyber and risk management. She also speaks domestically and internationally about risk management and cybersecurity. Dr. Jones previously served as a graduate school Associate Dean of Student Academic Affairs.

Irem Kose has professional experience in international patient services, healthcare market analysis, and digital transformation in healthcare organizations. Her academic interests include digitalization in health tourism, process innovation, and data-driven decision-making in hospital management. She has contributed to research on the application of multi-criteria decision-making methods in evaluating healthcare management challenges.

Eugene Lewis is a leading International Program Manager with an academic background in Information Systems, having taught in some Business areas such as Marketing, Management, and Information Systems. Dr. Lewis manages a portfolio of over 250 million dollars in the Foreign Military Sales (FMS) industry for the US Army. He is a leading expert in security assistance program management processes and competencies in total package approach (TPA) from hardware, training, case, financials, programmatic, contracting, technical support, logistics, and supply chain management. Currently, Dr. Lewis serves in the Aviation & Defense Industry, working in the Cargo CH-47 Helicopters Foreign Military Sales Program Office. Previously, he served as the Assistant Professor of Marketing at Oakwood

University as a full-time and part-time instructor in Marketing, Management (Leadership), and Computer Information Systems. Moreover, lecturing students in areas such as advertising management, consumer behavior, global marketing, marketing management, organizational dynamics, project management, sales management, project management, logistics, and supply chain management integrating marketing methodologies. Dr. Lewis is a Practitioner and Consultant with education and experience in real-world examples of strategic integration within the business field. As an International Program Manager, he gives several briefings to dignitaries and foreign leadership around the world on logistics and supply chain management. Much of his employment experience consists of incorporating new innovative concepts to provide consumers with the requisite knowledge necessary to understand the value of emotional (EQ) and business intelligence (BI) by incorporating technology into the organizational workflow of operational planning. Dr. Lewis enjoys giving back to the community, spending much of his free time assisting young people in underprivileged areas and instructing them in Boxing. Dr. Lewis has been a Boxing Coach and Instructor for over 15 years. He is a registered Coach with the United States Boxing Association for the Southeastern Region. Dr. Lewis believes in the efficacy of developing young people. He is originally from Seattle, Washington. His family still resides in the Pacific Northwest. He enjoys every opportunity available to travel, write poetry, play guitar, and meet inspiring people.

Adina Lundy is the associate director of undergraduate psychology at a small New England Institution. A leading authority on the lasting effects of chattel slavery on poverty and the contemporary structure of the Black family; she brings deep historical insight and interdisciplinary expertise to her work as an expert witness for the Public Defender's Office of the New Jersey Family Court, a professor, and a multiculturalism scholar. A dedicated researcher on racism and systemic inequity, Dr. Lundy's work focuses on how structural barriers—including the lack of social capital, institutional bias, and uneven resource allocation—impact both students of color and faculty of color. Her scholarship illuminates how these inequities limit opportunity, influence academic pathways, and contribute to long-term disparities across disciplines. Through her research, workshops (with institutions such as Duke Medical school and the New Jersey Judiciary) testimony, and community advocacy, Dr. Lundy provides critical expertise at the intersection of history, policy, and lived experience, advancing efforts to create equitable systems and strengthen outcomes for children, families, and people of color. Dr. Lundy's research program centers the experiences of marginalized populations seeking to improve their socioeconomic status, with a particular focus on the transformative role of social capital and the strategic engagement of white allies committed to meaningful, equity-driven change. Her work interrogates how access to education, resources, and informed accomplices shapes the trajectories of communities striving for liberation and opportunity. Grounded in qualitative inquiry, Dr. Lundy employs phenomenological methods rooted in community building, relational trust, and deep listening. Her approach prioritizes participants' lived experiences as essential sources of knowledge, informing policy recommendations and institutional practices.

Calvin Nobles is a cybersecurity professional and human factors engineer with more than 25 years of experience spanning military service, industry practice, and higher-education leadership. Dr. Nobles serves as Portfolio Vice President and Dean of the School of Cybersecurity and Information Technology at the University of Maryland Global Campus (UMGC), providing strategic and operational leadership for a large, workforce-aligned academic portfolio that advances cybersecurity education, applied partnerships, and innovation at scale. Dr. Nobles previously served as a Department Chair and Associate Professor at the Illinois Institute of Technology and completed a full career in the United States Navy before transitioning into roles within the financial services sector. Dr. Nobles is the author of a book on the integration of technologically advanced aircraft into general aviation and maintains active national service through advisory and professional bodies, including service on the National Cryptologic Foundation Board and the Cybersecurity Association, Inc. Board. Dr. Nobles is also a former Cybersecurity Fellow at Harvard University.

Jessica Parker is a consultant and researcher with a DBA in Business Intelligence from Marymount University, where her research focused on cybersecurity and Artificial Intelligence. She has over 25 years of Information Technology experience in executive leadership, financial management, project management, software quality assurance, business analysis, and data visualization. From a global perspective, she has worked with various Fortune 500 companies and multinational teams in countries including Australia, Brazil, China, England, France, Germany, India, Ireland, Italy, the Philippines, Switzerland, and Uruguay.

Chitra Perera is a financial executive with more than 15 years of experience in budgeting, financial analysis, forecasting, and fund management. As Director of Business Services at the Collins College of Hospitality Management, she oversees multi-million-dollar budgets and ensures compliance with CSU policies, accreditation standards, and strategic financial planning objectives. She is recognized for implementing robust internal controls and data-driven methodologies that improve transparency, cost efficiency, and interdepartmental collaboration. Her leadership supports the advancement of academic programs, revenue-generating operations, donor-supported initiatives, and major events, thereby strengthening financial accountability and operational efficiency. Previously, Chitra held senior positions in procurement and operations at sports and entertainment venues and in high-volume hospitality settings. She earned an MSc in Finance and an MSc in Accounting from the University of La Verne, as well as a BS in Hospitality Management from CSU Pomona. She is currently pursuing a DBA with a concentration in Business Intelligence at Marymount University.

KerriLaine Prunella is an executive with experience in leadership development, organizational transformation, and human capital strategy across public sector, consulting, and nonprofit entities. As Founder and CEO of VINA79Group, she brings entrepreneurial vision and a deep commitment to leadership excellence, serving as a strategic advisor. As a board member, Ms. Prunella values collaboration and duty of care. She leads risk management and strategic oversight emphasizing opportunities for operational excellence. She is adept at CEO mentorship, guiding organizational transformations, fostering inclusivity, and aligning human capital strategies. She has demonstrated strong governance expertise to maximize stakeholder value through her leadership roles such as Healthcare Access Maryland, where she currently serves as Vice Chair of the Board, former chair of the Nominating Committee, and the Penn State Alumni Association, serving as Chair of Audit and Risk, and former Vice Chair for Governance and Finance Committees. Additionally, she served as a founding Board Member and Executive Secretary of the Management Advisory Council at George Mason University, Costello College of Business. Ms. Prunella formerly served as Chief Human Capital Officer with oversight of strategy and operations for Talent Management, Talent Development, and Executive Compensation across all office locations. As an Acting Assistant Inspector General for Administration, she led six divisions and a $65 million budget. A Big Four Alumni, she led strategies for human capital process re-engineering. Ms. Prunella earned a Bachelor's degree in International Politics from The Pennsylvania State University, a Master's degree in Business Administration from Trinity Washington University, and a Graduate Certificate in Organization Development from Marymount University. She is currently pursuing her Doctorate in Business Intelligence at Marymount.

Célia Ramos holds a PhD in Quantitative Methods Applied to Economics and Management from Faculty of Economics, University of Algarve, obtained her Master in Electrical and Computers Engineering from the Higher Technical Institute, Lisbon University, and is graduated in Computer Engineering from the University of Coimbra, Portugal. She is Coordinator Professor with Habilitation at School for Management, Hospitality and Tourism, also in the UALG, where she lectures mainly Information Systems. Current research interests includes, information systems, data analytics, electronic tourism, business intelligence, digital marketing, panel data models. Has published in the fields of information systems, marketing and tourism,

She is reviewer of several scientific journals. At the level of applied research, she has participated in several funded projects. Researcher at the Centre for Tourism, Sustainability and Well-being (CinTurs).

Revati Rautrao, academic leader, researcher, author, and passionate about driving impact through research, innovation, and education. 20+ years of teaching experience, 60+ research papers, 12 patents, 11 copyrights, recipient of national & international awards including the Shiksha Padmabhushan Sanman & Global Teacher Excellence Award. Chief Editor, Conference Chair, Member of Editorial & Reviewer Boards for international journals. Research Areas: HRM, Sustainability, AI in Workforce Management, Indian Knowledge Systems, Literature & Culture Studies .#Education #Research #Leadership #Sustainability #AIinHR #WomenInAcademia

Amalisha Sabie-Aridi, Ph.D., MBA, CMP-UNGGA, is an esteemed scholar, educator, and executive leader whose career spans over a decade of exceptional contributions to higher education, project management, organizational strategy, and leadership development. As a distinguished graduate professor and doctoral committee chair at multiple institutions, including Marymount University and Washington University of Science & Technology, she has guided countless doctoral candidates and shaped innovative curricula across disciplines such as technology, business, leadership, and project management. Dr. Sabie-Aridi's prolific scholarly output includes over 80 peer-reviewed publications covering diverse fields such as organizational behavior, artificial intelligence strategy, healthcare management, strategic leadership, diversity and inclusion, and cybersecurity. She is renowned for integrating research excellence with practical expertise drawn from her tenure as a senior project manager and executive operations leader, notably in high-stakes environments like NASA Engineering Inc. Internationally recognized, she holds honors including UN Global Goodwill Ambassador.

Kevser Şahin is a research assistant at Istanbul Medipol University's Department of Health Management. She is pursuing a PhD in Health Management.

K.R.S. Sambasiva Rao, presently working as Vice-Chancellor of Mangalayatan University at Jabalpur, Madhya Pradesh, since September, 2023. He is formerly the Vice-Chancellor of Mizoram Central University during 2017- 2022 and Rector (Pro-Vice Chancellor) and Vice-Chancellor (i/c) of Acharya Nagarjuna University during 2014-2017. He did D.Sc. in Biotechnology from Berhampur University, in the field of Microbial Biotechnology; Ph.D. in Zoology from Sri Venkateswara University, in the field of Toxicology and Physiology; PhD. in Pharmaceutical Science from Andhra University, on Dosage formulation and Drug delivery systems. Prof.Rao had Post Graduation in M.Sc. Zoology, M.Sc. Psychology,

M.Sc. Information Technology, MBA Big Data Analytics and MA Sociology. Prof. Rao earlier worked as a Professor of Zoology and Biotechnology at Acharya Nagarjuna University over 29 years during 1988 to 2017, and worked in the areas of Aquaculture Biotechnology, Microbial Biotechnology and Molecular Biology. He is instrumental in establishing the Biotechnology department at Acharya Nagarjuna University, and initiated the courses such as MSc Biotechnology, MTech Biotechnology and MSc Nanobiotechnology. Prof. Rao published over 275 research papers in various national and international journals and filed eight patents and edited/authored 20 books. His research work has very good citations (Google Scholar – 22111; Scopus – 2739; WoS – 1338; H- index - 60). During his tenure, the Mizoram University has been glorified as one of the best University in North-East region and in the country with a consistent good ranking in NIRF and NAAC assessment. Under his leadership, Mizoram University has achieved many milestones and all round development of the University. Prof.Rao is the convener for implementation of National Educational Policy in North- Eastern and Eastern region. Prof. Rao worked as a Visiting Scientist at Emory University School of Medicine, Atlanta, USA during 1999-2000, at University of Gottingen, Germany during 2004 and Virginia Commonwealth University Medical College, Richmond, USA in the Division of Physiology during 2005 and worked as Adjunct Professor at INDICASAT, Republic of Panama during 2010 to 2018; at University of Malaysia, Terranganu and UCSI University, Kulalampur during 2020 to 2022. He-2- has been invited as Visiting Scientist to Technical University of Liberec, Czech Republic during June, 2022. He visited CSIRO and University of Adelaide, Australia and University of Singapore during the year 2001. Dr.Rao has visited China, Ethiopia and Tanzania as University representative during 2015-2017. He visited Pokhra University, Nepal as a visiting faculty during the year 2002. Prof. Rao has guided 26 PhD students and 21 MPhil students. Prof.Rao is instrumental in establishing the Association of Biotechnology and Pharmacy in the year 2007 and he is the founder General Secretary of Association of Biotechnology and Pharmacy; Editor, Current Trends in Biotechnology and Pharmacy; Elected Fellow of Andhra Pradesh Akademi of Sciences, Fellow of Association of Biotechnology and Pharmacy; Honorary Fellow of Society of Regenerative Medicine; and Honorary Fellow of Society of Biotechnology. Prof.Rao has received Panama National Award for Educational Development in Republic of Panama; Talented Biotechnologist award at Colombo University; Best Paper award from Polivet Journal. He is the member of several scientific and professional organizations and recipient of AP Best Teacher Award in 2013. He is the editorial board member for Indian Journal of Biosciences, Journal of Pharmacology & Toxicology, Journal of Applied Biosciences, Bioscan and Editor of Current Trends in Biotechnology and Pharmacy and reviewer for many journals.

Gregory Smith is a Senior Principal Operations Research Analyst and Simulation Developer at Raytheon (RTX), leveraging more than thirty years of Department of Defense experience to support complex modernization, modeling, and decision-analysis efforts. A West Point graduate with a Master's in Operations Research and Systems Analysis from the Naval Postgraduate School, he previously served in senior Army analytic and leadership roles across the Pentagon, Joint Staff, and deployed environments, including as the Military Deputy to the Director of the U.S. Army Combat Capabilities Development Command Analysis Center (DAC). He is currently a Doctor of Business Administration (DBA) candidate whose dissertation explores how professionals experience Generative AI pilot programs in capture and bid content management, and he remains active in the Central Arkansas community as President of the Douglas MacArthur Central Arkansas West Point Society and a lifetime member of Alpha Phi Alpha Fraternity, Inc.

Vijender Kumar Solanki, Ph.D. is an Associate Professor in Computer Science & Engineering, Stanley College of Engineering and Technology For Women (Autonomous), Hyderabad, TS, India. He has more than 15 years of academic experience in network security, IoT, Big Data, Smart City and IT. Prior to his current role, he was associated with Apeejay Institute of Technology, Greater Noida, UP, KSRCE (Autonomous) Institution, Tamilnadu, India & Institute of Technology & Science, Ghaziabad, UP, India. He has authored or co-authored more than 125 research articles that are published in journals, books and conference proceedings. He has edited or co-edited 24 books in the area of Information Technology.. He received Ph.D in Computer Science and Engineering from Anna University, Chennai, India in 2017 He is Editor in International Journal of Machine Learning and Networked Collaborative Engineering (IJMLNCE) ISSN 2581-3242, Associate Editor in International Journal of Information Retrieval Research (IJIRR), IGI-GLOBAL, USA, ISSN: 2155-6377 | E-ISSN: 2155-6385 . He is guest editor with IGI-Global, USA, InderScience & Many more publishers. He can be contacted at spesinfo@yahoo.com .

Won Song, Ph.D., DBA, MSN, RN, is a multidisciplinary researcher and practitioner specializing in the intersection of healthcare informatics, engineering management, and organizational development. He earned his Ph.D. in Engineering Management from Capitol Technology University (2025) and holds a Doctor of Business Administration (DBA) from Walden University. With a clinical foundation as a Registered Nurse (MSN, Liberty University; BSN, Rutgers University), Dr. Song has extensive experience in acute care and bedside nursing. His current research focuses on implementing Agentic AI and robotic systems to improve patient adherence and safety in chronic disease management. As the developer of the MADA (Medi-Assist IV System), he integrates reliability engineering with clinical practice to optimize nocturnal therapy. A US citizen and fluent in Korean, Dr. Song is dedicated to fostering cyberliterate organizational cultures that bridge the gap between complex technological adoption and high-quality patient care.

Samantha Thibodeau is a Software Engineering Leader and DBA Candidate at Marymount University. She began her career in medical device design and transitioned to Software Engineering. For the last 10 years, she has held industry roles across Quality Engineering, Software Development, and Engineering Management, blending her academic interests within her roles. Academically, Samantha holds a Bachelor of Science in Biomedical Engineering from Florida Gulf Coast University and a Master of Engineering in Engineering Management from Ohio University. She is currently a DBA candidate with Marymount University. She also holds University certifications in Lean Six Sigma and Engineering Leadership, as well as technical certifications including AWS Certified Cloud Practitioner, ICAgile Certified Professional - Foundations of Artificial Intelligence, and the ISTQB Foundation Level Exam (CTFL) with distinction. Her research interests include Engineering Leadership, Leadership Cultural Shifts, Change Management, Cybersecurity, Operational and Business Risk, and Software Quality. Samantha's publications span a breadth of peer-reviewed journals, including those focused on business resiliency and operations, cybersecurity risk management, and engineering leadership.

Duc-Tan Tran is an Associate professor and Vice Dean of Faculty of Electrical and Electronic Engineering, Phenikaa University. He has published over 150 research papers. His publications received the "Best Paper Award" at the 9th International Conference on Multimedia and Ubiquitous Engineering (MUE-15), and International Conference on Green and Human Information Technology (ICGHIT-2015). He was the recipient of the award for the excellent young researcher from Vietnam National University in 2008, Hanoi and the third prize in the contest "Vietnamese Talents" in 2008. His main research interests include the representation, processing, analysis, and communication of information embedded in signals and datasets. He serves as a TP Co-chair, technical committee program member, track chair, session chair and reviewer of many international conferences and journals.

Jason West heads Synalysis Group, advancing the state of knowledge through research synthesis and interpretive analysis. As Chief Strategy Officer, Jason has a talent and passion for conducting research and an insatiable curiosity about why people behave and feel the way they do. Jason's knowledge base is substantive, encompassing countless research studies and consultation engagements that helped solve clients' complex problems to determine content, marketing, advertising, and new product development strategies. Jason holds a BA in Communication Studies from the University of California, Santa Barbara, an MBA from Loyola Marymount University, and is a Business Doctoral student at Marymount University. He has received qualitative research/moderator training from the RIVA Institute. Additionally, Jason serves as a Lecturer for the MA in Communication Program at Johns Hopkins University.

Index

A

B

C

D

P

Q

R

S

T

V

W

www.ingramcontent.com/pod-product-compliance
Lightning Source LLC
LaVergne TN
LVHW062340300926
849161LV00007B/1765
* 9 7 9 8 2 6 0 0 1 0 2 6 6 *